Stem Cell Aging: Mechanisms, Consequences, Rejuvenation

Hartmut Geiger • Heinrich Jasper
Maria Carolina Florian
Editors

Stem Cell Aging: Mechanisms, Consequences, Rejuvenation

Springer

Editors
Hartmut Geiger
Dermatology and Allergic Diseases
Ulm University
Ulm
Germany

Maria Carolina Florian
Dermatology and Allergic Diseases
Ulm University
Ulm
Germany

Heinrich Jasper
Department of Biology
University of Rochester
Rochester, NY
USA

ISBN 978-3-7091-1231-1 ISBN 978-3-7091-1232-8 (eBook)
DOI 10.1007/978-3-7091-1232-8

Library of Congress Control Number: 2015945166

Springer Vienna Heidelberg New York Dordrecht London

Printed on acid-free paper

Springer-Verlag GmbH Wien is part of Springer Science+Business Media (www.springer.com)

Preface

In societies all around the globe, the population is aging. This fact will pose important challenges to healthcare and social systems in industrialized and emerging economies alike. It is therefore critical to devise strategies to extend the healthy years of life and thus limit the exponential increase in healthcare cost associated with growing old and frail. Understanding the aging process at a fundamental level will help do so, extending healthspan and maybe even increasing lifespan.

Stem cells are critical for maintaining tissue homeostasis, and age-related dysfunction of stem cells is likely the underlying cause of degeneration, regenerative dysfunction, and aging-associated disease in a great variety of organisms and tissues. Knowledge on the molecular mechanisms driving stem cell aging is thus expected to inspire rational interventions for a range of age-related pathologies.

This book summarizes our current understanding of cellular and molecular mechanisms of stem cell aging in a great variety of model systems while highlighting promising approaches to attenuate stem cell aging. We hope that this book will serve as a reference in the field and will also, in the long run, contribute to healthy aging.

Novato, CA, USA — Henri Jasper

Ulm, Germany — Maria Carolina Florian

Ulm, Germany — Hartmut Geiger

Contents

Contributors

Benoit Biteau Department of Biomedical Genetics, University of Rochester Medical Center, Rochester, NY, USA

Petra Boukamp Genetics of Skin Carcinogenesis, German Cancer Research Center, Heidelberg, Germany

Jose A. Cancelas, MD, PhD Hoxworth Blood Center, University of Cincinnati College of Medicine, Cincinnati, OH, USA

Kyung-Hee Chang, PhD Division of Experimental Hematology and Cancer Biology, Cincinnati Children's Hospital Medical Center, University of Cincinnati College of Medicine, Cincinnati, OH, USA

UC Department of Pediatrics, Hoxworth Blood Center, University of Cincinnati College of Medicine, Cincinnati, OH, USA

Gerald de Haan Department of Ageing Biology and Stem Cells, European Research Institute for the Biology of Aging (ERIBA), University Medical Center Groningen, University of Groningen, Groningen, The Netherlands

James DeGregori Department of Biochemistry and Molecular Genetics, University of Colorado Anschutz Medical Campus, Aurora, CO, USA

Department of Pediatrics, University of Colorado Anschutz Medical Campus, Aurora, CO, USA

Integrated Department of Immunology, National Jewish Health, Denver, CO, USA

Dounia Djeghloul INSERM U1126, Institut Universitaire d'Hématologie, Hôpital Saint-Louis, Paris, France

Maria Carolina Florian Institute for Molecular Medicine, University of Ulm, Ulm, Germany

Department of Experimental Hematology and Cancer Biology, Cincinnati Children Hospital Medical Centre (CCHMC), Cincinnati, OH, USA

David Garrick INSERM U1126, Institut Universitaire d'Hématologie, Hôpital Saint-Louis, Paris, France

Hartmut Geiger Institute for Molecular Medicine, University of Ulm, Ulm, Germany

Department of Experimental Hematology and Cancer Biology, Cincinnati Children Hospital Medical Centre (CCHMC), Cincinnati, OH, USA

Michele Goodhardt INSERM U1126, Institut Universitaire d'Hématologie, Hôpital Saint-Louis, Paris, France

Sabrina Gundermann Genetics of Skin Carcinogenesis, German Cancer Research Center, Heidelberg, Germany

Mark Gutierrez Department of Biology, Buck Institute for Research on Aging, Novato, CA, USA

Chong He Buck Institute for Research on Aging, Novato, CA, USA

Curtis J. Henry Department of Biochemistry and Molecular Genetics, University of Colorado Anschutz Medical Campus, Aurora, CO, USA

Henri Jasper Buck Institute for Research on Aging, Novato, CA, USA

Dongsheng Jiang Division of Dermatology and Allergic Diseases, Ulm University, Ulm, Germany

Dana Leanne Jones Department of Molecular, Cell, and Developmental Biology, Eli and Edythe Broad Center of Regenerative Medicine and Stem Cell Research, University of California Los Angeles, Los Angeles, CA, USA

Brian K. Kennedy Buck Institute for Research on Aging, Novato, CA, USA

Klaudia Kuranda INSERM U1126, Institut Universitaire d'Hématologie, Hôpital Saint-Louis, Paris, France

Deepak A. Lamba Department of Biology, Buck Institute for Research on Aging, Novato, CA, USA

Günter Lepperdinger Department of Cell Biology and Physiology, Stem Cell and Healthy Longevity Research, University Salzburg, Salzburg, Austria

Department of Extracellular Matrix Research, Institute for Biomedical Aging Research, Innsbruck, Austria

Institute for Biomedical Aging Research, Austrian Academy of Sciences, Innsbruck, Austria

Chih-chun Janet Lin Department of Molecular and Human Genetics, Huffington Center on Aging, Baylor College of Medicine, Houston, TX, USA

Bryon R. McKay, PhD Departments of Kinesiology, McMaster University, Hamilton, ON, Canada

Medical Physics and Applied Radiation Sciences, Faculty of Science, McMaster University, Hamilton, ON, Canada

Gianni Parise Department of Kinesiology, Ivor Wynne Centre, Toronto, ON, Canada

Medical Physics and Applied Radiation Sciences, Faculty of Science, McMaster University, Hamilton, ON, Canada

Departments of Kinesiology, McMaster University, Hamilton, ON, Canada

Stephan Reitinger Department of Extracellular Matrix Research, Institute for Biomedical Aging Research, Universität Innsbruck, Innsbruck, Austria

Institute for Biomedical Aging Research, Austrian Academy of Sciences, Innsbruck, Austria

Andrii I. Rozhok Department of Biochemistry and Molecular Genetics, University of Colorado Anschutz Medical Campus, Aurora, CO, USA

Lenhard Rudolph Division of Stem cell aging, Leibniz Institute for Age Research, Fritz Lipmann Institute e.V. (FLI), Jena, Germany

Karin Scharfetter-Kochanek Divison of Dermatology and Allergic Diseases, Ulm University, Ulm, Germany

Tobias Sperka Division of Stem cell aging, Leibniz Institute for Age Research, Fritz Lipmann Institute e.V. (FLI), Jena, Germany

Hans-Jürgen Stark Genetics of Skin Carcinogenesis, German Cancer Research Center, Heidelberg, Germany

Hila Toledano Department of Human Biology, Faculty of Natural Sciences, University of Haifa, Haifa, Israel

Evgenia Verovskaya Department of Ageing Biology and Stem Cells, European Research Institute for the Biology of Aging (ERIBA), University Medical Center Groningen, University of Groningen, Groningen, The Netherlands

Meng Carla Wang Department of Molecular and Human Genetics, Huffington Center on Aging, Baylor College of Medicine, Houston, TX, USA

Edyta Wójtowicz Department of Ageing Biology and Stem Cells, European Research Institute for the Biology of Aging (ERIBA), University Medical Center Groningen, University of Groningen, Groningen, The Netherlands

Part I

Introduction

Stem Cell Aging: An Overview

1

Heinrich Jasper

Abstract

Stem cell dysfunction is associated with age-related degenerative and proliferative diseases. Recent work in a range of model organisms has focused on characterizing the causes and consequences of stem cell aging. The insight obtained from these studies is likely to impact our ability to promote healthy aging and to develop new therapies against age-related diseases. This book provides an overview of such studies, aiming to present a comprehensive assessment of the current status of the field.

Loss of tissue homeostasis is a hallmark of aging, resulting in degenerative as well as proliferative diseases like cancer. A decline in stem cell (SC) function is a likely cause for these pathologies, and recent work in model organisms has focused on characterizing the causes and consequences of stem cell aging. Limiting these consequences of stem cell aging is likely to impact our ability to promote healthy aging and to develop new therapies against age-related diseases. Furthermore, the promise of regenerative medicine implies that mastering stem cell biology may open new avenues for truly rejuvenating therapies. For this promise to become reality, we need to understand not only the stem cell-intrinsic changes occurring as the organism ages but also the changes in local and systemic conditions that impact stem cell function and regenerative capacity in a variety of tissues. A wide range of studies in a variety of model systems have begun dissecting the mechanisms and consequences of stem cell aging, resulting in a coalescence of new concepts and models that promise to provide a comprehensive understanding of the decline of regenerative capacity in aging tissues (Jones and Rando 2011; Rando 2006; Sharpless and DePinho 2007).

H. Jasper
Buck Institute for Research on Aging,
8001 Redwood Boulevard, Novato, CA 94945-1400, USA
e-mail: hjasper@buckinstitute.org

H. Geiger et al. (eds.), *Stem Cell Aging: Mechanisms, Consequences, Rejuvenation*,
DOI 10.1007/978-3-7091-1232-8_1

This book provides an overview of such mechanisms, aiming to present a comprehensive assessment of the current status of the field. To achieve an accurate and exhaustive account of our understanding of stem cell aging, we have assembled a group of authors that study aging processes in a variety of model systems, including yeast, worms, and flies, as well as various mouse models of regenerative processes. The work of these authors thus represents a broad overview of stem cell aging ranging from cell-intrinsic mechanisms of replicative aging in yeast, over germline and somatic stem cell aging and niche-stem cell interactions in flies and worms to complex tissue/stem cell interactions in various vertebrate tissues, and strategies to improve regeneration and potentially design rejuvenating therapies in vertebrate models. We believe that this comprehensive approach is likely to achieve the most informative and timely overview of the field and allows identifying similarities and differences between individual stem cell systems. Here, I provide a broad overview of themes and insights that are elaborated on in more detail in the corresponding chapters.

1.1 SC-Intrinsic Changes in the Aging Organism: SC Dysfunction as Anticancer Mechanism

Age-related dysfunction of SCs is caused by both cell-intrinsic and cell-extrinsic mechanisms. While the germline stem cell lineage is in principle immortal, and should therefore have acquired mechanisms to prevent age-related decline, their proliferative activity is affected by changes in local and systemic support processes that decay in aging animals, leading to reproductive senescence. Somatic SCs, in turn, while being affected by systemic changes, also seem to age due to intrinsic mechanisms that limit their replicative lifespan. These include telomere dysfunction, DNA damage-induced cellular senescence, and age-related increases in the expression of cell cycle inhibitors (Gunes and Rudolph 2013; Jones and Rando 2011; Rando 2006; Sharpless and DePinho 2007). It has been proposed that these age-related changes in somatic stem cells are part of anticancer mechanisms that prevent deregulation of SC proliferation. The inability of somatic SCs to sustain efficient tissue regeneration in aging organisms would thus be a trade-off of processes required for cancer prevention (Campisi 2013; Sharpless and DePinho 2007). Supporting this view, recent studies confirm that deregulation of stem cell proliferation can contribute to tumor formation (Barker et al. 2009; Lapouge et al. 2011; White et al. 2011; Youssef et al. 2010). Furthermore, the identification of molecular similarities between cancer stem cells and tissue-specific stem cells highlights the importance of preventing deregulation of these cells to ensure long-term tissue homeostasis (Merlos-Suarez et al. 2011). This insight suggests that to promote tissue health and ultimately organismal lifespan, it is key to find strategies to modulate somatic SC function and increase their regenerative capacity without promoting excessive and deregulated proliferation.

1.2 Diverse Aging Processes in SCs

While stereotypical age-related processes that limit regeneration can thus be defined, somatic SCs exhibit a large proliferative diversity, reflecting the specific requirement of individual tissues. Specific modes of SC proliferation include (i) continuously cycling stem cells of high-turnover tissues, such as intestinal stem cells (Li and Clevers 2010; Simons and Clevers 2011; van der Flier and Clevers 2009) and short-term hematopoietic stem cells (HSCs) (Fuchs 2009); (ii) SCs whose proliferation is induced by injury, including airway basal epithelial stem cells and muscle satellite cells (Abou-Khalil and Brack 2010; Dhawan and Rando 2005; Rock et al. 2011); and (iii) stem cells with alternate quiescent and proliferative periods, such as hair follicle stem cells (Fuchs 2009). This diversity results in distinct age-associated deficiencies that contribute to regenerative decline of individual tissues and highlights the need to study aging of distinct SC systems separately to achieve a comprehensive understanding of the regenerative decline in aging organisms. Here, we take account of this diversity of aging processes in different stem cell populations by focusing on the age-associated dysfunction of individual stem cell populations separately in individual chapters.

1.3 SC Aging Through Evolutionarily Conserved Stress Responses

While cell-intrinsic age-related changes in somatic stem cells often specifically impact cell cycle regulation (and can thus be defined as anticancer "programs" that contribute to tissue aging), somatic SCs are also impacted by the random age-related accumulation of molecular damage, as well as by changes in local and systemic conditions in the aging organism. This includes inflammatory conditions and oxidative stress, as described in the hematopoietic system of the mouse and in the posterior midgut of flies (Biteau et al. 2011; Tothova and Gilliland 2007). The resulting accumulation of damage to DNA and other macromolecules is thus likely an important accelerator of SC aging (Biteau et al. 2011; Jones and Rando 2011; Rudolph et al. 2009; Tothova and Gilliland 2007). Mechanisms that allow cells to cope with these challenges and processes that contribute to the development of proliferative dysfunction in aging SCs are likely to be conserved from yeast to humans, and studies exploring the regulation of replicative lifespan in yeast are thus expected to provide important insight into somatic SC aging. One interesting example is the observation that budding yeast ensures rejuvenation of newly formed cells by asymmetrically segregating damaged and old macromolecules, providing a new "young copy" of such molecules to daughter cells (Aguilaniu et al. 2003; Shcheprova et al. 2008). Recent studies suggest that similar processes occur in human embryonic stem cells and in germline and somatic stem cells of flies (Bufalino et al. 2013; Fuentealba et al. 2008).

The complexity of somatic stem cell lineages in vertebrates often causes difficulties in definitively characterizing age-associated dysfunctions at the stem cell level in vivo. Characterization of stem cell aging in simpler model organisms in which lineage relationships are clearly defined, and which allow lineage tracing with relative ease to study stem cell proliferation and pluripotency, is thus necessary to provide a conceptual framework of stem cell aging and the loss of regenerative capacity. Work in flies and worms has provided such insight in recent years. This work initially focused on the effects of aging on germline stem cells (GSCs), and a large body of literature has examined mechanisms of stem cell aging in the worm and fly germline, as well as identified endocrine mechanisms by which GSCs influence metabolism and longevity of the animal (Arantes-Oliveira et al. 2002; Crawford et al. 2007; Drummond-Barbosa and Spradling 2001; Hsin and Kenyon 1999; Hsu et al. 2008; Jones and Rando 2011; LaFever and Drummond-Barbosa 2005; Mair et al. 2010; Pan et al. 2007; Pinkston et al. 2006; Toledano et al. 2012; Ueishi et al. 2009; Wang et al. 2008). These studies have established that in the aging animal, GSC maintenance is affected by a decline in specific niche factors, by metabolic imbalances, as well as by oxidative stress.

Adult somatic stem cells have been identified in the fly gonad, the intestine, and the Malpighian tubules (Decotto and Spradling 2005; Fox and Spradling 2009; Gonczy and DiNardo 1996; Margolis and Spradling 1995; Micchelli and Perrimon 2006; Ohlstein and Spradling 2006; Singh et al. 2007; Takashima et al. 2008). Among these stem cell populations, intestinal stem cells (ISCs) have served as a model to characterize processes and signaling mechanisms that regulate regenerative responses and whose deregulation contributes to the age-related decline in regenerative capacity (Biteau et al. 2011; Casali and Batlle 2009). These include signaling pathways required for homeostatic epithelial renewal (such as EGF and insulin signaling), stem cell maintenance (such as Wnt and Tor signaling), as well as stem cell stress responses (JNK, Jak/Stat, and Hippo signaling). Characterizing these signaling mechanisms is likely to not only provide insight into basic mechanisms of stem cell regulation but also elucidate the molecular etiology of tissue dysfunction, including age-related degeneration and cancer (Radtke and Clevers 2005; Rossi et al. 2008; Sharpless and DePinho 2007). It is important to note that the processes driving stem cell aging appear to be conserved from flies to mammals. FoxO transcription factors, for example, promote stem cell maintenance and function in both mice and flies (Biteau et al. 2010, 2011; Tothova and Gilliland 2007).

1.4 Oxidative Stress and Diet: Major Environmental Parameters Promoting Stem Cell Aging

1.4.1 Redox State and Stem Cell Function

The intracellular redox state has an important influence on stem cell function, and age-related changes in redox state or redox homeostasis have been implicated in age-related stem cell dysfunction of a wide range of stem cells. In mice, elevated ROS levels result in reduced regenerative potential and self-renewal in a wide range of stem cell populations, including neuronal and glial progenitors and HSCs (Diehn et al. 2009; Ito et al. 2004; Le Belle et al. 2011; Liu et al. 2009; Miyamoto et al. 2007; Smith et al. 2000; Tothova and Gilliland 2007; Tothova et al. 2007; Tsatmali et al. 2005). In flies, oxidative stress has been implicated in the age-associated loss of germline stem cells and in the age-associated deregulation of intestinal stem cell proliferation (Hochmuth et al. 2011; Pan et al. 2007). A low intracellular concentration of reactive oxygen species (ROS) is thus increasingly recognized as a critical condition for stemness, self-renewal, and pluripotency of stem cells.

Work in flies over the last decade has begun unraveling the signaling mechanisms mediating this effect. Increased ROS concentration primes hematopoietic progenitors of the larval lymph gland for differentiation (Owusu-Ansah and Banerjee 2009) and promotes ISC proliferation in the adult gut (Biteau et al. 2008; Buchon et al. 2009; Choi et al. 2008; Hochmuth et al. 2011). In the intestine, a Duox-mediated innate immune response in ECs produces ROS to control commensal and pathogenic bacteria, stimulating ISC proliferation (Biteau et al. 2008; Buchon et al. 2009; Choi et al. 2008; Ha et al. 2005; Hochmuth et al. 2011; Yang et al. 2009). The *Drosophila* homologue of the Nrf2 transcription factor, CncC, which regulates the expression of antioxidant enzymes, regulates ISC proliferation rates and limits ISC hyper-proliferation in aging flies (Hochmuth et al. 2011). It is likely that this mechanism is conserved in vertebrates, as mice deficient in Keap1, the negative regulator of Nrf2, show significant hyperkeratosis of the esophageal epithelium (Wakabayashi et al. 2003). Redox regulation by the Keap1/CncC regulatory module thus emerges as central to the control of SC proliferation and epithelial regeneration.

1.4.2 Metabolic/Dietary Effects on Stem Cell Function

Dietary conditions significantly influence lifespan, as exemplified by the robust lifespan extension observed in most tested organisms that are maintained on low-calorie food (dietary restriction, DR) (Jasper and Jones 2010; Kapahi et al. 2010; Katewa and Kapahi 2010). Somatic (as well as germline) stem cell function is strongly affected by nutritional conditions, suggesting that the longevity effects of DR are, at least in part, mediated by the enhanced regenerative capacity of a large number of tissues (Arantes-Oliveira et al. 2002; Mair et al. 2010). Many studies elucidating the interaction of dietary conditions, stem cell function, and lifespan have been performed in invertebrate model organisms, but recent work has also extended this inquiry into vertebrates.

In flies and worms, GSC proliferation is regulated by the nutrient-sensing insulin/IGF signaling (IIS) pathway (Drummond-Barbosa and Spradling 2001; Hsu et al. 2008; LaFever and Drummond-Barbosa 2005; Pinkston et al. 2006; Ueishi et al. 2009). IIS activity acts cell autonomously to maintain male GSCs (Drummond-Barbosa and Spradling 2001; Hsu et al. 2008; LaFever and Drummond-Barbosa 2005; McLeod et al. 2010; Ueishi et al. 2009), and DR conditions that extend lifespan delay the age-associated loss of GSCs (Mair et al. 2010). At the same time, poor diets can increase the rate of GSC loss during aging in females, where IIS activity also influences GSC maintenance in a non-autonomous manner by regulating the interaction between GSCs and the niche (Hsu and Drummond-Barbosa 2009). The

Tor signaling pathway contributes to the regulation of GSC and somatic stem cell maintenance in the *Drosophila* ovary, as it promotes germline and somatic stem cell proliferation independently of insulin signaling, and needs to be suppressed by the tuberous sclerosis complex (TSC) to prevent premature differentiation of GSCs (LaFever et al. 2010; Sun et al. 2010; Voog and Jones 2010).

Dietary conditions also influence somatic SC populations, affecting their proliferative activity, lineage commitment, and maintenance. Not surprisingly, the IIS and Tor signaling pathways play important roles in influencing SC function in these conditions. FoxO transcription factors, which are activated in conditions of low IIS activity, promote quiescence, long-term maintenance, and regenerative capacity of various somatic SC populations in flies and mice: In the *Drosophila* midgut, insulin signaling and FoxO regulate proliferation of intestinal stem cells, influencing the proliferative response of ISCs to tissue damage and the extent of age-related intestinal dysplasia (Amcheslavsky et al. 2009; Biteau et al. 2010). Accordingly, moderate reduction of IIS activity in the ISC lineage promotes intestinal homeostasis and extends fly lifespan (Biteau et al. 2010). The role of FoxO in the control of SC quiescence is conserved in mammalian hematopoietic stem cells (HSCs) and neuronal progenitor cells (Renault et al. 2009; Tothova and Gilliland 2007; Tothova et al. 2007), where it affects SC proliferation, in part, by its ability to regulate antioxidant gene expression (Renault et al. 2009; Tothova and Gilliland 2007; Tothova et al. 2007). The Tor signaling pathway plays a critical role in promoting stem cell maintenance in the intestinal epithelium of flies, where its activity has to be suppressed by TSC proteins in SCs to prevent premature differentiation and loss of these cells (Amcheslavsky et al. 2011; Kapuria et al. 2012).

It can be anticipated that the role of reducing IIS/Tor signaling in the lifespan extension by DR is mediated, at least in part, by its beneficial consequences for somatic stem cell maintenance and proliferative control (Biteau et al. 2010; Kapuria et al. 2012). Whether DR can promote somatic stem cell function by influencing the activity of IIS in mammalian systems remains unclear. However, the decreased cancer incidence in dietary-restricted mice suggests that SC maintenance and quiescence are improved in DR (Kalaany and Sabatini 2009).

It is likely that other nutrient-sensing signaling pathways also influence stem cell function and regenerative capacity. This includes signaling through the Sir2/SIRT1 family of NAD+-dependent deacetylases (Firestein et al. 2008; Guarente and Picard 2005).

Understanding the signaling mechanisms translating nutritional status to stem cell state and regenerative capabilities is likely to not only provide new insight into avenues to improve stem cell maintenance in aging animals but also to provide new therapies for proliferative diseases associated with metabolic dysfunction.

1.5 Age-Related Niche Dysfunction

A central cause of stem cell and regenerative dysfunction in aging organisms appears to be the loss of support from the local microenvironment, the "niche" (Jones and Rando 2011). Age-related niche dysfunction has been characterized primarily in the male and female germline of *Drosophila*, but studies in mice support the notion that similar processes are conserved in vertebrates. The interaction between GSCs and their niche is disrupted in aging females due to increased oxidative stress (Pan et al. 2007), while a decline in trophic support from the niche drives GSC decline in males (Jones and Rando 2011; Toledano et al. 2012). Interestingly, this decline is caused by an increase in the expression of the microRNA let-7, which negatively regulates the RNA-binding protein Imp (IGFII mRNA-binding protein), which in turn is required to stabilize the mRNA of the IL-6 like ligand Upd in the niche. Highlighting the potential conservation of processes driving age-related stem cell dysfunction, the let-7 microRNA has been implicated in stem cell self-renewal and differentiation in vertebrates, and deregulation of IL-6 expression by inhibition of let-7 promotes cell transformation into cancer stem cells in a breast cancer paradigm (Iliopoulos et al. 2009; Melton et al. 2010).

The influence of the local and systemic environment on somatic SC function has been explored extensively in mice. Studies using heterochronic parabiosis and assessing the regenerative capacity of satellite cells (muscle stem cells) suggest that changes in the exposure to growth and differentiation factors, including Wnt ligands, play a critical role in age-related stem cell dysfunction (Brack et al. 2007, 2008; Conboy et al. 2005; Conboy and Rando 2005; Jones and Rando 2011; Wagers and Conboy 2005).

A recent study further shows the importance of niche-derived factors for long-term maintenance of intestinal stem cells in mice and for the effects of nutritional status on stem cell maintenance: Paneth cells express the ligands Wnt3, Wnt11, and EGF to promote stem cell function in vitro and in vivo and can be considered "niche" cells for these SCs (Sato et al. 2011). Nutrient-responsive Tor signaling in Paneth cells influences their ability to promote SC maintenance, most likely by regulating the expression of bone stromal antigen 1 (Bst1), an ectoenzyme that produces the paracrine factor cyclic ADP ribose (Yilmaz et al. 2012).

1.6 SCs and Lifespan

While it is clear that improved regenerative capacity of stem cells promotes tissue health, the effects of stem cell function on overall lifespan of the organism have only recently begun to be understood. Studies in worms and flies have uncovered significant effects of germline stem cells on lifespan, primarily through endocrine effects on metabolism. But also somatic stem cell function in the fly intestine has recently been shown to affect lifespan, primarily by influencing the integrity of this critical tissue (Biteau et al. 2010; Rera et al. 2011, 2012). Strikingly, these studies revealed that stem cell activity has to be maintained within a critical window to improve tissue function and extend lifespan, highlighting the delicate balance between maintaining regenerative capacity and promoting dysplasia or cancer when stem cell proliferation is enhanced. Interestingly, IIS activity influences ISC function, suggesting that the lifespan effects of modulating IIS activity are mediated, at least in part, by its effects on ISC maintenance and activity. Similar effects may be conserved in vertebrates: Similar to flies, where dysplasia in the intestine is triggered by hyper-activation of IIS (Biteau et al. 2010), loss of FoxO negatively impacts NSC and HSC behavior in mammals (Renault et al. 2009; Tothova and Gilliland 2007; Tothova et al. 2007). It is likely that an optimal level of IIS is achieved during DR and contributes to SC maintenance (Mair et al. 2010).

The relationship between stem cell deregulation and tumor formation suggests that modulating somatic stem cell function in vertebrates should have significant effects on lifespan (Radtke and Clevers 2005). In the intestinal epithelium of mice, for example, intestinal stem cell proliferation and intestinal cancer have been linked: stress signaling can induce cell proliferation in the intestinal crypt and increases tumor incidence and growth in an inflammation-induced colon cancer model (Sancho et al. 2009). The precise role of stem cell-specific stress signaling in the development of tumors remains unclear, however, and the detailed analysis of stem cell activity in the *Drosophila* intestine is providing important new insight. Chronic and excessive stress signaling in the ISC lineage causes intestinal dysplasia (Biteau et al. 2008), while overexpression of tumorigenic RasV12 predisposes ISCs to tumor formation when exposed to a stimulus (Apidianakis et al. 2009). From such studies, a detailed picture of stem cell regulation under stress, infection, and aging conditions is emerging that has clear implications for our understanding of stem cell regulation in vertebrates (Biteau et al. 2008, 2011; Buchon et al. 2009; Jiang et al. 2009; Shea et al. 2010).

1.7 GSCs and Longevity

While somatic SCs thus seem to influence lifespan of the organism primarily by promoting tissue health, studies in worms and flies indicate that germline SCs influence longevity by endocrine mechanisms that affect somatic tissues and coordinate reproduction and somatic maintenance according to nutritional conditions (Arantes-Oliveira et al. 2002).

Ablation of GSCs, but not the entire gonad, extends lifespan, suggesting that GSCs promote aging of the organism, while somatic cells in the gonad contribute to the lifespan extension observed in GSC-ablated animals (Arantes-Oliveira et al. 2002; Hsin and Kenyon 1999), most likely by directly influencing metabolic homeostasis (Wang et al. 2008). However, the exact consequences of GSC regulation for the lifespan of the organism can be complex: while germline ablation extends lifespan robustly under normal conditions, under DR conditions GSC ablation can have positive or negative consequences for lifespan, depending on genetic backgrounds (Crawford et al. 2007). In the fly, ablation of the germline extends lifespan by inducing insulin resistance – like phenotypes in somatic cells (Flatt et al. 2008). However, these phenotypes are dependent on the specific method of GSC ablation, indicating that, as in the worm, lifespan extension by loss of GSCs is dependent on complex variables, such as timing of germline loss, presence of somatic cells in the gonad, genetic background, and diet (Barnes et al. 2006). Ovary transplantation studies indicate that the described effects of the germline on lifespan may be conserved in vertebrates (Cargill et al. 2003; Mason et al. 2009). The role of developing oocytes (as opposed to the ovarian soma) in this effect remains unclear, yet these studies confirm that maintenance of ovarian tissue also benefits life and healthspan of vertebrates.

1.8 Rejuvenation and Tissue Repair

The significant expansion of our understanding of stem cell regulation and of mechanisms promoting stem cell maintenance that is afforded by research in invertebrate model organisms is expected to inform strategies for stem cell rejuvenation and tissue repair in aging humans. The feasibility of interventions aimed at restoring regenerative capacity of old tissues has been demonstrated by experiments in which exposure to a young systemic environment restored regenerative capacity of muscle stem cells and neural stem cells (Conboy et al. 2005; Ruckh et al. 2012).

Whether such interventions, which aim at reactivating endogenous regenerative capacity of tissue stem cells, have the potential to also extend lifespan of the organism remains to be established. It is likely, however, that the systemic signals identified in these studies will enhance the efficacy of cell replacement strategies using patient-derived induced pluripotent stem cells (Lamba et al. 2009; Lindvall et al. 2012; Robinton and Daley 2012). While recent successes in engrafting isolated stem cells into mouse tissues highlight the potential of using lineage-restricted precursors or multipotent stem cells in such approaches (Yui et al. 2012), managing the aged systemic environment into which such cells are transplanted is likely to be required to ensure their efficacy.

Acknowledgments The author is supported by the National Institute on Aging (NIH RO1 AG028127) and the National Institute of General Medical Sciences (NIH RO1 GM100196).

References

Abou-Khalil R, Brack AS (2010) Muscle stem cells and reversible quiescence: the role of sprouty. Cell Cycle 9:2575–2580

Aguilaniu H, Gustafsson L, Rigoulet M, Nystrom T (2003) Asymmetric inheritance of oxidatively damaged proteins during cytokinesis. Science 299:1751–1753

Amcheslavsky A, Jiang J, Ip YT (2009) Tissue damage-induced intestinal stem cell division in Drosophila. Cell Stem Cell 4:49–61

Amcheslavsky A, Ito N, Jiang J, Ip YT (2011) Tuberous sclerosis complex and Myc coordinate the growth and division of Drosophila intestinal stem cells. J Cell Biol 193:695–710

Apidianakis Y, Pitsouli C, Perrimon N, Rahme L (2009) Synergy between bacterial infection and genetic predisposition in intestinal dysplasia. Proc Natl Acad Sci U S A 106:20883–20888

Arantes-Oliveira N, Apfeld J, Dillin A, Kenyon C (2002) Regulation of life-span by germ-line stem cells in Caenorhabditis elegans. Science 295:502–505

Barker N, Ridgway RA, van Es JH, van de Wetering M, Begthel H, van den Born M, Danenberg E, Clarke AR, Sansom OJ, Clevers H (2009) Crypt stem cells as the cells-of-origin of intestinal cancer. Nature 457:608–611

Barnes AI, Boone JM, Jacobson J, Partridge L, Chapman T (2006) No extension of lifespan by ablation of germ line in Drosophila. Proc Biol Sci 273:939–947
Biteau B, Hochmuth CE, Jasper H (2008) JNK activity in somatic stem cells causes loss of tissue homeostasis in the aging Drosophila gut. Cell Stem Cell 3:442–455
Biteau B, Karpac J, Supoyo S, DeGennaro M, Lehmann R, Jasper H (2010) Lifespan extension by preserving proliferative homeostasis in Drosophila. PLoS Genet 6:e1001159
Biteau B, Hochmuth CE, Jasper H (2011) Maintaining tissue homeostasis: dynamic control of somatic stem cell activity. Cell Stem Cell 9:402–411
Brack AS, Conboy MJ, Roy S, Lee M, Kuo CJ, Keller C, Rando TA (2007) Increased Wnt signaling during aging alters muscle stem cell fate and increases fibrosis. Science 317:807–810
Brack AS, Conboy IM, Conboy MJ, Shen J, Rando TA (2008) A temporal switch from notch to Wnt signaling in muscle stem cells is necessary for normal adult myogenesis. Cell Stem Cell 2:50–59
Buchon N, Broderick NA, Chakrabarti S, Lemaitre B (2009) Invasive and indigenous microbiota impact intestinal stem cell activity through multiple pathways in Drosophila. Genes Dev 23:2333–2344
Bufalino MR, DeVeale B, van der Kooy D (2013) The asymmetric segregation of damaged proteins is stem cell-type dependent. J Cell Biol 201:523–530
Campisi J (2013) Aging, cellular senescence, and cancer. Annu Rev Physiol 75:685–705
Cargill SL, Carey JR, Muller HG, Anderson G (2003) Age of ovary determines remaining life expectancy in old ovariectomized mice. Aging Cell 2:185–190
Casali A, Batlle E (2009) Intestinal stem cells in mammals and Drosophila. Cell Stem Cell 4:124–127
Choi NH, Kim JG, Yang DJ, Kim YS, Yoo MA (2008) Age-related changes in Drosophila midgut are associated with PVF2, a PDGF/VEGF-like growth factor. Aging Cell 7:318–334
Conboy IM, Rando TA (2005) Aging, stem cells and tissue regeneration: lessons from muscle. Cell Cycle 4:407–410
Conboy IM, Conboy MJ, Wagers AJ, Girma ER, Weissman IL, Rando TA (2005) Rejuvenation of aged progenitor cells by exposure to a young systemic environment. Nature 433:760–764
Crawford D, Libina N, Kenyon C (2007) Caenorhabditis elegans integrates food and reproductive signals in lifespan determination. Aging Cell 6:715–721
Decotto E, Spradling AC (2005) The Drosophila ovarian and testis stem cell niches: similar somatic stem cells and signals. Dev Cell 9:501–510
Dhawan J, Rando TA (2005) Stem cells in postnatal myogenesis: molecular mechanisms of satellite cell quiescence, activation and replenishment. Trends Cell Biol 15:666–673
Diehn M, Cho RW, Lobo NA, Kalisky T, Dorie MJ, Kulp AN, Qian D, Lam JS, Ailles LE, Wong M et al (2009) Association of reactive oxygen species levels and radioresistance in cancer stem cells. Nature 458:780–783
Drummond-Barbosa D, Spradling AC (2001) Stem cells and their progeny respond to nutritional changes during Drosophila oogenesis. Dev Biol 231:265–278
Firestein R, Blander G, Michan S, Oberdoerffer P, Ogino S, Campbell J, Bhimavarapu A, Luikenhuis S, de Cabo R, Fuchs C et al (2008) The SIRT1 deacetylase suppresses intestinal tumorigenesis and colon cancer growth. PLoS One 3:e2020
Flatt T, Min KJ, D'Alterio C, Villa-Cuesta E, Cumbers J, Lehmann R, Jones DL, Tatar M (2008) Drosophila germ-line modulation of insulin signaling and lifespan. Proc Natl Acad Sci U S A 105:6368–6373
Fox DT, Spradling AC (2009) The Drosophila hindgut lacks constitutively active adult stem cells but proliferates in response to tissue damage. Cell Stem Cell 5:290–297
Fuchs E (2009) The tortoise and the hair: slow-cycling cells in the stem cell race. Cell 137:811–819
Fuentealba LC, Eivers E, Geissert D, Taelman V, De Robertis EM (2008) Asymmetric mitosis: unequal segregation of proteins destined for degradation. Proc Natl Acad Sci U S A 105:7732–7737
Gonczy P, DiNardo S (1996) The germ line regulates somatic cyst cell proliferation and fate during Drosophila spermatogenesis. Development 122:2437–2447
Guarente L, Picard F (2005) Calorie restriction–the SIR2 connection. Cell 120:473–482

Gunes C, Rudolph KL (2013) The role of telomeres in stem cells and cancer. Cell 152:390–393

Ha EM, Oh CT, Bae YS, Lee WJ (2005) A direct role for dual oxidase in Drosophila gut immunity. Science 310:847–850

Hochmuth CE, Biteau B, Bohmann D, Jasper H (2011) Redox regulation by Keap1 and Nrf2 controls intestinal stem cell proliferation in Drosophila. Cell Stem Cell 8:1–12

Hsin H, Kenyon C (1999) Signals from the reproductive system regulate the lifespan of C. elegans. Nature 399:362–366

Hsu HJ, Drummond-Barbosa D (2009) Insulin levels control female germline stem cell maintenance via the niche in Drosophila. Proc Natl Acad Sci U S A 106:1117–1121

Hsu HJ, LaFever L, Drummond-Barbosa D (2008) Diet controls normal and tumorous germline stem cells via insulin-dependent and -independent mechanisms in Drosophila. Dev Biol 313:700–712

Iliopoulos D, Hirsch HA, Struhl K (2009) An epigenetic switch involving NF-kappaB, Lin28, Let-7 MicroRNA, and IL6 links inflammation to cell transformation. Cell 139:693–706

Ito K, Hirao A, Arai F, Matsuoka S, Takubo K, Hamaguchi I, Nomiyama K, Hosokawa K, Sakurada K, Nakagata N et al (2004) Regulation of oxidative stress by ATM is required for self-renewal of haematopoietic stem cells. Nature 431:997–1002

Jasper H, Jones DL (2010) Metabolic regulation of stem cell behavior and implications for aging. Cell Metab 12:561–565

Jiang H, Patel PH, Kohlmaier A, Grenley MO, McEwen DG, Edgar BA (2009) Cytokine/Jak/Stat signaling mediates regeneration and homeostasis in the Drosophila midgut. Cell 137:1343–1355

Jones DL, Rando TA (2011) Emerging models and paradigms for stem cell ageing. Nat Cell Biol 13:506–512

Kalaany NY, Sabatini DM (2009) Tumours with PI3K activation are resistant to dietary restriction. Nature 458:725–731

Kapahi P, Chen D, Rogers AN, Katewa SD, Li PW, Thomas EL, Kockel L (2010) With TOR, less is more: a key role for the conserved nutrient-sensing TOR pathway in aging. Cell Metab 11:453–465

Kapuria S, Karpac J, Biteau B, Hwangbo D, Jasper H (2012) Notch-mediated suppression of TSC2 expression regulates cell differentiation in the Drosophila intestinal stem cell lineage. PLoS Genet 8(11):e1003045

Katewa SD, Kapahi P (2010) Dietary restriction and aging, 2009. Aging Cell 9:105–112

LaFever L, Drummond-Barbosa D (2005) Direct control of germline stem cell division and cyst growth by neural insulin in Drosophila. Science 309:1071–1073

LaFever L, Feoktistov A, Hsu HJ, Drummond-Barbosa D (2010) Specific roles of Target of rapamycin in the control of stem cells and their progeny in the Drosophila ovary. Development 137:2117–2126

Lamba DA, Gust J, Reh TA (2009) Transplantation of human embryonic stem cell-derived photoreceptors restores some visual function in Crx-deficient mice. Cell Stem Cell 4: 73–79

Lapouge G, Youssef KK, Vokaer B, Achouri Y, Michaux C, Sotiropoulou PA, Blanpain C (2011) Identifying the cellular origin of squamous skin tumors. Proc Natl Acad Sci U S A 108:7431–7436

Le Belle JE, Orozco NM, Paucar AA, Saxe JP, Mottahedeh J, Pyle AD, Wu H, Kornblum HI (2011) Proliferative neural stem cells have high endogenous ROS levels that regulate self-renewal and neurogenesis in a PI3K/Akt-dependant manner. Cell Stem Cell 8:59–71

Li L, Clevers H (2010) Coexistence of quiescent and active adult stem cells in mammals. Science 327:542–545

Lindvall O, Barker RA, Brustle O, Isacson O, Svendsen CN (2012) Clinical translation of stem cells in neurodegenerative disorders. Cell Stem Cell 10:151–155

Liu J, Cao L, Chen J, Song S, Lee IH, Quijano C, Liu H, Keyvanfar K, Chen H, Cao LY et al (2009) Bmi1 regulates mitochondrial function and the DNA damage response pathway. Nature 459:387–392

Mair W, McLeod CJ, Wang L, Jones DL (2010) Dietary restriction enhances germline stem cell maintenance. Aging Cell 9(5):916–918

Margolis J, Spradling A (1995) Identification and behavior of epithelial stem cells in the Drosophila ovary. Development 121:3797–3807
Mason JB, Cargill SL, Anderson GB, Carey JR (2009) Transplantation of young ovaries to old mice increased life span in transplant recipients. J Gerontol A Biol Sci Med Sci 64:1207–1211
McLeod CJ, Wang L, Wong C, Jones DL (2010) Stem cell dynamics in response to nutrient availability. Curr Biol 20:2100–2105
Melton C, Judson RL, Blelloch R (2010) Opposing microRNA families regulate self-renewal in mouse embryonic stem cells. Nature 463:621–626
Merlos-Suarez A, Barriga FM, Jung P, Iglesias M, Cespedes MV, Rossell D, Sevillano M, Hernando-Momblona X, da Silva-Diz V, Munoz P et al (2011) The intestinal stem cell signature identifies colorectal cancer stem cells and predicts disease relapse. Cell Stem Cell 8:511–524
Micchelli CA, Perrimon N (2006) Evidence that stem cells reside in the adult Drosophila midgut epithelium. Nature 439:475–479
Miyamoto K, Araki KY, Naka K, Arai F, Takubo K, Yamazaki S, Matsuoka S, Miyamoto T, Ito K, Ohmura M et al (2007) Foxo3a is essential for maintenance of the hematopoietic stem cell pool. Cell Stem Cell 1:101–112
Ohlstein B, Spradling A (2006) The adult Drosophila posterior midgut is maintained by pluripotent stem cells. Nature 439:470–474
Owusu-Ansah E, Banerjee U (2009) Reactive oxygen species prime Drosophila haematopoietic progenitors for differentiation. Nature 461(7263):537–541
Pan L, Chen S, Weng C, Call G, Zhu D, Tang H, Zhang N, Xie T (2007) Stem cell aging is controlled both intrinsically and extrinsically in the Drosophila ovary. Cell Stem Cell 1:458–469
Pinkston JM, Garigan D, Hansen M, Kenyon C (2006) Mutations that increase the life span of C. elegans inhibit tumor growth. Science 313:971–975
Radtke F, Clevers H (2005) Self-renewal and cancer of the gut: two sides of a coin. Science 307:1904–1909
Rando TA (2006) Stem cells, ageing and the quest for immortality. Nature 441:1080–1086
Renault VM, Rafalski VA, Morgan AA, Salih DA, Brett JO, Webb AE, Villeda SA, Thekkat PU, Guillerey C, Denko NC et al (2009) FoxO3 regulates neural stem cell homeostasis. Cell Stem Cell 5:527–539
Rera M, Bahadorani S, Cho J, Koehler CL, Ulgherait M, Hur JH, Ansari WS, Lo T Jr, Jones DL, Walker DW (2011) Modulation of longevity and tissue homeostasis by the Drosophila PGC-1 homolog. Cell Metab 14:623–634
Rera M, Clark RI, Walker DW (2012) Intestinal barrier dysfunction links metabolic and inflammatory markers of aging to death in Drosophila. Proc Natl Acad Sci U S A 109:21528–21533
Robinton DA, Daley GQ (2012) The promise of induced pluripotent stem cells in research and therapy. Nature 481:295–305
Rock JR, Gao X, Xue Y, Randell SH, Kong YY, Hogan BL (2011) Notch-dependent differentiation of adult airway basal stem cells. Cell Stem Cell 8:639–648
Rossi DJ, Jamieson CH, Weissman IL (2008) Stems cells and the pathways to aging and cancer. Cell 132:681–696
Ruckh JM, Zhao JW, Shadrach JL, van Wijngaarden P, Rao TN, Wagers AJ, Franklin RJ (2012) Rejuvenation of regeneration in the aging central nervous system. Cell Stem Cell 10: 96–103
Rudolph KL, Hartmann D, Opitz OG (2009) Telomere dysfunction and DNA damage checkpoints in diseases and cancer of the gastrointestinal tract. Gastroenterology 137:754–762
Sancho R, Nateri AS, de Vinuesa AG, Aguilera C, Nye E, Spencer-Dene B, Behrens A (2009) JNK signalling modulates intestinal homeostasis and tumourigenesis in mice. EMBO J 28:1843–1854
Sato T, van Es JH, Snippert HJ, Stange DE, Vries RG, van den Born M, Barker N, Shroyer NF, van de Wetering M, Clevers H (2011) Paneth cells constitute the niche for Lgr5 stem cells in intestinal crypts. Nature 469:415–418

Sharpless NE, DePinho RA (2007) How stem cells age and why this makes us grow old. Nat Rev Mol Cell Biol 8:703–713

Shcheprova Z, Baldi S, Frei SB, Gonnet G, Barral Y (2008) A mechanism for asymmetric segregation of age during yeast budding. Nature 454:728–734

Shea KL, Xiang W, LaPorta VS, Licht JD, Keller C, Basson MA, Brack AS (2010) Sprouty1 regulates reversible quiescence of a self-renewing adult muscle stem cell pool during regeneration. Cell Stem Cell 6:117–129

Simons BD, Clevers H (2011) Strategies for homeostatic stem cell self-renewal in adult tissues. Cell 145:851–862

Singh SR, Liu W, Hou SX (2007) The adult Drosophila malpighian tubules are maintained by multipotent stem cells. Cell Stem Cell 1:191–203

Smith J, Ladi E, Mayer-Proschel M, Noble M (2000) Redox state is a central modulator of the balance between self-renewal and differentiation in a dividing glial precursor cell. Proc Natl Acad Sci U S A 97:10032–10037

Sun P, Quan Z, Zhang B, Wu T, Xi R (2010) TSC1/2 tumour suppressor complex maintains Drosophila germline stem cells by preventing differentiation. Development 137:2461–2469

Takashima S, Mkrtchyan M, Younossi-Hartenstein A, Merriam JR, Hartenstein V (2008) The behaviour of Drosophila adult hindgut stem cells is controlled by Wnt and Hh signalling. Nature 454:651–655

Toledano H, D'Alterio C, Czech B, Levine E, Jones DL (2012) The let-7-Imp axis regulates aging of the Drosophila testis stem cell niche. Nature 485(7400):605–610

Tothova Z, Gilliland DG (2007) FoxO transcription factors and stem cell homeostasis: insights from the hematopoietic system. Cell Stem Cell 1:140–152

Tothova Z, Kollipara R, Huntly BJ, Lee BH, Castrillon DH, Cullen DE, McDowell EP, Lazo-Kallanian S, Williams IR, Sears C et al (2007) FoxOs are critical mediators of hematopoietic stem cell resistance to physiologic oxidative stress. Cell 128:325–339

Tsatmali M, Walcott EC, Crossin KL (2005) Newborn neurons acquire high levels of reactive oxygen species and increased mitochondrial proteins upon differentiation from progenitors. Brain Res 1040:137–150

Ueishi S, Shimizu H, H Inoue Y (2009) Male germline stem cell division and spermatocyte growth require insulin signaling in Drosophila. Cell Struct Funct 34:61–69

van der Flier LG, Clevers H (2009) Stem cells, self-renewal, and differentiation in the intestinal epithelium. Annu Rev Physiol 71:241–260

Voog J, Jones DL (2010) Stem cells and the niche: a dynamic duo. Cell Stem Cell 6:103–115

Wagers AJ, Conboy IM (2005) Cellular and molecular signatures of muscle regeneration: current concepts and controversies in adult myogenesis. Cell 122:659–667

Wakabayashi N, Itoh K, Wakabayashi J, Motohashi H, Noda S, Takahashi S, Imakado S, Kotsuji T, Otsuka F, Roop DR et al (2003) Keap1-null mutation leads to postnatal lethality due to constitutive Nrf2 activation. Nat Genet 35:238–245

Wang MC, O'Rourke EJ, Ruvkun G (2008) Fat metabolism links germline stem cells and longevity in C. elegans. Science 322:957–960

White AC, Tran K, Khuu J, Dang C, Cui Y, Binder SW, Lowry WE (2011) Defining the origins of Ras/p53-mediated squamous cell carcinoma. Proc Natl Acad Sci U S A 108:7425–7430

Yang J, Kalhan SC, Hanson RW (2009) What is the metabolic role of phosphoenolpyruvate carboxykinase? J Biol Chem 284:27025–27029

Yilmaz ÖH, Katajisto P, Lamming DW, Gültekin Y, Bauer-Rowe KE, Sengupta S, Birsoy K, Dursun A, Yilmaz VO, Selig M et al (2012) mTORC1 in the Paneth cell niche couples intestinal stem-cell function to calorie intake. Nature 486(7404):490–495

Youssef KK, Van Keymeulen A, Lapouge G, Beck B, Michaux C, Achouri Y, Sotiropoulou PA, Blanpain C (2010) Identification of the cell lineage at the origin of basal cell carcinoma. Nat Cell Biol 12:299–305

Yui S, Nakamura T, Sato T, Nemoto Y, Mizutani T, Zheng X, Ichinose S, Nagaishi T, Okamoto R, Tsuchiya K et al (2012) Functional engraftment of colon epithelium expanded in vitro from a single adult Lgr5(+) stem cell. Nat Med 18(4):618–623

Part II

Mechanisms of Stem Cell Aging: Lessons from Model Organisms

Aging in the Single-Celled Eukaryote, *S. cerevisiae*

2

Chong He and Brian K. Kennedy

Abstract

In the last few decades, the budding yeast *Saccharomyces cerevisiae* has emerged as a simple and powerful model organism to study aging. Replicative aging and chronological aging are the two major models that have been established in yeast. In this chapter, we review the two aging model systems, focusing on genes and pathways that modulate replicative and chronological aging. The purpose of this chapter is to provide an overall understanding of the aging process in the single-celled yeast and a basis by which to generate models of molecular mechanisms that may affect aging stem cell populations in adult tissues, as well as the multicellular eukaryotes they inhabit.

2.1 Introduction

Why include a chapter about yeast in a book on stem cells and aging? The question revolves around whether aging is something that occurs at the level of the organism or whether the pathologies we describe as part of the aging process are aggregates of aging in single cells. Certainly, dividing and nondividing cells in the aging organism, including adult stem cells, have altered features with age and evidence points to both cell autonomous and non-autonomous processes.

One difficult aspect of studying mammalian aging is the lack of an accepted cell-based system to assess properties of aging. Certainly, cell senescence in fibroblasts and other cell types has been the subject of intense investigation, yielding important findings that may be relevant to aging of the organism. However, there are very likely aspects of aging that cannot be modeled in cell culture. Yeast lives as a

C. He • B.K. Kennedy (✉)
Buck Institute for Research on Aging, 8001 Redwood Blvd.,
Novato, CA 94945, USA
e-mail: bkennedy@buckinstitute.org

H. Geiger et al. (eds.), *Stem Cell Aging: Mechanisms, Consequences, Rejuvenation*,
DOI 10.1007/978-3-7091-1232-8_2

unicellular species. Thus, the cell and the organism are aligned. Moreover, it either replicates or enters a stress-resistant nonproliferative mode depending on environmental conditions, and aging models have been derived for both states. Simplistically, it has been speculated that the replicative aging model may provide a window to aging in dividing mammalian cells and the nonproliferative model to aging quiescent cells.

Over the last few decades, the budding yeast *Saccharomyces cerevisiae* has developed into one of the most prominent model organisms for aging-related research (Longo et al. 2012). The use of the yeast as an aging model organism is preferred due to several advantages: (1) yeasts are unicellular eukaryotic organisms with a short lifespan; (2) they have a relatively small genome, which is completely sequenced and mapped and for which a set of strains exists with each gene deleted individually (Goffeau et al. 1996; Winzeler et al. 1999); (3) the metabolic and regulatory mechanisms of the yeast are highly conserved within higher eukaryotic systems (Kaeberlein 2010). A large portion of genes in yeast have been shown to be orthologous in higher eukaryotes that have been implicated in human diseases (Ploger et al. 2000). (4) Lastly, the ease of genetic/environmental manipulation, maintenance and storage, low relative cost, and comprehensive integrated biological information allow for genome-wide studies and comparisons, leading to a more integrative understanding of aging-related pathways (McCormick and Kennedy 2012). To date, numerous genetic manipulations have been found to modulate aging in yeast, and although some are specific to this organism, many of the important pathways appear to play a conserved role in the aging of multicellular eukaryotes (Kaeberlein 2010).

Yeasts have also been used as a model system to unravel the molecular basis for several aging-related diseases, such as Alzheimer's disease, Parkinson's disease, Huntington's disease, and cancer (Outeiro and Giorgini 2006; Rubel et al. 2013; Willingham et al. 2003). Moreover, with the large amount of data collected in yeast, systems biology approaches are easily applied to this organism for quantitative description of complex phenotypes such as aging (Lorenz et al. 2009; Matecic et al. 2010; Yizhak et al. 2013).

There are two major lifespan assays that have been established in the study of yeast aging: the replicative lifespan assay and the chronological lifespan assay (Kaeberlein 2006; Longo et al. 2012). Replicative aging refers to a model of mitotically active cells, in which the replicative lifespan is defined by the number of daughter cells a mother cell can produce prior to senescence (Fig. 2.1) (Mortimer and Johnston 1959; Steffen et al. 2009). Chronological aging, in contrast, refers to an aging model of cells in a quiescent state, in which the chronological lifespan is defined by the length of time that a population of nondividing yeast cells can maintain viability or more often the capacity to resume cell division when restored to an environment conducive to proliferation (Fabrizio and Longo 2007; Murakami and Kaeberlein 2009). In this chapter, the main questions we address are the following: (1) What mechanisms underlie replicative and chronological aging? (2) To what extent are the two aging models related? (3) What do studies of aging in yeast tell us about aging in mammals? These studies of aging in a single-celled organism may presage important mechanisms driving the aging of adult stem cell populations.

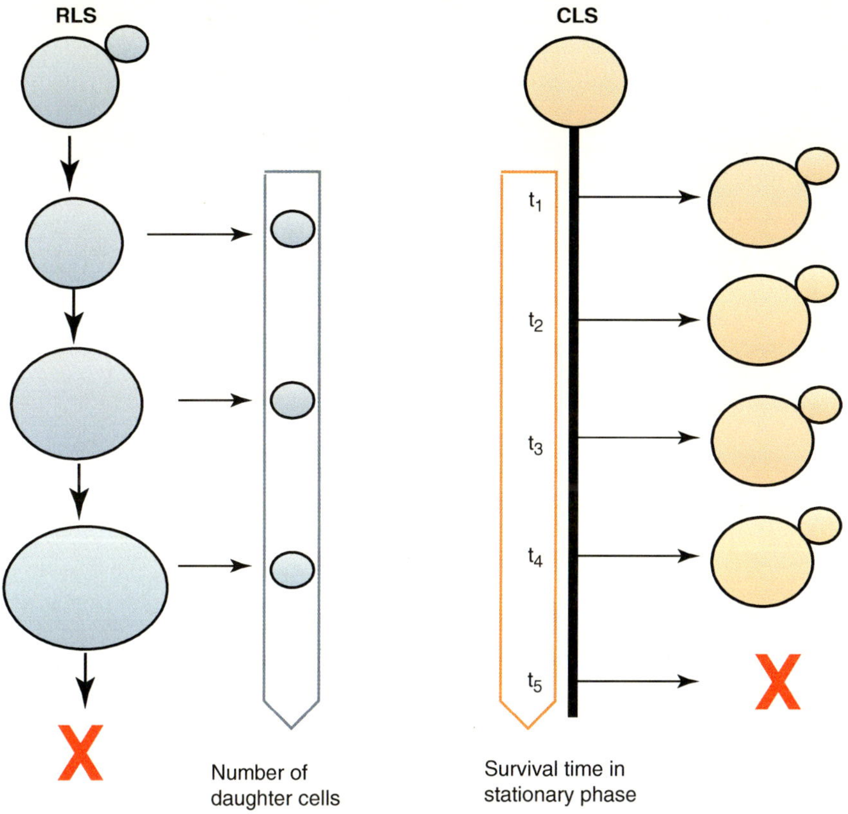

Fig. 2.1 Schematic for yeast replicative lifespan (*RLS*) and chronological lifespan (*CLS*). Replicative lifespan is defined by the number of daughter cells a mother cell can produce prior to senescence. Chronological lifespan is defined by the length of time that a population of nondividing yeast cells can maintain replicative potential when restored to rich media

2.2 Replicative Lifespan

The budding yeast *Saccharomyces cerevisiae* divides asymmetrically with the original "mother" cell budding to give rise to a "daughter" cell (Barker and Walmsley 1999; Jazwinski 1990). The first replicative lifespan studies (RLS) were published more than 60 years ago by Mortimer and Johnston, who designed a method to measure the number of buds or daughters produced by a single mother cell (Mortimer and Johnston 1959). Based on continual observation of individual cells and separation of the morphologically asymmetric buds, they determined that the number of buds produced by one "mother" cell is limited, defined as its lifespan. After undergoing a certain number of mitotic divisions, cells cease to divide and enter a short post-replicative state followed eventually by cell lysis.

The RLS assay is traditionally performed using a standard dissection microscope with a micromanipulator. From a logarithmically growing culture, individual newborn daughter cells are isolated onto a solid-media substrate.

Subsequent budded daughter cells are separated manually from mother cells by microdissection, and the number of divisions each mother cell has produced is then recorded, the totals of which result in the mean RLS (Kaeberlein and Kennedy 2005; Steffen et al. 2009). The typical mean lifespan of laboratory wild-type *Saccharomyces cerevisiae* is around 25 generations, with the maximum being around 40 (Gillespie et al. 2004). It is important to note that strain background, nutritional conditions, and temperature can have major effects on the outcome of RLS experiments (Kirchman et al. 1999; Kaeberlein and Powers 2007; Smeal et al. 1996; Kaeberlein 2010; Shama et al. 1998). While this technique may be the most accurate for determining lifespan, it has limitations including being labor intensive and suffers from an inability to generate large populations of old cells for biochemical analysis.

Lindstrom and Gottschling developed a new method, the Mother Enrichment Program, which blocks genetically the proliferative potential of daughter cells (Lindstrom and Gottschling 2009). This method can provide a large number of aged mother cell cohorts at a late stage of their lifespan, allowing the subsequent analysis of age-associated phenotypes, such as gene expression or protein levels. Recently, microfluidic devices have been developed to study yeast aging, providing another high-throughput method for automated analyses of single cells. These microfluidic systems use tiny transparent chambers to trap single mother cells while automatically removing daughter cells by continuous medium flowing through the device and allow for high-resolution microscopic live-cell imaging system to capture a cell throughout the aging process (Lee et al. 2012; Xie et al. 2012; Zhang et al. 2012). Together, these techniques now provide the capacity for a wide range of experimental approaches to study the yeast replicative aging process.

An early hypothesis to explain the limited lifespan of yeast cells was based on the observation that permanent, non-overlapping bud scars remained on the cell surface after each division (Barton 1950). In addition, cells increase in size as they get older, which reduces surface-to-volume ratios. Thus, a reduced relative surface area may eventually impair division of aging cells (Bacon et al. 1966; Seichertova et al. 1975). A more recent version, the "hypertrophy" hypothesis, has been proposed whereby yeast cells stop dividing after reaching a maximum size (Yang et al. 2011; Bilinski 2012; Bilinski et al. 2012). Longer-lived mutants would start out smaller as buds or increase in size at a reduced rate during the budding process. However, this theory has been disputed and remains controversial (Ganley et al. 2012; Kaeberlein 2012).

Since the early 1990s, when it was demonstrated that genetic manipulation were sufficient to alter yeast replicative lifespan (D'Mello et al. 1994; Kennedy et al. 1995; Sun et al. 1994), the yeast aging field has become a prominent model for aging research, serving as a platform for the discovery and characterization of genes, pathways, and molecular mechanisms involved in aging. As a result, there is growing interest in understanding which characteristics of replicative aging in yeast are shared with mammals, either with respect to the lifespan of the organism or the survival and proliferative capacity of adult cells within the mammalian organism.

2.3 Dietary Restriction and Yeast Replicative Lifespan

Dietary restriction (also referred to as calorie restriction) is defined as a reduction in nutrient intake without malnutrition. Even though the first demonstration of lifespan extension by dietary restriction (DR) was 80 years ago in rats (McCay et al. 1989), the mechanism behind its effect is still only partially understood (Gallinetti et al. 2013; Gems and Partridge 2013; Guarente 2013). DR is the only environmental intervention that can extend lifespan in all common model organisms, including yeast, worm, flies, mice, primates, and several others (Fontana et al. 2010; Anderson and Weindruch 2010; Masoro 2005; Weindruch et al. 1986; Colman et al. 2009; Mattison et al. 2012). In yeast, DR is typically invoked by reducing levels of glucose in the media from 2 % to either 0.5 % or 0.05 % during the lifespan experiment (Lin et al. 2000; Kaeberlein et al. 2004a). DR has been shown to increase both replicative and chronological lifespan in yeast (Kaeberlein 2010; Kennedy et al. 2007); however, the mechanism(s) proposed to explain the effects of DR in yeast remains only partly understood (Steinkraus et al. 2008). *HXK2* encodes the glycolytic enzyme hexokinase II, which converts glucose to glucose-6-phosphate and leads to glucose entering the glycolytic pathway (Walsh et al. 1983). Genetic models of DR, such as deletion of *HXK2* and other nutrient-responsive genes, also lead to RLS extension (Lin et al. 2000). Amino acid restriction, while maintaining normal glucose levels, has also been shown to extend lifespan (Jiang et al. 2000). However, the contribution of each amino acid component to aging is still a mystery. Methionine restriction has been shown to improve lifespan from yeast to rodents (Orentreich et al. 1993; Richie et al. 1994; Johnson and Johnson, 2014). A recent study showed that specific amino acid and glucose affect aging through different pathways (Mirisola et al. 2014). They demonstrated that threonine and valine promoted aging and sensitized yeast cells to stress mainly by activating the Tor/S6K pathway, while serine promoted aging by activating sphingolipid-mediated serine protein kinase (Pkh1/2); glucose promoted aging and sensitized yeast cells to stress through a Ras-dependent mechanism (Mirisola et al. 2014). Moreover, the function of amino acid transporter showed to link with longevity in yeast. A recent study reported that the inhibition of tryptophan uptake and destabilization of tryptophan permease Tat2 are benefit for increasing yeast RLS (He et al. 2014).

Much of the focus on the mechanisms by which DR extends lifespan has centered on the silent information regulator 2 (SIR2), which belongs to the deacetylase sirtuin protein family. SIR2 and its role in RLS are discussed in greater detail later in this chapter. It was originally proposed that DR increased RLS by activation of Sir2p (Lin et al. 2000); however, this hypothesis has since been called into question as DR was found to extend lifespan via Sir2-independent mechanisms (Kaeberlein et al. 2004b). Whether other Sirtuins play a redundant role with Sir2p in DR-mediated lifespan extension has also been debated (Lamming et al. 2005; Tsuchiya et al. 2006). Understanding the role of SIR2 in yeast aging is important given the findings implicating SIRT1, the mammalian ortholog, in aging, stem cell function, and cellular senescence (Giblin et al. 2014; Imai and Guarente 2014).

More recently, there has been a growing recognition of the possibility that lifespan extension by DR is mediated via several partially redundant nutrient-responsive signal transduction pathways, including PKA, TOR, and SCH9. Mutations with reduced PKA, TOR, or SCH9 signaling are replicatively long lived, and lifespans of these mutants cannot be further extended by DR (Kaeberlein et al. 2005c; Lin et al. 2000; Fabrizio et al. 2004). While the TOR pathway has been linked to Sir2 function (Ha and Huh 2011), the totality of evidence suggests that reduced signaling through these nutrient-responsive pathways occurs through multiple mechanisms (Longo et al. 2012).

Yeast cAMP-dependent protein kinase (PKA) is a conserved serine/threonine kinase complex containing three catalytic subunits encoded by *TPK1*, *TPK2*, and *TPK3* (Toda et al. 1987a, b). In yeast, cAMP/PKA is a major signaling pathway for growth on different carbon sources. The basal level of cAMP is higher in yeast growing on fermentable carbon sources (glucose, fructose, and sucrose), compared to growing on non-fermentable carbon sources (glycerol, ethanol, and acetate) (Robertson et al. 2000). PKA is regulated by at least two upstream sensing pathways via the induction of cAMP (Santangelo 2006). One sensing pathway involves RAS activation (Busti et al. 2010). G-protein Ras stimulates acetylate cyclase (Cyr1), which leads to increased cAMP levels in yeast (Thevelein 1994). RAS activation is upregulated by guanine nucleotide exchange factors such as Cdc25p (Martegani et al. 1986). Prevention of activation of this pathway by deletion of *CDC25* leads to lifespan extension (Lin et al. 2000). The other upstream sensing pathway is a G-protein-coupled receptor system encoded by *GPA2* and *GPR1*. Deletion of either *GPA2* or *GPR1* leads to RLS extension, and these mutants are commonly used as models for reducing PKA activity in lifespan experiments (Lin et al. 2000).

Another major nutrient-responsive pathway implicated in yeast replicative aging is the target of rapamycin (TOR) network (Heitman et al. 1991). The TOR protein is a serine/threonine kinase conserved in all eukaryotes (De Virgilio and Loewith 2006). Unlike higher eukaryotes, yeast has two TOR proteins, Tor1p and Tor2p, which function in two distinct multiprotein complexes: TOR complex 1 (TORC1) and TOR complex 2 (TORC2) (Dann and Thomas 2006; Loewith et al. 2002). TORC1 contains either Tor1p or Tor2p and is sensitive to rapamycin (Loewith et al. 2002; Stan et al. 1994). TORC2 contains only Tor2 (Loewith et al. 2002). TORC2 regulates polarization of the actin cytoskeleton in a rapamycin-insensitive manner (Zheng et al. 1995; Loewith et al. 2002). Aside from rapamycin, evidence demonstrates that the TOR signaling pathway is regulated by the presence of nitrogen sources such as glutamine availability; however, the sensing mechanism is still unclear (Crespo et al. 2002; Carvalho and Zheng 2003).

TOR was proposed as a primary conduit in RLS regulation based on the analysis of replicative lifespan in 564 single-gene deletion strains of yeast (Kaeberlein et al. 2005c). Among the ten single-gene deletions with increased RLS, which were identified from the 564 randomly selected deletion strains, at least five of these correspond to gene-encoding components of the TOR pathway (Kaeberlein et al. 2005c). Reduced TOR signaling by either rapamycin treatment or in a *tor1Δ* mutant strain leads to RLS extension in yeast (Kaeberlein et al. 2005c; Medvedik et al. 2007a). DR treatment to the *tor1Δ* mutant fails to further increase lifespan, indicating that TOR regulation of lifespan is through an overlapping pathway with DR (Kaeberlein et al. 2005c). Reduced TOR activity also leads to lifespan extension in *C. elegans* (Vellai et al. 2003; Jia et al. 2004; Hansen et al. 2008), *D. melanogaster* (Kapahi et al. 2004; Bjedov et al. 2010), and mice (Komarova et al. 2012; Harrison et al. 2009; Anisimov et al. 2011), suggesting that TOR signaling not only plays an important role in mediating the longevity effects of DR but also is a conserved nutrient-responsive pathway throughout divergent eukaryotic species.

Sch9p is also a nutrient-responsive kinase that can be phosphorylated by Tor1p and is the functional ortholog of mammalian ribosomal S6 kinase (Powers 2007; Urban et al. 2007). *SCH9* is required for TORC1-mediated regulation of ribosome biogenesis, translation initiation, and entry into G_0 phase (di Blasi et al. 1993; Hay and Sonenberg 2004). Sch9p also shows sequence homology to the kinase Akt, a central component of insulin/IGF-1-like signaling pathways (Paradis and Ruvkun 1998; Burgering and Coffer 1995). Even though yeast lacks a formal insulin/IGF-1-like signaling pathway, Sch9 may have a comparable role to both Akt and S6 kinases (Steinkraus et al. 2008). Deletion of *SCH9* extends both mean and maximum RLS (Fabrizio et al. 2004; Kaeberlein et al. 2005c) in yeast, and consistent with this observation, deletion of either Akt homologs or S6 kinase extends lifespan in *C. elegans* (Pan et al. 2007; Hansen et al. 2007; Paradis and Ruvkun 1998). Mice lacking S6 kinase 1 (S6K1) also have enhanced longevity (Selman et al. 2009). Interestingly, a recent report suggested that independently of TORC1, Sch9p can be independently phosphorylated and activated by the Snf1 kinase, which is the yeast ortholog of the mammalian AMPK kinase, through the acetylation of the Snf1 complex component, Sip2p (Lu et al. 2011). Enhanced Sip2p acetylation extends lifespan, likely because it results in downregulation of Snf1p-mediated phosphorylation of Sch9p (Lu et al. 2011).

Increasing attention has been paid to the downstream mediators of the PKA, TOR, and SCH9 pathways, specifically with regard to their importance in yeast RLS. Recently, a comprehensive, computational network of TORC1 has been constructed (Mohammadi et al. 2013). The functional map of TOR downstream effectors can be used to predict transcriptional changes and posttranslational modifications in response to TOR inhibition, which helps to identify new targets for antiaging therapy (Mohammadi et al. 2013).

Regulation of mRNA translation has been shown to be one of the most relevant downstream mediators required for RLS extension. It has been repeatedly shown that decreased mRNA translation, either by downregulating ribosomal protein biosynthesis or translation initiation factors, leads to lifespan extension (Steffen et al. 2008). The yeast ribosome has a small (40S) and a large (60S) ribosomal subunit which contains 78 ribosomal proteins (RP) encoded by 137 RP genes (Kaeberlein 2010). Although most RP subunits are essential in yeast, the majority of RP genes are duplicated, which allows for a viable deletion of one of the paralogs (Komili et al. 2007; McIntosh and Warner 2007). In the initial yeast deletion collection screen for replicative lifespan, the *rpl31aΔ* and *rpl6bΔ* strains, which lack genes coding for ribosomal large subunit proteins, were identified as long lived, although not every RP deletion extends lifespan (Kaeberlein et al. 2005c). Follow-up work demonstrated that among the 20 RP gene deletions which have been shown to increase RLS, 18 genes encode the 60S-subunit proteins, indicating genes that encode the components of the 60S subunit appear to more robustly regulate yeast RLS (Chiocchetti et al. 2007; Kaeberlein et al. 2005c; Managbanag et al. 2008; Steffen et al. 2008, 2012). Deletion of any one of three mRNA translation initiation factors (*tif1Δ*, *tif2Δ*, and *tif4631Δ*) has also been found to confer RLS extension (Steffen et al. 2008). Again, lifespan regulation by ribosomal proteins can also be found in another eukaryotic species. In *C. elegans*, an RNAi screen demonstrated that decreased mRNA translation level is associated with enhanced lifespan (Pan et al. 2007; Chen and Contreras 2007; Curran and Ruvkun 2007; Hansen et al. 2007), suggesting that modulation of mRNA translation by reducing the TOR signaling pathway is a conserved mechanism throughout divergent eukaryotic species (Smith et al. 2008).

Gcn4p is another downstream mediator of the nutrition signaling pathway. Gcn4p is one of the main transcriptional activators of amino acid biosynthetic genes, specifically in response to amino acid starvation (Hinnebusch 2005; Hinnebusch and Natarajan 2002). Gcn4p abundance is translationally regulated by four small inhibitory upstream open reading frames (uORFs) in the 5' untranslated region of the *GCN4* mRNA (Hinnebusch 2005). Due to the uORFs, *GCN4* is largely not translated under normal growth conditions. Nevertheless, alterations in 60S ribosomal subunit levels lead to increased translation of GCN4, despite an overall decrease in protein synthesis. This evidence suggests that decreased mRNA translation affects aging at least in part by differentially altering the translation of specific mRNA targets. As a result of these findings, Gcn4 has been identified as a translational regulator responsible for the role of the 60S subunit with regard to lifespan regulation, although GCN4-independent effects are also apparent (Steffen et al. 2008). In addition, yeast RLS extension resulting from DR, *tor1Δ*, and *sch9Δ* can be partially blocked by *gcn4Δ*, indicating that *GCN4* is a downstream mediator of DR, TOR, and *SCH9* in a conserved aging pathway. This pathway might also be conserved in mammals as recent evidence indicates that protein levels of the mammalian ortholog of *GCN4*, *ATF4*, are increased in fibroblasts from liver tissue from long-lived mouse models (Li and Miller 2015).

2.4 Sirtuins

Sirtuins are a highly conserved family of nicotinamide adenine dinucleotide (NAD^+)-dependent protein deacetylases. Sirtuins have been well studied in recent years for their potential roles in the regulation of yeast lifespan and because of their highly conserved nature (Finkel et al. 2009; Imai et al. 2000; Landry et al. 2000; Smith et al. 2000). Sirtuins are involved in multiple physiological processes including cellular stress resistance, genomic stability, energy metabolism, and longevity (Wierman and Smith 2014). Moreover, reports have shown that sirtuin activators are potentially beneficial in the treatment of mammalian disease models (Finkel et al. 2009; Michan and Sinclair 2007; Guarente 2006; Imai and Guarente 2010).

The silent information regulator (SIR) genes, including *SIR2*, *SIR3*, and *SIR4*, were first identified in *S. cerevisiae* as being required for silencing at the cryptic mating-type loci, HML and HMR (Ivy et al. 1986; Rine and Herskowitz 1987). In *S. cerevisiae*, Sir2p is a vital component of multiple protein complexes that are required for silencing chromatin regions in telomeres and ribosomal DNA (rDNA; Aparicio et al. 1991; Guarente 1999; Smith and Boeke 1997). It has been well established that Sir2p is recruited to the telomeres by Sir3p and Sir4p, as well as other factors. Disruption of these proteins leads to loss of silencing of the telomeres (Koch and Pillus 2009). Also, Sir2p protein abundance declines with age (Dang et al. 2009), which is consistent with findings that an extra genomic copy of *SIR2* leads to lifespan extension (Kaeberlein et al. 1999). Moreover, the decrease in Sir2 protein abundance correlates with an increase in histone H4 lysine 16 acetylation and loss of histones near telomeres. When histones are lost from the telomeres, transcriptional silencing near the telomeres is also lost. Overexpression of histones extends lifespan (Dang et al. 2009; Feser et al. 2010). Together, this data suggests that the Sir2p-dependent stabilization of telomere chromatin could have a critical role in RLS. Due to the potential relationship between telomeric dysfunction, cell senescence, and aging in humans, the hypothesis relating Sir2p to telomere function in yeast is appealing (Shawi and Autexier 2008; Blackburn et al. 2006).

S. cerevisiae contains 100–200 copies of an rDNA repeat unit, which are located in tandem on chromosome XII (Johnston et al. 1997; Petes 1979). The rDNA tandem array is unstable, with repeat excision from the genome occurring by homologous recombination and leading to formation of extrachromosomal rDNA circles (ERCs) (Sinclair and Guarente 1997). ERCs replicate during S-phase, but are largely maintained in mother cells during cell division because they do not have a centromeric element. As a result, they accumulate to toxic levels in the mother cell with age and can limit their lifespan (Sinclair and Guarente 1997). It has been hypothesized that Sir2p activity promotes replicative aging by suppressing ERC formation (Sinclair and Guarente 1997). Mutants lacking *SIR2* display a six- to tenfold increase in rDNA recombination and have a 50 % reduced lifespan. On the other hand, overexpressing SIR2, through integration of a second *SIR2* copy, extends replicative lifespan (Kaeberlein et al. 1999, 2004a; Kennedy et al. 1995). Additionally, Fob1 promotes ERC formation as an indirect consequence of its role in regulating replication fork traffic in the rDNA (Kobayashi and Horiuchi 1996). Mutants lacking *FOB1* display a dramatic reduction in rDNA recombination and ERC levels and a significant increase in replicative lifespan (Defossez et al. 1999). *fob1Δ* rescues the short replicative lifespan seen in *sir2Δ* (Kaeberlein et al. 1999; Defossez et al. 1999). Both *sir2Δ fob1Δ* and *fob1Δ* strains have low ERC levels, suggesting that *FOB1* is epistatic to *SIR2* for rDNA recombination (Kaeberlein et al. 1999). However, the lifespan of the *fob1Δsir2Δ* strain is similar to wild type, and shorter than that of the *fob1Δ* strain, indicating that Sir2 has a second function for regulating lifespan which is independent of both Fob1 and ERC accumulation (Kaeberlein et al. 1999). A recent study also showed that age-associated ERC formation can be

FOB1 independent using Mother Enrichment Program (Lindstrom et al. 2011). This study reported that increased rate of homologous recombination of rDNA, rather than ERCs, contributes to aging in yeast (Lindstrom et al. 2011). It is also suggested that Sir2 protein levels are not the major factor regulating rDNA recombination in old cells.

Inhibition of TOR signaling by DR or by rapamycin treatment increases rDNA stability by enhancing the association of Sir2p with rDNA (Ha and Huh 2011). Recently, a quantitative trait locus (QTL) study shows that SIR2 and rDNA loci play predominant roles in RLS (Stumpferl et al. 2012). Another QTL study also indicates that polymorphisms of rDNA, independent of SIR2 and FOB1, contribute to the lifespan extension due to modulating genome replication dynamics (Kwan et al. 2013). It has also been reported that Sir2 modulates lifespan through repression of E-pro, an rDNA noncoding promoter (Saka et al. 2013). When E-pro is repressed by an inducible promoter, it shows increased rDNA stability, decreased ERC levels, and increased RLS. Moreover, Sir2 is dispensable for lifespan extension when E-pro is repressed, indicating that Sir2 regulates lifespan through the repression of E-pro noncoding transcription in the rDNA loci. Although ERCs have been reported in higher eukaryotes including human tissues, their role in aging process is unknown (Cohen et al. 2010; Cohen and Segal 2009). Another possibility in mammals is that SIRT1 might regulate recombination in other repeated regions of the human genome, which could impact both cell senescence and aging (Oberdoerffer et al. 2008).

There is also evidence that SIR2 has functions outside of modulating rDNA recombination and telomere dynamics. First, it has been reported that Sir2 prevents daughter cells from inheriting oxidative damage that accumulates in aging mothers (Erjavec and Nystrom 2007; Aguilaniu et al. 2003). This hypothesis is further supported by the finding that SIR2 overexpression rescues the short lifespan of yeast treated with H_2O_2 (Oberdoerffer et al. 2008). Overexpression of Hsp104, a stress tolerance factor that functions with chaperones to dissipate protein aggregation, extends the lifespan of *sir2Δ* strains (Erjavec et al. 2007). Increased Hsp104 activity also rescues the missegregation of damaged proteins to daughters in the *sir2Δ* strain. In 2010, Liu et al. showed that Hsp104-recruiting protein aggregates move back from the bud to the mother cell along actin cables in an Sir2-dependent manner (Liu et al. 2010). Because oxidative damage has been implicated in aging process in invertebrates and mammals, there has been a growing interest in studying the link between sirtuins and oxidative stress and their roles in mediating lifespan. In addition, it has been showed that SIRT1, one of the sirtuin members that have been identified in mammals, regulates an oxidative stress response by deacetylating several transcription factors that regulate antioxidant genes (Brunet et al. 2004; Motta et al. 2004; van der Horst et al. 2004). Identifying yet another potential function of Sir2 relevant to aging, a recent study showed that Sir2 exhibits co-enrichment to various chromatin targets involved in metabolism and protein translation (Li et al. 2013).

It has been suggested that DR mediates lifespan in a sirtuin-dependent manner in yeast, worms, flies, and mice, although the topic is still highly debated (Guarente 2005; Rizki et al. 2011; Viswanathan and Guarente 2011; Burnett et al. 2011). One complicating factor is that DR can be induced in different ways and might require different downstream factors in different protocols, as appears to be the case in worms.

In mammals, there are seven sirtuins (SIRT1-7) that function as NAD-dependent deacetylases, deacylases, and/or ADP-ribosyl-transferases (Houtkooper et al. 2012). In mice, SIRT1 is the closest ortholog to yeast SIR2, and it has been suggested to be required for lifespan extension by DR (Dryden et al. 2003; Finkel et al. 2009; Cohen et al. 2004; Nemoto et al. 2004). However, it was reported recently that overexpression of SIRT1 in mice whole body does not influence lifespan (Herranz et al. 2010). In addition, recent evidence shows that overexpression of SIRT6 has a significantly increased lifespan in male mice (Kanfi et al. 2012). Nevertheless, there has been a growing recognition that sirtuins contribute to healthy lifespan by protecting against several age-associated diseases (Guarente 2011).

2.5 Mitochondria and Oxidative Damage in Yeast Replicative Lifespan

Mitochondria are semiautonomous organelles that contain their own genome (mtDNA) and have essential roles in energy production, metabolism, intracellular signaling, and apoptosis (Cheng et al. 2008; Braun and Westermann 2011; Suen et al. 2008). In addition, mitochondria are the major intracellular source of reactive oxygen species (ROS), which cause damage to cell structures and may regulate aging. However, in *S. cerevisiae*, the role of mitochondrial function in replicative aging is still poorly understood. In *S. cerevisiae*, yeasts lacking mtDNA have been associated with varying replicative lifespans depending on the strain background (Kaeberlein et al. 2005b; Kirchman et al. 1999). It has also been debated whether lifespan extension by DR requires enhanced respiration (Easlon et al. 2007; Kaeberlein et al. 2005a; Lin and Guarente 2006; Lin et al. 2000).

Similar to the lifespan phenotypes of strains lacking cytoplasmic ribosomal proteins, deletion of mitochondrial ribosomal proteins, for example, *MRPL25*, which encodes a component of the large subunit of the mitochondrial ribosome, leads to increased oxidative stress resistance and RLS (Heeren et al. 2009). A later study showed that strains lacking nuclear-encoded mitochondrial translation complex genes, such as SOV1, cause an Sir2-dependent RLS extension without affecting oxidative damage (Caballero et al. 2011). These studies indicate that mitochondrial translation may impinge on yeast RLS by multiple mechanisms.

Mitochondria exist in vivo as a dynamic organelle, which continually undergo fission and fusion (Westermann 2010). Recent work has shown that mitochondrial dynamics also play important roles in yeast replicative aging. A recent Mother Enrichment Program study confirmed and extended prior findings using microdissection that mitochondrial morphology changes dramatically through lifetime: mitochondria are tubular in young cells, fragmented in the early aging process, and form large aggregates in aged cells (Easlon et al. 2007; Hughes and Gottschling 2012). Mitochondria also lose membrane potential as cells age (Hughes and Gottschling

2012). The relationship between mitochondrial dynamics and aging is not fully understood. One study showed that strains lacking the mitofusin proteins Dnm1p or Fis1p, results in reduced fragmentation and extended RLS (Scheckhuber et al. 2007). Subsequently, deletion of the dynamin-related GTPases Mgm1p, results in increased fragmentation and reduced RLS and CLS (Scheckhuber et al. 2011). A recent additional study demonstrated that the double deletion of DNM1 and MGM1 exhibited a normal mitochondrial morphotype, but a decrease in mitophagy and a striking reduced RLS, indicating that a balanced mitochondrial dynamics, but not the filamentous mitochondrial morphotype might be benefit for longevity (Bernhardt et al. 2015).

The effects of mitochondrial uncouplers on yeast RLS have been studied. Partial uncoupling of oxidative phosphorylation by chemical uncoupling agent carbonyl cyanide p-trifluoromethoxyphenylhydrazone (FCCP) results in increased levels of ROS and decreased lifespan (Stockl et al. 2007). Conversely, cells treated with dinitrophenol (DNP), which causes proton leak and uncoupling of the electron transport, results in decreased ROS and increased lifespan (Barros et al. 2004). The relationship between ROS and DR is still not clear. Recent study suggested that elevated ROS levels caused by DR result in mild stress response, which promotes lifespan extension through a hormetic mechanism (Sharma et al. 2011; Calabrese et al. 2011). Higher reactive oxygen species has also been linked to lifespan extension in worms (Yang and Hekimi 2010).

The retrograde signaling pathway mediates stress signals from the mitochondria to the nucleus. One key output of the retrograde signaling pathway is an alteration in nuclear gene expression of mitochondrial proteins, such as *CIT2*, in response to mitochondrial dysfunction (Epstein et al. 2001; Traven et al. 2001). It has been demonstrated that genetic and environmental manipulations that induce retrograde response pathway can increase RLS in certain strain backgrounds (Kirchman et al. 1999). Three genes have been shown to mediate the signaling pathway. *RTG1* and *RTG3* encode the subunits of a leucine zipper transcription factor, which activates the transcription of many retrograde response target nuclear genes for mitochondrial proteins (such as *CIT2*) (Rothermel et al. 1997; Jazwinski 2005a, b). *RTG2* encodes part of SLIK histone acetyltransferase complex that triggers the translocation of Rtg1–Rtg3 transcription complex to the nucleus (Rothermel et al. 1995). It has been proposed that lifespan regulation by retrograde response occurs through both *RTG2*-independent and *RTG2*-dependent processes (Kirchman et al. 1999; Liu and Butow 2006). Ras-PKA and TOR pathways have been shown as upstream regulators for the retrograde signaling (Tate and Cooper 2003; Kirchman et al. 1999; Borghouts et al. 2004; Komeili et al. 2000). However, another study indicated that activation of retrograde response by mitochondria leads to increased ERC accumulation (Borghouts et al. 2004). Overall, the mechanisms by which the retrograde response regulates lifespan in yeast remain unclear; however, it is of interest because the mitochondria-to-nucleus signaling pathway may have an evolutionarily conserved role in the aging process in higher organisms (Kujoth et al. 2007; Sedensky and Morgan 2006).

Recent studies on mitochondria have elucidated a link between mitochondrial dysfunction and vacuolar acidification (Hughes and Gottschling 2012). Vacuolar acidity decreases during the aging process, which leads to mitochondrial dysfunction through reducing pH-dependent amino acid storage in the vacuolar lumen

(Hughes and Gottschling 2012). Preventing the decline of vacuolar acidity leads to a stabilization of mitochondrial function and increased lifespan (Hughes and Gottschling 2012). Along similar lines, it has been illustrated that increased vacuolar fusion, by overexpression of *OSH6*, leads to increased lifespan (Gebre et al. 2012). With multiple lines of evidence pointing to a linked function of the vacuole and mitochondria related to aging in yeast (Hughes and Gottschling 2012), it will be important to determine which of these events are conserved in mammalian aging, as well as adult stem cell function.

2.6 The Ubiquitin/Proteasome System in Yeast Replicative Lifespan

It has also been reported that the ubiquitin/proteasome system (UPS) may play a crucial role in RLS (Dange et al. 2011; Kruegel et al. 2011). Importantly, proteasome activity acts as a central regulator of oxidative stress and aging in different organisms, providing strong evidence for a conserved potential effect on longevity from yeast to humans (Vernace et al. 2007; Carrard et al. 2002; Ghazi et al. 2007). Long-lived organisms such as the naked mole rat display higher levels of proteasome activity when compared to similar shorter-lived species (Perez et al. 2009). In addition, a recent study showed that germline ablation in *C. elegans* results in increased somatic UPS activity and lifespan extension (Vilchez et al. 2012).

The function of UPS decreases during aging (Shringarpure and Davies 2002; Baraibar and Friguet 2012). Importantly, increased UPS activity results in a significantly increased RLS and resistance to proteotoxic stress (Kruegel et al. 2011). It has been proposed that increased proteasome activity is beneficial for longevity and is due to improved elimination of damaged proteins which are asymmetrically segregated to the mother cell during cell division (Zhou et al. 2011; Kruegel et al. 2011). Recently, a study tried to answer why UPS activity declines as cells age. If the level of damaged proteins reaches the proteasome capacity, then protein aggregates will occur. However, as cells age, aggregated proteins appear to accumulate beyond a threshold whereby they interfere directly with proteasome function creating a negative feedback loop (Andersson et al. 2013). A new study demonstrated that proteasomes regulates the AMPK pathway by targeting Mig1, a transcription factor involved in glucose repression (Yao et al. 2015). Increased proteasome activity caused reduced Mig1 level and incorrect localization of Mig1 (Yao et al. 2015). These findings might impact not only mammalian aging, but a wide range of protein aggregation-based diseases.

A recent study reported that yeast metacaspase (Mca1) play a critical role in aggregate management and RLS (Hill et al. 2014). MCA1 showed to help degrade misfolded proteins that left into the daughter cells. MCA1 deletion didn't affect lifespan in wild type strains, but decreased RLS in strains lacking Ydj1, the Hsp70 co-chaperone. Subsequently, overexpressing MCA1 increased RLS in YDJ1 deletion background. These findings not only challenges the idea that caspases execute cellular suicide in an altruistic mechanism, but also might shed lights on protein degradation and rejuvenation mechanisms in stem cells (Kampinga, 2014) (Hill and Nystrom, 2015).

2.7 The Methodology of Chronological Lifespan and New Variants

Chronological lifespan (CLS) is the length of time that a population of nondividing yeast cells can maintain viability or replicative potential (Longo et al. 1996). The chronological lifespan assay was developed as an alternative to the highly tedious replicative lifespan assays. Rather than measuring lifespan in a mitotic state, the chronological lifespan assay does so in a postmitotic or quiescent state. This has been argued to be a better representation of nondividing cells in higher organisms, but also has drawbacks that are addressed later in this section. Unlike the RLS studies, the CLS assay is more conducive to biochemical studies since survival is assessed in a population of cells.

Yeast cells are facultative anaerobes, generating energy mostly through fermentation in rich media with high glucose levels, but switching to respiration when nutrients become scarce. After nutrient depletion, cells enter a nonproliferative stress-resistant state where they can retain viability for a period from days to weeks depending on media conditions. While diploid cells can sporulate under certain conditions, the CLS assay is performed typically in haploid cells or in diploids in a non-sporulation environment. The most common method for CLS is to culture cells in liquid synthetic defined media for a period of 2–3 days where they progress through a postdiauxic, respiratory state and reach maximum cell density before stopping proliferation and ultimately stationary phase. Then, the percentage of cells that retain viability are monitored by periodically plating serial dilutions onto rich media and monitoring colony formation (Kaeberlein 2006; Fabrizio and Longo 2007). Readers can refer to these citations for detailed written and video protocols for a high-throughput method using a Bioscreen machine to obtain outgrowth curves from aged cells (Murakami et al. 2008; Murakami and Kaeberlein 2009). In addition, tools such as Yeast Outgrowth Data Analysis (YODA) have been developed in order to assist researchers (Olsen et al. 2010).

There are several variations of the chronological lifespan protocol, with each having its own strengths and weaknesses. All the methods are variations on the concept of placing cells in a postmitotic state and measuring how long the cells are able to maintain viability (Longo et al. 1996). This method attempts to mimic the conditions that yeast may commonly encounter in their environment, but slight variations in the preparation of the cultures such as the type of caps used on the culture tubes and the media/flask volume ratios can drastically alter the aeration of the culture, directly altering viability and influencing the pathways that modulate lifespan (Hu et al. 2014; Longo et al. 2012).

Several other complementary CLS methods have been developed. One involves measuring viability after starting cultures in a calorie-restricted state and another starting cells in normal conditions, but transferring them from spent media to water after nutrients are depleted. Both methods drastically extend chronologic lifespan. Depending on growth conditions, buildup of acetic acid during the growth phase as a result of fermentation can acidify media and accelerate cell death in the nonproliferative phase (Burtner et al. 2009). This observation has also been demonstrated to be true in several strain backgrounds, as well as wild vineyard strains (Murakami et al. 2011). This is one reason why starting cells in reduced glucose or transferring them to water after proliferation ceases extends lifespan. However, by altering growth conditions, the impact of acetic acid can be reduced (Hu et al. 2014; Longo et al. 2012). One additional way to reduce effects of acetic acid is to grow yeast in buffered media, so that the pH does not drop with acetic acid accumulation and secretion (Burtner et al. 2009).

Another method has been developed where cells in a tryptophan auxotrophic background are plated onto solid media lacking tryptophan. At every time point, tryptophan is added to the plate, and the number of colonies that form is counted (Wei et al. 2009). This method, however, is less conducive to high-throughput. In addition, measuring lifespan by this method slightly reduces chronological lifespan compared to the liquid media approach. The differences can likely be accounted for by the differential response of yeast cells depending on the nutrients for which they are limited. Cells starving for carbohydrates respond differently than tryptophan, for instance. An analysis of the response of yeast under a variety of conditions indicates that the cellular response depends on the first nutrient that becomes depleted (Gresham et al. 2011). Cells encountering starvation in a condition for which they are adapted to respond (e.g., a nutrient for which they would normally become restricted in the wild) have extended survival and evoke a common stress response that protects against a wide range of stressing agents. However, starvation in an unnatural condition leads to an ineffective response and rapid loss of viability.

2.8 Genes and Pathways Modulating Chronological Lifespan

By using a combination of several CLS methods, it has been possible to identify genetic and environmental manipulations that regulate chronological lifespan. Dietary restriction, which for CLS is defined as growth under reduced glucose, typically at a concentration of 0.05 % of the culture medium, has been reliably demonstrated to extend chronological aging significantly (Smith et al. 2007). Similarly, transfer of postmitotic cells to water also extends chronological lifespan. DR has been found to increase respiration (Lin et al. 2002), as a part of a global restructuring of the metabolic state of the cells. Consistent with this hypothesis, cells grown under caloric restriction result in a more basic medium, whereas cells grown under ordinary culture conditions (2 % w/v glucose) actually acidify their medium modestly. The difference in pH in the media likely is a result of byproducts of the metabolic restructuring, as Burtner et al. demonstrated that acetic acid is produced during growth and expelled into the extracellular environment (Burtner et al. 2009). Similarly, Burtner et al. demonstrated that acetic acid is a likely candidate because in cultures grown under calorie restriction or under 3 % glycerol, which forces the cells to respire, because the culture medium was not acidified and acetic acid did not accumulate under these conditions. Interestingly, a similar phenomenon has been reported for growth of mammalian cells in culture. Cells grown in culture acidify their medium by secretion of lactic acid, which reduces their long-term viability. Interestingly, rapamycin extends survival by reducing lactic acid production (Leontieva and Blagosklonny 2011).

Consistent with the findings that altered metabolism extends chronological lifespan, the two main pathways that have been well characterized for extending chronological lifespan are the nutrient sensing PKA and TOR/SCH9 pathways. Genetic perturbations in both have been demonstrated to reliably extend chronological lifespan in a variety of chronological methods, and these effects have been pinpointed to be a result of Rim15 and its downstream effectors Msn2/4 (Burtner et al. 2009). In accordance, reduced Tor signaling increases mitochondrial translation, oxidative phosphorylation (OXPHOS), and mitochondrial respiration, as well as mitochondrial reactive oxygen species (Bonawitz et al. 2007; Pan and Shadel 2009). These results taken together suggest that chronological aging is a result of an adaptive mitochondrial response, especially since stimulation of ROS during the growth phase is sufficient to extend chronological lifespan. In addition, pre-growth on glycerol and *HAP4* overexpression, which activates respiration, both extends CLS (Piper et al. 2006).

One of the most studied members of the Tor pathway is *SCH9*. Consistent with its direct role in the Tor pathway, several studies have demonstrated that deletion of *SCH9* extends chronological lifespan. Also consistent with the metabolic perturbations observed of cells under calorie restriction, *sch9Δ* demonstrates metabolic alterations, specifically enhanced glycerol biosynthesis. Glycerol biosynthetic genes are upregulated in *sch9Δ*, and these genes are required for the chronological lifespan extension in *sch9Δ* (Wei et al. 2009). They also proposed that glycerol is an important stress protectant during chronological aging. Cells lacking *SCH9* are also protected from loss of viability due to extracellular acetic acid (Burtner et al. 2009). *tor1D* cells also exhibit enhanced glycerol biosynthesis (Wei et al. 2009).

Another mechanism that has been directly linked to chronological lifespan is the role that membrane potential of the mitochondria plays in chronological aging. It is documented that as cells age by replication, their mitochondria display signs of altered morphology and of decreased membrane potential (Easlon et al. 2007; Hughes and Gottschling 2012). Respiration is also required for normal chronological lifespan; however, up to a 60 % reduction in respiration does not affect lifespan (Pan et al. 2011; Ocampo et al. 2012). Overexpression of superoxide dismutases curtails the extended chronological lifespan of Tor mutants (Pan et al. 2011), consistent with the finding that reactive oxygen species can promote enhanced CLS (Piper et al. 2006). These observations, along with the fact that addition of trehalose to the media is sufficient to restore chronological aging in respiratory-deficient mutants, suggest that there is an elevated stress response that is dependent on reactive oxygen species. Although yet to be fully elucidated, this mitochondrial reactive oxygen species signaling could prove a fruitful pathway to study given the multitude of diseases associated with aberrant mitochondrial function.

2.9 RLS and CLS: From Yeast to Mammalian Aging

Since the earliest chronological lifespan studies, it has been known that several interventions that affect chronological aging overlap with those that also affect replicative aging, such as DR, TOR/Sch9 pathway, and PKA pathway (Lin et al. 2000; Fabrizio et al. 2001, 2004). Mitochondria function and oxidative damage are also associated with both replicative and chronological aging (Kaeberlein 2010). However, some mutations have been reported to have conflicting effects in RLS and CLS assays (Harris et al. 2001, 2003; Fabrizio et al. 2004). rDNA instability is relevant to RLS, but not CLS since rDNA instability by *sir2Δ* results does not affect CLS (Smith et al. 2007; Wu et al. 2011; Kennedy et al. 2005).

Recently, a direct link between the CLS and RLS assays has been elucidated. Delaney et al. studied the effects chronological aging on the subsequent RLS of mother cells (Delaney et al. 2013). They demonstrated, consistent with an earlier report (Ashrafi et al. 1999), that cells aged chronologically display a reduced replicative potential and that this is exacerbated under conditions that give rise to acetic acid toxicity (Delaney et al. 2013).

Chronological aging and replicative aging have both shed light on similar pathways and have been insightful on their mechanisms. The first quantitative analysis study to address if the longevity genes are conserved between *S. cerevisiae* and *C. elegans* was demonstrated by Smith et al. (Smith et al. 2008). They found that the yeast orthologs of worm long-lived genes are more likely to impact yeast RLS, indicating that genes modulating aging have been conserved between divergent eukaryotic species (Smith et al. 2008). On an evolutionary timeline, the unicellular yeast *S. cerevisiae* and multicellular worm *C. elegans* are separated by 1.5 billion years, which is greater than worm and mammals (1 billion years) (Wang et al. 1999). This indicates that the conserved genes between yeast and worm are likely to have similar overlap in mammals. In contrast to RLS, and in spite of the fact that many of the major pathways affecting RLS, CLS, and worm aging are conserved, there has been no successful quantitative demonstration of conservation in aging pathways between yeast CLS and worm aging or even between CLS and RLS. The efforts to date have centered on CLS conditions where acidification is a prime determinant of aging (Burtner et al. 2011), and it may be the case that other methods of CLS may lead to greater correspondence with aging determinants in the RLS assay and multicellular eukaryotes.

Both the replicative aging and chronological aging assays are meant to establish aging mechanisms in relatively simple organisms that can be tested in metazoans. Chronological aging is proposed as a model for postmitotic cells such as neurons, while replicative aging can be a model for dividing populations of cells, such as adult stem cells. Like budding yeast, it has been reported that damaged proteins in stem cells are asymmetrically distributed during cell division (Fuentealba et al. 2008; Rujano et al. 2006), although it remains unclear whether the mechanisms by which this segregation occurs are conserved in mammals. We predict that the budding yeast replicative aging studies will provide further insights into key regulators for stem cell development and senescence (Thorpe et al. 2008).

In addition, yeast is increasingly used to model aging-related disease, such as neurodegenerative diseases and cancer (we refer readers with specific interest in

Table 2.1 Compounds increase lifespan in *S. cerevisiae*

Compound	Model	Type of compound	Potential mechanism	References
Rapamycin	RLS, CLS	Natural product, immunosuppressant	TOR	Kaeberlein et al. (2005c), Medvedik et al. (2007b), Powers et al. (2006)
Resveratrol	RLS	Polyphenol, natural product in red wine	Stimulate Sir2, increase DNA stability	Howitz et al. (2003)
Ibuprofen	RLS	nonsteroidal anti-inflammatory drug	Tryptophan import	He et al. (2014)
Spermidine	CLS	Natural product, polyamine	Induction of autophagy	Eisenberg et al. (2009)
Lithocholic acid	CLS	Bile acid	Regulate checkpoint, TOR independent	Goldberg et al. (2010), Burstein et al. (2012)
Methionine sulfoximine	CLS	Glutamine synthetase inhibitor	Increase glycogen accumulation, reduce TOR signaling by reducing intracellular glutamine	Powers et al. (2006)
Myriocin	CLS	Serine palmitoyltransferase inhibitor	Reduce sphingolipid levels	Powers et al. (2006)
Tanshinones	CLS	Natural product	TOR signaling pathway	Wu et al. (2014)

disease models to a dedicated review (Tenreiro et al. 2013; Pereira et al. 2012)). For example, Werner syndrome is an autosomal recessive disorder characterized by the appearance of premature aging and cancer predisposition (Epstein et al. 1966). Werner syndrome is caused by mutation of the WRN gene, which functions in telomere maintenance, DNA replication, and DNA repair (Yu et al. 1996). It has been shown that deletion of WRN ortholog in yeast, *SGS1*, results in decreased lifespan and elevated ERC accumulation (McVey et al. 2001). Many other chronic diseases have also been modeled.

The yeast model not only has provided a valuable model to discover compounds that impact aging. Many compounds have now been identified that extend yeast lifespan (Table 2.1). Examples include rapamycin and resveratrol, which were first identified and characterized as antiaging compounds in yeast and have been studied for a number of aging-related diseases in humans (Lamming et al. 2013; Kennedy and Pennypacker 2014). Novel targets such as serine palmitoyltransferase (the first enzyme in the de novo sphingolipid biosynthesis pathway) have also been identified that play important roles in aging, providing further promising pharmacological targets for antiaging interventions (Huang et al. 2014). Moreover, a systems biology approach has been applied to yeast to identify perturbations that can potentially transform an "aged" metabolic state to the desired "young" metabolic state, as well as predict new antiaging drug targets (Yizhak et al. 2013).

As the last few decades of aging research have shown us, yeasts are not only a viable model for studying aging, but they have been critical in discovering novel pathways that are involved in aging throughout different eukaryotic species. Replicative and chronological lifespan studies have unraveled sirtuins, TOR, and PKA, as conserved modulators of longevity, which have been the basis for numerous successful studies in higher organisms. More importantly, we are reaching the era when systems biology and high-throughput technology approaches can be applied to this organism in an attempt to understand aging holistically.

An interesting area of future studies will be to compare the yeast RLS assay to studies of adult stem cells from mammalian tissues. Indeed, while speculation has been that yeast RLS mimics aging in replicatively active adult stem cells, the evidence to support this is minimal. In fact, current studies would indicate that yeast aging corresponds more closely to organismal aging in higher eukaryotes. However, as the pathways modulating adult stem cell aging become further elucidated, it will be possible to perform more direct comparisons of aging mechanisms to that of yeast, and the answers to this question may become apparent. Suffice it to say that we have learned more about aging in mammals from yeast studies than anyone would have predicted, and the best may be yet to come.

References

Aguilaniu H, Gustafsson L, Rigoulet M, Nystrom T (2003) Asymmetric inheritance of oxidatively damaged proteins during cytokinesis. Science 299:1751–1753

Anderson RM, Weindruch R (2010) Metabolic reprogramming, caloric restriction and aging. Trends Endocrinol Metab 21:134–141

Andersson V, Hanzen S, Liu B, Molin M, Nystrom T (2013) Enhancing protein disaggregation restores proteasome activity in aged cells. Aging 5:802–812

Anisimov VN, Zabezhinski MA, Popovich IG, Piskunova TS, Semenchenko AV, Tyndyk ML, Yurova MN, Rosenfeld SV, Blagosklonny MV (2011) Rapamycin increases lifespan and inhibits spontaneous tumorigenesis in inbred female mice. Cell Cycle 10:4230–4236

Aparicio OM, Billington BL, Gottschling DE (1991) Modifiers of position effect are shared between telomeric and silent mating-type loci in S. cerevisiae. Cell 66:1279–1287

Ashrafi K, Sinclair D, Gordon JI, Guarente L (1999) Passage through stationary phase advances replicative aging in Saccharomyces cerevisiae. Proc Natl Acad Sci U S A 96: 9100–9105

Bacon JS, Davidson ED, Jones D, Taylor IF (1966) The location of chitin in the yeast cell wall. Biochem J 101:36C–38C

Baraibar MA, Friguet B (2012) Changes of the proteasomal system during the aging process. Prog Mol Biol Transl Sci 109:249–275

Barker MG, Walmsley RM (1999) Replicative ageing in the fission yeast Schizosaccharomyces pombe. Yeast 15:1511–1518

Barros MH, Bandy B, Tahara EB, Kowaltowski AJ (2004) Higher respiratory activity decreases mitochondrial reactive oxygen release and increases life span in Saccharomyces cerevisiae. J Biol Chem 279:49883–49888

Barton AA (1950) Some aspects of cell division in saccharomyces cerevisiae. J Gen Microbiol 4:84–86

Bernhardt D, Muller M, Reichert AS, Osiewacz HD (2015) Simultaneous impairment of mitochondrial fission and fusion reduces mitophagy and shortens replicative lifespan. Scientific reports 5, 7885

Bilinski T (2012) Hypertrophy, replicative ageing and the ageing process. FEMS Yeast Res 12:739–740

Bilinski T, Zadrag-Tecza R, Bartosz G (2012) Hypertrophy hypothesis as an alternative explanation of the phenomenon of replicative aging of yeast. FEMS Yeast Res 12:97–101

Bjedov I, Toivonen JM, Kerr F, Slack C, Jacobson J, Foley A, Partridge L (2010) Mechanisms of life span extension by rapamycin in the fruit fly Drosophila melanogaster. Cell Metab 11:35–46

Blackburn EH, Greider CW, Szostak JW (2006) Telomeres and telomerase: the path from maize, Tetrahymena and yeast to human cancer and aging. Nat Med 12:1133–1138

Bonawitz ND, Chatenay-Lapointe M, Pan Y, Shadel GS (2007) Reduced TOR signaling extends chronological life span via increased respiration and upregulation of mitochondrial gene expression. Cell Metab 5:265–277

Borghouts C, Benguria A, Wawryn J, Jazwinski SM (2004) Rtg2 protein links metabolism and genome stability in yeast longevity. Genetics 166:765–777

Braun RJ, Westermann B (2011) Mitochondrial dynamics in yeast cell death and aging. Biochem Soc Trans 39:1520–1526

Brunet A, Sweeney LB, Sturgill JF, Chua KF, Greer PL, Lin Y, Tran H, Ross SE, Mostoslavsky R, Cohen HY et al (2004) Stress-dependent regulation of FOXO transcription factors by the SIRT1 deacetylase. Science 303:2011–2015

Burgering BM, Coffer PJ (1995) Protein kinase B (c-Akt) in phosphatidylinositol-3-OH kinase signal transduction. Nature 376:599–602

Burnett C, Valentini S, Cabreiro F, Goss M, Somogyvari M, Piper MD, Hoddinott M, Sutphin GL, Leko V, McElwee JJ et al (2011) Absence of effects of Sir2 overexpression on lifespan in C. elegans and Drosophila. Nature 477:482–485

Burstein MT, Kyryakov P, Beach A, Richard VR, Koupaki O, Gomez-Perez A, Leonov A, Levy S, Noohi F, Titorenko VI (2012) Lithocholic acid extends longevity of chronologically aging yeast only if added at certain critical periods of their lifespan. Cell Cycle 11:3443–3462

Burtner CR, Murakami CJ, Kennedy BK, Kaeberlein M (2009) A molecular mechanism of chronological aging in yeast. Cell Cycle 8:1256–1270

Burtner CR, Murakami CJ, Olsen B, Kennedy BK, Kaeberlein M (2011) A genomic analysis of chronological longevity factors in budding yeast. Cell Cycle 10:1385–1396

Busti S, Coccetti P, Alberghina L, Vanoni M (2010) Glucose signaling-mediated coordination of cell growth and cell cycle in Saccharomyces cerevisiae. Sensors (Basel) 10:6195–6240

Caballero A, Ugidos A, Liu B, Oling D, Kvint K, Hao X, Mignat C, Nachin L, Molin M, Nystrom T (2011) Absence of mitochondrial translation control proteins extends life span by activating sirtuin-dependent silencing. Mol Cell 42:390–400

Calabrese V, Cornelius C, Cuzzocrea S, Iavicoli I, Rizzarelli E, Calabrese EJ (2011) Hormesis, cellular stress response and vitagenes as critical determinants in aging and longevity. Mol Aspects Med 32:279–304

Carrard G, Bulteau AL, Petropoulos I, Friguet B (2002) Impairment of proteasome structure and function in aging. Int J Biochem Cell Biol 34:1461–1474

Carvalho J, Zheng XF (2003) Domains of Gln3p interacting with karyopherins, Ure2p, and the target of rapamycin protein. J Biol Chem 278:16878–16886

Chen C, Contreras R (2007) Identifying genes that extend life span using a high-throughput screening system. Methods Mol Biol 371:237–248

Cheng WC, Leach KM, Hardwick JM (2008) Mitochondrial death pathways in yeast and mammalian cells. Biochim Biophys Acta 1783:1272–1279

Chiocchetti A, Zhou J, Zhu H, Karl T, Haubenreisser O, Rinnerthaler M, Heeren G, Oender K, Bauer J, Hintner H et al (2007) Ribosomal proteins Rpl10 and Rps6 are potent regulators of yeast replicative life span. Exp Gerontol 42:275–286

Cohen S, Segal D (2009) Extrachromosomal circular DNA in eukaryotes: possible involvement in the plasticity of tandem repeats. Cytogenet Genome Res 124:327–338

Cohen HY, Miller C, Bitterman KJ, Wall NR, Hekking B, Kessler B, Howitz KT, Gorospe M, de Cabo R, Sinclair DA (2004) Calorie restriction promotes mammalian cell survival by inducing the SIRT1 deacetylase. Science 305:390–392

Cohen S, Agmon N, Sobol O, Segal D (2010) Extrachromosomal circles of satellite repeats and 5S ribosomal DNA in human cells. Mob DNA 1:11

Colman RJ, Anderson RM, Johnson SC, Kastman EK, Kosmatka KJ, Beasley TM, Allison DB, Cruzen C, Simmons HA, Kemnitz JW et al (2009) Caloric restriction delays disease onset and mortality in rhesus monkeys. Science 325:201–204

Crespo JL, Powers T, Fowler B, Hall MN (2002) The TOR-controlled transcription activators GLN3, RTG1, and RTG3 are regulated in response to intracellular levels of glutamine. Proc Natl Acad Sci U S A 99:6784–6789

Curran SP, Ruvkun G (2007) Lifespan regulation by evolutionarily conserved genes essential for viability. PLoS Genet 3:e56

D'Mello NP, Childress AM, Franklin DS, Kale SP, Pinswasdi C, Jazwinski SM (1994) Cloning and characterization of LAG1, a longevity-assurance gene in yeast. J Biol Chem 269:15451–15459

Dang W, Steffen KK, Perry R, Dorsey JA, Johnson FB, Shilatifard A, Kaeberlein M, Kennedy BK, Berger SL (2009) Histone H4 lysine 16 acetylation regulates cellular lifespan. Nature 459:802–807

Dange T, Smith D, Noy T, Rommel PC, Jurzitza L, Cordero RJ, Legendre A, Finley D, Goldberg AL, Schmidt M (2011) Blm10 protein promotes proteasomal substrate turnover by an active gating mechanism. J Biol Chem 286:42830–42839

Dann SG, Thomas G (2006) The amino acid sensitive TOR pathway from yeast to mammals. FEBS Lett 580:2821–2829

De Virgilio C, Loewith R (2006) The TOR signalling network from yeast to man. Int J Biochem Cell Biol 38:1476–1481

Defossez PA, Prusty R, Kaeberlein M, Lin SJ, Ferrigno P, Silver PA, Keil RL, Guarente L (1999) Elimination of replication block protein Fob1 extends the life span of yeast mother cells. Mol Cell 3:447–455

Delaney JR, Murakami C, Chou A, Carr D, Schleit J, Sutphin GL, An EH, Castanza AS, Fletcher M, Goswami S et al (2013) Dietary restriction and mitochondrial function link replicative and chronological aging in Saccharomyces cerevisiae. Exp Gerontol 8(10):1006–1013

di Blasi F, Carra E, de Vendittis E, Masturzo P, Burderi E, Lambrinoudaki I, Mirisola MG, Seidita G, Fasano O (1993) The SCH9 protein kinase mRNA contains a long 5' leader with a small open reading frame. Yeast 9:21–32

Dryden SC, Nahhas FA, Nowak JE, Goustin AS, Tainsky MA (2003) Role for human SIRT2 NAD-dependent deacetylase activity in control of mitotic exit in the cell cycle. Mol Cell Biol 23:3173–3185

Easlon E, Tsang F, Dilova I, Wang C, Lu SP, Skinner C, Lin SJ (2007) The dihydrolipoamide acetyltransferase is a novel metabolic longevity factor and is required for calorie restriction-mediated life span extension. J Biol Chem 282:6161–6171

Eisenberg T, Knauer H, Schauer A, Buttner S, Ruckenstuhl C, Carmona-Gutierrez D, Ring J, Schroeder S, Magnes C, Antonacci L et al (2009) Induction of autophagy by spermidine promotes longevity. Nat Cell Biol 11:1305–1314

Epstein CJ, Martin GM, Schultz AL, Motulsky AG (1966) Werner's syndrome a review of its symptomatology, natural history, pathologic features, genetics and relationship to the natural aging process. Medicine 45:177–221

Epstein CB, Waddle JA, Hale W 4th, Dave V, Thornton J, Macatee TL, Garner HR, Butow RA (2001) Genome-wide responses to mitochondrial dysfunction. Mol Biol Cell 12:297–308

Erjavec N, Nystrom T (2007) Sir2p-dependent protein segregation gives rise to a superior reactive oxygen species management in the progeny of Saccharomyces cerevisiae. Proc Natl Acad Sci U S A 104:10877–10881

Erjavec N, Larsson L, Grantham J, Nystrom T (2007) Accelerated aging and failure to segregate damaged proteins in Sir2 mutants can be suppressed by overproducing the protein aggregation-remodeling factor Hsp104p. Genes Dev 21:2410–2421

Fabrizio P, Longo VD (2007) The chronological life span of Saccharomyces cerevisiae. Methods Mol Biol 371:89–95

Fabrizio P, Pozza F, Pletcher SD, Gendron CM, Longo VD (2001) Regulation of longevity and stress resistance by Sch9 in yeast. Science 292:288–290

Fabrizio P, Pletcher SD, Minois N, Vaupel JW, Longo VD (2004) Chronological aging-independent replicative life span regulation by Msn2/Msn4 and Sod2 in Saccharomyces cerevisiae. FEBS Lett 557:136–142

Feser J, Truong D, Das C, Carson JJ, Kieft J, Harkness T, Tyler JK (2010) Elevated histone expression promotes life span extension. Mol Cell 39:724–735

Finkel T, Deng CX, Mostoslavsky R (2009) Recent progress in the biology and physiology of sirtuins. Nature 460:587–591

Fontana L, Partridge L, Longo VD (2010) Extending healthy life span–from yeast to humans. Science 328:321–326

Fuentealba LC, Eivers E, Geissert D, Taelman V, De Robertis EM (2008) Asymmetric mitosis: Unequal segregation of proteins destined for degradation. Proc Natl Acad Sci U S A 105:7732–7737

Gallinetti J, Harputlugil E, Mitchell JR (2013) Amino acid sensing in dietary-restriction-mediated longevity: roles of signal-transducing kinases GCN2 and TOR. Biochem J 449:1–10

Ganley AR, Breitenbach M, Kennedy BK, Kobayashi T (2012) Yeast hypertrophy: cause or consequence of aging? Reply to Bilinski et al. FEMS Yeast Res 12:267–268

Gebre S, Connor R, Xia Y, Jawed S, Bush JM, Bard M, Elsalloukh H, Tang F (2012) Osh6 overexpression extends the lifespan of yeast by increasing vacuole fusion. Cell Cycle 11:2176–2188

Gems D, Partridge L (2013) Genetics of longevity in model organisms: debates and paradigm shifts. Annu Rev Physiol 75:621–644

Ghazi A, Henis-Korenblit S, Kenyon C (2007) Regulation of Caenorhabditis elegans lifespan by a proteasomal E3 ligase complex. Proc Natl Acad Sci U S A 104:5947–5952

Giblin W, Skinner ME, Lombard DB (2014) Sirtuins: guardians of mammalian healthspan. Trends Genet 30(7):271–286

Gillespie CS, Proctor CJ, Boys RJ, Shanley DP, Wilkinson DJ, Kirkwood TB (2004) A mathematical model of ageing in yeast. J Theor Biol 229:189–196

Goffeau A, Barrell BG, Bussey H, Davis RW, Dujon B, Feldmann H, Galibert F, Hoheisel JD, Jacq C, Johnston M et al (1996) Life with 6000 genes. Science 274:546, 563–547

Goldberg AA, Richard VR, Kyryakov P, Bourque SD, Beach A, Burstein MT, Glebov A, Koupaki O, Boukh-Viner T, Gregg C et al (2010) Chemical genetic screen identifies lithocholic acid as

an anti-aging compound that extends yeast chronological life span in a TOR-independent manner, by modulating housekeeping longevity assurance processes. Aging 2:393–414
Gresham D, Boer VM, Caudy A, Ziv N, Brandt NJ, Storey JD, Botstein D (2011) System-level analysis of genes and functions affecting survival during nutrient starvation in Saccharomyces cerevisiae. Genetics 187:299–317
Guarente L (1999) Diverse and dynamic functions of the Sir silencing complex. Nat Genet 23:281–285
Guarente L (2005) Calorie restriction and SIR2 genes–towards a mechanism. Mech Ageing Dev 126:923–928
Guarente L (2006) Sirtuins as potential targets for metabolic syndrome. Nature 444:868–874
Guarente L (2011) Franklin H. Epstein Lecture: Sirtuins, aging, and medicine. N Engl J Med 364:2235–2244
Guarente L (2013) Calorie restriction and sirtuins revisited. Genes Dev 27:2072–2085
Ha CW, Huh WK (2011) Rapamycin increases rDNA stability by enhancing association of Sir2 with rDNA in Saccharomyces cerevisiae. Nucleic Acids Res 39:1336–1350
Hansen M, Taubert S, Crawford D, Libina N, Lee SJ, Kenyon C (2007) Lifespan extension by conditions that inhibit translation in Caenorhabditis elegans. Aging Cell 6:95–110
Hansen M, Chandra A, Mitic LL, Onken B, Driscoll M, Kenyon C (2008) A role for autophagy in the extension of lifespan by dietary restriction in C elegans. PLoS Genet 4:e24
Harris N, MacLean M, Hatzianthis K, Panaretou B, Piper PW (2001) Increasing Saccharomyces cerevisiae stress resistance, through the overactivation of the heat shock response resulting from defects in the Hsp90 chaperone, does not extend replicative life span but can be associated with slower chronological ageing of nondividing cells. Mol Genet Genomics 265:258–263
Harris N, Costa V, MacLean M, Mollapour M, Moradas-Ferreira P, Piper PW (2003) Mnsod overexpression extends the yeast chronological (G(0)) life span but acts independently of Sir2p histone deacetylase to shorten the replicative life span of dividing cells. Free Radic Biol Med 34:1599–1606
Harrison DE, Strong R, Sharp ZD, Nelson JF, Astle CM, Flurkey K, Nadon NL, Wilkinson JE, Frenkel K, Carter CS et al (2009) Rapamycin fed late in life extends lifespan in genetically heterogeneous mice. Nature 460:392–395
Hay N, Sonenberg N (2004) Upstream and downstream of mTOR. Genes Dev 18:1926–1945
He C, Tsuchiyama SK, Nguyen QT, Plyusnina EN, Terrill SR, Sahibzada S, Patel B, Faulkner AR, Shaposhnikov MV, Tian R et al (2014) Enhanced longevity by ibuprofen, conserved in multiple species, occurs in yeast through inhibition of tryptophan import. PLoS genetics 10, e1004860.
Heeren G, Rinnerthaler M, Laun P, von Seyerl P, Kossler S, Klinger H, Hager M, Bogengruber E, Jarolim S, Simon-Nobbe B et al (2009) The mitochondrial ribosomal protein of the large subunit, Afo1p, determines cellular longevity through mitochondrial back-signaling via TOR1. Aging 1:622–636
Heitman J, Movva NR, Hall MN (1991) Targets for cell cycle arrest by the immunosuppressant rapamycin in yeast. Science 253:905–909
Herranz D, Munoz-Martin M, Canamero M, Mulero F, Martinez-Pastor B, Fernandez-Capetillo O, Serrano M (2010) Sirt1 improves healthy ageing and protects from metabolic syndrome-associated cancer. Nat Commun 1:3
Hill SM, Hao X, Liu B, Nystrom T (2014) Life-span extension by a metacaspase in the yeast Saccharomyces cerevisiae. Science 344, 1389–1392
Hill SM, Nystrom T (2015) The dual role of a yeast metacaspase: What doesn't kill you makes you stronger. BioEssays : news and reviews in molecular, cellular and developmental biology.
Hinnebusch AG (2005) Translational regulation of GCN4 and the general amino acid control of yeast. Annu Rev Microbiol 59:407–450
Hinnebusch AG, Natarajan K (2002) Gcn4p, a master regulator of gene expression, is controlled at multiple levels by diverse signals of starvation and stress. Eukaryot Cell 1:22–32
Houtkooper RH, Pirinen E, Auwerx J (2012) Sirtuins as regulators of metabolism and healthspan. Nat Rev Mol Cell Biol 13:225–238

Howitz KT, Bitterman KJ, Cohen HY, Lamming DW, Lavu S, Wood JG, Zipkin RE, Chung P, Kisielewski A, Zhang LL et al (2003) Small molecule activators of sirtuins extend Saccharomyces cerevisiae lifespan. Nature 425:191–196

Hu J, Wei M, Mirzaei H, Madia F, Mirisola M, Amparo C, Chagoury S, Kennedy B, Longo VD (2014) Tor-Sch9 deficiency activates catabolism of the ketone body-like acetic acid to promote trehalose accumulation and longevity. Aging Cell 13:457–467

Huang X, Liu J, Dickson RC (2012) Down-regulating sphingolipid synthesis increases yeast lifespan. PLoS Genet 8:e1002493

Huang X, Withers BR, Dickson RC (2014) Sphingolipids and lifespan regulation. Biochim Biophys Acta 1841:657–664

Hughes AL, Gottschling DE (2012) An early age increase in vacuolar pH limits mitochondrial function and lifespan in yeast. Nature 492:261–265

Imai S, Guarente L (2010) Ten years of NAD-dependent SIR2 family deacetylases: implications for metabolic diseases. Trends Pharmacol Sci 31:212–220

Imai SI, Guarente L (2014) NAD and sirtuins in aging and disease. Trends Cell Biol 24(8):464–471

Imai S, Armstrong CM, Kaeberlein M, Guarente L (2000) Transcriptional silencing and longevity protein Sir2 is an NAD-dependent histone deacetylase. Nature 403:795–800

Ivy JM, Klar AJ, Hicks JB (1986) Cloning and characterization of four SIR genes of Saccharomyces cerevisiae. Mol Cell Biol 6:688–702

Jazwinski SM (1990) An experimental system for the molecular analysis of the aging process: the budding yeast Saccharomyces cerevisiae. J Gerontol 45:B68–B74

Jazwinski SM (2005a) The retrograde response links metabolism with stress responses, chromatin-dependent gene activation, and genome stability in yeast aging. Gene 354:22–27

Jazwinski SM (2005b) Rtg2 protein: at the nexus of yeast longevity and aging. FEMS Yeast Res 5:1253–1259

Jia K, Chen D, Riddle DL (2004) The TOR pathway interacts with the insulin signaling pathway to regulate C. elegans larval development, metabolism and life span. Development 131:3897–3906

Jiang JC, Jaruga E, Repnevskaya MV, Jazwinski SM (2000) An intervention resembling caloric restriction prolongs life span and retards aging in yeast. FASEB J 14:2135–2137

Johnson JE, Johnson FB (2014) Methionine restriction activates the retrograde response and confers both stress tolerance and lifespan extension to yeast, mouse and human cells. PloS one 9, e97729.

Johnston M, Hillier L, Riles L, Albermann K, Andre B, Ansorge W, Benes V, Bruckner M, Delius H, Dubois E et al (1997) The nucleotide sequence of Saccharomyces cerevisiae chromosome XII. Nature 387:87–90

Kaeberlein M (2006) Longevity and aging in the budding yeast. In: Conn PM (ed) Handbook of models for human aging. Elsevier Press, Boston, pp 109–120

Kaeberlein M (2010) Lessons on longevity from budding yeast. Nature 464:513–519

Kaeberlein M (2012) Hypertrophy and senescence factors in yeast aging. A reply to Bilinski et al. FEMS Yeast Res 12:269–270

Kaeberlein M, Kennedy BK (2005) Large-scale identification in yeast of conserved ageing genes. Mech Ageing Dev 126:17–21

Kaeberlein M, Powers RW 3rd (2007) Sir2 and calorie restriction in yeast: a skeptical perspective. Ageing Res Rev 6:128–140

Kaeberlein M, McVey M, Guarente L (1999) The SIR2/3/4 complex and SIR2 alone promote longevity in Saccharomyces cerevisiae by two different mechanisms. Genes Dev 13:2570–2580

Kaeberlein M, Kirkland KT, Fields S, Kennedy BK (2004a) Sir2-independent life span extension by calorie restriction in yeast. PLoS Biol 2:E296

Kaeberlein M, Kirkland KT, Fields S, Kennedy BK (2004b) Sir2-independent life span extension by calorie restriction in yeast. PLoS Biol 2:1381–1387

Kaeberlein M, Hu D, Kerr EO, Tsuchiya M, Westman EA, Dang N, Fields S, Kennedy BK (2005a) Increased life span due to calorie restriction in respiratory-deficient yeast. PLoS Genet 1:e69

Kaeberlein M, Kirkland KT, Fields S, Kennedy BK (2005b) Genes determining yeast replicative life span in a long-lived genetic background. Mech Ageing Dev 126:491–504

Kaeberlein M, Powers RW 3rd, Steffen KK, Westman EA, Hu D, Dang N, Kerr EO, Kirkland KT, Fields S, Kennedy BK (2005c) Regulation of yeast replicative life span by TOR and Sch9 in response to nutrients. Science 310:1193–1196

Kampinga HH (2014) Cell biology. A cell death avenue evolved from a life-saving path. Science 344, 1341–1342

Kanfi Y, Naiman S, Amir G, Peshti V, Zinman G, Nahum L, Bar-Joseph Z, Cohen HY (2012) The sirtuin SIRT6 regulates lifespan in male mice. Nature 483:218–221

Kapahi P, Zid BM, Harper T, Koslover D, Sapin V, Benzer S (2004) Regulation of lifespan in Drosophila by modulation of genes in the TOR signaling pathway. Curr Biol 14:885–890

Kennedy BK, Pennypacker JK (2014) Drugs that modulate aging: the promising yet difficult path ahead. Transl Res 163:456–465

Kennedy BK, Austriaco NR Jr, Zhang J, Guarente L (1995) Mutation in the silencing gene SIR4 can delay aging in S. cerevisiae. Cell 80:485–496

Kennedy BK, Smith ED, Kaeberlein M (2005) The enigmatic role of Sir2 in aging. Cell 123:548–550

Kennedy BK, Steffen KK, Kaeberlein M (2007) Ruminations on dietary restriction and aging. Cell Mol Life Sci 64:1323–1328

Kirchman PA, Kim S, Lai CY, Jazwinski SM (1999) Interorganelle signaling is a determinant of longevity in Saccharomyces cerevisiae. Genetics 152:179–190

Kobayashi T, Horiuchi T (1996) A yeast gene product, Fob1 protein, required for both replication fork blocking and recombinational hotspot activities. Genes Cells 1:465–474

Koch MR, Pillus L (2009) The glucanosyltransferase Gas1 functions in transcriptional silencing. Proc Natl Acad Sci U S A 106:11224–11229

Komarova EA, Antoch MP, Novototskaya LR, Chernova OB, Paszkiewicz G, Leontieva OV, Blagosklonny MV, Gudkov AV (2012) Rapamycin extends lifespan and delays tumorigenesis in heterozygous p53+/– mice. Aging 4:709–714

Komeili A, Wedaman KP, O'Shea EK, Powers T (2000) Mechanism of metabolic control. Target of rapamycin signaling links nitrogen quality to the activity of the Rtg1 and Rtg3 transcription factors. J Cell Biol 151:863–878

Komili S, Farny NG, Roth FP, Silver PA (2007) Functional specificity among ribosomal proteins regulates gene expression. Cell 131:557–571

Kruegel U, Robison B, Dange T, Kahlert G, Delaney JR, Kotireddy S, Tsuchiya M, Tsuchiyama S, Murakami CJ, Schleit J et al (2011) Elevated proteasome capacity extends replicative lifespan in Saccharomyces cerevisiae. PLoS Genet 7:e1002253

Kujoth GC, Bradshaw PC, Haroon S, Prolla TA (2007) The role of mitochondrial DNA mutations in mammalian aging. PLoS Genet 3:e24

Kwan EX, Foss EJ, Tsuchiyama S, Alvino GM, Kruglyak L, Kaeberlein M, Raghuraman MK, Brewer BJ, Kennedy BK, Bedalov A (2013) A natural polymorphism in rDNA replication origins links origin activation with calorie restriction and lifespan. PLoS Genet 9:e1003329

Lamming DW, Latorre-Esteves M, Medvedik O, Wong SN, Tsang FA, Wang C, Lin SJ, Sinclair DA (2005) HST2 mediates SIR2-independent life-span extension by calorie restriction. Science 309:1861–1864

Lamming DW, Ye L, Sabatini DM, Baur JA (2013) Rapalogs and mTOR inhibitors as anti-aging therapeutics. J Clin Invest 123:980–989

Landry J, Sutton A, Tafrov ST, Heller RC, Stebbins J, Pillus L, Sternglanz R (2000) The silencing protein SIR2 and its homologs are NAD-dependent protein deacetylases. Proc Natl Acad Sci U S A 97:5807–5811

Lee SS, Avalos Vizcarra I, Huberts DH, Lee LP, Heinemann M (2012) Whole lifespan microscopic observation of budding yeast aging through a microfluidic dissection platform. Proc Natl Acad Sci U S A 109:4916–4920

Leontieva OV, Blagosklonny MV (2011) Yeast-like chronological senescence in mammalian cells: phenomenon, mechanism and pharmacological suppression. Aging (Albany NY) 3:1078–1091

Li W, Miller RA (2015) Elevated ATF4 function in fibroblasts and liver of slow-aging mutant mice. J Gerontol A Biol Sci Med Sci 70(3):263–272

Li M, Valsakumar V, Poorey K, Bekiranov S, Smith JS (2013) Genome-wide analysis of functional sirtuin chromatin targets in yeast. Genome Biol 14:R48

Lin SJ, Guarente L (2006) Increased life span due to calorie restriction in respiratory-deficient yeast. PLoS Genet 2:e33; author reply e34

Lin SJ, Defossez PA, Guarente L (2000) Requirement of NAD and SIR2 for life-span extension by calorie restriction in Saccharomyces cerevisiae. Science 289:2126–2128

Lin SJ, Kaeberlein M, Andalis AA, Sturtz LA, Defossez PA, Culotta VC, Fink GR, Guarente L (2002) Calorie restriction extends Saccharomyces cerevisiae lifespan by increasing respiration. Nature 418:344–348

Lindstrom DL, Gottschling DE (2009) The mother enrichment program: a genetic system for facile replicative life span analysis in Saccharomyces cerevisiae. Genetics 183:413–422, 411SI–413SI

Lindstrom DL, Leverich CK, Henderson KA, Gottschling DE (2011) Replicative age induces mitotic recombination in the ribosomal RNA gene cluster of Saccharomyces cerevisiae. PLoS Genet 7:e1002015

Liu Z, Butow RA (2006) Mitochondrial retrograde signaling. Annu Rev Genet 40:159–185

Liu B, Larsson L, Caballero A, Hao X, Oling D, Grantham J, Nystrom T (2010) The polarisome is required for segregation and retrograde transport of protein aggregates. Cell 140:257–267

Loewith R, Jacinto E, Wullschleger S, Lorberg A, Crespo JL, Bonenfant D, Oppliger W, Jenoe P, Hall MN (2002) Two TOR complexes, only one of which is rapamycin sensitive, have distinct roles in cell growth control. Mol Cell 10:457–468

Longo VD, Gralla EB, Valentine JS (1996) Superoxide dismutase activity is essential for stationary phase survival in Saccharomyces cerevisiae. Mitochondrial production of toxic oxygen species in vivo. J Biol Chem 271:12275–12280

Longo VD, Shadel GS, Kaeberlein M, Kennedy B (2012) Replicative and chronological aging in Saccharomyces cerevisiae. Cell Metab 16:18–31

Lorenz DR, Cantor CR, Collins JJ (2009) A network biology approach to aging in yeast. Proc Natl Acad Sci U S A 106:1145–1150

Lu JY, Lin YY, Sheu JC, Wu JT, Lee FJ, Chen Y, Lin MI, Chiang FT, Tai TY, Berger SL et al (2011) Acetylation of yeast AMPK controls intrinsic aging independently of caloric restriction. Cell 146:969–979

Managbanag JR, Witten TM, Bonchev D, Fox LA, Tsuchiya M, Kennedy BK, Kaeberlein M (2008) Shortest-path network analysis is a useful approach toward identifying genetic determinants of longevity. PLoS One 3:e3802

Martegani E, Baroni MD, Frascotti G, Alberghina L (1986) Molecular cloning and transcriptional analysis of the start gene CDC25 of Saccharomyces cerevisiae. EMBO J 5:2363–2369

Masoro EJ (2005) Overview of caloric restriction and ageing. Mech Ageing Dev 126:913–922

Matecic M, Smith DL, Pan X, Maqani N, Bekiranov S, Boeke JD, Smith JS (2010) A microarray-based genetic screen for yeast chronological aging factors. PLoS Genet 6:e1000921

Mattison JA, Roth GS, Beasley TM, Tilmont EM, Handy AM, Herbert RL, Longo DL, Allison DB, Young JE, Bryant M et al (2012) Impact of caloric restriction on health and survival in rhesus monkeys from the NIA study. Nature 489:318–321

McCay CM, Crowell MF, Maynard LA (1989) The effect of retarded growth upon the length of life span and upon the ultimate body size. 1935. Nutrition 5:155–171; discussion 172

McCormick MA, Kennedy BK (2012) Genome-scale studies of aging: challenges and opportunities. Curr Genomics 13:500–507

McIntosh KB, Warner JR (2007) Yeast ribosomes: variety is the spice of life. Cell 131:450–451

McVey M, Kaeberlein M, Tissenbaum HA, Guarente L (2001) The short life span of Saccharomyces cerevisiae sgs1 and srs2 mutants is a composite of normal aging processes and mitotic arrest due to defective recombination. Genetics 157:1531–1542

Medvedik O, Lamming DW, Kim KD, Sinclair DA (2007a) *MSN2* and *MSN4* link calorie restriction and TOR to Sirtuin-mediated lifespan extension in *Saccharomyces cerevisiae*. PLoS Biol 5:230–241

Medvedik O, Lamming DW, Kim KD, Sinclair DA (2007b) MSN2 and MSN4 link calorie restriction and TOR to sirtuin-mediated lifespan extension in Saccharomyces cerevisiae. PLoS Biol 5:e261

Michan S, Sinclair D (2007) Sirtuins in mammals: insights into their biological function. Biochem J 404:1–13

Mirisola MG, Taormina G, Fabrizio P, Wei M, Hu J, Longo VD (2014) Serine- and threonine/valine-dependent activation of PDK and Tor orthologs converge on Sch9 to promote aging. PLoS genetics 10, e1004113

Mohammadi S, Subramaniam S, Grama A (2013) Inferring the effective TOR-dependent network: a computational study in yeast. BMC Syst Biol 7:84

Mortimer RK, Johnston JR (1959) Life span of individual yeast cells. Nature 183:1751–1752

Motta MC, Divecha N, Lemieux M, Kamel C, Chen D, Gu W, Bultsma Y, McBurney M, Guarente L (2004) Mammalian SIRT1 represses forkhead transcription factors. Cell 116:551–563

Murakami C, Kaeberlein M (2009) Quantifying yeast chronological life span by outgrowth of aged cells. J Vis Exp (27). pii:1156

Murakami CJ, Burtner CR, Kennedy BK, Kaeberlein M (2008) A method for high-throughput quantitative analysis of yeast chronological life span. J Gerontol A Biol Sci Med Sci 63:113–121

Murakami CJ, Wall V, Basisty N, Kaeberlein M (2011) Composition and acidification of the culture medium influences chronological aging similarly in vineyard and laboratory yeast. PLoS One 6:e24530

Nemoto S, Fergusson MM, Finkel T (2004) Nutrient availability regulates SIRT1 through a forkhead-dependent pathway. Science 306:2105–2108

Oberdoerffer P, Michan S, McVay M, Mostoslavsky R, Vann J, Park SK, Hartlerode A, Stegmuller J, Hafner A, Loerch P et al (2008) SIRT1 redistribution on chromatin promotes genomic stability but alters gene expression during aging. Cell 135:907–918

Ocampo A, Liu J, Schroeder EA, Shadel GS, Barrientos A (2012) Mitochondrial respiratory thresholds regulate yeast chronological life span and its extension by caloric restriction. Cell Metab 16:55–67

Olsen B, Murakami CJ, Kaeberlein M (2010) YODA: software to facilitate high-throughput analysis of chronological life span, growth rate, and survival in budding yeast. BMC Bioinformatics 11:141

Orentreich N, Matias JR, DeFelice A, Zimmerman JA (1993) Low methionine ingestion by rats extends life span. J Nutr 123, 269–274.

Outeiro TF, Giorgini F (2006) Yeast as a drug discovery platform in Huntington's and Parkinson's diseases. Biotechnol J 1:258–269

Pan Y, Shadel GS (2009) Extension of chronological life span by reduced TOR signaling requires down-regulation of Sch9p and involves increased mitochondrial OXPHOS complex density. Aging (Albany NY) 1:131–145

Pan KZ, Palter JE, Rogers AN, Olsen A, Chen D, Lithgow GJ, Kapahi P (2007) Inhibition of mRNA translation extends lifespan in Caenorhabditis elegans. Aging Cell 6:111–119

Pan Y, Schroeder EA, Ocampo A, Barrientos A, Shadel GS (2011) Regulation of yeast chronological life span by TORC1 via adaptive mitochondrial ROS signaling. Cell Metab 13:668–678

Paradis S, Ruvkun G (1998) Caenorhabditis elegans Akt/PKB transduces insulin receptor-like signals from AGE-1 PI3 kinase to the DAF-16 transcription factor. Genes Dev 12:2488–2498

Pereira C, Leao M, Soares J, Bessa C, Saraiva L (2012) New therapeutic strategies for cancer and neurodegeneration emerging from yeast cell-based systems. Curr Pharm Des 18:4223–4235

Perez VI, Buffenstein R, Masamsetti V, Leonard S, Salmon AB, Mele J, Andziak B, Yang T, Edrey Y, Friguet B et al (2009) Protein stability and resistance to oxidative stress are determinants of longevity in the longest-living rodent, the naked mole-rat. Proc Natl Acad Sci U S A 106:3059–3064

Petes TD (1979) Yeast ribosomal DNA genes are located on chromosome XII. Proc Natl Acad Sci U S A 76:410–414

Piper PW, Harris NL, MacLean M (2006) Preadaptation to efficient respiratory maintenance is essential both for maximal longevity and the retention of replicative potential in chronologically ageing yeast. Mech Ageing Dev 127:733–740
Ploger R, Zhang J, Bassett D, Reeves R, Hieter P, Boguski M, Spencer F (2000) XREFdb: cross-referencing the genetics and genes of mammals and model organisms. Nucleic Acids Res 28:120–122
Powers T (2007) TOR signaling and S6 kinase 1: Yeast catches up. Cell Metab 6:1–2
Powers RW 3rd, Kaeberlein M, Caldwell SD, Kennedy BK, Fields S (2006) Extension of chronological life span in yeast by decreased TOR pathway signaling. Genes Dev 20:174–184
Richie JP, Leutzinger Y, Parthasarathy S, Malloy V, Orentreich N, Zimmerman JA (1994) Methionine restriction increases blood glutathione and longevity in F344 rats. FASEB journal: official publication of the Federation of American Societies for Experimental Biology 8, 1302–1307
Rine J, Herskowitz I (1987) Four genes responsible for a position effect on expression from HML and HMR in Saccharomyces cerevisiae. Genetics 116:9–22
Rizki G, Iwata TN, Li J, Riedel CG, Picard CL, Jan M, Murphy CT, Lee SS (2011) The evolutionarily conserved longevity determinants HCF-1 and SIR-2.1/SIRT1 collaborate to regulate DAF-16/FOXO. PLoS Genet 7:e1002235
Robertson LS, Causton HC, Young RA, Fink GR (2000) The yeast A kinases differentially regulate iron uptake and respiratory function. Proc Natl Acad Sci U S A 97:5984–5988
Rothermel BA, Shyjan AW, Etheredge JL, Butow RA (1995) Transactivation by Rtg1p, a basic helix-loop-helix protein that functions in communication between mitochondria and the nucleus in yeast. J Biol Chem 270:29476–29482
Rothermel BA, Thornton JL, Butow RA (1997) Rtg3p, a basic helix-loop-helix/leucine zipper protein that functions in mitochondrial-induced changes in gene expression, contains independent activation domains. J Biol Chem 272:19801–19807
Rubel AA, Ryzhova TA, Antonets KS, Chernoff YO, Galkin A (2013) Identification of PrP sequences essential for the interaction between the PrP polymers and Abeta peptide in a yeast-based assay. Prion 7:469–476
Rujano MA, Bosveld F, Salomons FA, Dijk F, van Waarde MA, van der Want JJ, de Vos RA, Brunt ER, Sibon OC, Kampinga HH (2006) Polarised asymmetric inheritance of accumulated protein damage in higher eukaryotes. PLoS Biol 4:e417
Saka K, Ide S, Ganley AR, Kobayashi T (2013) Cellular senescence in yeast is regulated by rDNA noncoding transcription. Curr Biol 23:1794–1798
Santangelo GM (2006) Glucose signaling in Saccharomyces cerevisiae. Microbiol Mol Biol Rev 70:253–282
Scheckhuber CQ, Erjavec N, Tinazli A, Hamann A, Nystrom T, Osiewacz HD (2007) Reducing mitochondrial fission results in increased life span and fitness of two fungal ageing models. Nat Cell Biol 9:99–105
Scheckhuber CQ, Wanger RA, Mignat CA, Osiewacz HD (2011) Unopposed mitochondrial fission leads to severe lifespan shortening. Cell cycle 10, 3105-3110.
Sedensky MM, Morgan PG (2006) Mitochondrial respiration and reactive oxygen species in C. elegans. Exp Gerontol 41:957–967
Seichertova O, Beran K, Holan Z, Pokorny V (1975) The chitin-glucan complex of Saccharomyces cerevisiae III. Electron-microscopic study of the prebudding stage. Folia Microbiol (Praha) 20:371–378
Selman C, Tullet JM, Wieser D, Irvine E, Lingard SJ, Choudhury AI, Claret M, Al-Qassab H, Carmignac D, Ramadani F et al (2009) Ribosomal protein S6 kinase 1 signaling regulates mammalian life span. Science 326:140–144
Shama S, Lai CY, Antoniazzi JM, Jiang JC, Jazwinski SM (1998) Heat stress-induced life span extension in yeast. Exp Cell Res 245:379–388
Sharma PK, Agrawal V, Roy N (2011) Mitochondria-mediated hormetic response in life span extension of calorie-restricted Saccharomyces cerevisiae. Age 33:143–154
Shawi M, Autexier C (2008) Telomerase, senescence and ageing. Mech Ageing Dev 129:3–10

Shringarpure R, Davies KJ (2002) Protein turnover by the proteasome in aging and disease. Free Radic Biol Med 32:1084–1089
Sinclair DA, Guarente L (1997) Extrachromosomal rDNA circles–a cause of aging in yeast. Cell 91:1033–1042
Smeal T, Claus J, Kennedy B, Cole F, Guarente L (1996) Loss of transcriptional silencing causes sterility in old mother cells of S. cerevisiae. Cell 84:633–642
Smith JS, Boeke JD (1997) An unusual form of transcriptional silencing in yeast ribosomal DNA. Genes Dev 11:241–254
Smith JS, Brachmann CB, Celic I, Kenna MA, Muhammad S, Starai VJ, Avalos JL, Escalante-Semerena JC, Grubmeyer C, Wolberger C et al (2000) A phylogenetically conserved NAD+-dependent protein deacetylase activity in the Sir2 protein family. Proc Natl Acad Sci U S A 97:6658–6663
Smith DL Jr, McClure JM, Matecic M, Smith JS (2007) Calorie restriction extends the chronological lifespan of Saccharomyces cerevisiae independently of the Sirtuins. Aging Cell 6:649–662
Smith ED, Tsuchiya M, Fox LA, Dang N, Hu D, Kerr EO, Johnston ED, Tchao BN, Pak DN, Welton KL et al (2008) Quantitative evidence for conserved longevity pathways between divergent eukaryotic species. Genome Res 18:564–570
Stan R, McLaughlin MM, Cafferkey R, Johnson RK, Rosenberg M, Livi GP (1994) Interaction between FKBP12-rapamycin and TOR involves a conserved serine residue. J Biol Chem 269:32027–32030
Steffen KK, MacKay VL, Kerr EO, Tsuchiya M, Hu D, Fox LA, Dang N, Johnston ED, Oakes JA, Tchao BN et al (2008) Yeast life span extension by depletion of 60s ribosomal subunits is mediated by Gcn4. Cell 133:292–302
Steffen KK, Kennedy BK, Kaeberlein M (2009) Measuring replicative life span in the budding yeast. J Vis Exp (28). pii:1209
Steffen KK, McCormick MA, Pham KM, MacKay VL, Delaney JR, Murakami CJ, Kaeberlein M, Kennedy BK (2012) Ribosome deficiency protects against ER stress in Saccharomyces cerevisiae. Genetics 191:107–118
Steinkraus KA, Kaeberlein M, Kennedy BK (2008) Replicative aging in yeast: the means to the end. Annu Rev Cell Dev Biol 24:29–54
Stockl P, Zankl C, Hutter E, Unterluggauer H, Laun P, Heeren G, Bogengruber E, Herndler-Brandstetter D, Breitenbach M, Jansen-Durr P (2007) Partial uncoupling of oxidative phosphorylation induces premature senescence in human fibroblasts and yeast mother cells. Free Radic Biol Med 43:947–958
Stumpferl SW, Brand SE, Jiang JC, Korona B, Tiwari A, Dai J, Seo JG, Jazwinski SM (2012) Natural genetic variation in yeast longevity. Genome Res 22:1963–1973
Suen DF, Norris KL, Youle RJ (2008) Mitochondrial dynamics and apoptosis. Genes Dev 22:1577–1590
Sun J, Kale SP, Childress AM, Pinswasdi C, Jazwinski SM (1994) Divergent roles for *RAS1* and *RAS2* in yeast longevity. J Biol Chem 269:18638–18645
Tate JJ, Cooper TG (2003) Tor1/2 regulation of retrograde gene expression in Saccharomyces cerevisiae derives indirectly as a consequence of alterations in ammonia metabolism. J Biol Chem 278:36924–36933
Tenreiro S, Munder MC, Alberti S, Outeiro TF (2013) Harnessing the power of yeast to unravel the molecular basis of neurodegeneration. J Neurochem 127:438–452
Thevelein JM (1994) Signal transduction in yeast. Yeast 10:1753–1790
Thorpe PH, Bruno J, Rothstein R (2008) Modeling stem cell asymmetry in yeast. Cold Spring Harb Symp Quant Biol 73:81–88
Toda T, Cameron S, Sass P, Zoller M, Scott JD, McMullen B, Hurwitz M, Krebs EG, Wigler M (1987a) Cloning and characterization of BCY1, a locus encoding a regulatory subunit of the cyclic AMP-dependent protein kinase in Saccharomyces cerevisiae. Mol Cell Biol 7:1371–1377
Toda T, Cameron S, Sass P, Zoller M, Wigler M (1987b) Three different genes in S. cerevisiae encode the catalytic subunits of the cAMP-dependent protein kinase. Cell 50:277–287

Traven A, Wong JM, Xu D, Sopta M, Ingles CJ (2001) Interorganellar communication. Altered nuclear gene expression profiles in a yeast mitochondrial DNA mutant. J Biol Chem 276:4020–4027

Tsuchiya M, Dang N, Kerr EO, Hu D, Steffen KK, Oakes JA, Kennedy BK, Kaeberlein M (2006) Sirtuin-independent effects of nicotinamide on lifespan extension from calorie restriction in yeast. Aging Cell 5:505–514

Urban J, Soulard A, Huber A, Lippman S, Mukhopadhyay D, Deloche O, Wanke V, Anrather D, Ammerer G, Riezman H et al (2007) Sch9 is a major target of TORC1 in Saccharomyces cerevisiae. Mol Cell 26:663–674

van der Horst A, Tertoolen LG, de Vries-Smits LM, Frye RA, Medema RH, Burgering BM (2004) FOXO4 is acetylated upon peroxide stress and deacetylated by the longevity protein hSir2(SIRT1). J Biol Chem 279:28873–28879

Vellai T, Takacs-Vellai K, Zhang Y, Kovacs AL, Orosz L, Muller F (2003) Genetics: influence of TOR kinase on lifespan in C elegans. Nature 426:620

Vernace VA, Schmidt-Glenewinkel T, Figueiredo-Pereira ME (2007) Aging and regulated protein degradation: who has the UPPer hand? Aging Cell 6:599–606

Vilchez D, Morantte I, Liu Z, Douglas PM, Merkwirth C, Rodrigues AP, Manning G, Dillin A (2012) RPN-6 determines C. elegans longevity under proteotoxic stress conditions. Nature 489:263–268

Viswanathan M, Guarente L (2011) Regulation of Caenorhabditis elegans lifespan by sir-2.1 transgenes. Nature 477:E1–E2

Walsh RB, Kawasaki G, Fraenkel DG (1983) Cloning of genes that complement yeast hexokinase and glucokinase mutants. J Bacteriol 154:1002–1004

Wang DY, Kumar S, Hedges SB (1999) Divergence time estimates for the early history of animal phyla and the origin of plants, animals and fungi. Proc Biol Sci 266:163–171

Wei M, Fabrizio P, Madia F, Hu J, Ge H, Li LM, Longo VD (2009) Tor1/Sch9-regulated carbon source substitution is as effective as calorie restriction in life span extension. PLoS Genet 5:e1000467

Weindruch R, Walford RL, Fligiel S, Guthrie D (1986) The retardation of aging in mice by dietary restriction: longevity, cancer, immunity and lifetime energy intake. J Nutr 116:641–654

Westermann B (2010) Mitochondrial fusion and fission in cell life and death. Nat Rev Mol Cell Biol 11:872–884

Wierman MB, Smith JS (2014) Yeast sirtuins and the regulation of aging. FEMS Yeast Res 14(1):73–88

Willingham S, Outeiro TF, DeVit MJ, Lindquist SL, Muchowski PJ (2003) Yeast genes that enhance the toxicity of a mutant huntingtin fragment or alpha-synuclein. Science 302:1769–1772

Winzeler EA, Shoemaker DD, Astromoff A, Liang H, Anderson K, Andre B, Bangham R, Benito R, Boeke JD, Bussey H et al (1999) Functional characterization of the S. cerevisiae genome by gene deletion and parallel analysis. Science 285:901–906

Wu Z, Song L, Liu SQ, Huang D (2011) A high throughput screening assay for determination of chronological lifespan of yeast. Exp Gerontol 46:915–922

Wu Z, Song L, Liu SQ, Huang D (2014) Tanshinones extend chronological lifespan in budding yeast Saccharomyces cerevisiae. Applied microbiology and biotechnology 98, 8617-8628

Xie Z, Zhang Y, Zou K, Brandman O, Luo C, Ouyang Q, Li H (2012) Molecular phenotyping of aging in single yeast cells using a novel microfluidic device. Aging Cell 11:599–606

Yang W, Hekimi S (2010) A mitochondrial superoxide signal triggers increased longevity in Caenorhabditis elegans. PLoS Biol 8:e1000556

Yang J, Dungrawala H, Hua H, Manukyan A, Abraham L, Lane W, Mead H, Wright J, Schneider BL (2011) Cell size and growth rate are major determinants of replicative lifespan. Cell Cycle 10:144–155

Yao Y, Tsuchiyama S, Yang C, Bulteau AL, He C, Robison B, Tsuchiya M, Miller D, Briones V, Tar K et al (2015) Proteasomes, Sir2, and Hxk2 form an interconnected aging network that

impinges on the AMPK/Snf1-regulated transcriptional repressor Mig1. PLoS genetics 11, e1004968.
Yizhak K, Gabay O, Cohen H, Ruppin E (2013) Model-based identification of drug targets that revert disrupted metabolism and its application to ageing. Nat Commun 4:2632
Yu CE, Oshima J, Fu YH, Wijsman EM, Hisama F, Alisch R, Matthews S, Nakura J, Miki T, Ouais S et al (1996) Positional cloning of the Werner's syndrome gene. Science 272:258–262
Zhang Y, Luo C, Zou K, Xie Z, Brandman O, Ouyang Q, Li H (2012) Single cell analysis of yeast replicative aging using a new generation of microfluidic device. PLoS One 7:e48275
Zheng XF, Florentino D, Chen J, Crabtree GR, Schreiber SL (1995) TOR kinase domains are required for two distinct functions, only one of which is inhibited by rapamycin. Cell 82:121–130
Zhou C, Slaughter BD, Unruh JR, Eldakak A, Rubinstein B, Li R (2011) Motility and segregation of Hsp104-associated protein aggregates in budding yeast. Cell 147:1186–1196

Germline Stem Cells and Their Roles in the Regulation of Organism Longevity

3

Chih-chun Janet Lin and Meng Carla Wang

Abstract

Aging is a genetically regulated process that happens in all organisms. Lifespan extension can be achieved through several mechanisms, including regulatory signaling from germline stem cells, which will be the focus of this chapter. The free-living nematode *Caenorhabditis elegans* has become the standard workhorse for aging studies due to its fast life cycle, short lifespan, and powerful functional genomics. In this chapter, we will first introduce germline organization and germline stem cell maintenance in *C. elegans*. Next, we will review the knowledge achieved by *C. elegans* research on how gonadal signaling pathways regulate organism longevity. Lastly, the current model of lipid metabolic reprogramming as the link between germline and longevity will be discussed as well.

3.1 Introduction

Aging is a fundamental biological process occurring in the whole animal kingdom. It refers to the gradual loss of normal functions in organs, physiological systems, and ultimately the organism as a whole. Although this phenomenon is a familiar experience for all, it is only recently that people has started to realize aging is not just a simple process of wear and tear but rather a series of complex regulatory processes under tight control. Over the last two decades, groundbreaking work in model organisms has identified many genes and signaling pathways that regulate organism lifespan, including insulin/IGF-1 receptor, forkhead transcription factor, TOR kinase, AMP kinase, and Sirtuin (Blagosklonny et al. 2010; Kenyon 2010b). Strikingly, most of those molecular mechanisms were first linked to lifespan

C.-C.J. Lin • M.C. Wang (✉)
Department of Molecular and Human Genetics, Huffington Center on Aging,
Baylor College of Medicine, Houston, TX 77030, USA
e-mail: chihchul@bcm.edu; wmeng@bcm.edu

H. Geiger et al. (eds.), *Stem Cell Aging: Mechanisms, Consequences, Rejuvenation*,
DOI 10.1007/978-3-7091-1232-8_3

regulation by using a tiny free-living nematode, *Caenorhabditis elegans*, as a genetic model organism and lately shown to be well conserved across other species.

C. elegans is small (~1 mm in length) and transparent and has a fast life cycle, passing through four larval stages to adulthood in approximately 3 days. It also has short lifespan (~2–3 weeks), which is remarkably shorter than 2–3 months in *Drosophila* or 2–3 years in mice. The entire cell lineage of *C. elegans* has been completely mapped and found to be invariant among generations and populations (Sulston et al. 1983). Therefore, the developmental fate of each single cell in those tiny worms is clear throughout their lives. Furthermore, with fully sequenced genome (~97 megabases) (Consortium 1998), and the ease of RNA interference (RNAi) and transgenic techniques, *C. elegans* is highly amenable for genetic manipulation. Together, those features have made *C. elegans* a powerful animal model in studies of development, reproduction, metabolism, evolution, and aging. *C. elegans* is a multicellular organism like human beings, in which the functions of different cells and organs integrate and cooperate to support various biological processes, including aging. *C. elegans* lifespan can be modulated via a cell nonautonomous manner with the nervous system, intestine (the worm fat storage tissue), and germline stem cells (GSCs) serving as a major endocrine signaling network (Russell and Kahn 2007). In this chapter, we will focus on the endocrine function of GSCs and review the knowledge known to date on how those stem cells systemically modulate the aging process of the whole organism in *C. elegans*.

3.2 GSC Maintenance in *C. elegans*

Before we start to discuss the endocrine functions of GSCs in the regulation of organism aging, it is necessary to first overview their organization in the germline, and the signaling events that determine their stem cell fate and functions.

3.2.1 Germline Development in C. elegans

The reproductive system in adult *C. elegans* is consisting of two symmetrical U-shaped arms, connected by a common uterus. In each germline arm of adult worms, all stages of germ cells, from GSCs to differentiated gametes, are present at one time but finely regulated to specialize their positions. Germline specification occurs during embryogenesis and early L1 larval stage (Strome 2005). This process relies on the distribution of P granule at early embryogenesis (Updike and Strome 2010). P granule contains maternal RNAs and is asymmetrically segregated into P blastomeres, which eventually grow into Z2 and Z3 primordial germ cells (Fig. 3.1a). In early L1 worms, the Z2 and Z3 primordial germ cells (PGCs), together with Z1 and Z4 somatic gonad precursors (SGPs), are surrounded within a basement membrane (Fig. 3.1b). These four cells will give rise to the complete reproductive system during the postembryonic development.

Germline proliferation, survival, and differentiation continue through L1 to L3 larval stages, followed by gametogenesis at L4 and early adulthood (Fig. 3.1b) (Hubbard and Greenstein 2005). When nutritional condition is favorable, PGCs undergo MES-mediated chromatin modification to activate proliferation genes, divide symmetrically to expand syncytial GSC population, and ultimately give rise to germline tissues. Meanwhile, each SGP differentiates asymmetrically into five proximal cells and one distal tip cell (DTC). The proximal cells eventually turn into

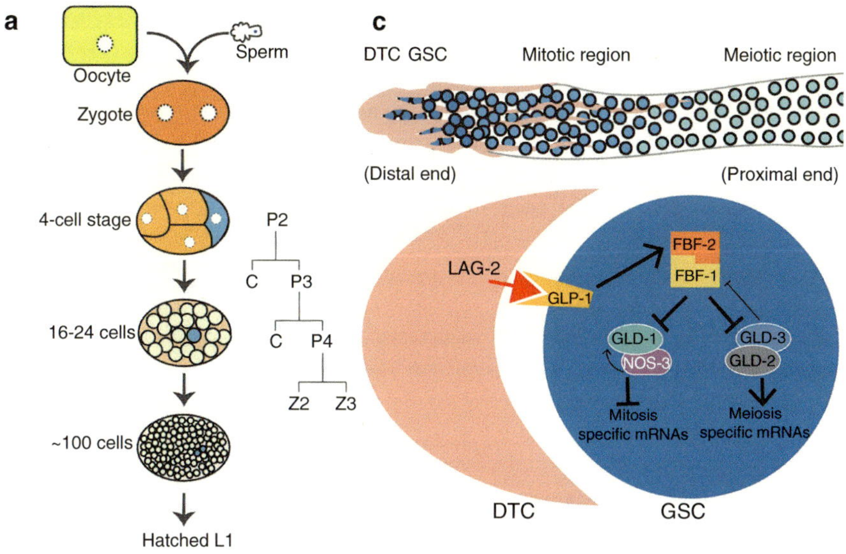

Fig. 3.1 Germline organization in *C. elegans*. (**a**) Germline specification relies on maternally inherited germ granules, called P granules. P granules contain RNA and proteins required for germline determination and are progressively segregated into a series of P blastomeres (highlighted as turquoise cells) during embryonic development. P4 will eventually divide into two PGCs (Z2 and Z3, turquoise) at the ~100-cell stage. (**b**) In postembryonic development, the Z2 and Z3 PGCs will continue to proliferate and comprise the future germ cell population after hatch, which are supported by the DTCs (*red*) and sheath cells developed from Z1 and Z4 SGPs (*white*). The DTCs locate at the distal ends of gonad arms, acting as the niche for GSC maintenance and proliferation. Germ cells proliferate mitotically (*turquoise*) in the distal region; when progressing into the meiotic region, germ cells become meiotic (*light blue*) and undergo sperm (*purple dot*) differentiation at the L4 stage and oogenesis (*yellow*) later during adulthood. The matured oocyte goes through spermatheca and becomes self-fertilized egg (*orange*) in the uterus. (**c**) The DTC projects processes and surrounds the distal end of the gonad arm. Those germ cells adjacent to the DTC remain mitotic; otherwise, they turn into meiotic differentiation program. This is controlled via LAG-2/GLP-1 Notch-like signaling pathway. In brief, the DTC-expressed LAG-2 triggers the activation of its receptor GLP-1 and downstream FEB-1/2 signaling in GSCs. FEB-1/2 maintain mitotic proliferation fate by promoting mitosis-specific mRNA translation via inhibiting GLD-1/NOS-3 and by suppressing meiosis-specific mRNA translation via inhibiting GLD-2/3. When germ cells progress toward the proximal end, they lose the DTC signaling; thus, GLD-1/NOS-3 and GLD-2/3 take the lead to switch on meiotic program. Abbreviation: *PGC* primordial germ cell, *SGP* somatic gonad precursors, *DTC* distal tip cell, *GSC* germline stem cell

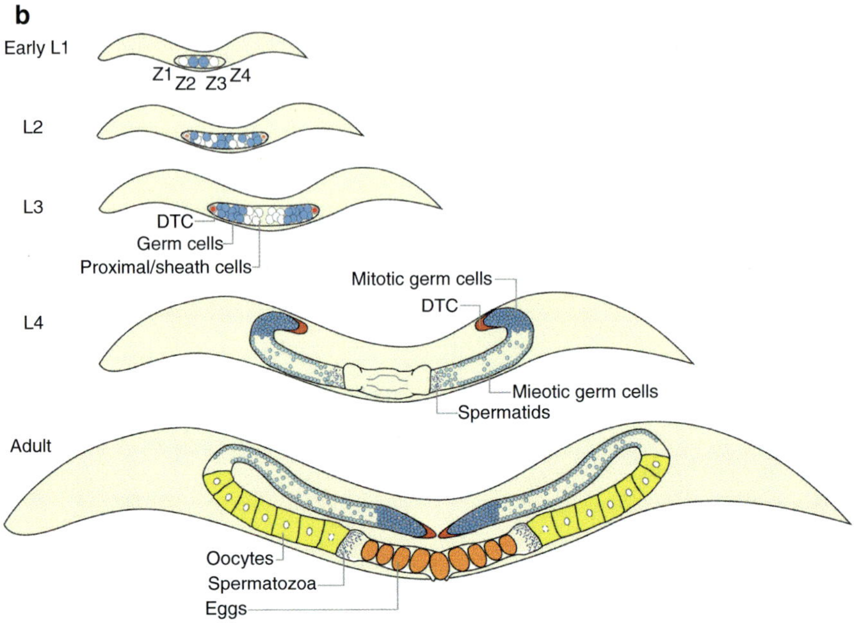

Fig. 3.1 (continued)

sheath cells, forming single layer intimately surrounding the germ cells. The DTC, at the distal end of each germline arm, forms processes that stretch out proximally along germline surface and functions as a niche to maintain the adjacent GSCs (Fig. 3.1c). During the L4 larval stage, the polarity of the germline forms. As new cells continue proliferating at the distal end, the other germ cells are pushed toward the proximal side and farther away from the DTC. Only GSCs and early germ cells at the distal end remain mitotic. In contrast, those germ cells progressing to the proximal end switch fate from mitotic proliferation to meiotic differentiation and undergo spermatogenesis during the L4 stage, followed by oogenesis starting at adulthood (Fig. 3.1b). In adulthood, oogenesis can continue throughout the whole reproductive period until sperm exhaustion.

3.2.2 Molecular Mechanism Controlling GSC Maintenance and Activity

In *C. elegans*, GSCs are located at the most distal end of the germline. The somatic DTC extends thin cytoplasmic processes that encircle the most distal end germ cells and serves as a niche for adjacent GSCs (Fig. 3.1c) (Li and Xie 2005; Xie 2008). When DTC is killed by laser ablation, the germline mitotic region diminishes (Kimble and White 1981). Conversely, DTC relocation results in the development of ectopic proliferating germ cells at the corresponding position, and DTC duplication leads to doubling of germ cell pools (Kidd et al. 2005; Kipreos et al. 2000;

Lam et al. 2006). The DTC promotes mitotic proliferation of GSCs via GLP-1/Notch signaling. LAG-2, a notch-like signaling ligand, is expressed in the DTC and activates its receptor GLP-1, located on the surface of GSCs (Fig. 3.1c) (Crittenden et al. 1994; Fitzgerald and Greenwald 1995; Henderson et al. 1994). The activation of GLP-1/Notch signaling is necessary and sufficient for GSC maintenance and proliferation. In the loss-of-function mutants of *glp-1*, all of the germ cells progress into meiotic differentiation, leading to a complete loss of the GSC population (Fitzgerald and Greenwald 1995). Conversely, gain-of-function mutations of *glp-1* promote GSC overproliferation and the formation of germline tumor (Berry et al. 1997).

In response to the activation of GLP-1/Notch signaling, a battery of downstream RNA regulators cooperate to promote GSC self-renewal and prevent cell differentiation (Fig. 3.1c) (Byrd and Kimble 2009; Kimble and Crittenden 2005). In brief, GLP-1 can transcriptionally activate *fbf-2* (along with other unidentified targets) (Lamont et al. 2004). FBF-2 is a pumilio-like RNA-binding translational repressor that can work together with another pumilio-like protein FBF-1 to repress the expression of *gld-1* and *gld-3* via binding to their 3′-untranslated regions (UTRs) (Crittenden et al. 2002; Eckmann et al. 2004). GLD-1 together with NOS-3 and GLD-3 together with GLD-2 form two parallel regulatory branches to control meiotic entry in the germline. In one branch, GLD-1 functions as a translational repressor that acts through regulatory elements present in the 3′-UTR of mitosis-promoting genes and represses their expression posttranscriptionally. NOS-3, a Nanos family of RNA-binding proteins, physically interacts with FBF-1 and promotes GLD-1 accumulation (Hansen et al. 2004). In the other branch, GLD-2, a cytoplasmic poly(A) polymerase and translational activator, activates meiosis-promoting genes while GLD-3, a member of the bicaudal-C family of RNA-binding protein, enhances the poly(A) polymerase activity of GLD-2 and antagonizes the FBF-mediated repression (Eckmann et al. 2002, 2004; Wang et al. 2002). Together, GLP-1/Notch signaling and FBF-GLD proteins constitute the control hub for GSC maintenance in adults (Fig. 3.1c).

3.3 GSC Arrest Promotes Longevity in *C. elegans*: Trade-Off Between Reproduction and Longevity?

As mentioned in the introduction, we now know that the aging process is under the control of various regulatory pathways (Blagosklonny et al. 2010; Kenyon 2010b). Altered activity of those pathways by genetic or pharmacological methods can effectively influence organism lifespan. The first of such pathways was characterized in *C. elegans* and named the insulin/IGF-1 signaling (IIS) pathway. Mutations of *age-1* or *daf-2*, the worm homolog of phosphoinositide 3-kinase (PI3K) or insulin/IGF-1 receptor, respectively, were shown to increase *C. elegans* adult lifespan by two-folds (Friedman and Johnson 1988; Kimura et al. 1997; Samuelson et al. 2007). IIS activation is transduced through PDK-1 (PIP3-dependent kinase 1) and AKT-1/2 (AKT/protein kinase B) and inhibits the transcriptional activity of the FOXO forkhead transcription factor, DAF-16. Loss-of-function mutations of *pdk-1* or *akt-1/2* increase *C. elegans* lifespan (Paradis et al. 1999; Paradis and Ruvkun 1998) while the increased lifespan associated with *daf-2* or *age-1* mutants is suppressed by mutations in *daf-16* (Ogg et al. 1997). Following *C. elegans* studies, the IIS pathway has been shown to regulate lifespan also in yeast, fruit flies, and mice (Longo et al. 2005). In human, several insulin receptor and IGF-1 receptor variants are linked to extremely long life in Ashkenazi Jewish centenarian and Japanese centenarian, respectively (Kojima et al. 2004; Suh et al. 2008). Furthermore, a number of AKT and FOXO3 variants have been implicated to be responsible for human exceptional longevity, as shown in multiple independent genetic association studies (Anselmi et al. 2009; Flachsbart et al. 2009; Pawlikowska et al. 2009; Willcox et al. 2008).

To date, the IIS pathway has been well recognized as a remarkable conserved mechanism to modulate the aging process across species. However, when first identified, its life-extending effect was suspected to be a result of fertility trade-off. Could it be possible that the *daf-2* and *age-1* mutants live longer simply because of their comprised reproductive abilities and consequent allocation of energy resources from reproduction to somatic maintenance? Several lines of evidences do not agree with this assumption. First of all, sterility alone, either by physical removal of whole gonad or using non-reproducing mutants, is not sufficient to ensure longevity (Kenyon et al. 1993). Secondly, *fer-15* rescue can restore normal fertility in the *age-1* mutants but has no effect on the lifespan extension (Johnson et al. 1993). In addition, the *daf-2* mutants remain long lived at 15 °C, where the animals have similar reproductive ability as wild type (Tissenbaum and Ruvkun 1998). Last but not least, knockdown of *daf-2* by RNAi only during adulthood extends lifespan without affecting reproduction while *daf-2* inactivation during development reduces fertility with no lifespan extension (Dillin et al. 2002). Therefore, decreased fertility in the *daf-2* mutants is likely due to its requirement during development, but not related to its longevity effects at adulthood.

So, what is the relationship between reproduction and longevity, if there is no simple trade-off between them? The merging view is that the reproductive system can produce regulatory signals that actively coordinate the metabolic state of the organism toward reproduction or toward survival. In the following sections, we will focus on the mechanistic details of how reproductive signaling might regulate lifespan in *C. elegans*.

3.4 Germline, GSCs, and Longevity

The first line of evidence that the reproductive system produces signals to regulate lifespan came from the classic experiments by the Kenyon laboratory (Hsin and Kenyon 1999). They showed that removal of the germline precursors Z2/3 PGCs by laser microsurgery results in germline-less animals that live 60 % longer than the mock control, while additional removal of the somatic precursors Z1/4 SGPs, which eliminates somatic gonad surrounding and supporting the germ cells, abrogates this lifespan extension (Hsin and Kenyon 1999). These suggest that the longevity effect conferred by germline ablation is not a simple consequence of sterility, but rather arises from certain regulatory signals. Removal of somatic gonad may antagonize those longevity-promoting signals (Hsin and Kenyon 1999). In addition, the longevity phenotype in the germline-less animals resulting from physical ablation can also be recapitulated using genetic mutants. Loss-of-function mutations in *mes-1* that have no germ cells can live twofold longer, and the *glp-1* mutants that arrest germ cell proliferation can live 1.5-fold longer (Arantes-Oliveira et al. 2002). In both studies of *mes-1* and *glp-1*, the presence of somatic gonad is required for the longevity effects.

In the adult germline, different stages of proliferative and differentiated germ cells, including GSCs, mitotic germ cells, meiotic germ cells, oocytes, and sperms, are present at one time (Fig. 3.1b). Which of those specific germ cell groups contribute to the lifespan regulation? Surprisingly, several different sperm-deficient (*fem-1*, *fog-1*, *fog-2*, or *fog-3*) or oocyte-deficient (*daz-1*) mutants all exhibit normal lifespan without extension (Arantes-Oliveira et al. 2002). It indicates the negligible functions of gametogenesis in the regulation of organism lifespan. This is particularly unexpected for the oocyte-deficient mutants, given that vast energy is invested during oocyte formation. These observations further support that the lifespan extension conferred by germline deficiency is not a simple passive result of energy reallocation. On the other hand, both GLP-1/Notch signaling and FBF-GLD proteins are required for GSC maintenance and proliferation. Interestingly, loss-of-function mutations of *glp-1* that arrest GSCs promote longevity, while *glp-1* gain-of-function mutations or *gld-1* loss-of-function mutations that cause GSC overproliferation in contrast shorten lifespan (Arantes-Oliveira et al. 2002). These observations reveal the crucial function of GCSs in regulating longevity. Furthermore, in the temperature-sensitive *glp-1* mutants, GSC arresting upon temperature shift can always lead to increased longevity, even at day-1 adulthood when the whole germline development is completely accomplished (Arantes-Oliveira et al. 2002). Thus, GSC arrest itself is sufficient to promote longevity, and the lifespan regulatory signals could directly associate with the presence of GSCs.

Importantly, this GSC longevity regulatory mechanism is evolutionarily conserved. In *Drosophila*, adult flies lacking GSCs live 50 % longer than the controls (Flatt et al. 2008). In mice, ovary transplantation from young females increases the lifespan of older female recipients (Cargill et al. 2003), suggesting that unknown lifespan-enhancing endocrine signals are produced from mammalian ovary tissues.

3.5 Signaling Pathways Regulating Longevity in Response to Gonadal Signals

3.5.1 Steroid Hormone Control from the Reproductive System

Reproductive tissues are crucial functional parts of the endocrine system. Steroid hormones synthesized and released from the reproductive system can systemically coordinate the whole body physiology. Consistently, steroidal signaling has been implicated as a key factor contributing to the regulation of GSC longevity (Fig. 3.2).

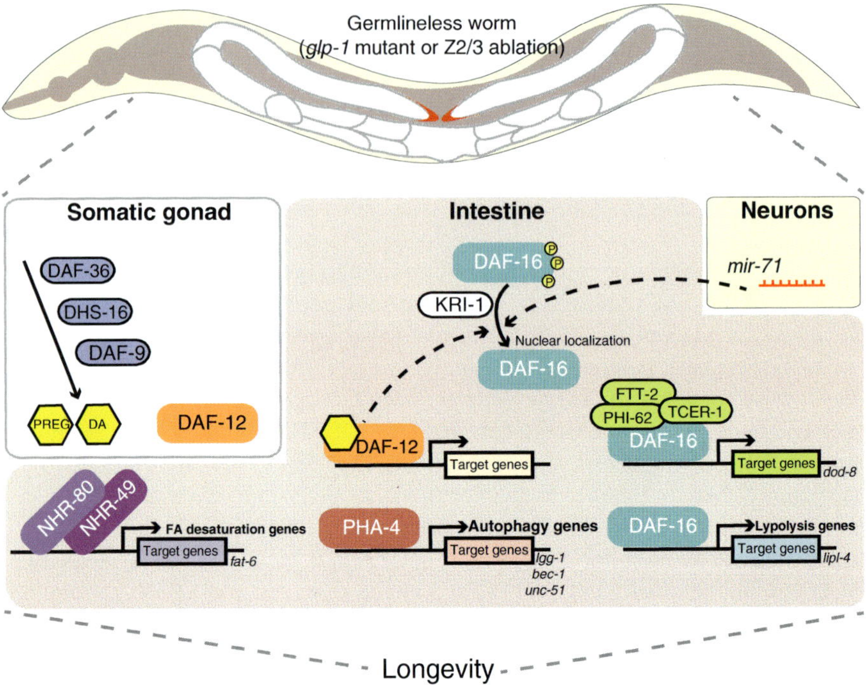

Fig. 3.2 Longevity regulation in germline-deficient *C. elegans*. Germ cell loss works in concert with somatic gonad, intestine, and neuronal cells to regulate lifespan extension. In somatic gonad, steroid synthesis via DAF-36, DHS-16, and DAF-9 generates hormone ligands, like DA and PREG, for DAF-12 activation. DAF-12 promotes longevity via regulating downstream target gene expression and via assisting FOXO/DAF-16 nuclear localization in the intestine. DAF-16 activation also requires KRI-1 and neuronal *mir-71* activity. Once DAF-16 enters nucleus, it activates assorted downstream targets through forming distinct complexes with different factors. When DAF-16 interacts with TCER-1, PHI-62, and FTT-2, it is known to regulate the gene *dod-8*. DAF-16 also plays a central role in regulating *lipl-4* and other lipolysis genes. With the aid of *lipl-4*, FOXA/PHA-4 regulates autophagy gene expression, e.g., *lgg-1*, *bec-1*, and *unc-51*. On the other hand, NHR-80 and NHR-49 activate the expression of SCD, including *fat-6*, to alter lipid composition by increasing oleic acid synthesis. These subsets of regulatory modules do not work self-sufficiently; instead, they belong to a closely linked network and may have partial dependency on others. These findings also suggest there is cross-tissue endocrine signaling to communicate the whole organism in the germline-depleted scenario and to achieve lifespan extension. Abbreviation: *DA* dafachronic acid, *PREG* pregnenolone, *SCD* stearoyl-CoA-Δ9-desaturases

This includes the bile acid-like steroids (dafachronic acids (DA)), its receptor DAF-12, and several enzymes involved in DA synthesis, such as DAF-36/Rieske-like oxygenase, DHS-16/3-hydroxysteroid dehydrogenase, and DAF-9/cytochrome P450 (Lee and Schroeder 2012). Loss-of-function mutations of *daf-12*, *daf-36*, *dhs-16*, or *daf-9* completely abrogate the longevity in the germline-less animals conferred by the *glp-1* mutations or Z2/3 laser ablation (Beckstead and Thummel 2006; Gerisch et al. 2001; Wollam et al. 2012). Moreover, DA supplementation restores longevity in the *daf-9*;*glp-1*, *dhs-16*;*glp-1*, and *daf-36*;*glp-1* double mutants back to that of *glp-1* but has no effect in extending the shortened lifespan of *daf-12;glp-1* (Gerisch et al. 2007; Wollam et al. 2012). Thus, DA biosynthesis and its transcriptional activation of DAF-12 are both required in promoting longevity of the germline-less animals.

As mentioned above, an intact somatic gonad is required for the longevity of the germline-less animals, suggesting somatic gonad as a source of life-extending signals. Notably, DA supplementation can restore longevity of the whole gonad-deficient worms lacking both germline and somatic gonad, which requires the presence of *daf-12* (Gerisch et al. 2007). In addition, DAF-12 transcriptional activities are stimulated by germline loss, but further removal of somatic gonad diminishes this induction, which can be rescued by supplying DA exogenously (Yamawaki et al. 2010). These indicate that DA biosynthesis and DAF-12 signaling are both involved in mediating life-extending signals from the somatic gonad. In wild-type animals, those signals may be hindered by the presence of germline. Besides DA, *C. elegans* also contain several other hormonal steroids that are present in humans, including pregnenolone (3β-hydroxy-pregn-5-en-20-one; PREG) and other pregnane and androstane derivatives (Broue et al. 2007). Some of those steroids are also involved in the regulation of longevity in the germline-less animals. For example, in the *glp-1* mutants, PREG levels are elevated in a DAF-9-dependent fashion. Moreover, PREG supplementation restores longevity in the *daf-9;glp-1* double mutants, but fails to do so in the *daf-12;glp-1* mutants, suggesting the involvement of PREG in mediating gonadal longevity signals (Broue et al. 2007). However, neither DA nor PREG supplementation is sufficient to extend lifespan in wild type or further increase longevity of germline-deficient animals. Therefore, there must be also other signaling components mediating the longevity effects conferred by germline loss.

3.5.2 Regulation of Intestinal FOXO Activity Upon Germline Loss

The FOXO forkhead transcription factor DAF-16 serves as another central regulator of longevity by germline deficiency (Fig. 3.2). Null mutation of *daf-16* completely abolishes the longevity of the germline-ablated animals (Hsin and Kenyon 1999) and the *glp-1* germline-defective mutants (Arantes-Oliveira et al. 2002; Berman and Kenyon 2006). Despite ubiquitous distribution in all worm tissues, DAF-16 activities in the intestine may be particularly critical to mediate the longevity effect brought about by germline loss. Upon laser microsurgery of germline precursor cells or in the *glp-1* germline-deficient mutants, DAF-16 is translocalized to the nucleus in intestinal cells during the first day of adulthood, but its subcellular localization remains unchanged in other tissues (Berman and Kenyon 2006; Lin et al. 2001). Similar to mammalian FOXO transcription factors, translocation of DAF-16 from the cytoplasm to the nucleus triggers its activation, and three conserved AKT phosphorylation sites are required for its cytoplasmic retention. Substitution of those three consensus sites with alanines generates a constitutively nuclear DAF-16 protein (referred to as DAF-16 AM) (Lin et al. 2001). Importantly, expression of this mutated form of DAF-16 AM within the intestine fully restores the longevity of the *glp-1;daf-16* double mutants, showing the functional significance of DAF-16 nuclear localization in this tissue (Berman and Kenyon 2006). The worm intestine is not only a digestive organ, but also a key metabolic and endocrine organ to store lipid and release humoral signals to the rest of the organism. Thus, activation of DAF-16 in the intestine by germline signals may generate secondary effects and hence have global impacts on organism longevity.

Multiple factors genetically act upstream of DAF-16 in response to germline loss. First of all, DA/DAF-12 signaling assists DAF-16 nuclear localization in the germline-less animals. Loss-of-function mutations of either *daf-9*, *daf-36*, *dhs-16*, or *daf-12* substantially compromise the nuclear translocation of DAF-16 in the intestine, and DA supplementation restores DAF-16 nuclear localization in the *glp-1* double mutants with those DA biosynthesis genes (Gerisch et al. 2007; Libina et al. 2003; Wollam et al. 2012). Noteworthily, expression of constitutively nuclear localized DAF-16 AM fails to restore longevity in *glp-1*; *daf-16*; *daf-12* mutants (Berman and Kenyon 2006). It suggests that DAF-12 may not only facilitate DAF-16 nuclear translocalization, but also aid its transcriptional activation or have other unidentified downstream pathways. Secondly, scientists also apply unbiased genetic screens to search for other components involved in the germline longevity signaling. *kri-1* (Berman and Kenyon 2006) and *tcer-1* (Ghazi et al. 2009) are two genes identified in large-scale RNAi screens that remarkably suppress the longevity of the *glp-1* mutants but have no effect on wild-type lifespan. KRI-1, a worm homolog of the human KRIT/CCM1 protein, is expressed in the intestine and is required for the intestinal nuclear localization of DAF-16 in the *glp-1* mutants (Berman and Kenyon 2006). In the presence of DAF-16 AM, *kri-1* loss-of-function mutations cannot abrogate the longevity of the *glp-1* mutants (Berman and Kenyon 2006). Thus, KRI-1 is likely a downstream intestinal factor of the germline longevity signaling pathway, and its primary function is to promote DAF-16 nuclear localization in this tissue. As a transcription elongation factor, TCER-1 works together with DAF-16 in regulating the expression of specific target genes in response to germline depletion. *tcer-1* is specifically upregulated in the nuclei of intestine and neurons in the germline-less animals, and its intestinal upregulation is dependent on *kri-1*, but not *daf-12* (Ghazi et al. 2009). Although loss-of-function mutations of *tcer-1* do not affect DAF-16 nuclear localization in the germline-less animals, they reduce the induction of several DAF-16 target genes and abrogate the longevity (Ghazi et al. 2009). Conversely, TCER-1 overexpression significantly triggers specific DAF-16 target gene expressions, along with lifespan extension by 15 %, which is dependent on DAF-16 activity (Ghazi et al. 2009). The mammalian homolog of TCER-1, TCERG-1, associates with RNA polymerase II and regulates transcription elongation and pre-mRNA splicing (Carty et al. 2000; Goldstrohm et al. 2001; Lin et al. 2004; Pearson et al. 2008; Smith et al. 2004; Sune and Garcia-Blanco 1999). A plausible model is that germline removal triggers the formation of specialized transcriptional complexes including TCER-1, DAF-16, and other unidentified factors, which modulate the expression of a specific subset of genes that are crucial for organism longevity. This is supported by the findings that TCER-1 and DAF-16 physically interact with FTT-2 (14-3-3 protein) and PHI-62 (RNA-binding protein) to regulate *dod-8* (putative steroid dehydrogenase) expression, and both *ftt-2* and *phi-62* are required for the longevity in germline-less animals (Li et al. 2007; McCormick et al. 2012). However, the expression of *sod-3* (Mn++ superoxide dismutase), another *daf-16* downstream target, does not require *ftt-2* nor *phi-62* (McCormick et al. 2012). This reveals that DAF-16 may form a variety of complexes with different factors to exert diverse regulatory functions on multiple subsets of downstream genes and together create a network that promotes organism longevity (Fig. 3.2).

Interestingly, recent studies have also implicated neuronal microRNA *mir-71* in the regulation of DAF-16 nuclear localization and longevity in the germline-less animals (Boulias and Horvitz 2012). In a large screen of microRNA loss-of-function mutants, *mir-71* loss-of-function was identified to shorten wild-type lifespan by 40 %. More importantly, loss of *mir-71* fully abrogates the longevity in the *glp-1* mutants or the germline-ablated animals and also blocks intestinal nuclear localization of DAF-16 and its target gene expressions in those germline-less animals (Boulias and Horvitz 2012). Conversely, *mir-71* overexpression modestly increases lifespan in wild type and further enhances the *glp-1* longevity. This increased longevity fully requires DAF-16 and TCER-1 activities within the intestine but is not affected by DAF-12. Although broadly expressed in multiple tissues, *mir-71* activity within neurons is particularly important to mediate the longevity effects by germline loss. Thus, *mir-71* acts in a cell nonautonomous manner to regulate DAF-16 activity in the intestine. A complex crosstalk among the reproductive system, neurons, and intestine systemically regulates the organism aging process (Fig. 3.2).

3.5.3 Interaction with the IIS Pathway in the Regulation of Longevity

As mentioned above, the IIS pathway is the first identified, the best characterized, and the most conserved longevity mechanism to date. Whether and how does the germline longevity signaling interact with the IIS pathway? So far, the evidences suggest that these two mechanisms largely work independently (Kenyon 2010a; Panowski and Dillin 2009). First, germline deficiency and reduced *daf-2* activity have synergistic effects genetically in promoting longevity. Ablation of germ cells in the *daf-2* loss-of-function mutants further enhances already increased lifespan in the mutants (Hsin and Kenyon 1999). Second, DAF-16 responds to germline loss signals differently from reduced IIS. Removal of germline causes DAF-16 localization in intestinal nuclei at adulthood, while *daf-2* loss-of-function mutants display DAF-16 nuclear accumulation in all somatic cells throughout life (Henderson and Johnson 2001; Lee et al. 2001; Lin et al. 2001). Third, as discussed above, multiple factors are required to promote longevity in the germline-less animals, including *daf-12*, *kri-1*, *tcer-1*, and *mir-71*. However, none of those components are necessary for the longevity conferred by reduced *daf-2* activity (Berman and Kenyon 2006; Gems et al. 1998; Ghazi et al. 2009; Hsin and Kenyon 1999).

Nevertheless, the IIS pathway is likely involved in mediating the lifespan extending signals from the somatic gonad. In wild-type animals, removal of somatic gonad precursors abrogates the longevity brought by ablating germline precursors. However, in the strong *daf-2* loss-of-function mutants, longevity remains even with whole gonad depletion (Hsin and Kenyon 1999), indicating that the somatic gonad promotes longevity via reducing IIS. Furthermore, a subset of DAF-16 target genes is induced by both *daf-2* reduction and germline removal, such as *sod-3*, *gpd-2* (glyceraldehyde 3-phosphate dehydrogenase), and *dod-8* (Murphy et al. 2003). Interestingly, the inductions of *sod-3* and *gpd-2* both require the presence of somatic gonad (Yamawaki et al. 2008). Thus, the somatic gonad may exert influence through a regulatory network that at least partially overlaps with the IIS pathway, while germ cells act in parallel with the somatic reproductive tissue to regulate lifespan in a counterbalancing manner.

3.6 Metabolic Reprogramming in Mediating GSC Longevity

Following the removal of germline and activation of a series of downstream signaling pathways, what might be the mechanisms that confer longevity? Although not conclusive, recent studies reveal that the germline removal actively modulates the organism metabolic state, especially the lipid metabolism. This metabolic reprogramming may play a crucial role in promoting longevity.

3.6.1 LIPL-4 Lysosomal Lipid Signaling

The germline-less *glp-1* mutants have substantial fat accumulation in the intestine and also in the hypodermis, another major fat storage tissue in *C. elegans* (Ashrafi 2007; Yokota et al. 2002). On the other hand, they have reduced levels of Nile red staining signals, which labels worm lysosomal-like organelles involved in lipid metabolism (Greenspan and Fowler 1985; Wang et al. 2008). To understand the mechanisms by which germline signals regulate lipid metabolism, scientists screened through an RNAi library targeting various key metabolic genes and identified a group of genes whose inactivation specifically increase Nile red staining levels in the *glp-1* mutants (Wang et al. 2008). Interestingly, most of those genes are involved in catalyzing lipid hydrolysis, fatty acid transport and β-oxidation, and citric acid cycle, revealing an altered lipid metabolic state in the germline-less animals (Wang et al. 2008). Importantly, further characterization of one candidate, *lipl-4*, showed that its knockdown completely abrogates longevity of *glp-1* and its overexpression robustly extends lifespan in gonad intact wild-type animals (Wang et al. 2008).

lipl-4 expression is upregulated in the germline-less animals in a *daf-16-* and *kri-1-dependent* manner (Wang et al. 2008) (Fig. 3.2). LIPL-4 is a lysosomal acid lipase, and its overexpression alone sufficiently promotes longevity through triggering the nuclear localization of LBP-8, a lysosomal lipid chaperone. This allows LBP-8 to deliver a lipid messenger oleoylethanolamide (OEA) to the nucleus and to consequently activate nuclear hormone receptors NHR-80 and NHR-49 (Folick et al. 2015). Activation of this lysosome-to-nucleus lipid-signaling pathway promotes a metabolic shift toward lipid catabolism (Folick et al. 2015). Further studies to fully understand the signaling role of lysosomes will be crucial to unveil the metabolic link between GSCs and longevity.

3.6.2 NHR-80/NHR-49 Mediated Fatty Acid Metabolic Changes

Not only total lipid content increases in the germline-less animals, but also lipid composition is altered in those animals. The monounsaturated oleic acid (OA, C18:1(n-9)) levels and the ratio of OA/stearic acid (SA, C18:0) are specifically increased in the germline-less animals (Goudeau et al. 2011). In *C. elegans*, OA is synthesized from SA by stearoyl-CoA-Δ9-desaturases (SCDs), *fat-6*, and *faf-7* and is the key precursor for a variety of polyunsaturated fatty acids (PUFAs) (Watts and Browse 2002). Loss-of-function mutations of both *fat-6* and *fat-7* abrogate longevity of *glp-1*, which can be perfectly rescued if OA is supplied exogenously (Goudeau et al. 2011). This suggests that elevation of OA synthesis is necessary for the longevity in the germline-less animals. However, OA dietary supplementation is not sufficient to extend lifespan in gonad intact wild-type animals or further enhance longevity in the germline-less animals (Goudeau et al. 2011). The induction of OA synthesis is not mediated by *daf-16* or *daf-12* but requires other nuclear hormone receptors, including *nhr-80* and *nhr-49* (Brock et al. 2006; Goudeau et al. 2011) (Fig. 3.2).

Knockdown of *nhr-80* by RNAi or mutation completely blocks *glp-1* longevity, but has no effect on the lifespan of wild type or other long-lived conditions, such as *daf-2* mutants, dietary restricted worms, or worms with reduced mitochondrial functions (Goudeau et al. 2011). NHR-80 is expressed in both neurons and intestine, and only its intestinal levels are dramatically increased in germline-less animals, suggesting the intestine as its functional site in mediating the germline signals (Goudeau et al. 2011). Interestingly, *nhr-80* overexpression leads to further lifespan extension in the *glp-1* mutants, but is not sufficient to prolong wild-type lifespan (Goudeau et al. 2011). Thus, activating ligands of NHR-80 are likely generated in the *glp-1* mutants, which are absent in wild type.

On the other hand, *nhr-49* expression is also upregulated in the *glp-1* mutant and is required for its longevity (Ratnappan et al. 2014). NHR-49 and NHR-80 form a nuclear complex and regulate fatty acid desaturation coordinately (Pathare et al. 2012; Ratnappan et al. 2014). *nhr-49* also controls the induction of a variety of genes involved in fatty acid beta-oxidation in the *glp-1* mutant, including *acs-2*, *acs-22*, *acdh-11*, and *hacd-1* (Ratnappan et al. 2014). Those fatty acid beta-oxidation genes are required for the longevity of the *glp-1* mutant, suggesting an increased lipid catabolism in germline-less animals despite their increased fat content levels.

3.6.3 Autophagy

Autophagy is a cellular process where cells deliver unnecessary or dysfunctional cellular organelles and protein aggregates to the lysosomes for degradation and recycling. This process is crucial for a healthy turnover of cellular organelles and proteins and also produces energy from unwanted waste under nutrient starvation. Interestingly, autophagy has been revealed as a crucial link between lipid metabolism and germline-loss-associated longevity. Autophagy is induced in the germline-less animals, as seen by the accumulation of autophagic particles in the intestine and hypodermis (Lapierre et al. 2011). Consistently, several autophagy genes, like *unc-51* (ULK1 homolog), *bec-1* (Beclin1 homolog), and *lgg-1* (LC3 homolog), are upregulated in the *glp-1* mutant (Lapierre et al. 2011). Importantly, inactivation of those autophagy genes by RNAi specifically abrogates longevity of the *glp-1* mutant, but does not affect wild-type lifespan (Lapierre et al. 2011). The induction of autophagy in the germline-less animals requires the activity of the FOXA forkhead transcription factor, PHA-4 (Lapierre et al. 2011). PHA-4 has been previously identified as required for dietary restriction-mediated longevity (Panowski et al. 2007). Adult-specific knockdown of *pha-4* abrogates *glp-1* longevity with minor shortening effects on wild-type lifespan (Goudeau et al. 2011). Autophagy has also been linked to lipid catabolism where it stimulates lipolysis in lysosomal compartments (lipophagy) (Singh et al. 2009). Interestingly, overexpression of the LIPL-4 lipase alone leads to increased autophagic particles in the hypodermal seam cells. Knockdown of *pha-4* or of autophagy genes blocks the lifespan extension conferred by *lipl-4* overexpression (Goudeau et al. 2011). These findings suggest a possible induction of lipophagy mechanisms following germline removal and its crucial roles in regulating longevity (Fig. 3.2).

How can lipid metabolism influence lifespan? Current hypotheses include lipophilic signaling, alteration of lipid composition or distribution, and lipotoxicity (Ackerman and Gems 2012). A study found that lipolytic products and intermediates participate in specific cellular signaling and can systemically influence the whole organism activity (Haemmerle et al. 2011), which unveils a previously underappreciated aspect of lipolysis. Interestingly, a lipolytic product OEA has been identified to activate NHR-49/NHR-80 nuclear hormone receptor complex and consequently promote longevity (Folick et al. 2015). Lipid composition of saturated versus unsaturated fat may decide oxidative stress tolerance capacity; the ratio of short- versus long-chain length, or the lipid turnover rate, may all affect fat profile contributed by healthy or unhealthy lipid (Shmookler Reis et al. 2011). Yet, more supporting biochemical evidence is required to back up this model. On the other hand, in the mammalian system, it has long been supposed that good fat or bad fat could also be defined by where it locates. Indeed, visceral fat or fat in non-adipocyte is considered deleterious and may cause metabolic syndromes while subcutaneous fat or adipocyte fat is recognized fine (Huffman and Barzilai 2009; Schaffer 2003). In *C. elegans*, yolk lipoprotein ectopic accumulation has been observed post-reproductively and may lead to age-related tissue dysfunction. Long-lived *daf-2* mutants display lower yolk levels after reproduction; and RNAi knockdown of yolk genes extends worm lifespan (Murphy et al. 2003). Hence, ectopic accumulation of fat may lead to lipotoxicity and thus trigger aging. Up to now, it is still under debate which of these hypotheses is correct, notwithstanding that they might not be mutually exclusive with each other.

3.7 Perspectives

During evolution, the maximum reproductive success is the ultimate purpose of all organisms. In the wild, the animals that lack the germline will not continue living, so why does this germline-mediated regulatory mechanism exist? Interestingly, germline proliferation is under tight control of nutrient availability. Several nutrient-sensing pathways regulate germline proliferation and maintenance, such as IIS, TOR, and dietary restriction (Narbonne and Roy 2006). Under unfavorable environmental conditions, germline proliferation is expected to stall. This is carried out by several developmental and reproductive arrest checkpoints in *C. elegans*. In particular, *C. elegans* can enter into an adult reproductive diapause in response to starvation, during which only a few GSCs are maintained in the germline. Animals can survive in this diapause state for extremely long time until nutrient becomes available and will reconstitute the germline and resume full reproductive capacity. We could expect that the germline-mediated longevity mechanism might be responsible for the long-term survival under diapause. In the wild, these mechanisms could assist animals survive through harsh environment and ensure the maximum reproductive success.

In the past, lifespan extension sounded like an unrealistic dream. Through decades of efforts, scientists have demonstrated that lifespan can be prolonged via various genetic mechanisms. Nevertheless, there are still many questions in the field of germline-regulated longevity. Loss of germ cells drives signaling responses in neurons, intestine, and somatic gonads that lead to enhanced somatic maintenance and healthy aging. How do different tissues crosstalk with each other? The lipid metabolism in intestine is altered in germline-less animals. What are the intestinal receptors interacting with the germline-less signals? What are the downstream mechanisms of lipid metabolism that directly affect lifespan? Somatic gonad regulates lifespan via *daf-12* lipophilic signaling. How does *daf-2* IIS signaling pathway intervene in *daf-12* signaling? All these interesting questions (and many more to come) remain to be answered. There is always more knowledge of aging waiting for us to be explored and amazed.

Acknowledgment We would like to acknowledge all the scientists in the *C. elegans* research community for their contributions to the field of reproductive system-regulated aging. We also thank Cheng-lin Li and Timothy R. Mahoney for helpful comments and valuable advice on the manuscript. The work in the authors' laboratory is supported by the National Institutes of Health grant RO0 AG034988 and RO1 AG045183 and the Ellison Medical Foundation grant AG-NS-0756-11.

References

Ackerman D, Gems D (2012) The mystery of C. elegans aging: an emerging role for fat. Distant parallels between C. elegans aging and metabolic syndrome? Bioessays 34:466–471

Anselmi CV, Malovini A, Roncarati R, Novelli V, Villa F, Condorelli G, Bellazzi R, Puca AA (2009) Association of the FOXO3A locus with extreme longevity in a southern Italian centenarian study. Rejuvenation Res 12:95–104

Arantes-Oliveira N, Apfeld J, Dillin A, Kenyon C (2002) Regulation of life-span by germ-line stem cells in Caenorhabditis elegans. Science 295:502–505

Ashrafi K (2007) Obesity and the regulation of fat metabolism. WormBook, ed. The *C. elegans* Research Community, WormBook 1–20, doi:10.1895/wormbook.1.130.1, http://www.wormbook.org

Beckstead RB, Thummel CS (2006) Indicted: worms caught using steroids. Cell 124:1137–1140

Berman JR, Kenyon C (2006) Germ-cell loss extends C. elegans life span through regulation of DAF-16 by kri-1 and lipophilic-hormone signaling. Cell 124:1055–1068

Berry LW, Westlund B, Schedl T (1997) Germ-line tumor formation caused by activation of glp-1, a Caenorhabditis elegans member of the Notch family of receptors. Development 124:925–936

Blagosklonny MV, Campisi J, Sinclair DA, Bartke A, Blasco MA, Bonner WM, Bohr VA, Brosh RM Jr, Brunet A, Depinho RA et al (2010) Impact papers on aging in 2009. Aging (Albany NY) 2:111–121

Boulias K, Horvitz HR (2012) The C. elegans microRNA mir-71 acts in neurons to promote germline-mediated longevity through regulation of DAF-16/FOXO. Cell Metab 15:439–450

Brock TJ, Browse J, Watts JL (2006) Genetic regulation of unsaturated fatty acid composition in C. elegans. PLoS Genet 2, e108

Broue F, Liere P, Kenyon C, Baulieu EE (2007) A steroid hormone that extends the lifespan of Caenorhabditis elegans. Aging Cell 6:87–94

Byrd DT, Kimble J (2009) Scratching the niche that controls Caenorhabditis elegans germline stem cells. Semin Cell Dev Biol 20:1107–1113

C. elegans Sequencing Consortium (1998) Genome sequence of the nematode C. elegans: a platform for investigating biology. Science 282:2012–2018

Cargill SL, Carey JR, Muller HG, Anderson G (2003) Age of ovary determines remaining life expectancy in old ovariectomized mice. Aging Cell 2:185–190

Carty SM, Goldstrohm AC, Sune C, Garcia-Blanco MA, Greenleaf AL (2000) Protein-interaction modules that organize nuclear function: FF domains of CA150 bind the phosphoCTD of RNA polymerase II. Proc Natl Acad Sci U S A 97:9015–9020

Crittenden SL, Troemel ER, Evans TC, Kimble J (1994) GLP-1 is localized to the mitotic region of the C. elegans germ line. Development 120:2901–2911

Crittenden SL, Bernstein DS, Bachorik JL, Thompson BE, Gallegos M, Petcherski AG, Moulder G, Barstead R, Wickens M, Kimble J (2002) A conserved RNA-binding protein controls germline stem cells in Caenorhabditis elegans. Nature 417:660–663

Dillin A, Crawford DK, Kenyon C (2002) Timing requirements for insulin/IGF-1 signaling in C. elegans. Science 298:830–834

Eckmann CR, Kraemer B, Wickens M, Kimble J (2002) GLD-3, a bicaudal-C homolog that inhibits FBF to control germline sex determination in C. elegans. Dev Cell 3:697–710

Eckmann CR, Crittenden SL, Suh N, Kimble J (2004) GLD-3 and control of the mitosis/meiosis decision in the germline of Caenorhabditis elegans. Genetics 168:147–160

Fitzgerald K, Greenwald I (1995) Interchangeability of Caenorhabditis elegans DSL proteins and intrinsic signalling activity of their extracellular domains in vivo. Development 121:4275–4282

Flachsbart F, Caliebe A, Kleindorp R, Blanche H, von Eller-Eberstein H, Nikolaus S, Schreiber S, Nebel A (2009) Association of FOXO3A variation with human longevity confirmed in German centenarians. Proc Natl Acad Sci U S A 106:2700–2705

Flatt T, Min KJ, D'Alterio C, Villa-Cuesta E, Cumbers J, Lehmann R, Jones DL, Tatar M (2008) Drosophila germ-line modulation of insulin signaling and lifespan. Proc Natl Acad Sci U S A 105:6368–6373

Folick A, Oakley HD, Yu Y, Armstrong EH, Kumari M, Sanor L, Moore DD, Ortlund EA, Zechner R, Wang MC (2015) Aging. Lysosomal signaling molecules regulate longevity in Caenorhabditis elegans. Science 347:83–86

Friedman DB, Johnson TE (1988) A mutation in the age-1 gene in Caenorhabditis elegans lengthens life and reduces hermaphrodite fertility. Genetics 118:75–86

Gems D, Sutton AJ, Sundermeyer ML, Albert PS, King KV, Edgley ML, Larsen PL, Riddle DL (1998) Two pleiotropic classes of daf-2 mutation affect larval arrest, adult behavior, reproduction and longevity in Caenorhabditis elegans. Genetics 150:129–155

Gerisch B, Weitzel C, Kober-Eisermann C, Rottiers V, Antebi A (2001) A hormonal signaling pathway influencing C. elegans metabolism, reproductive development, and life span. Dev Cell 1:841–851

Gerisch B, Rottiers V, Li D, Motola DL, Cummins CL, Lehrach H, Mangelsdorf DJ, Antebi A (2007) A bile acid-like steroid modulates Caenorhabditis elegans lifespan through nuclear receptor signaling. Proc Natl Acad Sci U S A 104:5014–5019

Ghazi A, Henis-Korenblit S, Kenyon C (2009) A transcription elongation factor that links signals from the reproductive system to lifespan extension in Caenorhabditis elegans. PLoS Genet 5, e1000639

Goldstrohm AC, Albrecht TR, Sune C, Bedford MT, Garcia-Blanco MA (2001) The transcription elongation factor CA150 interacts with RNA polymerase II and the pre-mRNA splicing factor SF1. Mol Cell Biol 21:7617–7628

Goudeau J, Bellemin S, Toselli-Mollereau E, Shamalnasab M, Chen Y, Aguilaniu H (2011) Fatty acid desaturation links germ cell loss to longevity through NHR-80/HNF4 in C. elegans. PLoS Biol 9:e1000599

Greenspan P, Fowler SD (1985) Spectrofluorometric studies of the lipid probe, nile red. J Lipid Res 26:781–789

Haemmerle G, Moustafa T, Woelkart G, Buttner S, Schmidt A, van de Weijer T, Hesselink M, Jaeger D, Kienesberger PC, Zierler K et al (2011) ATGL-mediated fat catabolism regulates cardiac mitochondrial function via PPAR-alpha and PGC-1. Nat Med 17:1076–1085

Hansen D, Wilson-Berry L, Dang T, Schedl T (2004) Control of the proliferation versus meiotic development decision in the C. elegans germline through regulation of GLD-1 protein accumulation. Development 131:93–104

Henderson ST, Johnson TE (2001) daf-16 integrates developmental and environmental inputs to mediate aging in the nematode Caenorhabditis elegans. Curr Biol 11:1975–1980

Henderson ST, Gao D, Lambie EJ, Kimble J (1994) lag-2 may encode a signaling ligand for the GLP-1 and LIN-12 receptors of C. elegans. Development 120:2913–2924

Hsin H, Kenyon C (1999) Signals from the reproductive system regulate the lifespan of C. elegans. Nature 399:362–366

Hubbard EJA, Greenstein D (2005) Introduction to the germ line. WormBook, ed. The *C. elegans* Research Community, WormBook 1–4, doi:10.1895/wormbook.1.18.1, http://www.wormbook.org

Huffman DM, Barzilai N (2009) Role of visceral adipose tissue in aging. Biochim Biophys Acta 1790:1117–1123

Johnson TE, Tedesco PM, Lithgow GJ (1993) Comparing mutants, selective breeding, and transgenics in the dissection of aging processes of Caenorhabditis elegans. Genetica 91: 65–77

Kenyon C (2010a) A pathway that links reproductive status to lifespan in Caenorhabditis elegans. Ann N Y Acad Sci 1204:156–162

Kenyon CJ (2010b) The genetics of ageing. Nature 464:504–512

Kenyon C, Chang J, Gensch E, Rudner A, Tabtiang R (1993) A C. elegans mutant that lives twice as long as wild type. Nature 366:461–464

Kidd AR 3rd, Miskowski JA, Siegfried KR, Sawa H, Kimble J (2005) A beta-catenin identified by functional rather than sequence criteria and its role in Wnt/MAPK signaling. Cell 121:761–772

Kimble J, Crittenden SL (2005) Germline proliferation and its control. WormBook, ed. The *C. elegans* Research Community, WormBook 1–14, doi:10.1895/wormbook.1.13.1, http://www.wormbook.org

Kimble JE, White JG (1981) On the control of germ cell development in Caenorhabditis elegans. Dev Biol 81:208–219

Kimura KD, Tissenbaum HA, Liu Y, Ruvkun G (1997) daf-2, an insulin receptor-like gene that regulates longevity and diapause in Caenorhabditis elegans. Science 277:942–946

Kipreos ET, Gohel SP, Hedgecock EM (2000) The C. elegans F-box/WD-repeat protein LIN-23 functions to limit cell division during development. Development 127:5071–5082

Kojima T, Kamei H, Aizu T, Arai Y, Takayama M, Nakazawa S, Ebihara Y, Inagaki H, Masui Y, Gondo Y et al (2004) Association analysis between longevity in the Japanese population and polymorphic variants of genes involved in insulin and insulin-like growth factor 1 signaling pathways. Exp Gerontol 39:1595–1598

Lam N, Chesney MA, Kimble J (2006) Wnt signaling and CEH-22/tinman/Nkx2.5 specify a stem cell niche in C. elegans. Curr Biol 16:287–295

Lamont LB, Crittenden SL, Bernstein D, Wickens M, Kimble J (2004) FBF-1 and FBF-2 regulate the size of the mitotic region in the C. elegans germline. Dev Cell 7:697–707

Lapierre LR, Gelino S, Melendez A, Hansen M (2011) Autophagy and lipid metabolism coordinately modulate life span in germline-less C. elegans. Curr Biol 21:1507–1514

Lee SS, Schroeder FC (2012) Steroids as central regulators of organismal development and lifespan. PLoS Biol 10, e1001307

Lee RY, Hench J, Ruvkun G (2001) Regulation of C. elegans DAF-16 and its human ortholog FKHRL1 by the daf-2 insulin-like signaling pathway. Curr Biol 11:1950–1957

Li L, Xie T (2005) Stem cell niche: structure and function. Annu Rev Cell Dev Biol 21:605–631

Li J, Tewari M, Vidal M, Lee SS (2007) The 14-3-3 protein FTT-2 regulates DAF-16 in Caenorhabditis elegans. Dev Biol 301:82–91

Libina N, Berman JR, Kenyon C (2003) Tissue-specific activities of C. elegans DAF-16 in the regulation of lifespan. Cell 115:489–502

Lin K, Hsin H, Libina N, Kenyon C (2001) Regulation of the Caenorhabditis elegans longevity protein DAF-16 by insulin/IGF-1 and germline signaling. Nat Genet 28:139–145

Lin KT, Lu RM, Tarn WY (2004) The WW domain-containing proteins interact with the early spliceosome and participate in pre-mRNA splicing in vivo. Mol Cell Biol 24:9176–9185

Longo VD, Mitteldorf J, Skulachev VP (2005) Programmed and altruistic ageing. Nat Rev Genet 6:866–872

McCormick M, Chen K, Ramaswamy P, Kenyon C (2012) New genes that extend Caenorhabditis elegans' lifespan in response to reproductive signals. Aging Cell 11:192–202

Murphy CT, McCarroll SA, Bargmann CI, Fraser A, Kamath RS, Ahringer J, Li H, Kenyon C (2003) Genes that act downstream of DAF-16 to influence the lifespan of Caenorhabditis elegans. Nature 424:277–283

Narbonne P, Roy R (2006) Regulation of germline stem cell proliferation downstream of nutrient sensing. Cell Div 1:29

Ogg S, Paradis S, Gottlieb S, Patterson GI, Lee L, Tissenbaum HA, Ruvkun G (1997) The Fork head transcription factor DAF-16 transduces insulin-like metabolic and longevity signals in C. elegans. Nature 389:994–999

Panowski SH, Dillin A (2009) Signals of youth: endocrine regulation of aging in Caenorhabditis elegans. Trends Endocrinol Metab 20:259–264

Panowski SH, Wolff S, Aguilaniu H, Durieux J, Dillin A (2007) PHA-4/Foxa mediates diet-restriction-induced longevity of C. elegans. Nature 447:550–555

Paradis S, Ruvkun G (1998) Caenorhabditis elegans Akt/PKB transduces insulin receptor-like signals from AGE-1 PI3 kinase to the DAF-16 transcription factor. Genes Dev 12:2488–2498

Paradis S, Ailion M, Toker A, Thomas JH, Ruvkun G (1999) A PDK1 homolog is necessary and sufficient to transduce AGE-1 PI3 kinase signals that regulate diapause in Caenorhabditis elegans. Genes Dev 13:1438–1452

Pathare PP, Lin A, Bornfeldt KE, Taubert S, Van Gilst MR (2012) Coordinate regulation of lipid metabolism by novel nuclear receptor partnerships. PLoS Genet 8, e1002645

Pawlikowska L, Hu D, Huntsman S, Sung A, Chu C, Chen J, Joyner AH, Schork NJ, Hsueh WC, Reiner AP et al (2009) Association of common genetic variation in the insulin/IGF1 signaling pathway with human longevity. Aging Cell 8:460–472

Pearson JL, Robinson TJ, Munoz MJ, Kornblihtt AR, Garcia-Blanco MA (2008) Identification of the cellular targets of the transcription factor TCERG1 reveals a prevalent role in mRNA processing. J Biol Chem 283:7949–7961

Ratnappan R, Amrit FR, Chen SW, Gill H, Holden K, Ward J, Yamamoto KR, Olsen CP, Ghazi A (2014) Germline signals deploy NHR-49 to modulate fatty-acid beta-oxidation and desaturation in somatic tissues of C. elegans. PLoS Genet 10:e1004829

Russell SJ, Kahn CR (2007) Endocrine regulation of ageing. Nat Rev Mol Cell Biol 8:681–691

Samuelson AV, Klimczak RR, Thompson DB, Carr CE, Ruvkun G (2007) Identification of Caenorhabditis elegans genes regulating longevity using enhanced RNAi-sensitive strains. Cold Spring Harb Symp Quant Biol 72:489–497

Schaffer JE (2003) Lipotoxicity: when tissues overeat. Curr Opin Lipidol 14:281–287

Shmookler Reis RJ, Xu L, Lee H, Chae M, Thaden JJ, Bharill P, Tazearslan C, Siegel E, Alla R, Zimniak P et al (2011) Modulation of lipid biosynthesis contributes to stress resistance and longevity of C. elegans mutants. Aging (Albany NY) 3:125–147

Singh R, Kaushik S, Wang Y, Xiang Y, Novak I, Komatsu M, Tanaka K, Cuervo AM, Czaja MJ (2009) Autophagy regulates lipid metabolism. Nature 458:1131–1135

Smith MJ, Kulkarni S, Pawson T (2004) FF domains of CA150 bind transcription and splicing factors through multiple weak interactions. Mol Cell Biol 24:9274–9285

Strome S (2005) Specification of the germ line. WormBook, ed. The *C. elegans* Research Community, WormBook 1–10, doi:10.1895/wormbook.1.9.1, http://www.wormbook.org

Suh Y, Atzmon G, Cho MO, Hwang D, Liu B, Leahy DJ, Barzilai N, Cohen P (2008) Functionally significant insulin-like growth factor I receptor mutations in centenarians. Proc Natl Acad Sci U S A 105:3438–3442

Sulston JE, Schierenberg E, White JG, Thomson JN (1983) The embryonic cell lineage of the nematode Caenorhabditis elegans. Dev Biol 100:64–119

Sune C, Garcia-Blanco MA (1999) Transcriptional cofactor CA150 regulates RNA polymerase II elongation in a TATA-box-dependent manner. Mol Cell Biol 19:4719–4728

Tissenbaum HA, Ruvkun G (1998) An insulin-like signaling pathway affects both longevity and reproduction in Caenorhabditis elegans. Genetics 148:703–717

Updike D, Strome S (2010) P granule assembly and function in Caenorhabditis elegans germ cells. J Androl 31:53–60

Wang L, Eckmann CR, Kadyk LC, Wickens M, Kimble J (2002) A regulatory cytoplasmic poly(A) polymerase in Caenorhabditis elegans. Nature 419:312–316

Wang MC, O'Rourke EJ, Ruvkun G (2008) Fat metabolism links germline stem cells and longevity in C. elegans. Science 322:957–960

Watts JL, Browse J (2002) Genetic dissection of polyunsaturated fatty acid synthesis in Caenorhabditis elegans. Proc Natl Acad Sci U S A 99:5854–5859

Willcox BJ, Donlon TA, He Q, Chen R, Grove JS, Yano K, Masaki KH, Willcox DC, Rodriguez B, Curb JD (2008) FOXO3A genotype is strongly associated with human longevity. Proc Natl Acad Sci U S A 105:13987–13992

Wollam J, Magner DB, Magomedova L, Rass E, Shen Y, Rottiers V, Habermann B, Cummins CL, Antebi A (2012) A novel 3-hydroxysteroid dehydrogenase that regulates reproductive development and longevity. PLoS Biol 10, e1001305

Xie T (2008) Germline stem cell niches. StemBook, ed. The Stem Cell Research Community, StemBook, doi:10.3824/stembook.1.23.1, http://www.stembook.org

Yamawaki TM, Arantes-Oliveira N, Berman JR, Zhang P, Kenyon C (2008) Distinct activities of the germline and somatic reproductive tissues in the regulation of Caenorhabditis elegans' longevity. Genetics 178:513–526

Yamawaki TM, Berman JR, Suchanek-Kavipurapu M, McCormick M, Gaglia MM, et al (2010) The Somatic Reproductive Tissues of C. elegans Promote Longevity through Steroid Hormone Signaling. PLoS Biol 8(8): e1000468. doi:10.1371/journal.pbio.1000468

Yokota S, Togo SH, Maebuchi M, Bun-Ya M, Haraguchi CM, Kamiryo T (2002) Peroxisomes of the nematode Caenorhabditis elegans: distribution and morphological characteristics. Histochem Cell Biol 118:329–336

Age-Related Changes to *Drosophila m.* Male Germline Stem Cells

4

Hila Toledano and D. Leanne Jones

Abstract

A decline in the regenerative potential of adult tissues is one of the most apparent hallmarks of aging. As tissue regeneration is facilitated by resident stem cells, this age-related decline has been attributed to altered stem cell behavior. The male germline of *Drosophila melanogaster* has provided a valuable model system for studying the effects of aging on stem cells, due to the presence of both somatic and germline stem cells that reside within the same environment. Stem cells can be easily identified and manipulated genetically to allow for precise tracking of age-related changes in vivo. In this chapter we discuss the age-related decline in spermatogenesis in *Drosophila*. Specifically, we outline intrinsic changes both to stem cells and to the local microenvironment, known as the stem cell niche. Elucidation of mechanisms underlying these age-related changes has provided new paradigms that have been used to guide work in more complex mammalian stem cell systems.

H. Toledano
Department of Human Biology, Faculty of Natural Sciences,
University of Haifa, Mount Carmel, Haifa 3498838, Israel
e-mail: hila@univ.haifa.ac.il

D.L. Jones (✉)
Department of Molecular, Cell, and Developmental Biology,
Eli and Edythe Broad Center of Regenerative
Medicine and Stem Cell Research, University of California Los Angeles,
Los Angeles, CA 90095, USA
e-mail: leannejones@ucla.edu

H. Geiger et al. (eds.), *Stem Cell Aging: Mechanisms, Consequences, Rejuvenation*,
DOI 10.1007/978-3-7091-1232-8_4

4.1 Introduction

Adult stem cells are rare populations of tissue-specific cells that govern homeostasis and repair throughout the lifespan of an organism. These potent stem cells allow replenishment of gametes, skin, blood, and intestinal cells, for example, by continually providing new cells to replace the spent ones. In other tissues, such as the lung or skeletal muscle, stem cells lie dormant and are activated to replace damaged cells, following injury or stress. The regenerative potential of all adult stem cells relies on the ability to divide asymmetrically, with respect to cell fate, such that one of the cells will initiate differentiation along a specific lineage, while the other daughter cell maintains stem cell characteristics in a process that is referred to as stem cell self-renewal (Morrison and Spradling 2008). One of the hallmarks of the aging process is a gradual loss in the ability of tissues to maintain homeostasis and proper repair of damage (Jones and Rando 2011). In tissues maintained by stem cells, this strongly implies an age-related decline in adult stem cell functionality.

Germ cells are highly specialized cells that form gametes, and they are the only cells within an organism that contribute genes to offspring. Germline stem cells (GSCs) sustain gamete production, both oogenesis (egg production) and spermatogenesis (sperm production), in many organisms. Since the genetic information contained within germ cells is passed from generation to generation, the germline is often referred to as "immortal." Therefore, it is likely that germ cells possess unique strategies to protect and transmit the genetic information contained within them indefinitely. However, aging often leads to a dramatic decrease in gamete production and fecundity. In addition, single-gene mutations affecting longevity can have a converse effect on reproduction (Partridge et al. 2005). Studies examining age-related changes in GSC number and activity, as well as changes to the stem cell microenvironment, have provided insights into the mechanisms underlying the observed reduction in gametogenesis over the lifetime of the organism (Jones 2007).

4.2 The Stem Cell Niche

The dynamic balance between the production of new stem cells and differentiating progenitor cells is subject to the demands of the tissue and, ultimately, regulated by the integration of intrinsic factors and extrinsic cues. Such extrinsic signals are provided both by circulating systemic factors and by the local stem cell microenvironment, known as the stem cell *niche* (Schofield 1978).

Stem cell niches are composed of both cellular and acellular components that act together to regulate stem cell quiescence, activation, division, and survival, to precisely control the number of stem cells and differentiating progeny (Jones and Wagers 2008). Within the local niche, adjacent support cells and cell-cell/cell-extracellular matrix contacts supply growth factors and polarity cues to influence cell fates. In many tissues, the vasculature is also a critical component of the niche, which supports stem cell function by providing access to circulating growth factors and metabolites (Conboy et al. 2005; Mirzadeh et al. 2008; Shen et al. 2008; Tavazoie et al. 2008; Yoshida et al. 2007). In sum, the niche acts as rheostat that integrates local and systemic signals to determine stem cell proliferation rate stem cell and daughter cell fate, and to ensure tissue homeostasis and proper repair after tissue damage.

In aged organisms, the net age-related decline in stem cell function is the sum of changes that occur in all three compartments: stem cells, niche, and systemic circulating factors. Therefore, the age-related decline in stem cell functionality is best studied in a whole organism model, throughout its lifespan, and ideally would include analysis of the stem cells, the niche, and the systemic milieu, simultaneously.

The ability to conclusively identify stem cells within complex mammalian tissues, combined with a relatively long lifespan, has complicated a thorough analysis of the effects of aging on mammalian adult stem cells, the corresponding niche, and circulating factors. However, work in model organisms, such as the fruit fly *Drosophila melanogaster*, has provided an excellent foundation for probing the coordination of organism, tissue, and cellular aging for several reasons: (1) flies possess adult stem cells that are functionally analogous to their mammalian counterparts; (2) stem cells, niche components, and differentiated cells can be easily identified at a single-cell resolution in vivo; (3) a wide variety of genetic tools are available to manipulate gene expression in a cell-type and temporal fashion; and (4) conserved signaling pathways exist that regulate stem cell behavior and organism aging (Wang and Jones 2011).

4.3 The *Drosophila* Testis Stem Cell Niche

In *Drosophila*, the testis is a long coiled tube, and spermatogenesis proceeds in a spatiotemporal manner, from the apical tip to the base. Male germline stem cells (GSCs), somatic cyst stem cells (CySCs), and a cluster of 10–12 somatic cells called the hub are located at the apical tip of the testis (Fig. 4.1). GSCs and CySCs lie adjacent to and are in direct contact with hub cells, forming a rosette-like structure (Fig. 4.1) (Fuller 1993; Hardy et al. 1979). When a GSC divides, one daughter cell remains in contact with the hub and retains stem cell identity, while the other daughter cell is displaced away from the hub and initiates differentiation as a gonialblast (GB). GBs undergo four rounds of mitotic amplification divisions with incomplete cytokinesis to produce a cyst of 16 interconnected spermatogonia. On average, two CySCs encapsulate a each GSC and aids in regulating GSC self-renewal, and cyst cells derived from CySCs ensure differentiation of the developing spermatogonia (Kiger et al. 2000; Leatherman and Dinardo 2008; Matunis et al. 1997; Tran et al. 2000).

The architecture and function of the testis stem cell niche are influenced by spatially restricted production and secretion of the JAK-STAT ligand Unpaired (Upd), exclusively by hub cells (Fig. 4.1) (Harrison et al. 1998; Kiger et al. 2001; Tulina and Matunis 2001). JAK-STAT signaling acts intrinsically within CySCs to regulate CySC self-renewal and maintenance. In addition, activation of Stat92E, the single Stat orthologue in *Drosophila*, in CySCs is also important for regulating the behavior of adjacent GSCs in a non-autonomous manner (Flaherty et al. 2010; Leatherman and Dinardo 2008, 2010; Lim and Fuller 2012). Activation of Stat92E within GSCs,

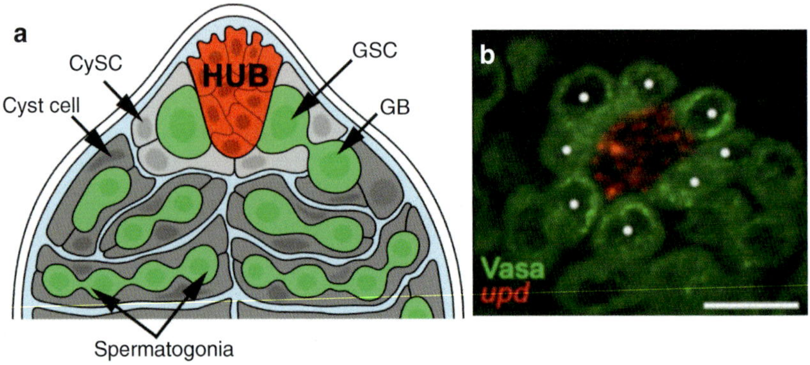

Fig. 4.1 The *Drosophila* male stem cell niche: (**a**) schematic representation of the stem cell niche at the apical tip of the testis. Hub cells (*red*). Germline stem cell (*GSC*)/progenitor gonialblast (*GB*)/spermatogonia cells (*green*). Cyst stem cell (*CySC*, *light gray*), cyst cells (*dark gray*). (**b**) Dual immunofluorescence image of the testis tip. Fluorescent in situ hybridization (FISH) to detect expression of the JAK-STAT ligand *upd* marks the hub (*red*). Antibody staining against Vasa (*green*) marks the germ cells. Eight GSCs, indicated by white dots, form a rosette around the apical hub (Adapted from Toledano et al. 2012)

however, appears to be important for regulating hub cell-GSC adhesion, rather than proliferation (Leatherman and Dinardo 2010). In addition to the JAK-STAT pathway, Hh (Amoyel et al. 2013; Michel et al. 2012; Zhang et al. 2013) and BMP (Kawase et al. 2004; Leatherman and Dinardo 2010; Michel et al. 2011; Shivdasani and Ingham 2003; Zheng et al. 2011) signaling also play important roles in regulating stem cell behavior within the testis stem cell niche. Therefore, successful spermatogenesis requires adequate signaling between hub cells, CySCs, and GSCs to coordinate proper function of each cell population.

In addition to GSCs that self-renew to maintain the stem cell pool, replacement of lost GSCs can also occur by symmetric GSC divisions, as observed by live cell imaging and lineage tracing (Salzmann et al. 2013; Sheng and Matunis 2011). Furthermore, spermatogonia can dedifferentiate to reacquire stem cell properties to contribute to tissue maintenance (Brawley and Matunis 2004; Sheng et al. 2009). Thus, germline maintenance relies on an intricate balance of renewal/differentiation/dedifferentiation of germ cells that is guided by the niche. This regeneration strategy in place in the testis is highly effective in young adults, but it gradually declines over time.

4.4 Intrinsic Age-Related Changes to Male GSCs

All aspects of spermatogenesis, including stem cell self-renewal, germline differentiation, and gamete maturation, appear to be affected by aging (Boyle et al. 2007; Cheng et al. 2008; Toledano et al. 2012; Wallenfang et al. 2006). Quantification of GSC number in testes of aged (30–35- and 50-day-old) males revealed a significant decrease in the average number of GSCs over time (Fig. 4.2), which appear to be lost as a consequence of detachment from the hub, followed by differentiation and/

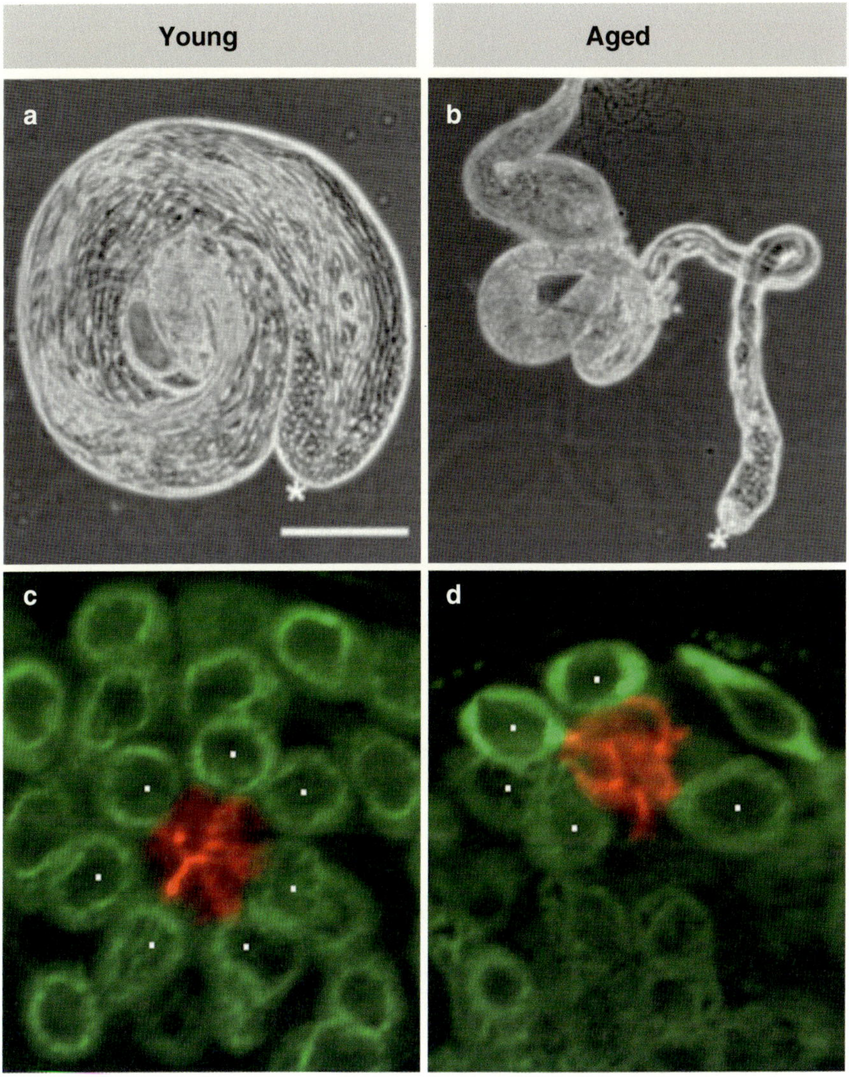

Fig. 4.2 Aging of the *Drosophila* testis. (**a**, **b**) Phase contrast images (in the same magnification, scale bar 250 μm) of testes dissected from (**a**) a young male (1 day old) or (**b**) an aged male (50 days old). Note the dramatic reduction in testis size. *Asterisks* mark apical tip of the testis. (**c**, **d**) Immunofluorescence images of testis tips from (**c**) a young or (**d**) aged male stained for Fas3 (*red*, hub) and Vasa (*green*, germ cells). Note the reduced number of GSCs (indicated by *white dots*) (Adapted from Boyle et al. 2007)

or cell death (Boyle et al. 2007; Wallenfang et al. 2006). The remaining GSCs progress through the cell cycle more slowly; therefore, both GSC loss and decreased cell division appear to contribute to the decrease in spermatogenesis observed in older males (Boyle et al. 2007; Wallenfang et al. 2006). Phase contrast analysis of testes from older males also shows increased defects in mitotic amplification divisions, although the mechanisms underlying this aspect of germline aging have not been investigated in depth (Boyle et al. 2007).

Using clonal marking strategies, it was demonstrated that GSCs have an approximate half-life of 14 days (Wallenfang et al. 2006). Therefore, one would predict that, given a starting pool of approximately 10 GSCs, testes should be devoid of germline stem cells by 7 weeks; however, testes of 50-day-old males contained an average of five GSCs/testes (Boyle et al. 2007) (Fig. 4.2). This observation indicated that a mechanism must be in place to maintain a minimum stem cell pool. One intriguing observation that was made by using clonal analysis was that the pool of GSCs appeared to progress toward clonality over time (Wallenfang et al. 2006). Therefore, either a few stem cells are more robust and maintained throughout the lifetime of the fly or, alternatively, dedifferentiating spermatogonia could aid in repopulation of the niche. Interestingly, studies have indicated that spermatogonial dedifferentiation does not decline with age and, therefore, could be a major mechanism supporting GSC maintenance during aging (Wong and Jones 2012).

Some insights were provided by studies examining centrosome positioning in male GSCs (Cheng et al. 2008; Yamashita et al. 2003). As mentioned above, *Drosophila* male GSCs typically divide in an asymmetric fashion with respect to the hub, with the daughter cell destined to differentiate being displaced away from the hub. During interphase, the mitotic spindle is oriented perpendicularly to the hub to facilitate this asymmetric outcome, and centrosome positioning regulates this stereotypical orientation of the mitotic spindle (Yamashita et al. 2003). The mother centrosome in the GSC is located adjacent to the hub, whereas the daughter centrosome migrates to the opposite side (Yamashita et al. 2007). In young adults, GSCs remain oriented perpendicularly to the hub throughout the cell cycle. However, as flies age, the percentage of GSCs with misoriented centrosomes, in which neither centrosome is located at the GSC-hub interface, increases to about 40 % of total GSCs in 30-day-old flies (Cheng et al. 2008). The age-related accumulation in GSCs bearing misoriented centrosomes was attributed primarily to the dedifferentiation process, as mitotic spindles in mitotically dividing spermatogonia are orientated randomly (Cheng et al. 2008). GSCs with misoriented centrosomes do not enter mitosis, but they can restart the cell division once the centrosomes restore their perpendicular positioning (Cheng et al. 2008).

The extra time required for reorientation of the mitotic spindle could be one contributing factor to the decline in the mitotic index observed in GSCs of aged males (Boyle et al. 2007; Wallenfang et al. 2006). In addition, studies have demonstrated that the level of cell cycle regulatory factors is altered in male germ cells (Boyle et al. 2007; Inaba et al. 2011). For example, the levels of cyclin E, a regulator of progression through the G1 phase of the cell cycle, were increased in GSCs of aged males, consistent with an arrest or extension of G1 (Boyle et al. 2007). In addition, String (Stg), the *Drosophila* homologue of the dual-specificity phosphatase Cdc25, which promotes cell cycle progression, is highly expressed in both germline and cyst stem cells and is rapidly downregulated in differentiating progeny. However, with increasing age, Stg expression declines specifically in GSCs, but not in the somatic CySCs. Loss and gain of function analysis showed that Stg is required for stem cell maintenance and division, and constitutive expression of Stg in the germline rescued the age-related decrease in GSC division and CySC number (Inaba et al. 2011). Thus, further investigation into the age-related decline in Stg expression might lead to additional insights into mechanisms regulating aging of the male germline.

Interestingly, a recent report noted that one of the core histones, histone H3, is asymmetrically segregated during male GSC divisions (Tran et al. 2012). With every cell cycle, newly synthesized H3 is segregated to the daughter gonialblast, while the original pool of H3, presumably carrying the appropriate posttranslational modifications, is maintained in the GSC. If increasing age were to alter the fidelity of H3 segregation, this could be an epigenetic mechanism leading to changes in gene expression and germline differentiation.

4.5 Aging of the Male Germline Stem Cell Niche

4.5.1 Somatic Cyst Cells

Communication between germ cells and somatic cyst cells is essential for regulating GSC proliferation and spermatogonial differentiation and dedifferentiation (Kiger et al. 2000; Leatherman and Dinardo 2008; Matunis et al. 1997; Sheng et al. 2009; Tran et al. 2000). Therefore, changes to CySC behavior over time could contribute to altered germ cell behavior in a non-cell autonomous manner. Staining for Zfh-1, a transcription factor expressed in cyst cells, including CySCs, revealed a significant decrease in the average number of early cyst cells in aged males, similar to what was observed for GSCs (Wong and Jones 2012). However, despite an age-related decrease in early cyst cell numbers, those cyst cells that remain are more active as revealed by an increased percentage of early somatic cells that are progressing through S-phase (Wong and Jones 2012). The increase in cyst cell activity is reminiscent of the behavior of somatic cells in the complete absence of germ cells, as in the case of agametic animals (Gönczy and DiNardo 1996), which may reflect a decline in antiproliferative signals normally emanating from germ cells. Alternatively, the increase in early cyst cell activity with age may represent a mechanism by remaining cyst cells to compensate for the age-related loss of somatic cells.

4.5.2 The Apical Hub

The apical hub cells are specified early in embryogenesis (Boyle and DiNardo 1995), and later during adulthood, hub cells appear to be nondividing and relatively stable with respect to gene expression (Gönczy and DiNardo 1996; Voog et al. 2008, 2014). However, the ability of hub cells to support stem cell behavior changes dramatically with age (Boyle et al. 2007; Toledano et al. 2012). Over time, the 10–12 hub cells appear to become disorganized, and a consistent decrease in the expression of the *Drosophila* homologue of E-cadherin (DE-cadherin) has been observed (Boyle et al. 2007). DE-cadherin is expressed highly at hub cell-hub cell and hub cell-GSC junctions and is required for maintenance of both male and female GSCs (Inaba et al. 2010; Song and Xie 2002; Song et al. 2002; Voog et al. 2008; Yamashita et al. 2003).

Remarkably, however, roughly 40 % of testes from aged (50-day-old) males showed no detectable expression of a key self-renewal signal, *upd*. As activation of the JAK-STAT pathways regulates the behavior of both CySCs and GSCs, the decreased expression of *upd* could contribute to loss of CySCs and/or decreased cell-cell adhesion between GSCs and hub cells directly, as well as decreased GSC proliferation indirectly, by way of altered STAT activity in CySCs. Constitutive re-expression of *upd* in hub cells suppressed the loss of GSCs of aged males, but had no effect in young males, indicating that loss of *upd* expression could be a major contributing factor to the age-related decline in spermatogenesis (Boyle et al. 2007). The effects of aging on Hh signaling or the BMP pathway in the *Drosophila* testis has not been analyzed to date. Interestingly, studies have suggested that decreased expression of self-renewal factors in the murine testis might also contribute to decreased spermatogenesis in this system (Oatley and Brinster 2012; Ryu et al. 2006; Zhang et al. 2006).

The decrease in *upd* expression in hub cells appears to be due, in part, to the activities of both RNA-binding proteins (RBPs) and small RNAs (both microRNAs (miRNAs) and endogenous small interfering RNAs (endo siRNAs)). The evolutionarily conserved RBP, Imp, possesses K homology (KH) domains that bind the 3′ untranslated region (3′UTR) of *upd* to stabilize the mRNA. In hub cells, Imp appears to counteract endo-siRNAs that target *upd* mRNA and, thus, contributes to the maintenance of niche function in young adults (Toledano et al. 2012). However, in testes of older males, the levels of Imp were significantly reduced specifically in hub cells (Toledano et al. 2012). The age-related decrease in Imp also appears to be due to the activity of a small RNA, in this case the conserved heterochronic miRNA *let-7*, which targets *Imp* mRNA (Ambros 2011; Toledano et al. 2012). Expression of *let-7* is faintly detected in testes from young males; however, in older males, *let-7* expression is strongly induced in hub cells (Table 4.1; Toledano et al. 2012). An age-related increase in *let-7* in hub cells leads to a decrease in Imp, which in turn exposes *upd* mRNA to targeting by endo-siRNAs. Therefore, elevated levels of the heterochronic miRNA *let-7* in the hub cells of aged males trigger a cascade of post-transcriptional events that result in decreased JAK-STAT signaling and impaired niche function over time (Toledano et al. 2012).

Notably, constitutive expression of Upd or Imp that is not susceptible to *let-7-mediated* degradation in hub cells suppresses GSC loss in older males (Boyle et al. 2007; Toledano et al. 2012). Nonetheless, forced Upd/Imp expression was not sufficient to fully rescue defects in spermatogonial proliferation and differentiation, confirming that aging independently modulates other stages of spermatogenesis.

4.5.3 Systemic Signals

Numerous studies have indicated that stem cells within a variety of tissues and species are influenced by insulin/IGF signaling (Shim et al. 2013). The insulin signaling pathway, which is conserved in *Drosophila*, encodes a single insulin/IGF-like receptor (*dInR*) and seven insulin-like peptides (*dILPs*) in its genome (Brogiolo et al. 2001; Ikeya et al. 2002; Puig et al. 2003). Experiments have clearly demonstrated that insulin signaling directly regulates the rate of GSC proliferation in the

Table 4.1 Deep sequencing data of *Drosophila melanogaster* (dme)-miRNAs in libraries from testis of wild-type (OregonR) young (1 day) and aged (30 day) old flies

miRNA	Sequence	1 day adult testis library		30 day adult testis library		Ratio 30 day/1 day
		No. of reads	% of lib.	No. of reads	% of lib.	
dme-let-7-5p	TGAGGTAGTAGGTTGTATAGT	1,17,568	0.98	2,17,661	1.86	1.90
dme-mir-100-5p	AACCCGTAAATCCGAACTTGT	1057	0.01	2049	0.02	1.99
dme-mir-125-5p	TCCCTGAGACCCTAACTTGTGA	2534	0.02	5053	0.04	2.05

The three miRNAs of let-7 cluster (dme-let-7C) that are co-transcribed are presented. Note that the ratio 30 day/1 day is similar for all three miRNAs

Drosophila ovary (Ikeya et al. 2002; LaFever and Drummond-Barbosa 2005; Yu et al. 2009) and regulates the maintenance of female GSCs indirectly by preservation of the niche (Hsu and Drummond-Barbosa 2009). Clonal analysis demonstrated that *dInR* function is also required cell autonomously for GSC maintenance in the *Drosophila* testis, suggesting that male GSCs are competent to receive and respond to insulin signaling directly (McLeod et al. 2010; Ueishi et al. 2009). Importantly, dInR appears to regulate GSC maintenance independently of the JAK-STAT pathway (McLeod et al. 2010; Wang et al. 2011).

Three dILPs, *dILP2*, *dILP3*, and *dILP5*, are expressed in insulin-producing cells (IPCs) in the brain, secreted into the circulating hemolymph, and act to regulate InR activity systemically (Brogiolo et al. 2001; Ikeya et al. 2002; Kramer et al. 2003; Puig et al. 2003). While a significant source of dILPs is in the brain, expression of *dILP5* and *dILP7* has been reported in the ovary and female reproductive tract, where they regulate the onset of vitellogenesis and coordinate the strong and swift response of the female reproductive program to nutrition (Slaidina et al. 2009; Yang et al. 2008). In contrast, *dILP2* and *dILP3* are expressed in somatic cyst cells in the testis that encapsulate spermatocytes and differentiating cysts at the basal end of the testis (McLeod et al. 2010; Wang et al. 2011). Therefore, decreased production of brain-derived and/or locally produced dILPs, as a consequence of aging, could contribute to a reduction of dInR signaling in GSCs in the testis, and consequently a loss of GSCs and a decrease in proliferation rate. Indeed, using a phosphoinositide 3-kinase reporter, Hsu et al. found a decrease in insulin signaling in the ovaries of aged females (Hsu and Drummond-Barbosa 2009). Studies have yet to dissect the relative effects of aging on local dILPs versus brain-derived dILPs and the subsequent contribution to aging-related changes in tissue homeostasis.

Conclusions

Studies performed in the *Drosophila* male germline have established important paradigms for how stem cell niches are established and maintained over time. Both autonomous and non-autonomous mechanisms have been elucidated that contribute to decreased stem cell activity in the male germline upon aging (Fig. 4.3). Importantly, changes such as decreased expression of self-renewal factors and decreased cell-cell adhesion appear to be conserved in more complex mammalian systems (Jones and Rando 2011). Future studies should focus on a thorough analysis of age-related changes in gene expression, due to both genetic and epigenetic mechanisms, and additional signaling pathways known to regulate the behavior of stem cells and support cells in the testis. In addition, given the integral role of metabolic pathways in regulating lifespan of organisms as diverse as yeast and nonhuman primates, it will be interesting to investigate whether key metabolic programs might be central players in coordinating organismal, tissue, and cellular aging.

Altogether, it is clear that the remarkable ability of stem cells to maintain tissue homeostasis and repair damage relies on a complex tuning of local cues and systemic signals. Compromised niche function over time may lead to the selection of stem cells that acquire the ability to self-renew independently of the niche

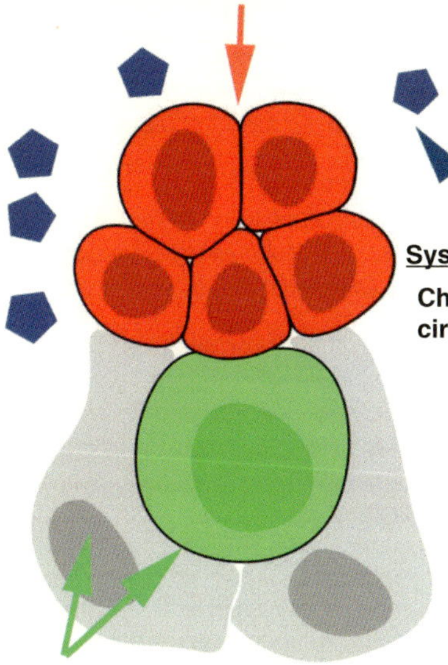

Fig. 4.3 Mechanisms contributing to age-related changes in stem cell behavior in the *Drosophila* testis

and/or progenitor cells that acquire self-renewal capabilities (Li and Neaves 2006). Such cells could be the precursors to cancer stem cells that arise to contribute to tumorigenesis in variety of tissues. Furthermore, investigating how the process of aging affects tissue homeostasis has important implications for stem cell-based therapies. The use of tissue replacement in regenerative medicine will require the isolation, propagation, and transplantation of stem cells. Expansion and manipulation of tissue stem cells from older individuals for autologous transplants may prove to be more difficult due to autonomous changes that occur with age (Rheinwald and Green 1975). In addition, it is not yet clear how age affects the reprogramming of tissues or the expansion of induced pluripotent stem cells from older patients (Isobe et al. 2014). Therefore, stem cells transplanted into older individuals may be unable to initiate self-renewing divisions and functionally replace damaged or diseased tissues without co-transplantation of "younger" niche cells or mobilization of endogenous stem cells from functional niches. Continuing studies to identify niche components and thoroughly characterize the effects of aging on both stem cells and the niche will be critical for the implementation of stem cell-based therapies to treat degenerative diseases, particularly in older individuals.

References

Ambros V (2011) MicroRNAs and developmental timing. Curr Opin Genet Dev 21:511–517
Amoyel M, Sanny J, Burel M, Bach EA (2013) Hedgehog is required for CySC self-renewal but does not contribute to the GSC niche in the Drosophila testis. Development 140:56–65
Boyle M, DiNardo S (1995) Specification, migration and assembly of the somatic cells of the Drosophila gonad. Development 121:1815–1825
Boyle M, Wong C, Rocha M, Jones DL (2007) Decline in self-renewal factors contributes to aging of the stem cell niche in the Drosophila testis. Cell Stem Cell 1:470–478
Brawley C, Matunis E (2004) Regeneration of male germline stem cells by spermatogonial dedifferentiation in vivo. Science 304:1331–1334
Brogiolo W, Stocker H, Ikeya T, Rintelen F, Fernandez R, Hafen E (2001) An evolutionarily conserved function of the Drosophila insulin receptor and insulin-like peptides in growth control. Curr Biol 11:213–221
Cheng J, Turkel N, Hemati N, Fuller MT, Hunt AJ, Yamashita YM (2008) Centrosome misorientation reduces stem cell division during ageing. Nature 456:599–604
Conboy IM, Conboy MJ, Wagers AJ, Girma ER, Weissman IL, Rando TA (2005) Rejuvenation of aged progenitor cells by exposure to a young systemic environment. Nature 433:760–764
Flaherty MS, Salis P, Evans CJ, Ekas LA, Marouf A, Zavadil J, Banerjee U, Bach EA (2010) Chinmo is a functional effector of the JAK/STAT pathway that regulates eye development, tumor formation, and stem cell self-renewal in Drosophila. Dev Cell 18:556–568
Fuller MT (1993) Spermatogenesis. In: Bate M, Martinez-Arias A (eds) The development of Drosophila melanogaster. Cold Spring Harbor Laboratory Press, Cold Spring Harbor
Gönczy P, DiNardo S (1996) The germ line regulates somatic cyst cell proliferation and fate during Drosophila spermatogenesis. Development 122:2437–2447
Hardy RW, Tokuyasu KT, Lindsley DL, Garavito M (1979) The germinal proliferation center in the testis of Drosophila melanogaster. J Ultrastruct Res 69:180–190
Harrison DA, McCoon PE, Binari R, Gilman M, Perrimon N (1998) Drosophila unpaired encodes a secreted protein that activates the JAK signaling pathway. Genes Dev 12:3252–3263
Hsu HJ, Drummond-Barbosa D (2009) Insulin levels control female germline stem cell maintenance via the niche in Drosophila. Proc Natl Acad Sci U S A 106:1117–1121
Ikeya T, Galic M, Belawat P, Nairz K, Hafen E (2002) Nutrient-dependent expression of insulin-like peptides from neuroendocrine cells in the CNS contributes to growth regulation in Drosophila. Curr Biol 12:1293–1300
Inaba M, Yuan H, Salzmann V, Fuller MT, Yamashita YM (2010) E-cadherin is required for centrosome and spindle orientation in Drosophila male germline stem cells. PLoS One 5, e12473
Inaba M, Yuan H, Yamashita YM (2011) String (Cdc25) regulates stem cell maintenance, proliferation and aging in Drosophila testis. Development 138:5079–5086
Isobe KI, Cheng Z, Nishio N, Suganya T, Tanaka Y, Ito S (2014) iPSCs, aging and age-related diseases. N Biotechnol 31:411–421
Jones DL (2007) Aging and the germ line: where mortality and immortality meet. Stem Cell Rev 3:192–200
Jones DL, Rando TA (2011) Emerging models and paradigms for stem cell ageing. Nat Cell Biol 13:506–512
Jones DL, Wagers AJ (2008) No place like home: anatomy and function of the stem cell niche. Nat Rev Mol Cell Biol 9:11–21
Kawase E, Wong MD, Ding BC, Xie T (2004) Gbb/Bmp signaling is essential for maintaining germline stem cells and for repressing bam transcription in the Drosophila testis. Development 131:1365–1375
Kiger AA, White-Cooper H, Fuller MT (2000) Somatic support cells restrict germline stem cell self-renewal and promote differentiation. Nature 407:750–754
Kiger AA, Jones DL, Schulz C, Rogers MB, Fuller MT (2001) Stem cell self-renewal specified by JAK-STAT activation in response to a support cell cue. Science 294:2542–2545
Kramer JM, Davidge JT, Lockyer JM, Staveley BE (2003) Expression of Drosophila FOXO regulates growth and can phenocopy starvation. BMC Dev Biol 3:5

LaFever L, Drummond-Barbosa D (2005) Direct control of germline stem cell division and cyst growth by neural insulin in Drosophila. Science 309:1071–1073

Leatherman JL, Dinardo S (2008) Zfh-1 controls somatic stem cell self-renewal in the Drosophila testis and nonautonomously influences germline stem cell self-renewal. Cell Stem Cell 3:44–54

Leatherman JL, Dinardo S (2010) Germline self-renewal requires cyst stem cells and stat regulates niche adhesion in Drosophila testes. Nat Cell Biol 12:806–811

Li L, Neaves WB (2006) Normal stem cells and cancer stem cells: the niche matters. Cancer Res 66:4553–4557

Lim JG, Fuller MT (2012) Somatic cell lineage is required for differentiation and not maintenance of germline stem cells in Drosophila testes. Proc Natl Acad Sci U S A 109:18477–18481

Matunis E, Tran J, Gönczy P, Caldwell K, DiNardo S (1997) punt and schnurri regulate a somatically derived signal that restricts proliferation of committed progenitors in the germline. Development 124:4383–4391

McLeod CJ, Wang L, Wong C, Jones DL (2010) Stem cell dynamics in response to nutrient availability. Curr Biol 20:2100–2105

Michel M, Raabe I, Kupinski AP, Perez-Palencia R, Bokel C (2011) Local BMP receptor activation at adherens junctions in the Drosophila germline stem cell niche. Nat Commun 2:415

Michel M, Kupinski AP, Raabe I, Bokel C (2012) Hh signalling is essential for somatic stem cell maintenance in the Drosophila testis niche. Development 139:2663–2669

Mirzadeh Z, Merkle FT, Soriano-Navarro M, Garcia-Verdugo JM, Alvarez-Buylla A (2008) Neural stem cells confer unique pinwheel architecture to the ventricular surface in neurogenic regions of the adult brain. Cell Stem Cell 3:265–278

Morrison SJ, Spradling AC (2008) Stem cells and niches: mechanisms that promote stem cell maintenance throughout life. Cell 132:598–611

Oatley JM, Brinster RL (2012) The germline stem cell niche unit in mammalian testes. Physiol Rev 92:577–595

Partridge L, Gems D, Withers DJ (2005) Sex and death: what is the connection? Cell 120:461–472

Puig O, Marr MT, Ruhf ML, Tjian R (2003) Control of cell number by Drosophila FOXO: downstream and feedback regulation of the insulin receptor pathway. Genes Dev 17:2006–2020

Rheinwald JG, Green H (1975) Serial cultivation of strains of human epidermal keratinocytes: the formation of keratinizing colonies from single cells. Cell 6:331–343

Ryu BY, Orwig KE, Oatley JM, Avarbock MR, Brinster RL (2006) Effects of aging and niche microenvironment on spermatogonial stem cell self-renewal. Stem Cells (Dayton, Ohio) 24:1505–1511

Salzmann V, Inaba M, Cheng J, Yamashita YM (2013) Lineage tracing quantification reveals symmetric stem cell division in Drosophila male germline stem cells. Cell Mol Bioeng 6:441–448

Schofield R (1978) The relationship between the spleen colony-forming cell and the haemopoietic stem cell. Blood Cells 4:7–25

Shen Q, Wang Y, Kokovay E, Lin G, Chuang SM, Goderie SK, Roysam B, Temple S (2008) Adult SVZ stem cells lie in a vascular niche: a quantitative analysis of niche cell-cell interactions. Cell Stem Cell 3:289–300

Sheng XR, Matunis E (2011) Live imaging of the Drosophila spermatogonial stem cell niche reveals novel mechanisms regulating germline stem cell output. Development 138:3367–3376

Sheng XR, Brawley CM, Matunis EL (2009) Dedifferentiating spermatogonia outcompete somatic stem cells for niche occupancy in the Drosophila testis. Cell Stem Cell 5:191–203

Shim J, Gururaja-Rao S, Banerjee U (2013) Nutritional regulation of stem and progenitor cells in Drosophila. Development 140:4647–4656

Shivdasani AA, Ingham PW (2003) Regulation of stem cell maintenance and transit amplifying cell proliferation by tgf-beta signaling in Drosophila spermatogenesis. Curr Biol 13:2065–2072

Slaidina M, Delanoue R, Gronke S, Partridge L, Leopold P (2009) A Drosophila insulin-like peptide promotes growth during nonfeeding states. Dev Cell 17:874–884

Song X, Xie T (2002) DE-cadherin-mediated cell adhesion is essential for maintaining somatic stem cells in the Drosophila ovary. Proc Natl Acad Sci U S A 99:14813–14818

Song X, Zhu CH, Doan C, Xie T (2002) Germline stem cells anchored by adherens junctions in the Drosophila ovary niches. Science 296:1855–1857

Tavazoie M, Van der Veken L, Silva-Vargas V, Louissaint M, Colonna L, Zaidi B, Garcia-Verdugo JM, Doetsch F (2008) A specialized vascular niche for adult neural stem cells. Cell Stem Cell 3:279–288

Toledano H, D'Alterio C, Czech B, Levine E, Jones DL (2012) The let-7-Imp axis regulates ageing of the Drosophila testis stem-cell niche. Nature 485:605–610

Tran J, Brenner TJ, DiNardo S (2000) Somatic control over the germline stem cell lineage during Drosophila spermatogenesis. Nature 407:754–757

Tran V, Lim C, Xie J, Chen X (2012) Asymmetric division of Drosophila male germline stem cell shows asymmetric histone distribution. Science 338:679–682

Tulina N, Matunis E (2001) Control of stem cell self-renewal in Drosophila spermatogenesis by JAK-STAT signaling. Science 294:2546–2549

Ueishi S, Shimizu H, H Inoue Y (2009) Male germline stem cell division and spermatocyte growth require insulin signaling in Drosophila. Cell Struct Funct 34:61–69

Voog J, D'Alterio C, Jones DL (2008) Multipotent somatic stem cells contribute to the stem cell niche in the Drosophila testis. Nature 454:1132–1136

Voog J, Sandall SL, Hime GR, Resende LP, Loza-Coll M, Aslanian A, Yates JR 3rd, Hunter T, Fuller MT, Jones DL (2014) Escargot restricts niche cell to stem cell conversion in the Drosophila testis. Cell Rep 7:722–734

Wallenfang MR, Nayak R, DiNardo S (2006) Dynamics of the male germline stem cell population during aging of Drosophila melanogaster. Aging Cell 5:297–304

Wang L, Jones DL (2011) The effects of aging on stem cell behavior in Drosophila. Exp Gerontol 46:340–344

Wang L, McLeod CJ, Jones DL (2011) Regulation of adult stem cell behavior by nutrient signaling. Cell Cycle 10:2628–2634

Wong C, Jones DL (2012) Efficiency of spermatogonial dedifferentiation during aging. PLoS One 7, e33635

Yamashita Y, Jones DL, Fuller MT (2003) Orientation of asymmetric stem cell division by the APC tumor suppressor and centrosome. Science 301:1547–1550

Yamashita YM, Mahowald AP, Perlin JR, Fuller MT (2007) Asymmetric inheritance of mother versus daughter centrosome in stem cell division. Science 315:518–521

Yang CH, Belawat P, Hafen E, Jan LY, Jan YN (2008) Drosophila egg-laying site selection as a system to study simple decision-making processes. Science 319:1679–1683

Yoshida S, Sukeno M, Nabeshima Y (2007) A vasculature-associated niche for undifferentiated spermatogonia in the mouse testis. Science 317:1722–1726

Yu JY, Reynolds SH, Hatfield SD, Shcherbata HR, Fischer KA, Ward EJ, Long D, Ding Y, Ruohola-Baker H (2009) Dicer-1-dependent Dacapo suppression acts downstream of Insulin receptor in regulating cell division of Drosophila germline stem cells. Development 136:1497–1507

Zhang X, Ebata KT, Robaire B, Nagano MC (2006) Aging of male germ line stem cells in mice. Biol Reprod 74:119–124

Zhang Z, Lv X, Jiang J, Zhang L, Zhao Y (2013) Dual roles of Hh signaling in the regulation of somatic stem cell self-renewal and germline stem cell maintenance in Drosophila testis. Cell Res 23:573–576

Zheng Q, Wang Y, Vargas E, DiNardo S (2011) magu is required for germline stem cell self-renewal through BMP signaling in the Drosophila testis. Dev Biol 357:202–210

Aging of Intestinal Stem Cells in *Drosophila Melanogaster*

5

Benoît Biteau

Abstract

Drosophila m. has a long history of major contributions to basic biology and biomedical research. Not surprisingly, the recent identification of several multipotent stem cell populations in the *Drosophila* fly digestive tract has generated an immense enthusiasm in the research community. This experimental model provides a unique opportunity to study adult somatic stem cells, using the power of fly genetics. Over the past few years, research in this field has focused on the characterization of the signaling pathways and mechanisms that control stem cell function and tissue repair in the intestine. Importantly, the rapid aging and short lifespan of *Drosophila* make this model ideal to investigate the impact of aging on stem cell populations and test the contribution of somatic stem cells to normal healthspan and lifespan. This chapter presents recent findings that elucidate the mechanisms causing age-related loss of tissue homeostasis in the fly intestine, as well as strategies of stem cell-specific genetic manipulation that significantly impact physiology in aging animals and can extend lifespan.

5.1 Introduction

With the identification of stem cell populations in the developing larval brain and the gonads, studies in *Drosophila* have been pivotal to the discovery of fundamental processes that regulate stem cell biology in many other stem/progenitor compartments. For example, many conserved mechanisms that control asymmetric cell division and cell specification were first identified in larval neuroblasts (Knoblich 2010), while studies of the germline stem cells in the fly testis and ovary

B. Biteau
Department of Biomedical Genetics, University of Rochester Medical Center, 633 Elmwood Avenue, Rochester, NY, USA
e-mail: benoit_biteau@urmc.rochester.edu

H. Geiger et al. (eds.), *Stem Cell Aging: Mechanisms, Consequences, Rejuvenation*,
DOI 10.1007/978-3-7091-1232-8_5

lead to the first demonstration of the concept of stem cell niches (Decotto and Spradling 2005; Fuller and Spradling 2007). Although these models have significantly contributed to our understanding of stem cell biology, they provided only limited insight into the impact of aging on somatic multipotent stem cell populations and the influence of these progenitors on the maintenance of healthy tissues and normal longevity.

The recent identification of multipotent stem cell populations in the fly gastrointestinal tract has revolutionized aging research in *Drosophila*. While previous studies had vastly focused on the mechanisms allowing the protection of somatic cells, the possibility that adult somatic cells are maintained by tissue-specific progenitors opened up new research avenues. Importantly, combined with the short lifespan and the power of fly genetics, this new experimental system provides a unique opportunity to directly address in vivo questions that were previously inaccessible. For example, what is the impact of aging on the mechanisms that control somatic stem cell proliferation, cell-fate decision, and differentiation of progenitors in this tissue? How do stem cell populations in the intestine influence metabolism and physiology in young and old animals? Is *Drosophila* longevity limited by the fitness of somatic stem cell populations? Can lifespan be extended by stem cell-specific manipulations?

This chapter presents recent studies that provide new answers to these questions and discuss the aspects of stem cell aging in the fly intestine that remain to be investigated.

5.2 Stem Cell Populations in the *Drosophila* Digestive Tract

Drosophila adult organs have been thought to be strictly postmitotic for decades. Surprisingly, while powerful lineage tracing methods have been developed in this genetic model to study progenitors in developing tissues and stem cell populations in gonads (Lee and Luo 1999; Perrimon 1998), these methods have only recently been used to investigate the possibility of the existence of other tissue-specific stem cells in the adult fly. These breakthrough studies have identified several multipotent stem cell populations in the fly digestive tract and established that many of these adult tissues are maintained by specialized progenitor pools (Fig. 5.1).

5.2.1 Cardia and Stomach

Multipotent stem cells (GaSCs) have been identified in the cardia (also called proventriculus – Fig. 5.1), a structure that connects the crop, the esophagus, and the anterior midgut (Singh et al. 2011). These progenitors were found to continuously give rise to cells that compose the cardia itself as well as the crop, under unstressed conditions, suggesting that they are multipotent and constantly proliferate.

The central section of the fly midgut generates an acidic environment, analogous to the mammalian stomach (Shanbhag and Tripathi 2009). The epithelium of this

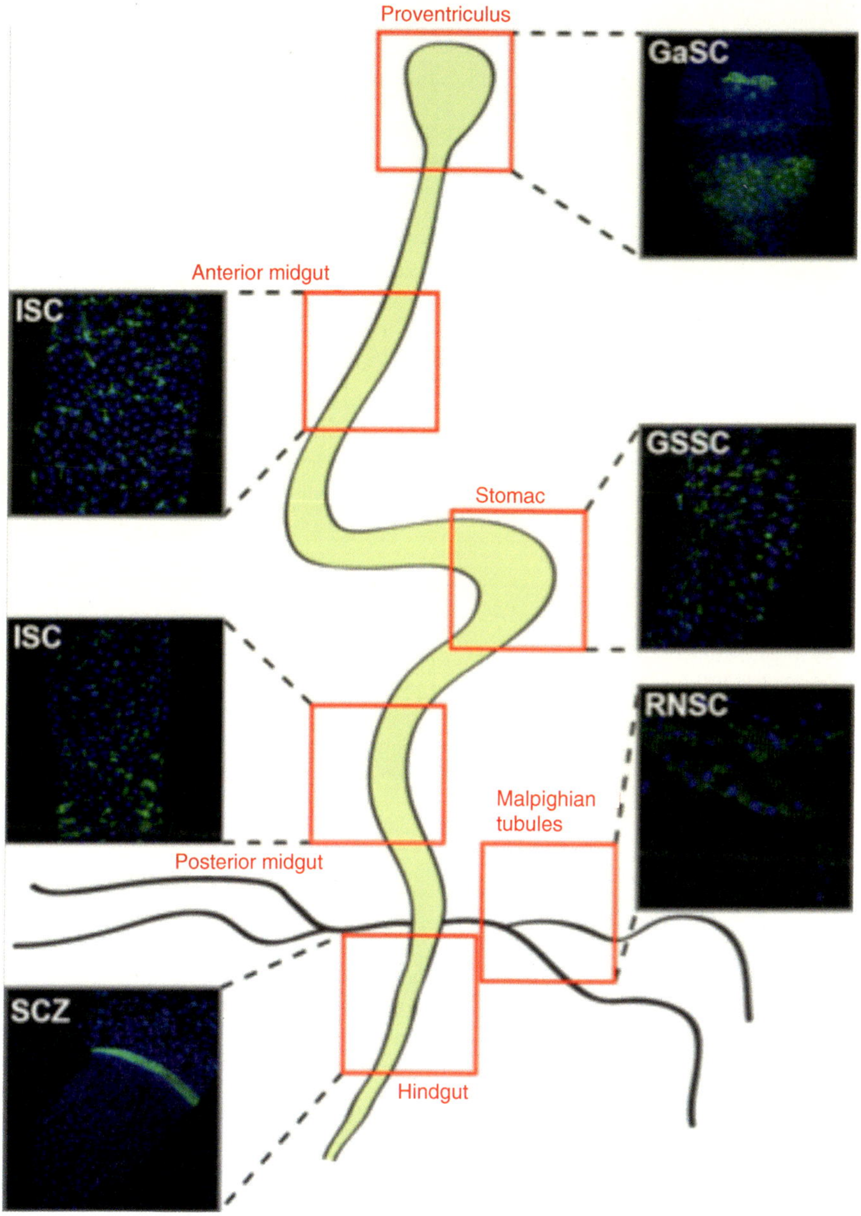

Fig. 5.1 Stem cell populations in the *Drosophila m.* digestive tract

region, including the copper cells responsible for the acidification, interstitial cells, and endocrine cells, is maintained by a specialized population of multipotent gastric stem cells (GSSC – Fig. 5.1) (Strand and Micchelli 2011). These cells are relatively quiescent but rapidly proliferate in response to specific tissue damage. The induction of GSSC proliferation by stress suggests that the behavior of this cell population may change over time. However, potential age-related changes in these tissues and effect of aging on the GaSC and GSSC populations remain to be tested.

5.2.2 Malpighian Tubules

Similar to the kidneys in mammals, the excretory function in *Drosophila* is fulfilled by specialized organs, the Malpighian tubules. It has been recently recognized that this tissue is maintained by a multipotent stem cell population, the renal and nephric stem cells (RNSCs – Fig. 5.1) (Singh et al. 2007). These cells are capable of self-renewal and give rise to progenitors that can differentiate in all the cell types that compose the tubule (i.e., renalcytes, principal cells, and stellate cells). The mechanisms that control SC proliferation and differentiation in this lineage are starting to be elucidated (Singh et al. 2007; Zeng et al. 2010). However, the impact of aging on the maintenance and turnover of this tissue remains unknown.

5.2.3 Hindgut

The most posterior portion of the intestine is formed by the hindgut. Recently, proliferating cells have been identified in the most anterior segment of the hindgut, the hindgut proliferation zone (HPZ) (Takashima et al. 2008). The anterior portion of the HPZ is composed of a stem cell population, the spindle cell zone (SCZ), which maintains the hindgut epithelium (Fig. 5.1). Contrary to all other stem cell lineages identified in the adult fly digestive tract, these stem cells give rise to a clearly identified transit-amplifying progenitor population.

While these cells were first described as continuously proliferating (Takashima et al. 2008), additional studies have demonstrated that the SCZ stem cell population is mostly quiescent in healthy young animals (Fox and Spradling 2009). Interestingly, proliferation of stem cells and transit-amplifying cells is dramatically increased in response to tissue damage (Fox and Spradling 2009). As the aging intestine is characterized by a chronic exposure to a damaging environment (see below), it is expected that aging can affect the proliferative behavior of these hindgut progenitor cells. In addition, the *Drosophila* hindgut has recently been described as a model of infection- and oncogene-mediated cell dissemination and metastasis (Bangi et al. 2012; Christofi and Apidianakis 2012). However, this process remains to be investigated in the context of the aging hindgut epithelium.

5.2.4 Posterior and Anterior Midgut

The intestinal stem cells (ISCs – Fig. 5.1) population, present in the posterior and anterior midgut, was the first multipotent stem cell population described in the digestive tract (Micchelli and Perrimon 2006; Ohlstein and Spradling 2006). ISCs were first identified as small diploid cells interspaced between the large absorptive cells that compose most of the epithelium (enterocytes, ECs) and located in close contact with the basal membrane (Baumann 2001). However, only recent lineage tracing experiments have demonstrated that ISCs continuously proliferate and maintain the intestinal epithelium by producing new ECs and enteroendocrine cells (EEs) (Micchelli and Perrimon 2006; Ohlstein and Spradling 2006). Following the discovery of ISCs, the *Drosophila* midgut has rapidly emerged as an ideal model system to study the mechanisms that control intestinal homeostasis and all aspects of tissue-specific stem cells, under unchallenged conditions or during tissue repair (Biteau et al. 2011; Jiang and Edgar 2012; Wang and Hou 2010).

Interestingly, ISCs continuously proliferate at low rate, replacing the entire epithelium several times throughout the life of the animal, in particular in females that have a much higher nutritional demand to accommodate egg production (Jiang et al. 2009). Normal tissue turnover requires a variety of signals that control ISC self-renewal, proliferation, and differentiation, including the following pathways:

- Notch (Bardin et al. 2010; Ohlstein and Spradling 2007; Perdigoto et al. 2011)
- Insulin (Amcheslavsky et al. 2009; Biteau et al. 2010; Choi et al. 2011; Kapuria et al. 2012)
- Epidermal growth factor (EGF) (Biteau and Jasper 2011; Buchon et al. 2010; Jiang et al. 2011; Xu et al. 2011)
- PDGF/VEGF receptor (Pvr) (Bond and Foley 2012; Choi et al. 2008)
- Target of rapamycin (TOR) (Amcheslavsky et al. 2011; Kapuria et al. 2012)
- Janus kinase – signal transducer and activator of transcription (JAK-STAT) (Beebe et al. 2010; Lin et al. 2010; Liu et al. 2010)
- Hippo (Karpowicz et al. 2010; Ren et al. 2010; Shaw et al. 2010; Staley and Irvine 2010)
- Wnt/wingless and adenomatous polyposis coli (APC) (Cordero et al. 2012; Lee et al. 2009; Lin et al. 2008)

In addition, cell proliferation is rapidly induced in the intestine in response to stress and infection, in order to replace damaged differentiated cells and maintain the integrity of the epithelium. This proliferative response directly involves the activation of ISCs themselves, as no transit-amplifying cell population exists in the midgut, and relies on the induction of many conserved signaling pathways, such as:

- Jun N-terminal kinase (JNK) (Biteau et al. 2008; Buchon et al. 2009a; Jiang et al. 2009)
- JAK-STAT pathway (Buchon et al. 2009a, b; Cronin et al. 2009; Jiang et al. 2009; Zhou et al. 2013)
- p38 (Park et al. 2009)
- Nrf2 (Hochmuth et al. 2011)

While the impact of aging on RNSCs, GaSCs, GSSCs, and the SCZ remains to be investigated, the age-related changes that impair ISC function in older flies are starting to be elucidated. These mechanisms include extrinsic and intrinsic signals, which are largely related to the chronic inflammatory state that develops in the aging intestinal epithelium, and cause uncontrolled ISC proliferation and differentiation defects. In addition, this stem cell population has proven to be a valuable model to address critical questions regarding the influence of stem cell-mediated tissue homeostasis on physiology and longevity.

5.3 Age-Related ISC Hyper-proliferation

ISCs were first characterized as a slow cycling stem cell population. Only a few cells, out of the thousands of ISCs dispersed in the midgut, are undergoing mitosis at any given time in young animals. However, ISC proliferation dramatically increases with age (Biteau et al. 2008; Choi et al. 2008). This hyper-proliferation is reminiscent of the response to tissue damage, which relies on an increased ISC proliferation to replace damaged differentiated cells (Amcheslavsky et al. 2009; Buchon et al. 2009b; Choi et al. 2008; Cronin et al. 2009; Jiang et al. 2009). This insight led to the current model that proposes that the chronic activation of the mechanisms that control tissue repair is the primary cause of the changes observed in the aging intestine and age-related disruption of the ISC function.

The aging process in the posterior midgut includes changes in the intestinal environment, in particular in the gut microbial flora, and the activation of both non-cell-autonomous and intrinsic signals that ultimately drive uncontrolled ISC proliferation (Fig. 5.2). However, many questions remain unanswered. For example, the mechanisms that contribute to the changes in bacteria population are still not fully understood; the specific contribution of each signaling pathway remains unclear; and it has yet to be tested whether some aspects of the intestinal and ISC aging are independent of the gut flora and stress responses and directly related to an intrinsic aging process in ISCs.

5.4 Environmental Changes and Extrinsic Signals

5.4.1 Role of Age-Related Changes in the Intestinal Flora

Like the intestine in other animals, the *Drosophila* midgut hosts a variety of commensal, food-borne, and pathogenic bacterial populations (Hooper and Gordon 2001; Sansonetti 2004). Flies have evolved a variety of mechanisms to accommodate beneficial flora and defend the intestinal epithelium against pathogens (Ryu et al. 2010). Interestingly, dramatic changes in the intestinal flora occur during aging in mammals (Tiihonen et al. 2010; Woodmansey 2007). Supporting the notion that similar changes in gut flora strongly contribute to the aging process of the intestinal epithelium in *Drosophila*, maintaining flies in an axenic environment prevents age-induced ISC proliferation and intestinal dysplasia (Buchon et al. 2009a). However, it is still lacking a clear description of age-related changes in the

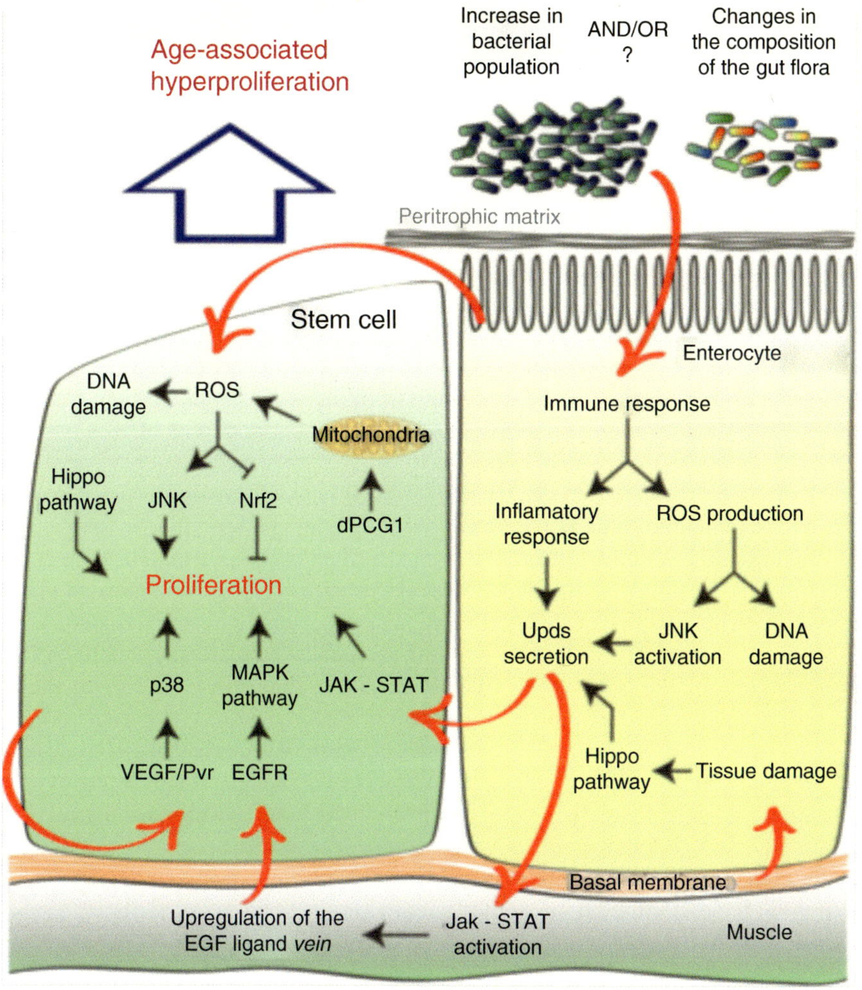

Fig. 5.2 Schematic representation of the numerous cell-autonomous mechanisms and extrinsic signals that may contribute to the ISC hyper-proliferation observed in the aging *Drosophila* intestine

composition and/or size of bacterial populations in the midgut of older flies. In addition, the reasons for these potential changes remain unclear. Several non-exclusive mechanisms may drive these alterations, such as defects in the immune system (i.e., immunosenescence); decay of the peritrophic membrane, a structure essential to protect the intestinal epithelium against bacterial infections (Kuraishi et al. 2011); changes in intestinal function and chemical properties; and selection or expansion of more virulent bacterial strains over time. Importantly, two recent studies have started to address this question and suggest that a collapse of proper immune signaling in the gut itself and the adipose tissue, another essential immune organ in flies, plays a fundamental role in dysbiotic changes occurring in older animals (Chen et al. 2014; Guo et al. 2014). Yet, further investigations will be required to test all the possible factors contributing to age-related changes in microbiota and fully understand the underlying age-related molecular changes.

5.4.2 Bacterial Infection and Tissue Damage: The ISC Response to Enterocyte- and Muscle-Derived Proliferative Signals

After establishing that bacteria that populate the intestine play a major role in the age-related degeneration of this tissue, one essential question emerges: how do changes in the gut flora influence intestinal homeostasis and ISC proliferation in older flies? Answers to this question came from the study of the mechanisms that control intestinal regeneration in response to bacterial infection and tissue damage (Fig. 5.2).

Production of reactive oxygen species (ROS) is one of the major responses of the intestinal epithelium to cope with bacterial infection, and specific mechanisms to fight against this oxidative stress have been identified in *Drosophila* (Ha et al. 2009, 2005a, b). Interestingly, antioxidant molecules, such as N-acetylcysteine and glutathione, prevent induction of ISC proliferation after bacterial infection (Buchon et al. 2009a), while exposure to ROS-generating molecules dramatically increases ISC proliferation (Biteau et al. 2008; Choi et al. 2008), indicating that this oxidative burst directly impacts ISC proliferation.

The increased ROS production and the recognition of bacteria themselves lead to the stimulation of multiple signaling pathways in ECs and the establishment of both an antimicrobial and an inflammatory response, which are essential for the survival of the organism (Buchon et al. 2009a, b; Cronin et al. 2009; Jiang et al. 2009). One major component of this response is the activation of the JNK signaling pathway in ECs, which induces the production of interleukin-like cytokines (unpaired, Upds). This secretion of Upds by ECs leads to the activation of the JAK-STAT signaling pathway in ISCs, directly promoting their proliferation (Buchon et al. 2009a, b; Jiang et al. 2009). At the same time, this inflammatory signal causes the activation of JAK-STAT in the visceral muscle that surrounds the intestinal epithelium. This results in the upregulation of the expression of the EGF ligand *vein*, which in turn further promotes stem cell proliferation through the activation of the EGFR-MAPK signaling pathway in ISCs (Buchon et al. 2010; Jiang et al. 2011).

Similar to the response to bacterial infection, damage to the intestinal epithelium, such as DNA damage or disruption of the basal membrane, promotes ISC proliferation and tissue repair through the activation of the same stress-responsive signaling pathways (Amcheslavsky et al. 2009).

Over the last few years, it has become clear that during normal aging, similar mechanisms are chronically activated in the intestine. For example, high levels of ROS and increased activity of the JNK pathway are detected in ECs of older flies (Biteau et al. 2008; Hochmuth et al. 2011), and the expression of proliferative signals such as EGF ligand and Upd cytokines is greatly elevated in the aging fly midgut. In addition, the artificial activation of the mechanisms that promote ISC proliferation in response to infection or tissue damage recapitulates the dysplastic phenotype observed in old animals. Finally, recent evidence suggests that the muscles that surround the gut epithelium deteriorate with age (Larson et al. 2012), a process that may indirectly affect ISCs, through the deregulation of muscle-derived proliferative signals (in particular the ligand *vein*).

Taken together, these studies strongly suggest that chronic activation of the non-cell-autonomous mechanisms that control tissue repair in young animals significantly contributes to the uncontrolled ISC proliferation and intestinal dysplasia observed in older flies. However, it is important to note that the current model remains to be conclusively validated, as the effects of stress signaling pathways in ECs on age-related changes in the ISC regulation and intestinal homeostasis have yet to be directly investigated.

5.4.3 Other Cell-Cell and Tissue-Tissue Interactions

Interestingly, other cell types and tissues have been shown to interact with ISCs and the midgut epithelium to control stem cell proliferation and the function of the intestine. While little is known about the potential effect of aging on these other interactions, it is important to take these into account when trying to elucidate the complex mechanisms that affect ISC in aging flies, as defects in these mechanisms may directly contribute to the ISC hyper-proliferation and differentiation defects observed in the intestine of old flies.

Within the intestinal epithelium, all the different cell types constantly communicate to maintain homeostasis. Undifferentiated EC-committed daughter cells (i.e., enteroblasts, EBs) are capable of repressing ISC proliferation through the maintenance of E-cadherin-mediated cell-cell interactions (Choi et al. 2011). This interaction is directly regulated by the activity of the insulin receptor pathway in EBs, a conserved signaling pathway that is strongly influenced by aging (see below). At the same time, secretory cells (i.e., enteroendocrine cells, EEs) indirectly control ISC proliferation by regulating muscle-derived production of dilp3 (insulin) and vein (EGF) (Amcheslavsky et al. 2014; Scopelliti et al. 2014). Hence, while age-related defects in these feedback mechanisms have not yet been investigated, impaired signaling in EBs and/or EEs may significantly contribute to the uncontrolled proliferation of aging ISCs.

The complex interactions between ISCs, ECs, and the surrounding visceral muscle were described before. In addition to this regulatory network, other neighboring tissues influence the cells of the intestinal epithelium. The tracheal system, which delivers oxygen to all tissues in the fly, also provides an essential factor that promotes EC survival (Li et al. 2013). Loss of this signal, dpp (*decapentaplegic*, a member of the bone morphogenetic protein family), induces a strong compensatory ISC proliferation. In addition, the intestine is innervated by a complex neuronal network. These neurons control the adaptation of the intestinal function and excretion to nutrient availability and reproduction (Cognigni et al. 2011). Age-related changes in these organs have not yet been identified, but it is conceivable that misregulation of the signals originating from these tissues participates to the decline in intestinal function in older animals.

Finally, ISCs are regulated by systemic factors, including the activity of the insulin signaling pathway and nutrient availability. In *Drosophila*, insulin-like peptides (dilps) are secreted by multiple tissues. Notably, dilp2, dilp3, and dilp5 are produced by the neurosecretory insulin-producing cells (IPCs) in the brain (Teleman 2010). Relevant to the regulation of ISCs, dilp3 is also expressed in the visceral muscle surrounding the intestine and regulates the proliferation and number of ISCs in the intestine in response to changing nutritional conditions (O'Brien et al. 2011). Importantly, modulation of the Insulin signaling pathways is the one of the most widely conserved mechanisms that affects aging and longevity, from worm to mammals (Fontana et al. 2010; Kenyon 2010; Partridge et al. 2011). Indeed, reducing insulin signaling systematically or specifically in ISCs prevents age-related intestinal dysplasia (Biteau et al. 2010). Similarly, changes in nutrient availability that are known to affect insulin signaling, aging, and longevity modulate ISC proliferation (McLeod et al. 2010). These observations thus support the notion that, in *Drosophila*, age-related changes in metabolism and insulin signaling impact the biology of ISCs and that alterations of insulin signaling influence aging and extend lifespan, at least in part, by controlling the function of the stem cell pool in the intestine in old flies.

5.5 Cell Intrinsic Induction of ISC Proliferation

In addition to signals received from ECs and other surrounding cells and organs in response to tissue damage and infection, ISCs are capable of directly sensing changes in their environment and intracellular state and adapt their proliferation rate to these challenging conditions. As a consequence, cell-intrinsic mechanisms also significantly contribute to the changes in ISC regulation that take place in the intestine of aging flies.

5.5.1 Redox State and Stress Signals in Aging ISCs

Supporting the notion that ISC themselves are stressed in the aging intestine, high levels of ROS are detected in old ISCs (Hochmuth et al. 2011), and ISCs accumulate damaged macromolecules with age, as shown by the activation of the DNA damage response (γH2AX) and presence of DNA lesions (8-oxo-deoxyguanosine) (Park et al. 2012). Furthermore, overexpression of antioxidant proteins (the peroxiredoxin Jafrac1 and Gclc, an enzyme in the glutathione biosynthetic pathway) in ISCs and EBs is sufficient to reduce ISC proliferation and delay the progression of the intestinal dysplastic phenotype (Biteau et al. 2010; Hochmuth et al. 2011). These intracellular prooxidant conditions lead to the cell-autonomous activation of multiple signaling pathways that promote ISC proliferation, including the JNK and MAPK p38 pathways (Biteau et al. 2008; Park et al. 2009). At the same time, the Nrf2 signaling pathway, which promotes ISC quiescence under normal conditions, is inactivated, further increasing ISC proliferation rate (Hochmuth et al. 2011). Importantly, genetic modulations of these signaling pathways that revert the influence of aging on their activity are sufficient to prevent age-related tissue degeneration in the posterior midgut (Biteau et al. 2010; Hochmuth et al. 2011; Park et al. 2009), demonstrating the critical role of the cell-autonomous activation of stress signaling in the aging of the intestinal stem cell population.

In addition to the response to oxidative stress, ISCs are capable of responding directly to changes in the epithelium integrity and cell-cell interactions, through the inactivation of the Hippo pathway. Decreased activity of this pathway in ISCs leads to the activation of the transcription factor Yorkie and results in increased ISC proliferation (Ren et al. 2010; Shaw et al. 2010). Therefore, age-related changes in the architecture of the intestinal epithelium are likely to lead to the activation of Yorkie and also participate to the establishment of a chronic ISC-proliferative state.

5.5.2 Role of Mitochondrial Function in ISC Aging

As discussed previously, the increasingly prooxidant ISC microenvironment and the intracellular redox imbalance appear to be the major causes of the increased proliferation of aging ISCs. In addition to the ROS produced by ECs as part of the immune response, it has recently been recognized that mitochondria in aging ISCs themselves are an additional source of ROS (Rera et al. 2011). This recent study suggests that the mitochondrial function in the intestine declines with age. Importantly, reverting these effects of age by increasing the expression of the fly homolog of the PGC1 transcription factors is sufficient to promote mitochondrial activity, decrease the accumulation of ROS in aging ISCs, and prevent ISC hyper-proliferation. These recent findings further confirm that age-related intracellular changes in ISCs directly affect their function in older animals.

5.6 Age-Associated Defects in the ISC Lineage

In mammalian organs, age-related changes in the proliferative behavior of stem/progenitor cells are often accompanied by defects in the mechanisms that control cell-fate decision and differentiation in aging tissues (Brack and Rando 2007; Kim et al. 2012; Rossi et al. 2008). This can generate an imbalance between the different cell types that compose the tissue and/or lead to the production of nonfunctional cells, significantly contributing to age-related loss of tissue function. Similarly, in the aging *Drosophila* intestine, ISC hyper-proliferation results in a defective lineage

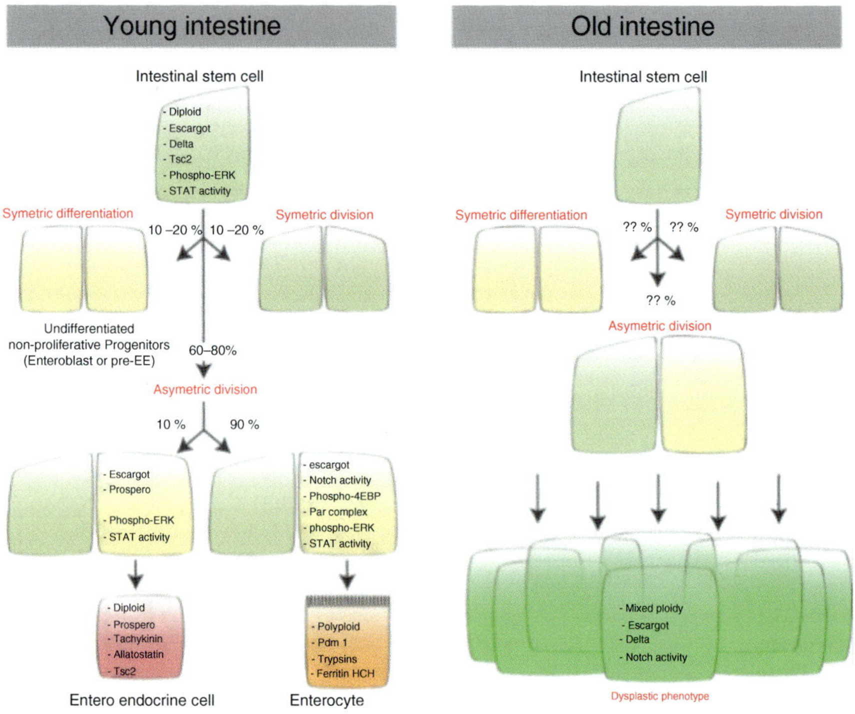

Fig. 5.3 The ISC lineage in young and old flies. Molecular markers of each cell type that composes the intestinal epithelium are indicated

that produces large numbers of abnormal cells (Biteau et al. 2008). Therefore, a better understanding of the mechanisms that control ISC self-renewal and cell-fate decision in young animals is essential to fully elucidate the mechanisms that drive dysplasia in the intestine of older flies.

5.6.1 The Posterior Midgut Lineage: Description of the Lineage

Lineage tracing analyses and the use of many cell-specific markers have allowed the elucidation of the ISC lineage in *Drosophila* posterior midgut (Fig. 5.3). These multipotent stem cells were first described as generating at least three different cell types, nonproliferating undifferentiated progenitors (enteroblasts, EBs), which differentiate in either enterocytes (ECs) or enteroendocrine cells (EEs) (Micchelli and Perrimon 2006; Ohlstein and Spradling 2006). However, it has recently been recognized that the ISC lineage is much more complex than what was originally thought. ISCs primarily asymmetrically divide to give rise to another ISC and a daughter that will undergo differentiation. Yet, ISCs have also been shown to symmetrically divide and generate two ISCs or two daughter cells (de Navascues et al. 2012; Goulas et al. 2012; O'Brien et al. 2011). In addition, it has recently been found that

ISCs can give rise to two distinct types of daughter cells: EBs, committed to the absorptive lineage (ECs), and pre-EEs, which differentiate into secretory cells (Biteau and Jasper 2014; Zeng and Hou 2015).

5.6.2 Molecular Mechanisms Controlling the ISC Lineage

While the intestinal lineage is relatively well characterized, the mechanisms that regulate the balance between self-renewal and differentiation remain largely unclear. In this regard, ISC seems to share some similarities with larval neuroblasts, as suggested by the asymmetric segregation of some components of the Par complex (Goulas et al. 2012), which may specify the EB cell fate during cell division. In addition, it has been suggested that the wingless/Wnt signaling pathway is required for ISC maintenance and may control self-renewal (Lin et al. 2008). Also, the balance between self-renewal and differentiation can be dynamically regulated, in particular during the regrowth of the intestine in response to re-feeding after a period of starvation, a process that may involve the insulin signaling pathway (O'Brien et al. 2011). Several models have been developed to describe the mechanisms that control the decision between self-renewal and differentiation (de Navascues et al. 2012; Kuwamura et al. 2012; Perdigoto et al. 2011). However, it has been recently found that endocrine cells directly control their own production, through the Slit/Robo2 signaling pathway, ensuring that Prospero-expressing daughter cells, which are committed to the secretory lineage, are only produced when EEs are missing in the immediate ISC vicinity (Biteau and Jasper 2014).

In addition to self-renewal signals, multiple conserved signaling pathways have been shown to be critical for the differentiation of ISC daughter cells. Notably, the Notch signaling pathway is absolutely required for the differentiation of EBs into ECs, while the JAK/STAT pathway is involved in the differentiation of EEs (Beebe et al. 2010; Lin et al. 2010; Ohlstein and Spradling 2007). Defects in these signaling pathways result in the accumulation of ISCs and/or an impaired balance between EEs and ECs. It is likely that all these signaling pathways form a complex regulatory system that precisely controls cell differentiation. For example, the Notch signaling pathway induces the TOR pathway in EBS to promote cell growth (Kapuria et al. 2012). However, most of the molecular interactions between the components of these pathways in ISCs remain to be identified, and the impact of aging or stress response on this network is not yet understood (see below).

5.6.3 Age-Related Defects in the ISC Lineage

As described in the previous section, the ISC lineage in young flies is characterized by the segregation of many molecular markers that clearly identify the different cell types that compose the midgut (Fig. 5.3). On the contrary, the architecture of the epithelium in the aging intestine is severely disturbed. In particular, a large number of nonfunctional cells accumulate in the epithelium, disrupting the monolayer organization. These cells express stem/progenitor cell makers and show characteristics specific of differentiated cells, such as increased ploidy, at the same time (Biteau et al. 2008) demonstrating that the cells produced by highly proliferating ISCs in older animals are not capable of properly differentiating. Interestingly, reducing the activity of the Notch signaling pathway is sufficient to prevent the accumulation of these misdifferentiated cells without preventing age-related ISC hyper-proliferation (Biteau et al. 2008), suggesting that perturbation of Notch signaling between ISCs and their daughter plays an important role in the age-related loss of tissue homeostasis in the aging intestine.

At least, three non-exclusive models can be proposed to explain the observed differentiation defects in the aging lineage:

- Aging may directly impact ISCs themselves and the intestinal lineage through the intrinsic activation of stress signals in ISCs and a direct effect on the mechanisms that govern cell-fate decision, differentiation, and the balance between symmetric and asymmetric ISC division. Thus, investigating the molecular interactions between the JNK/p38 pathways and the Notch/JAK pathways may be critical to understand the differentiation defects observed in the old intestine.
- It is also possible that uncontrolled ISC proliferation and the dramatic increase in the number of daughter cells produced indirectly affect cell-cell interactions that are essential for the proper differentiation of ISCs daughter cells. Supporting this hypothesis, correct E-cadherin-mediated interactions between ISCs and EBs are essential for the activation of the Notch signaling pathway in differentiating cells (Maeda et al. 2008).
- High activities of several stress signaling pathways have been reported in aging endocrine cells, including the JNK and Nrf2 pathways. Due to their role in maintaining the proper cellular balance in young animals (EEs versus ECs), it is plausible that age-related changes in cell signaling in EEs significantly contribute to the deterioration of the ISC lineage.

A careful analysis of the intestinal lineage in older flies will be required to distinguish between these models and provide information regarding the interactions between the mechanisms that control the intestinal lineage and the signaling pathways known to be influenced by the aging process.

5.7 Influence of Intestinal Stem Cell Function on Healthspan and Longevity

One of the most important questions regarding the relationship between stem cell function and aging is: can healthspan and/or lifespan be ameliorated by stem cell-specific manipulations that prevent or revert the effects of age on progenitor cells? The *Drosophila* midgut has proven to be an ideal experimental model to start addressing this problem.

Several characteristics specific to this *Drosophila* intestine have allowed these studies:

- *The GI tract is the only somatic tissue maintained by resident stem cell populations in Drosophila.*

Multiple progenitor pools have been identified in the intestine. While reports have suggested the existence of a stem cell population in the adult fly brain (Fernandez-Hernandez et al. 2013; von Trotha et al. 2009), it remains unclear if the presence of these proliferating cells is essential for the long-term maintenance of the central nervous system. Also, contrary to other invertebrates, germline stem cells have a controversial effect on longevity in *Drosophila* (Barnes et al. 2006; Flatt et al. 2008).

- *Optimal intestinal function is essential for metabolic and immune homeostasis.*

In *Drosophila*, the intestine is one of the major organs where metabolism, in particular cholesterol and lipid homeostasis, and immune response are regulated (Buchon et al. 2009b; Cronin et al. 2009; Karpac et al. 2013; Sieber and Thummel 2012). Accordingly, age-related intestinal dysplasia correlates with dramatic changes in the intestinal flora, as well as lipid and glucose metabolism (Biteau et al. 2008, 2010). In addition, the integrity of the intestinal epithelium correlates with the age and life expectancy of flies under many different culture conditions (Rera et al. 2012), confirming the central role of the gut in the maintenance of healthspan.

- *The function of ISCs can be easily and specifically manipulated in vivo.*

As discussed in previous sections, the major identified cause of tissue degeneration in the aging midgut is the uncontrolled ISC proliferation that results from the chronic activation of tissue repair mechanisms. Thus, using powerful genetic tools, it is relatively easy to specifically manipulate the ISC proliferation rate in the adult fly to influence tissue homeostasis.

For all these reasons, it has been possible to demonstrate that ISC function directly correlates with healthspan and longevity and that optimizing the ISC proliferation rate to preserve tissue homeostasis in the aging intestine is sufficient to extend lifespan (Fig. 5.4). Indeed, manipulations that strongly prevent ISC proliferation and impair tissue repair not only render animals sensitive to infection or tissue damage but also shorten lifespan under unchallenged conditions (Apidianakis et al. 2009; Biteau et al. 2010; Buchon et al. 2009a; Cronin et al. 2009; Jiang et al. 2009). On the other hand, conditions that chronically promote ISC proliferation also shorten lifespan (Apidianakis et al. 2009; Biteau et al. 2008). However, optimal conditions have been identified: ISC-specific manipulations that prevent tissue dysplasia without blocking tissue turnover (e.g., overexpression of antioxidant proteins, overexpression of the dPGC1, or the moderate inhibition of the insulin, EGF, or JNK signaling pathways) prevent age-related metabolic decay and result in long-lived flies (Biteau et al. 2010; Rera et al. 2011).

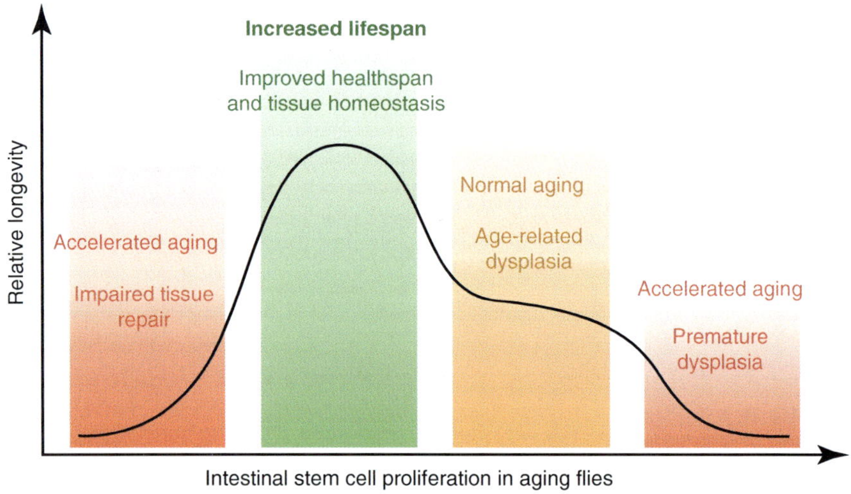

Fig. 5.4 Relationship between ISC proliferation rate in old flies and *Drosophila* longevity

Conclusions

In only a few years, the discovery of ISCs and studies exploring their influence on tissue homeostasis and longevity have had a major impact on aging research in *Drosophila*. The fly intestine has become an exceptionally powerful model to investigate the mechanisms by which aging affects stem cells, progenitors, differentiated cells, the stem cell niche, and the systemic environment in the same biological system.

These studies have shed new light on the role of the intestinal flora, immunity, mitochondrial function, and stress responses in the control of epithelial homeostasis throughout the life of an organism. In addition, for the first time, these studies have demonstrated that preventing the effects of age on a specific somatic stem cell population is a viable strategy to preserve healthspan and extend longevity of an invertebrate organism.

As for many discoveries first made using *Drosophila*, it can be anticipated that the mechanisms of stem cell aging which are rapidly being elucidated in the *Drosophila* intestine will provide new concepts valuable to the study of the influence of aging on stem cell function in aging organs throughout the animal kingdom, including in vertebrates.

References

Amcheslavsky A, Jiang J, Ip YT (2009) Tissue damage-induced intestinal stem cell division in Drosophila. Cell Stem Cell 4:49–61

Amcheslavsky A, Ito N, Jiang J, Ip YT (2011) Tuberous sclerosis complex and Myc coordinate the growth and division of Drosophila intestinal stem cells. J Cell Biol 193:695–710

Amcheslavsky A, Song W, Li Q, Nie Y, Bragatto I, Ferrandon D, Perrimon N, Ip YT (2014) Enteroendocrine cells support intestinal stem-cell-mediated homeostasis in Drosophila. Cell Rep 9:32–39

Apidianakis Y, Pitsouli C, Perrimon N, Rahme L (2009) Synergy between bacterial infection and genetic predisposition in intestinal dysplasia. Proc Natl Acad Sci U S A 106:20883–20888

Bangi E, Pitsouli C, Rahme LG, Cagan R, Apidianakis Y (2012) Immune response to bacteria induces dissemination of Ras-activated Drosophila hindgut cells. EMBO Rep 13:569–576

Bardin AJ, Perdigoto CN, Southall TD, Brand AH, Schweisguth F (2010) Transcriptional control of stem cell maintenance in the Drosophila intestine. Development 137:705–714

Barnes AI, Boone JM, Jacobson J, Partridge L, Chapman T (2006) No extension of lifespan by ablation of germ line in Drosophila. Proc Biol Sci/The Royal Society 273:939–947

Baumann O (2001) Posterior midgut epithelial cells differ in their organization of the membrane skeleton from other drosophila epithelia. Exp Cell Res 270:176–187

Beebe K, Lee WC, Micchelli CA (2010) JAK/STAT signaling coordinates stem cell proliferation and multilineage differentiation in the Drosophila intestinal stem cell lineage. Dev Biol 338:28–37

Biteau B, Jasper H (2011) EGF signaling regulates the proliferation of intestinal stem cells in Drosophila. Development 138:1045–1055

Biteau B, Jasper H (2014) Slit/Robo signaling regulates cell fate decisions in the intestinal stem cell lineage of Drosophila. Cell Rep 7:1867–1875

Biteau B, Hochmuth CE, Jasper H (2008) JNK activity in somatic stem cells causes loss of tissue homeostasis in the aging Drosophila gut. Cell Stem Cell 3:442–455

Biteau B, Karpac J, Supoyo S, Degennaro M, Lehmann R, Jasper H (2010) Lifespan extension by preserving proliferative homeostasis in Drosophila. PLoS Genet 6, e1001159

Biteau B, Hochmuth CE, Jasper H (2011) Maintaining tissue homeostasis: dynamic control of somatic stem cell activity. Cell Stem Cell 9:402–411

Bond D, Foley E (2012) Autocrine PDGF- VEGF- receptor related (Pvr) pathway activity controls intestinal stem cell proliferation in the adult Drosophila midgut. J Biol Chem 287:27359–27370

Brack AS, Rando TA (2007) Intrinsic changes and extrinsic influences of myogenic stem cell function during aging. Stem Cell Rev 3:226–237

Buchon N, Broderick NA, Chakrabarti S, Lemaitre B (2009a) Invasive and indigenous microbiota impact intestinal stem cell activity through multiple pathways in Drosophila. Genes Dev 23:2333–2344

Buchon N, Broderick NA, Poidevin M, Pradervand S, Lemaitre B (2009b) Drosophila intestinal response to bacterial infection: activation of host defense and stem cell proliferation. Cell Host Microbe 5:200–211

Buchon N, Broderick NA, Kuraishi T, Lemaitre B (2010) Drosophila EGFR pathway coordinates stem cell proliferation and gut remodeling following infection. BMC Biol 8:152

Chen H, Zheng X, Zheng Y (2014) Age-associated loss of lamin-B leads to systemic inflammation and gut hyperplasia. Cell 159:829–843

Choi NH, Kim JG, Yang DJ, Kim YS, Yoo MA (2008) Age-related changes in Drosophila midgut are associated with PVF2, a PDGF/VEGF-like growth factor. Aging Cell 7:318–334

Choi NH, Lucchetta E, Ohlstein B (2011) Nonautonomous regulation of Drosophila midgut stem cell proliferation by the insulin-signaling pathway. Proc Natl Acad Sci U S A 108:18702–18707

Christofi T, Apidianakis Y (2012) Ras-oncogenic Drosophila hindgut but not midgut cells use an inflammation-like program to disseminate to distant sites. Gut Microbes 4:54–59

Cognigni P, Bailey AP, Miguel-Aliaga I (2011) Enteric neurons and systemic signals couple nutritional and reproductive status with intestinal homeostasis. Cell Metab 13:92–104

Cordero JB, Stefanatos RK, Scopelliti A, Vidal M, Sansom OJ (2012) Inducible progenitor-derived Wingless regulates adult midgut regeneration in Drosophila. EMBO J 31:3901–3917

Cronin SJ, Nehme NT, Limmer S, Liegeois S, Pospisilik JA, Schramek D, Leibbrandt A, Simoes Rde M, Gruber S, Puc U et al (2009) Genome-wide RNAi screen identifies genes involved in intestinal pathogenic bacterial infection. Science (New York, NY) 325:340–343 de Navascues J, Perdigoto CN, Bian Y, Schneider MH, Bardin AJ, Martinez-Arias A, Simons BD (2012) Drosophila midgut homeostasis involves neutral competition between symmetrically dividing intestinal stem cells. EMBO J 31:2473–2485

Decotto E, Spradling AC (2005) The Drosophila ovarian and testis stem cell niches: similar somatic stem cells and signals. Dev Cell 9:501–510

Fernandez-Hernandez I, Rhiner C, Moreno E (2013) Adult neurogenesis in Drosophila. Cell Rep 3:1857–1865

Flatt T, Min KJ, D'Alterio C, Villa-Cuesta E, Cumbers J, Lehmann R, Jones DL, Tatar M (2008) Drosophila germ-line modulation of insulin signaling and lifespan. Proc Natl Acad Sci U S A 105:6368–6373

Fontana L, Partridge L, Longo VD (2010) Extending healthy life span–from yeast to humans. Science (New York, NY) 328:321–326

Fox DT, Spradling AC (2009) The Drosophila hindgut lacks constitutively active adult stem cells but proliferates in response to tissue damage. Cell Stem Cell 5:290–297

Fuller MT, Spradling AC (2007) Male and female Drosophila germline stem cells: two versions of immortality. Science (New York, NY) 316:402–404

Goulas S, Conder R, Knoblich JA (2012) The par complex and integrins direct asymmetric cell division in adult intestinal stem cells. Cell Stem Cell 11:529–540

Guo L, Karpac J, Tran SL, Jasper H (2014) PGRP-SC2 promotes gut immune homeostasis to limit commensal dysbiosis and extend lifespan. Cell 156:109–122

Ha EM, Oh CT, Bae YS, Lee WJ (2005a) A direct role for dual oxidase in Drosophila gut immunity. Science (New York, NY) 310:847–850

Ha EM, Oh CT, Ryu JH, Bae YS, Kang SW, Jang IH, Brey PT, Lee WJ (2005b) An antioxidant system required for host protection against gut infection in Drosophila. Dev Cell 8:125–132

Ha EM, Lee KA, Seo YY, Kim SH, Lim JH, Oh BH, Kim J, Lee WJ (2009) Coordination of multiple dual oxidase-regulatory pathways in responses to commensal and infectious microbes in drosophila gut. Nat Immunol 10:949–957

Hochmuth CE, Biteau B, Bohmann D, Jasper H (2011) Redox regulation by Keap1 and Nrf2 controls intestinal stem cell proliferation in Drosophila. Cell Stem Cell 8:188–199

Hooper LV, Gordon JI (2001) Commensal host-bacterial relationships in the gut. Science (New York, NY) 292:1115–1118

Jiang H, Edgar BA (2012) Intestinal stem cell function in Drosophila and mice. Curr Opin Genet Dev 22:354–360

Jiang H, Patel PH, Kohlmaier A, Grenley MO, McEwen DG, Edgar BA (2009) Cytokine/Jak/Stat signaling mediates regeneration and homeostasis in the Drosophila midgut. Cell 137:1343–1355

Jiang H, Grenley MO, Bravo MJ, Blumhagen RZ, Edgar BA (2011) EGFR/Ras/MAPK signaling mediates adult midgut epithelial homeostasis and regeneration in Drosophila. Cell Stem Cell 8:84–95

Kapuria S, Karpac J, Biteau B, Hwangbo D, Jasper H (2012) Notch-mediated suppression of TSC2 expression regulates cell differentiation in the Drosophila intestinal stem cell lineage. PLoS Genet 8, e1003045

Karpac J, Biteau B, Jasper H (2013) Misregulation of an adaptive metabolic response contributes to the age-related disruption of lipid homeostasis in Drosophila. Cell Rep 4:1250–1261

Karpowicz P, Perez J, Perrimon N (2010) The Hippo tumor suppressor pathway regulates intestinal stem cell regeneration. Development 137:4135–4145

Kenyon CJ (2010) The genetics of ageing. Nature 464:504–512

Kim M, Kim C, Choi YS, Park C, Suh Y (2012) Age-related alterations in mesenchymal stem cells related to shift in differentiation from osteogenic to adipogenic potential: implication to age-associated bone diseases and defects. Mech Ageing Dev 133:215–225

Knoblich JA (2010) Asymmetric cell division: recent developments and their implications for tumour biology. Nature Rev Mol Cell Biol 11:849–860

Kuraishi T, Binggeli O, Opota O, Buchon N, Lemaitre B (2011) Genetic evidence for a protective role of the peritrophic matrix against intestinal bacterial infection in Drosophila melanogaster. Proc Natl Acad Sci U S A 108:15966–15971

Kuwamura M, Maeda K, Adachi-Yamada T (2012) Mathematical modelling and experiments for the proliferation and differentiation of Drosophila intestinal stem cells II. J Biol Dyn 6:267–276

Larson K, Yan SJ, Tsurumi A, Liu J, Zhou J, Gaur K, Guo D, Eickbush TH, Li WX (2012) Heterochromatin formation promotes longevity and represses ribosomal RNA synthesis. PLoS Genet 8, e1002473

Lee T, Luo L (1999) Mosaic analysis with a repressible cell marker for studies of gene function in neuronal morphogenesis. Neuron 22:451–461

Lee WC, Beebe K, Sudmeier L, Micchelli CA (2009) Adenomatous polyposis coli regulates Drosophila intestinal stem cell proliferation. Development 136:2255–2264

Li Z, Zhang Y, Han L, Shi L, Lin X (2013) Trachea-derived dpp controls adult midgut homeostasis in Drosophila. Dev Cell 24:133–143

Lin G, Xu N, Xi R (2008) Paracrine Wingless signalling controls self-renewal of Drosophila intestinal stem cells. Nature 455:1119–1123

Lin G, Xu N, Xi R (2010) Paracrine unpaired signaling through the JAK/STAT pathway controls self-renewal and lineage differentiation of drosophila intestinal stem cells. J Mol Cell Biol 2:37–49

Liu W, Singh SR, Hou SX (2010) JAK-STAT is restrained by Notch to control cell proliferation of the Drosophila intestinal stem cells. J Cell Biochem 109:992–999

Maeda K, Takemura M, Umemori M, Adachi-Yamada T (2008) E-cadherin prolongs the moment for interaction between intestinal stem cell and its progenitor cell to ensure Notch signaling in adult Drosophila midgut. Genes Cells 13:1219–1227

McLeod CJ, Wang L, Wong C, Jones DL (2010) Stem cell dynamics in response to nutrient availability. Curr Biol 20:2100–2105

Micchelli CA, Perrimon N (2006) Evidence that stem cells reside in the adult Drosophila midgut epithelium. Nature 439:475–479

O'Brien LE, Soliman SS, Li X, Bilder D (2011) Altered modes of stem cell division drive adaptive intestinal growth. Cell 147:603–614

Ohlstein B, Spradling A (2006) The adult Drosophila posterior midgut is maintained by pluripotent stem cells. Nature 439:470–474

Ohlstein B, Spradling A (2007) Multipotent Drosophila intestinal stem cells specify daughter cell fates by differential notch signaling. Science (New York, NY) 315:988–992

Park JS, Kim YS, Yoo MA (2009) The role of p38b MAPK in age-related modulation of intestinal stem cell proliferation and differentiation in Drosophila. Aging (Albany NY) 1:637–651

Park JS, Lee SH, Na HJ, Pyo JH, Kim YS, Yoo MA (2012) Age- and oxidative stress-induced DNA damage in Drosophila intestinal stem cells as marked by Gamma-H2AX. Exp Gerontol 47:401–405

Partridge L, Alic N, Bjedov I, Piper MD (2011) Ageing in Drosophila: the role of the insulin/Igf and TOR signalling network. Exp Gerontol 46:376–381

Perdigoto CN, Schweisguth F, Bardin AJ (2011) Distinct levels of Notch activity for commitment and terminal differentiation of stem cells in the adult fly intestine. Development 138:4585–4595

Perrimon N (1998) Creating mosaics in Drosophila. Int J Dev Biol 42:243–247

Ren F, Wang B, Yue T, Yun EY, Ip YT, Jiang J (2010) Hippo signaling regulates Drosophila intestine stem cell proliferation through multiple pathways. Proc Natl Acad Sci U S A 107:21064–21069

Rera M, Bahadorani S, Cho J, Koehler CL, Ulgherait M, Hur JH, Ansari WS, Lo T Jr, Jones DL, Walker DW (2011) Modulation of longevity and tissue homeostasis by the Drosophila PGC-1 homolog. Cell Metab 14:623–634

Rera M, Clark RI, Walker DW (2012) Intestinal barrier dysfunction links metabolic and inflammatory markers of aging to death in Drosophila. Proc Natl Acad Sci U S A 109:21528–21533

Rossi DJ, Jamieson CH, Weissman IL (2008) Stems cells and the pathways to aging and cancer. Cell 132:681–696

Ryu JH, Ha EM, Lee WJ (2010) Innate immunity and gut-microbe mutualism in Drosophila. Dev Comp Immunol 34:369–376

Sansonetti PJ (2004) War and peace at mucosal surfaces. Nat Rev Immunol 4:953–964
Scopelliti A, Cordero JB, Diao F, Strathdee K, White BH, Sansom OJ, Vidal M (2014) Local control of intestinal stem cell homeostasis by enteroendocrine cells in the adult Drosophila midgut. Curr Biol 24:1199–1211
Shanbhag S, Tripathi S (2009) Epithelial ultrastructure and cellular mechanisms of acid and base transport in the Drosophila midgut. J Exp Biol 212:1731–1744
Shaw RL, Kohlmaier A, Polesello C, Veelken C, Edgar BA, Tapon N (2010) The Hippo pathway regulates intestinal stem cell proliferation during Drosophila adult midgut regeneration. Development 137:4147–4158
Sieber MH, Thummel CS (2012) Coordination of triacylglycerol and cholesterol homeostasis by DHR96 and the Drosophila LipA homolog magro. Cell Metab 15:122–127
Singh SR, Liu W, Hou SX (2007) The adult Drosophila Malpighian tubules are maintained by multipotent stem cells. Cell Stem Cell 1:191–203
Singh SR, Zeng X, Zheng Z, Hou SX (2011) The adult Drosophila gastric and stomach organs are maintained by a multipotent stem cell pool at the foregut/midgut junction in the cardia (proventriculus). Cell Cycle 10:1109–1120
Staley BK, Irvine KD (2010) Warts and Yorkie mediate intestinal regeneration by influencing stem cell proliferation. Curr Biol 20:1580–1587
Strand M, Micchelli CA (2011) Quiescent gastric stem cells maintain the adult Drosophila stomach. Proc Natl Acad Sci U S A 108:17696–17701
Takashima S, Mkrtchyan M, Younossi-Hartenstein A, Merriam JR, Hartenstein V (2008) The behaviour of Drosophila adult hindgut stem cells is controlled by Wnt and Hh signalling. Nature 454:651–655
Teleman AA (2010) Molecular mechanisms of metabolic regulation by insulin in Drosophila. Biochem J 425:13–26
Tiihonen K, Ouwehand AC, Rautonen N (2010) Human intestinal microbiota and healthy ageing. Ageing Res Rev 9:107–116
von Trotha JW, Egger B, Brand AH (2009) Cell proliferation in the Drosophila adult brain revealed by clonal analysis and bromodeoxyuridine labelling. Neural Dev 4:9
Wang P, Hou SX (2010) Regulation of intestinal stem cells in mammals and Drosophila. J Cell Physiol 222:33–37
Woodmansey EJ (2007) Intestinal bacteria and ageing. J Appl Microbiol 102:1178–1186
Xu N, Wang SQ, Tan D, Gao Y, Lin G, Xi R (2011) EGFR, Wingless and JAK/STAT signaling cooperatively maintain Drosophila intestinal stem cells. Dev Biol 354:31–43
Zeng X, Hou SX (2015) Enteroendocrine cells are generated from stem cells through a distinct progenitor in the adult Drosophila posterior midgut. Development 142:644–653
Zeng X, Singh SR, Hou D, Hou SX (2010) Tumor suppressors Sav/scrib and oncogene ras regulate stem-cell transformation in adult Drosophila malpighian tubules. J Cell Physiol 224:766–774
Zhou F, Rasmussen A, Lee S, Agaisse H (2013) The upd3 cytokine couples environmental challenge and intestinal stem cell division through modulation of JAK/STAT signaling in the stem cell microenvironment. Dev Biol 373:383–393

Part III

Mechanisms of Stem Cell Aging in Mammalian Tissues

Aging of Murine Hematopoietic Stem Cells

6

Edyta Wójtowicz, Evgenia Verovskaya, and Gerald de Haan

Abstract

Hematopoietic stem cells (HSCs) are unique in their ability to self-renew and differentiate into all mature blood lineages. The equilibrium between these processes is crucial for tissue maintenance during the lifetime of the organism. However, with age the functionality of HSCs declines, resulting in development of anemias, deficiencies of immune response, and increased risk of hematopoietic malignancies. Aged HSCs are characterized by preferential differentiation toward myeloid lineage, impaired self-renewal, and engraftment. Recent evidence provides clues to the understanding of these processes on cellular and molecular levels. Key components contributing to stem cell aging are shifts in the transcriptome and epigenome, accompanied by dysfunction of DNA repair pathways. In this chapter we will focus on studies and conceptual models of murine HSC aging.

6.1 Introduction

Organismal aging is associated with functional decline of many tissues. In the hematopoietic system, aging is associated with a decreased immune response, an increased incidence of hematopoietic cancers, anemia, and several autoimmune disorders (Balducci et al. 2008). A growing body of data supports the notion that at least some of these changes originate from dysfunctional hematopoietic stem cells (HSCs). Functional decline of HSCs is believed to be orchestrated by the interplay

E. Wójtowicz • E. Verovskaya • G. de Haan (✉)
Department of Ageing Biology and Stem Cells,
European Research Institute for the Biology of Aging (ERIBA),
University Medical Center Groningen, University of Groningen,
Antonius Deusinglaan 1, 9713 AV Groningen, The Netherlands
e-mail: g.de.haan@med.umcg.nl

H. Geiger et al. (eds.), *Stem Cell Aging: Mechanisms, Consequences, Rejuvenation*,
DOI 10.1007/978-3-7091-1232-8_6

between cell-intrinsic and cell-extrinsic factors. Insight into the mechanism underlying HSC aging can be obtained from molecular studies in mice. This chapter provides an overview of our current understanding of the mechanisms of murine HSC aging.

6.2 Characteristics of Murine HSCs

In the mouse, HSCs constitute a small proportion (<0.01 %) of all bone marrow cells. However, their extensive proliferative capacity ensures a daily production of over 10^{11} blood cells (Lansdorp 1998). All types of mature hematopoietic and immune cells are generated and regenerated by HSCs, in an intricate process of stepwise gradual differentiation (Fig. 6.1). The abilities to self-renew and to differentiate into multiple lineages are the defining properties of the HSCs (also see Box 6.1).

Box 6.1 *Hematopoietic stem cell* – adult multipotent stem cell that is able to self-renew and differentiate into all mature types of immune and hematopoietic cells.

Functionally defined HSC – a cell that can support long-term robust multilineage repopulation of a host, whose own hematopoietic system was ablated (e.g., by irradiation). This definition is retrospective. Criteria for defining stemness differ from lab to lab and are based on either total donor cell presence (chimerism) in blood of transplant recipient or chimerism in specific mature blood populations. The timescale for defining long-term repopulation varies but usually constitutes at least 4 months of activity after transplantation. A very common functional HSC definition is its ability to support production at least 1 % of myeloid and lymphoid cells for 4 months after transplantation.

Phenotypically defined HSC – hematopoietic cell expressing a panel of cell surface markers characteristic for HSCs. Phenotypic definitions of HSCs have become more and more stringent over the years, coinciding with discoveries of novel markers. Marker combinations that allow to achieve isolation of highly enriched HSC populations include among others:

$CD150^+$ $CD244^-$ $CD48^-$ (SLAM) (Kiel et al. 2005)
$CD45^{mid}$ Rho^- SP (Dykstra et al. 2007)
$lineage^-$ $Sca1^+$ $c\text{-}Kit^+$ (LSK) $CD48^-$ $CD34^-$ $EPCR^+$ $CD150^+$ (Dykstra et al. 2011)
$CD45^+$ $EPCR^+$ $CD48^-$ $CD150^+$ (ESLAM) (Benz et al. 2012)
$LSKCD34^-$ $CD150^+$ (Morita et al. 2010)
SP LSK (Challen et al. 2010)
CD45 is expressed on hematopoietic cells and allows to distinguish them from non-hematopoietic populations.

HSC clone – all progeny of a single HSC

Clone size – proportion of differentiated cells in blood generated by an HSC clone

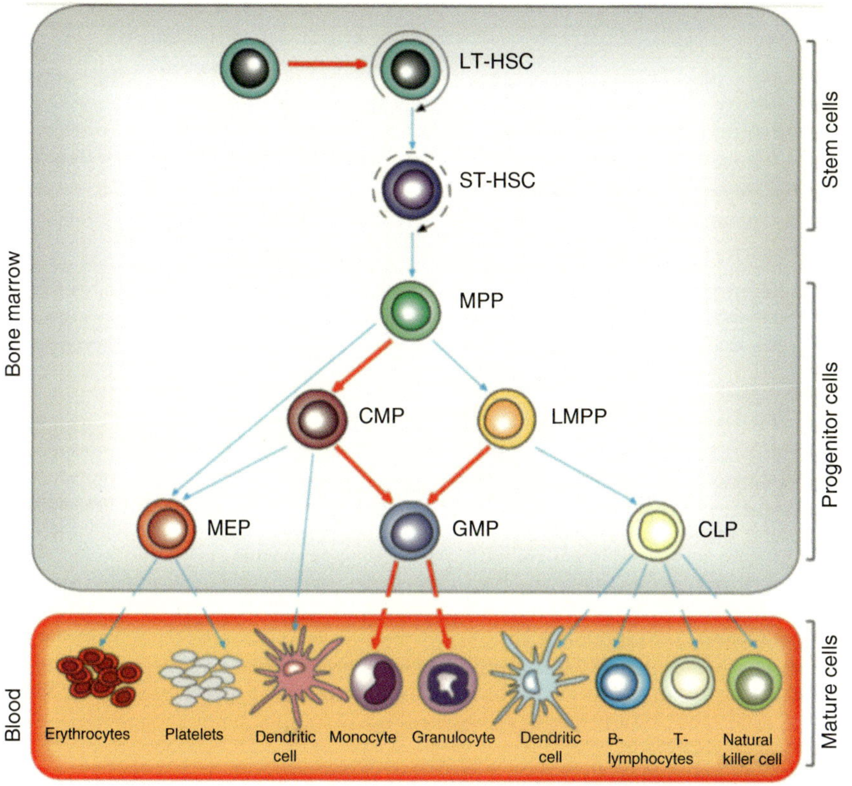

Fig. 6.1 Simplified representation of the hematopoietic hierarchy and the effects of aging (Adapted from Verovskaya and de Haan (2011)). HSCs residing in the bone marrow produce all the mature hematopoietic and immune cells that are found in the blood. The majority of HSCs resides in a quiescent G_0 state (*top left*) and becomes activated into cycling and differentiation. During the gradual process of multilineage differentiation, HSCs lose the ability to self-renew (as indicated by *round arrows*) and become more committed to a certain lineage. In aged mice, HSCs have been found to more actively cycle. Moreover, hematopoiesis becomes skewed toward production of myeloid cells, while lymphoid production becomes deficient. Regeneration of red blood cells and platelets is also decreased with lifespan. Processes activated with age are indicated with *red arrows*, while decreased activity is depicted with *blue arrows*. *LT-HSCs* indicate long-term repopulating HSC, *ST-HSCs* short-term repopulating HSCs. Several described intermediate progenitor states are shown on the figure: *MPP* multipotent progenitor cells, *LMPP* lymphoid-primed multipotent progenitors, *CMP* common myeloid progenitors, *CLP* common lymphoid progenitors, *MEP* megakaryocyte-erythroid progenitors, *GMP* granulocyte-macrophage progenitors

Morphologically, HSCs cannot be discriminated from more differentiated progenitor cells (Fig. 6.1). However, a panel of cell surface markers, characteristically expressed on HSCs, has been identified. These include expression of Sca1, c-Kit, CD150 (one of the SLAM markers), and CD201 (EPCR) and lack of the expression of lineage surface markers CD48, CD244, Flt3, and CD34. Alternatively, the ability of HSCs to efflux certain dyes (Rhodamine-123 (Rho) and Hoechst 33342) can be

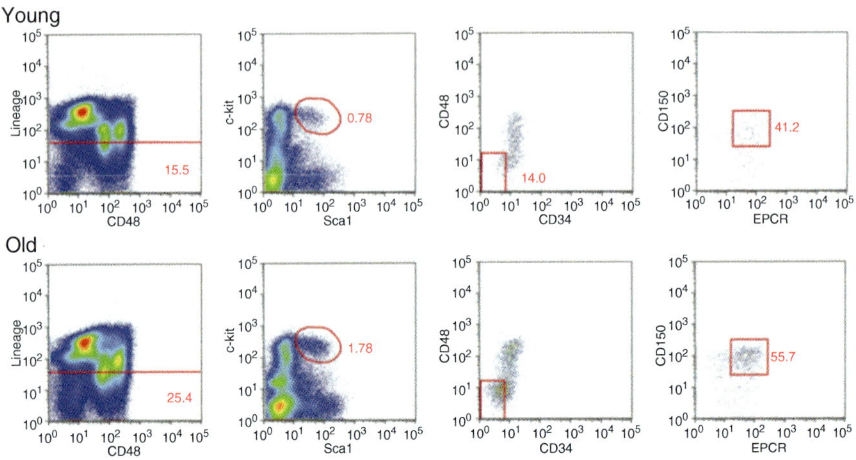

Fig. 6.2 Number of phenotypically defined HSCs increases with age. Here we show the gating strategy for isolation of LSK CD48$^-$ CD150$^+$ EPCR$^+$ CD34$^-$ cells (of young ~4-month-old and 1 of ~25-month-old animals) that have robust repopulating potential in vivo (Dykstra et al. 2011). Note that percentages of primitive cells are higher for an old mouse at all gating stages

also used for purification (Bertoncello et al. 1985; Wolf et al. 1993). This feature of HSC results from the expression of the ABC transporters P-glycoprotein and Abcg2 (Zhou et al. 2001). Hoechst 33342 exclusion results in a characteristic appearance of primitive cells during fluorescence-activated cell sorting (FACS), known as the "side population" (SP) (Goodell et al. 1996). A combination of multiple of these markers allows prospective isolation and characterization of HSCs with purity up to 50 %, defined by the ability of a single cell to reconstitute hematopoiesis of a lethally irradiated animal (Kiel et al. 2005; Dykstra et al. 2007, 2011; Challen et al. 2009; Kent et al. 2009).

Aging has been shown to increase both the relative frequency and the absolute number of HSCs in the bone marrow of C57BL/6 mice. Initial studies using unfractionated bone marrow cells had demonstrated a competitive advantage of old cells when co-transplanted with equal numbers of young cells (Harrison 1983), which was attributed to higher concentrations of repopulating cells in the aged population (Harrison et al. 1989). More recent studies, using different combinations of phenotypic markers, confirmed a remarkable increase in the pool size of HSC upon aging (Dykstra et al. 2011; Morrison et al. 1996; Rossi et al. 2005; Sudo et al. 2000). An example of such a phenotypic increase of primitive hematopoietic population upon aging is shown in Fig. 6.2. The extent of such expansion was seven- to 16-fold (Dykstra and de Haan 2008). Furthermore, while the frequency of HSCs is similar among individual young mice, HSC frequencies become highly variable in old mice, indicating a possible loss of control of HSC pool size with age (Dykstra et al. 2011).

Despite many advances that have been made in the identification of the phenotype of HSC, "stemness" of a (murine) HSC can only be demonstrated retrospectively by its ability for robust multilineage repopulation of a myeloablated host. An

HSC is often defined by the ability of a cell to support at least 1 % lymphomyeloid chimerism in a recipient animal for 4 months posttransplantation (Ema et al. 2005; Miller and Eaves 1997). Primitive cells contributing to peripheral blood cells less than 1 % are usually unable to sustain robust blood production in the long-term or upon secondary transplantation. Therefore, they are not considered *bona fide* stem cells. In contrast to studies where whole bone marrow cells were used, transplantations of purified aged HSCs demonstrated a two- to sevenfold decrease in functional stem cell frequency *in vivo* (Dykstra et al. 2011; Morrison et al. 1996; Rossi et al. 2005; Yilmaz et al. 2006). Moreover, compared to young, aged HSCs have been shown to exhaust more rapidly upon serial transplantation (Kamminga et al. 2005). Finally, the ability of HSCs to generate cells of all different lineages changes as a function of age and switches toward production of myeloid cells concurrent with decreased lymphopoiesis (Morrison et al. 1996; Sudo et al. 2000; Liang et al. 2005) (indicated in Fig. 6.1).

6.3 Not All Stem Cells Are Equal

Historically, it has been assumed that the HSC pool is homogeneous and all HSCs are uniform in their ability to produce cells of all lineages (as shown in Fig. 6.1). However, this concept was challenged by clonal studies demonstrating profound heterogeneity among HSCs regarding three of their key properties: (1) repopulation potential, (2) cycling activity, and (3) developmental program.

It was shown that individual HSCs display a marked diverse and unpredictable self-renewal capacity and repopulating activity upon transplantation into irradiated host mice (Dykstra et al. 2007; Ema et al. 2005). For instance, upon single-cell transplantation, purified cells can generate highly variable numbers of circulating white blood cells ranging from 1 to >80 % chimerism (Dykstra et al. 2007; Ema et al. 2005).

Further, the HSCs are highly heterogeneous in regard to their cycling rate (Wilson et al. 2008; Foudi et al. 2009). While HSCs have a high proliferative capacity, their turnover is slow, what protects HSCs from accumulating mutations and prevents their exhaustion (Dykstra et al. 2006; Orford and Scadden 2008; Arai and Suda 2007). Analysis of cell cycle in the primitive compartment has shown that more than 70 % of phenotypically defined HSCs are found in quiescent G_0 stage, opposed to less than 10 % of progenitor population (Wilson et al. 2008). Multiple other studies supported this notion (Wilson et al. 2008; Passegue et al. 2005; Cheshier et al. 1999). Several reports suggest presence of at least two populations within the HSC pool – "homeostatic" HSCs that have a turnover time of approximately 1 month and "dormant" HSCs that are entering the cell cycle once in around 170 days (Wilson et al. 2008; Foudi et al. 2009).

Moreover, a highly variable ability of HSCs to produce myeloid and lymphoid progeny is well documented. This was first reported in retroviral marking studies where different vector integration profiles were observed in myeloid and lymphoid tissues (Jordan and Lemischka 1990). Later, experiments in animals transplanted with limiting dilutions of bone marrow cells (Muller-Sieburg et al. 2002) and in single-cell transplant recipients (Dykstra et al. 2007) demonstrated the presence of at least three distinct developmental types of behavior of HSCs. These studies established that individual cells can preferentially support myeloid cell production (so-called α or myeloid-biased HSCs) and lymphoid cell development (γ or lymphoid-biased HSCs) or be equally prone to generate both lineages (β or balanced HSCs). High expression of CD150 has been demonstrated to correlate with a myeloid potential of HSCs (Beerman et al. 2010a). Lymphoid-biased HSCs have been shown to have a relatively short lifespan and to exhaust readily upon secondary transplantation (Muller-Sieburg et al. 2004). On the contrary, myeloid-biased HSCs had the longest lifespan and the highest self-renewal capacity of all HSC types (Muller-Sieburg et al. 2004).

Distinct gene expression profiles appear characteristic to HSCs with distinct developmental programs. CD150 (SLAM) and SP profiling in phenotypically isolated HSCs allowed their sub-fractionation into lymphoid-biased and myeloid-biased subsets (Challen et al. 2010) (Box 6.1). Serial transplantation experiments revealed that a lower SP^{LSK} $CD150^{high}$ population preferentially contributed to LT-HSC, MPP, and CMPs (Fig. 6.1), while upper SP^{LSK} $CD150^{low}$ cells produced significantly more CLPs. Microarray analysis carried out on those two populations revealed that pathways involved in cell cycling and metabolism were highly activated in lymphoid-biased HSCs (Challen et al. 2010).

During aging, the proportion of myeloid-biased HSCs in the bone marrow increases at the expense of lymphoid-biased and balanced HSCs (Challen et al. 2010; Beerman et al. 2010b; Cho et al. 2008; Morita et al. 2010). Such an expansion of myeloid-primed HSCs is likely to result from a combination of higher self-renewal potential (Beerman et al. 2010b), different response to cytokines, and lower metabolic activity (Challen et al. 2010) of this population. Moreover, while the pool of young HSCs is largely quiescent, a substantially higher fraction of aged HSCs is cycling (Morrison et al. 1996; Sudo et al. 2000; Yilmaz et al. 2006; Takizawa et al. 2011) (Fig. 6.1).

6.4 Changes in Homing and Migration of Stem Cells Upon Aging

The ability to engraft bone marrow and to migrate throughout the body is an essential property of HSCs that allows them to exit the fetal liver and repopulate bone marrow in the initial stages of development. Studies in the 1960s demonstrated that a small number of repopulating cells is also present in the blood of mice in steady-state conditions (Goodman and Hodgson 1962). These circulating cells are capable of reentering available HSC niches, as demonstrated by experiments with pairs of parabiotic mice that share their circulation (Wright et al. 2001) and by transplantation of bone marrow cells into nonirradiated recipients (Ramshaw et al. 1995). Homing of cells to myeloablated or unmanipulated niches is a prerequisite in experimental or clinical bone marrow transplantations, where intravenous cell injection is sufficient for grafting and regeneration of the hematopoietic system (van Os et al. 2010).

Several studies demonstrated that upon aging, the engraftment and homing ability of HSCs decreases ~2-fold. Liang et al. demonstrated diminished short-term (24 h) homing of old HSCs by limiting-dilution competitive transplantation of young and aged HSCs (Liang et al. 2005). Recently, our lab confirmed this observation by direct measurement of fluorescent cells in bone marrow after co-transplantation of HSCs isolated from old and young differentially colored transgenic donors (Dykstra et al. 2011).

On the contrary, levels of both mobilized progenitors and transplantable HSCs were ~5-fold higher in peripheral blood and spleen of granulocyte colony-stimulating factor (G-CSF)-treated 25-month-old mice compared to young 3-month-old animals (Xing et al. 2006). Xing et al. proposed that lower adhesion of aged HSCs to stromal cells and increased activity of small Rho GTPases that are involved in regulation thereof play a role in this change (Xing et al. 2006).

Interestingly, data generated in our lab demonstrates that on a clonal level young and aged HSCs are similar in their migratory behavior (Verovskaya et al. 2014). When purified HSCs were labeled with DNA barcodes and transplanted into irradiated hosts, barcoded clones of both young and old HSCs were unequally distributed across the skeleton. However, administration of G-CSF led to redistribution of HSC around the skeleton. At the same time, clonal analysis demonstrated that the pool of old HSCs was composed of multiple small clones, whereas the young HSC pool contained fewer, but larger clones (Verovskaya et al. 2013).

The observed functional decline in aged HSCs is reflected in their transcriptome and epigenome, which will be discussed in the next section. A final issue that we will discuss is how HSCs reside in an hypoxic environment and how oxidative metabolism, production of reactive oxygen species (ROS), and DNA repair pathways are linked to diminished self-renewal and proliferative abilities of old HSCs.

6.5 Gene Expression Changes

The increased number of HSCs in aged animals, accompanied by defects in their repopulating ability, as discussed above, suggests that changes in the transcriptional and epigenetic landscape occur with time. However, data describing the transcriptome of old HSCs are limited (Rossi et al. 2005; Chambers et al. 2007a; Noda et al. 2009; Sun et al. 2014) and highly variable. Our understanding of the epigenome is even more restricted, with only two studies recently been published (Beerman et al. 2013; Sun et al. 2014). In this section, we will discuss molecular mechanisms of HSC aging identified thus far.

To date, three groups have independently used microarray analysis to compare gene expression profile in young and aged HSCs. Categories of genes that were upregulated with age in HSCs included those involved in NO-mediated signal transduction, stress response (protein folding), and the inflammatory response. Age-dependent repressed genes were enriched for those involved in the

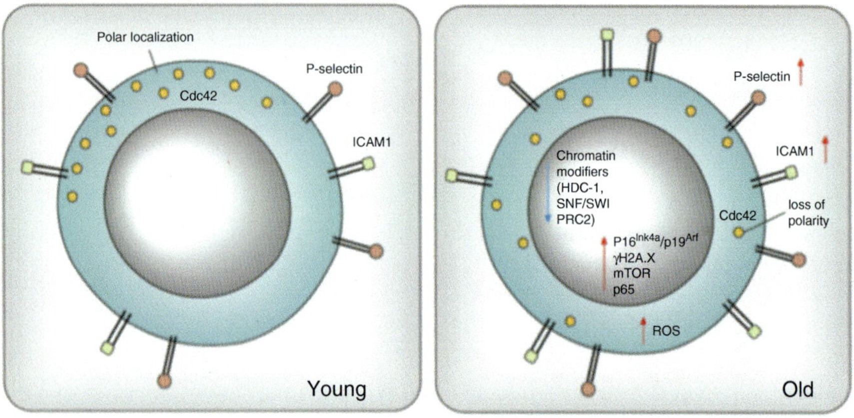

Fig. 6.3 Molecular mechanisms of murine HSC aging. The figure summarizes molecules/pathways reported to be changed in aged HSCs. Increased expression is shown with *red arrows* and decreased expression with *blue*. Changes in cellular localization of Cdc42 are depicted: while in young cells it is polarly distributed, in old HSC it is apolar

preservation of genomic integrity, e.g., chromatin remodeling and DNA repair (Rossi et al. 2005; Chambers et al. 2007b). Surprisingly, while results of all studies indicate that hundreds of genes are differentially expressed, only five of those genes were overlapping among these reports. Of note, all of these studies used pooled samples from multiple mice to initiate the analysis. While differences in the panel of markers used for selection of HSCs (Box 6.1) could contribute to such discrepancies, it is also likely that the pronounced heterogeneity of individual aged mice that we discussed above can explain these differences. In the future, analysis of individual mice will be instrumental for answering this question. Nevertheless, below we will further discuss molecular players that were indicated in those and other studies.

Genes that have been found to be involved in HSC aging thus far include receptors, chromatin modifiers, transcription factors, and genes involved in the DNA damage response and cell proliferation. Two of the highly upregulated genes involve the surface adhesion molecules P-selectin and intercellular adhesion molecule 1 (ICAM1) (Fig. 6.3) (Rossi et al. 2005; Chambers et al. 2007a). Increased levels of P-selectin and ICAM1 are implicated in a higher mobility of those cells and can interfere with HSCs engraftment in bone marrow, which is strictly dependent on adhesion. Of these genes, P-selectin has been shown to serve as a marker for physiological stress, including inflammation, and it is therefore interesting that it has now been shown to be activated upon aging. It mediates the leukocyte-vascular endothelium interaction involved in leukocyte extravasation and therefore appears important for HSCs migration (Karin and Ben-Neriah 2000). At the same time, a decreased expression level of chromatin modifiers (SWI/SNF and PRC2 core members) implicates reorganization of the chromatin structure (Chambers et al. 2007a; Beerman et al. 2013; Hidalgo et al. 2012). Epigenetic perturbations may propagate skewing toward the myeloid lineage and be involved in anemia in aged individuals, because of hypermethylation of promoter regions of genes regulating lymphopoiesis and erythropoiesis (Beerman et al. 2013). Further, aged HSCs are likely to display a more euchromatic genome state, since genes encoding chromatin compaction tend to be downregulated in old HSCs (besides the abovementioned SWI/SNF remodelers, expression of related chromatin remodeling genes such as Smarca4 and Smarcal, as well as histone deacetylases Hdac-1, Hdac-5, and Hdac-6 and a DNA methyltransferase Dnmt3b is decreased) (Chambers et al. 2007a).

Moreover, recent RNA-Seq-based report on epigenomic state of old HSCs documents an increased occurrence of the transcription activating mark-H3K4me3 across self-renewal genes and decreased methylation of HSC maintenance genes combined with hypermethylation of gene promoters associated with differentiation, reflecting phenotypical HSC aging (Sun et al. 2014). Interestingly ribosomal biogenesis seems to be a particular target of aging with hypomethylation of rRNA genes.

The expression of P-selectin, one of the few proteins to be consistently upregulated in aged HSCs, is regulated by NF-κB (the p65 isoform). In HSCs from aged mice, it has been observed that p65 accumulates in the nucleus, indicating transcriptional activation. Increased abundance of this protein in the nucleus of aged HSC may induce upregulation of Toll-like receptor 4 (TLR-4), as well as higher levels of proinflammatory molecules in the plasma of aged individuals (Bruunsgaard et al. 1999; Ershler and Keller 2000). Such higher levels of proinflammatory cytokines (TNFα, IL-1β), together with increased expression of inflammation-related genes, support the concept of inflammation playing a significant role in the aging process (Chambers et al. 2007a).

Two other genes that have often been associated with the aging process are $p16^{Ink4a}$ and $p19^{Arf}$, which are tumor suppressors that induce loss of HSCs without inducing apoptosis (Ito et al. 2004). Furthermore, $p16^{Ink4a}$ has been proposed as a biomarker of aging as its expression increases with age in mammalian tissues, including murine bone marrow (Krishnamurthy et al. 2004). The presence of $p16^{Ink4a}$ restricts the number of HSCs and affects their serial repopulation capacity, cell cycling, and apoptosis (Janzen et al. 2006). However, a causal role for $p16^{Ink4a}$ in the aging process is still under discussion since an effective repression of this gene by H3K27me3 mark in young and old HSCs was recently reported (Sun et al. 2014). Moreover, in the study from Attema et al., no increased levels of $p16^{Ink4a}$ transcripts in individual aged HSCs were found (Attema et al. 2009). The expression of $p16^{Ink4a}$ and $p19^{Arf}$ is mediated by the activity of Polycomb Repressive Complex 2 (PRC2). Recent studies showed upregulation of PRC2 core component in aged HSCs, suggesting in fact increased repression of this locus upon aging (Hidalgo et al. 2012; Norddahl et al. 2011). This is in line with findings that have shown hypermethylation of loci targeted by PRC2 during aging (Beerman et al. 2013).

Aside from the possible, if not likely, contribution of epigenetic mechanisms underlying stem cell aging, there are also many studies that indicate a causal role of genetic and genomic alterations in the aging process. A decline in the expression of genes involved in DNA repair, resulting in an increased rate of non-repaired DNA damage, is thought to be a driving factor in aging. The role of these pathways is described in the next section.

6.6 The Role of Oxidative Metabolism and DNA Repair in HSC Aging

HSCs reside in the bone marrow and are supposed to replenish blood cell production for the entire lifetime of the organism. Both cell-extrinsic localization and cell-intrinsic DNA damage repair mechanisms allow HSCs to maintain their functionality. HSCs appear not be lodging randomly in the bone marrow, but in fact preferentially reside in a low oxygen environment which: (1) reduces molecular damage that results from oxidation by ROS, (2) determines their energy status, (3) induces pathways that are active under hypoxic conditions, and (4) decreases the rate by which DNA mutations accumulate. Apart from preventing damage by avoiding exposure to oxygen, HSCs implement efficient mechanisms of DNA repair to remove deleterious mutations and prevent these to propagate to their progeny. To accomplish this ability, HSCs can employ two independent pathways, depending on the type of the DNA damage and their cell cycle status. We will discuss these parameters in more detail in this section.

One of the first concepts suggesting a link between ROS and the aging process is the "free radical theory of aging," proposed by Harman in the 1950s (Harman 1956). This theory argues that the effect of ROS on cellular components (proteins, nucleic acids, and lipids) may be key in determining the lifespan of the organism. Thus, if we consider that the lifespan of the organism is at least partially determined by the activity of stem cells to maintain tissue homeostasis, effective neutralization of ROS may be crucial for maintaining the stem cell pool. Indeed, several recent studies have demonstrated a role of ROS, and how managing ROS levels and, more general, metabolism affect proper HSC functioning.

The redox potential in HSCs is strictly regulated by intrinsic and extrinsic factors that will be described later in detail. The average O_2 concentration in BM is approximately three times lower compared to blood (Klebanov et al. 2000; Parmar et al. 2007). This suggests that HSCs are residing in an environment with limited availability of oxygen, which propagates glycolytic metabolism (Simsek et al. 2010). However, the content of mitochondria in HSC is relatively high compared to myeloid cells, which suggest that they are less active (Norddahl et al. 2011).

Although the net production from 2 glucose molecules is only 2 ATP (versus 36–38 ATPs during mitochondrial oxidation), this metabolic landscape is compatible with the low-energy requirements of HSC (Folmes et al. 2012) and significantly reduces the rate of ROS production. Whether ROS levels in HSCs are mainly determined by the niche location or by the stem cell-intrinsic activation status remains to be further investigated.

Several studies have shown that ROS levels increase with proliferation and during early steps of lineage commitment (Simsek et al. 2010; Jang and Sharkis 2007). These experiments have combined regular FACS staining for HSCs with DCF-DA, a probe enabling the measurement of ROS activity in viable cells. Using this approach allowed the identification of two populations of ROS^{low} and ROS^{high} stem cells. The population of ROS^{high} stem cells showed functional decline upon serial transplantation and a typical pattern of differentiation skewing toward the myeloid lineage. This clearly mimics characteristics of aged HSCs. Moreover, these ROS^{high} cells express activated p38 kinase and components of the mTOR pathway, both involved in stress response. Treatment of ROS^{high} HSCs with inhibitors of both of these pathways was able to restore functional activity (Jang and Sharkis 2007). The comparison of the ratio of ROS^{low} versus ROS^{high} populations in aged mice showed a strong decline of cells expressing low levels of ROS and only a moderate increase in the ROS^{high} population. Therefore, a decrease in ROS^{low} stem cells may be responsible for the skewing of myeloid differentiation and the impaired repopulating activity (Jang and Sharkis 2007).

The major impact of oxidation on HSCs discussed above seems to be reflected also by the presence of O_2 in the hematopoietic niche. Decreased oxygen availability activates hypoxia-inducible factor (HIF) signaling, which regulates the redox potential in HSCs. The HIF family of transcription factors comprises four proteins, among which HIF-1α is the best studied. HIF-1α levels are controlled by products of the Ink4a gene locus and appear to play a protective role against senescence-induced HSCs exhaustion. Bone marrow cells deficient of HIF-1α showed increased ROS levels (Takubo et al. 2010). Similar phenotype with HSC exhaustion and increase in ROS content was observed during deletion of subfamily of transcription factors-Forkhead O (FoxOs 1,2,3,4) (Tothova et al. 2007), which regulate the expression of genes involved in metabolism and oxidative stress (Salih and Brunet 2008). However, the treatment of those cells with antioxidants partially rescued defects in HSCs deficient for these proteins (Tothova et al. 2007). Thus, induction of cellular senescence may originate from ineffective ROS management, which on its turn may lead to DNA damage. Indeed, a recent study provides evidence directly linking cellular senescence with DNA damage (Di Micco et al. 2011).

In line with a detrimental influence of ROS on HSCs, it has been shown that endogenous DNA damage accumulates with age in normal stem cells. Multiple foci of phosphorylated γ-H2AX, a marker of DNA damage, were found in over 70 % of old HSCs (Rossi et al. 2005). However, the percentage of γ-H2AX-positive cells decreases as the cell differentiated to more committed progenitors. The alternative explanation would be that γ-H2AX is a marker for impaired replication and associated with stalled or collapsed replication forks. This hypothesis is in line with reduced expression of MDM helicase family genes that decline with age. Importantly, downregulation of members of this family in young HSCs led to their impaired engraftment – a feature characteristic of old HSCs (Flach et al. 2014). Furthermore, cellular mislocalization of PP4c (γ-H2AX phosphatase) in aged HSCs leads to delayed and ineffective dephosphorylation and accumulation of γ-H2AX (Flach et al. 2014). This indicates that with age, DNA damage accumulates preferentially in stem cells, while this is either repaired more quickly in proliferating progenitor cells or DNA damage-containing progenitors are eliminated. This also reiterates the notion of cell type-dependent efficiency of DNA damage repair. In the next paragraph we will further discuss how mechanisms for DNA repair are implicated in HSC aging.

To repair DNA damage, cells utilize a repair machinery that is either DNA template dependent or template independent. So far, it has been shown that several template-dependent mechanisms of DNA damage repair are active in HSCs. All of these are negatively affected by aging. Such repair mechanisms include non-homologous end joining (NHEJ) which occurs during G_0/G_1 stages of cell cycle and nucleotide-excision repair (NER) utilized to remove modified nucleotides. The effect of aging on these processes has been assessed using mutant mouse models with defects in these pathways (Blanpain et al. 2011). Both XPD^{TTD} mice, with deficiency of NER, and $Ku80^{-/-}$animals, where NHEJ is affected, generated numbers of old HSCs comparable to normal mice, but their functionality was severely affected upon serial transplantation (Rossi et al. 2007; Nijnik et al. 2007).

Interestingly, a similar phenotype was observed in $mTR^{-/-}$ mice, a mouse strain deficient for telomerase. Telomere shortening affected the intrinsic ability of HSCs to self-renew (Rossi et al. 2007) and also limited the capacity of niche cells to support HSC homing (Ju et al. 2007). The decline of HSCs activity in the absence of telomerase could be reverted by downregulation of the DNA damage checkpoint – Basic leucine zipper transcriptional factor (BATF), which simultaneously resulted in accumulation of DNA damage in HSCs (Wang et al. 2012).

In the case of double-stranded DNA breaks, template-independent DNA repair takes place. This requires the activity of the protein kinase ataxia telangiectasia mutated (ATM). The analysis of $Atm^{-/-}$ HSCs revealed a lack of long-term repopulating capacity (Ito et al. 2006). Moreover, on the molecular level, increased ROS levels (much higher than those observed in normal aged HSCs) contributed to upregulation of $p16^{Ink4a}$ and $p19^{Arf}$ in those cells (Ito et al. 2004; Maryanovich et al. 2012). Malfunctioning of $Atm^{-/-}$ HSCs could be rescued by the long-term treatment of mice with the antioxidative molecule N-acetyl cysteine (NAC). This highlights the antiaging properties of ROS-quenching antioxidants. Moreover, increased cellular ROS levels activate the p38 MAPK pathway (Jang and Sharkis 2007; Ito et al. 2006), which inhibits quiescence state in $Atm^{-/-}$ HSCs (Ito et al. 2006).

A schematic overview of the molecular events implicated in HSC aging is provided in Fig. 6.3. To summarize, adult HSCs accumulate DNA damage under physiological conditions *in vivo* due to a plethora of intrinsic and extrinsic factors. The mechanisms that we describe above are relevant for neutralizing DNA damage in order to maintain the pool of HSCs; dysfunctions in any of those pathways will inevitably cause HSCs exhaustion.

The functional and molecular changes that occur during aging in individual HSCs impact on the behavior of the entire hematopoietic stem cell pool, i.e., on how stem cells act in concert during the aging process. In the next section we will briefly discuss various currently prevailing stem cell aging models.

6.7 Models of HSC Aging

Originally it was considered that the age-related decline in functioning of a population of HSCs was caused by cellular aging of individual stem cells and involved a homogeneous HSC population (Rossi et al. 2007; Nijnik et al. 2007). Such aging model was dubbed the "population shift model." This model presumes that the HSC population is uniform and all cells gradually accumulate damages and lose their ability to produce lymphoid cells and to engraft bone marrow (Muller-Sieburg and Sieburg 2008). Later, the identification of distinct developmental programs of individual HSCs and observation of changes in the representation of these HSC subsets with age led to the establishment of an alternative hypothesis. The "clonal selection model" implied that changes in the clonal make-up of the HSC pool, rather than changes in individual HSCs, underlie the hematopoietic aging (Cho et al. 2008; Muller-Sieburg and Sieburg 2008). This theory also assumes that individual HSCs do not change with age (Muller-Sieburg and Sieburg 2008). The most recent conceptual view on aging combines both models: a changing pool of HSCs is simultaneously affected by cellular aging (Dykstra et al. 2011; Beerman et al. 2010a).

6.8 Perspectives

Several other mechanisms that may contribute to functional hematopoietic decline are currently being investigated. These include factors such as metabolism, epigenetic changes, polarity, the regulation of posttranscriptional gene expression, and hematopoietic microenvironment.

6.8.1 Metabolism

A recent paper reports an increased activation of the mTOR pathway in HSCs and progenitors from aged mice (Chen et al. 2009). Administration of rapamycin (the inhibitor of mTOR pathway) limited the age-related expansion of the HSC pool and also improved their functionality so that these aged HSCs became as effective as young HSCs (Chen et al. 2009). It has also been shown that autophagy is increased in old HSCs and ensures their survival (Warr et al. 2013). Additionally, loss of expression of the nutrient sensing protein SIRT7 with age has been associated with increased HSC proliferation and skewing of hematopoiesis toward production of myeloid cells (Mohrin et al. 2015).

6.8.2 Noncoding RNAs

Noncoding RNAs seem to play a pivotal role in the aging process. In the study from Boon et al., authors studied the role of microRNAs (miRNA) in aged cardiomyocytes. MiR-34a, by suppressing the expression of its targets, leads to inhibition of telomere erosion, DNA damage response, and apoptosis (Boon et al. 2013). It leaves the ground for speculation that also during HSC aging certain miRNAs play an important role in determining their functionality and moreover specifying the cellular program to preferentially differentiate into certain lineages. No data on miRNA functioning in aged HSCs are currently available. However, a recent report documents the importance and the potential of long noncoding (Lnc-RNAs) in regulating self-renewal and lineage commitment in old HSCs (Luo et al. 2015). Two lnc-RNAs (referred to as LncHSC-1 and LncHSC-2) affect lineage commitment. Whereas downregulation of LncHSC-1 led to myeloid skewing at the expense of B cells, LncHSC-2 promoted T lymphopoiesis (Luo et al. 2015).

6.8.3 Cell Polarity

Recent studies have demonstrated that polar localization of the small Rho-GTPase Cdc42 is characteristic for young HSC (but not more committed cells), and this polarity is lost with age due to increased activity of Cdc42 (Florian et al. 2012). The shift from canonical to non-canonical Wnt signaling has been associated with apolar Cdc42 cellular localization (Florian et al. 2013). Remarkably, administration of a chemical compound (a Cdc42 specific inhibitor, CASIN) allowed restoration of both polarity and functionality of old cells and resulted in their rejuvenation (Florian et al. 2012). Similarly reduced non-canonical Wnt5a signaling in old HSCs rejuvenates chronologically aged HSCs and restores polar localization of Cdc42 (Florian et al. 2013). The prospect that we may indeed be able to rejuvenate aged stem cells is of considerable interest for future clinical interventions.

6.8.4 Microenvironment

Finally, while this chapter focuses on cell-intrinsic mechanisms of HSC aging, recent findings indicate the potential contribution of the bone marrow microenvironment, or "niche," to the HSC aging phenotype. To support this notion, the level of inflammatory cytokine CCL5 that is shown to induce myelopoiesis at expense of lymphopoiesis was shown to increase upon aging (Ergen et al. 2012). It was also demonstrated that localization of hematopoietic progenitors in old bone changes further from endosteum and is associated with reduced cell polarity (Kohler et al. 2009). A recent report also highlights the initial role of $p19^{INK4d}$ in niche remodeling upon aging (Hilpert et al. 2014). The authors argue that loss of $p19^{INK4d}$ in the microenvironment with age results in expansion of myeloid-biased HSCs, possibly through changes of expressed levels of TGF-β1 by niche cells (Hilpert et al. 2014). Therefore, it is likely that HSC aging is a complex interplay between changes in the HSC pool itself and the aging environment they reside in.

Acknowledgements This work was supported by grants from the Netherlands Institute for Regenerative Medicine (NIRM), the European Union FP7 Marie Curie Initial Training Network "EuroCancer Stem Cell Training," the Netherlands Organization for Scientific Research (NWO), and the Mouse Clinic for Cancer and Aging, part of the large RoadMap Initiatives funded by the Dutch Ministry of Education, Culture, and Science.

References

Arai F, Suda T (2007) Maintenance of quiescent hematopoietic stem cells in the osteoblastic niche. Ann N Y Acad Sci 1106:41–53

Attema JL, Pronk CJ, Norddahl GL, Nygren JM, Bryder D (2009) Hematopoietic stem cell ageing is uncoupled from p16 INK4A-mediated senescence. Oncogene 28(22):2238–2243

Balducci L, Ershler W, Gaetano G (eds) (2008) Blood disorders in the elderly. Cambridge University Press, New York

Beerman I, Bhattacharya D, Zandi S, Sigvardsson M, Weissman IL, Bryder D et al (2010a) Functionally distinct hematopoietic stem cells modulate hematopoietic lineage potential during aging by a mechanism of clonal expansion. Proc Natl Acad Sci U S A 107(12):5465–5470

Beerman I, Maloney WJ, Weissmann IL, Rossi DJ (2010b) Stem cells and the aging hematopoietic system. Curr Opin Immunol 22(4):500–506

Beerman I, Bock C, Garrison BS, Smith ZD, Gu H, Meissner A et al (2013) Proliferation-dependent alterations of the DNA methylation landscape underlie hematopoietic stem cell aging. Cell Stem Cell 12:413–425

Benz C, Copley MR, Kent DG, Wohrer S, Cortes A, Aghaeepour N et al (2012) Hematopoietic stem cell subtypes expand differentially during development and display distinct lymphopoietic programs. Cell Stem Cell 10(3):273–283

Bertoncello I, Hodgson GS, Bradley TR (1985) Multiparameter analysis of transplantable hemopoietic stem cells: I. The separation and enrichment of stem cells homing to marrow and spleen on the basis of rhodamine-123 fluorescence. Exp Hematol 13(10):999–1006

Blanpain C, Mohrin M, Sotiropoulou PA, Passegue E (2011) DNA-damage response in tissue-specific and cancer stem cells. Cell Stem Cell 8(1):16–29

Boon RA, Iekushi K, Lechner S, Seeger T, Fischer A, Heydt S et al (2013) MicroRNA-34a regulates cardiac ageing and function. Nature 495:107–110

Bruunsgaard H, Pedersen AN, Schroll M, Skinhoj P, Pedersen BK (1999) Impaired production of proinflammatory cytokines in response to lipopolysaccharide (LPS) stimulation in elderly humans. Clin Exp Immunol 118(2):235–241

Challen GA, Boles N, Lin KK, Goodell MA (2009) Mouse hematopoietic stem cell identification and analysis. Cytometry A 75(1):14–24

Challen GA, Boles NC, Chambers SM, Goodell MA (2010) Distinct hematopoietic stem cell subtypes are differentially regulated by TGF-beta1. Cell Stem Cell 6(3):265–278

Chambers SM, Shaw CA, Gatza C, Fisk CJ, Donehower LA, Goodell MA (2007a) Aging hematopoietic stem cells decline in function and exhibit epigenetic dysregulation. PLoS Biol 5(8), e201

Chambers SM, Boles NC, Lin KY, Tierney MP, Bowman TV, Bradfute SB et al (2007b) Hematopoietic fingerprints: an expression database of stem cells and their progeny. Cell Stem Cell 1(5):578–591

Chen C, Liu Y, Liu Y, Zheng P (2009) mTOR regulation and therapeutic rejuvenation of aging hematopoietic stem cells. Sci Signal 2(98):ra75

Cheshier SH, Morrison SJ, Liao X, Weissman IL (1999) In vivo proliferation and cell cycle kinetics of long-term self-renewing hematopoietic stem cells. Proc Natl Acad Sci U S A 96(6):3120–3125

Cho RH, Sieburg HB, Muller-Sieburg CE (2008) A new mechanism for the aging of hematopoietic stem cells: aging changes the clonal composition of the stem cell compartment but not individual stem cells. Blood 111(12):5553–5561

Di Micco R, Sulli G, Dobreva M, Liontos M, Botrugno OA, Gargiulo G et al (2011) Interplay between oncogene-induced DNA damage response and heterochromatin in senescence and cancer. Nat Cell Biol 13(3):292–302

Dykstra B, de Haan G (2008) Hematopoietic stem cell aging and self-renewal. Cell Tissue Res 331(1):91–101

Dykstra B, Ramunas J, Kent D, McCaffrey L, Szumsky E, Kelly L et al (2006) High-resolution video monitoring of hematopoietic stem cells cultured in single-cell arrays identifies new features of self-renewal. Proc Natl Acad Sci U S A 103(21):8185–8190

Dykstra B, Kent D, Bowie M, McCaffrey L, Hamilton M, Lyons K et al (2007) Long-term propagation of distinct hematopoietic differentiation programs in vivo. Cell Stem Cell 1(2):218–229

Dykstra B, Olthof S, Schreuder J, Ritsema M, de Haan G (2011) Clonal analysis reveals multiple functional defects of aged murine hematopoietic stem cells. J Exp Med 208(13):2691–2703

Ema H, Sudo K, Seita J, Matsubara A, Morita Y, Osawa M et al (2005) Quantification of self-renewal capacity in single hematopoietic stem cells from normal and Lnk-deficient mice. Dev Cell 8(6):907–914

Ergen AV, Boles NC, Goodell MA (2012) Rantes/Ccl5 influences hematopoietic stem cell subtypes and causes myeloid skewing. Blood 119(11):2500–2509

Ershler WB, Keller ET (2000) Age-associated increased interleukin-6 gene expression, late-life diseases, and frailty. Annu Rev Med 51:245–270

Flach J, Bakker ST, Mohrin M, Conroy PC, Pietras EM et al (2014) Replicative stress is a potent driver of functional decline in ageing hematopoietic stem cells. Nature 512(7513):198–202

Florian MC, Dorr K, Niebel A, Daria D, Schrezenmeier H, Rojewski M et al (2012) Cdc42 activity regulates hematopoietic stem cell aging and rejuvenation. Cell Stem Cell 10(5):520–530

Florian CM, Nattamai KJ, Doerr K, Marka G, Überle B et al (2013) A canonical to non-canonical Wnt signaling switch in hematopoietic stem-cell ageing. Nature 503(7476):392–396

Folmes CD, Dzeja PP, Nelson TJ, Terzic A (2012) Metabolic plasticity in stem cell homeostasis and differentiation. Cell Stem Cell 11(5):596–606

Foudi A, Hochedlinger K, Van Buren D, Schindler JW, Jaenisch R, Carey V et al (2009) Analysis of histone 2B-GFP retention reveals slowly cycling hematopoietic stem cells. Nat Biotechnol 27(1):84–90

Goodell MA, Brose K, Paradis G, Conner AS, Mulligan RC (1996) Isolation and functional properties of murine hematopoietic stem cells that are replicating in vivo. J Exp Med 183(4):1797–1806

Goodman JW, Hodgson GS (1962) Evidence for stem cells in the peripheral blood of mice. Blood 19:702–714

Harman D (1956) Aging: a theory based on free radical and radiation chemistry. J Gerontol 11(3):298–300

Harrison DE (1983) Long-term erythropoietic repopulating ability of old, young, and fetal stem cells. J Exp Med 157(5):1496–1504

Harrison DE, Astle CM, Stone M (1989) Numbers and functions of transplantable primitive immunohematopoietic stem cells. Effects of age. J Immunol 142(11):3833–3840

Hidalgo I, Herrera-Merchan A, Ligos JM, Carramolino L, Nunez J, Martinez F et al (2012) Ezh1 is required for hematopoietic stem cell maintenance and prevents senescence-like cell cycle arrest. Cell Stem Cell 11(5):649–662

Hilpert M, Legrand C, Bluteau C, Balayn N, Betems A et al (2014) p19 INK4d controls hematopoietic stem cell in a cell-autonomous manner during genotoxic stress and through the microenvironment during aging. Stem Cell Reports 3(6):1085–1102

Ito K, Hirao A, Arai F, Matsuoka S, Takubo K, Hamaguchi I et al (2004) Regulation of oxidative stress by ATM is required for self-renewal of haematopoietic stem cells. Nature 431(7011):997–1002

Ito K, Hirao A, Arai F, Takubo K, Matsuoka S, Miyamoto K et al (2006) Reactive oxygen species act through p38 MAPK to limit the lifespan of hematopoietic stem cells. Nat Med 12(4):446–451

Jang YY, Sharkis SJ (2007) A low level of reactive oxygen species selects for primitive hematopoietic stem cells that may reside in the low-oxygenic niche. Blood 110(8):3056–3063

Janzen V, Forkert R, Fleming HE, Saito Y, Waring MT, Dombkowski DM et al (2006) Stem-cell ageing modified by the cyclin-dependent kinase inhibitor p16INK4a. Nature 443(7110): 421–426

Jordan CT, Lemischka IR (1990) Clonal and systemic analysis of long-term hematopoiesis in the mouse. Genes Dev 4(2):220–232

Ju Z, Jiang H, Jaworski M, Rathinam C, Gompf A, Klein C et al (2007) Telomere dysfunction induces environmental alterations limiting hematopoietic stem cell function and engraftment. Nat Med 13(6):742–747

Kamminga LM, van Os R, Ausema A, Noach EJ, Weersing E, Dontje B et al (2005) Impaired hematopoietic stem cell functioning after serial transplantation and during normal aging. Stem Cells 23(1):82–92

Karin M, Ben-Neriah Y (2000) Phosphorylation meets ubiquitination: the control of NF-[kappa]B activity. Annu Rev Immunol 18:621–663

Kent DG, Copley MR, Benz C, Wohrer S, Dykstra BJ, Ma E et al (2009) Prospective isolation and molecular characterization of hematopoietic stem cells with durable self-renewal potential. Blood 113(25):6342–6350

Kiel MJ, Yilmaz OH, Iwashita T, Yilmaz OH, Terhorst C, Morrison SJ (2005) SLAM family receptors distinguish hematopoietic stem and progenitor cells and reveal endothelial niches for stem cells. Cell 121(7):1109–1121

Klebanov S, Flurkey K, Roderick TH, Archer JR, Astle MC, Chen J et al (2000) Heritability of life span in mice and its implication for direct and indirect selection for longevity. Genetica 110(3):209–218

Kohler A, Schmithorst V, Filippi MD, Ryan MA, Daria D et al (2009) Altered cellular dynamics and endosteal location of aged early hematopoietic progenitor cells revealed by time-lapse intravital imaging in long bones. Blood 114(2):290–298

Krishnamurthy J, Torrice C, Ramsey MR, Kovalev GI, Al-Regaiey K, Su L et al (2004) Ink4a/Arf expression is a biomarker of aging. J Clin Invest 114(9):1299–1307

Lansdorp PM (1998) Stem cell biology for the transfusionist. Vox Sang 74(Suppl 2):91–94

Liang Y, Van Zant G, Szilvassy SJ (2005) Effects of aging on the homing and engraftment of murine hematopoietic stem and progenitor cells. Blood 106(4):1479–1487

Luo M, Jeong M, Sun D, Jung Park H, Rodrigues BAT et al (2015) Long non-coding RNAs control hematopoietic stem cell function. Cell Stem Cell 16(4):426–438

Maryanovich M, Oberkovitz G, Niv H, Vorobiyov L, Zaltsman Y, Brenner O et al (2012) The ATM-BID pathway regulates quiescence and survival of haematopoietic stem cells. Nat Cell Biol 14(5):535–541

Miller CL, Eaves CJ (1997) Expansion in vitro of adult murine hematopoietic stem cells with transplantable lympho-myeloid reconstituting ability. Proc Natl Acad Sci U S A 94(25):13648–13653

Mohrin M, Shin J, Liu Y, Brown K, Luo H et al (2015) A mitochondrial UPR-mediated checkpoint regulates hematopoietic stem cell aging. Science 347(6228):1347–1377

Morita Y, Ema H, Nakauchi H (2010) Heterogeneity and hierarchy within the most primitive hematopoietic stem cell compartment. J Exp Med 207(6):1173–1182

Morrison SJ, Wandycz AM, Akashi K, Globerson A, Weissman IL (1996) The aging of hematopoietic stem cells. Nat Med 2(9):1011–1016

Muller-Sieburg C, Sieburg HB (2008) Stem cell aging: survival of the laziest? Cell Cycle 7(24):3798–3804

Muller-Sieburg CE, Cho RH, Thoman M, Adkins B, Sieburg HB (2002) Deterministic regulation of hematopoietic stem cell self-renewal and differentiation. Blood 100(4): 1302–1309

Muller-Sieburg CE, Cho RH, Karlsson L, Huang JF, Sieburg HB (2004) Myeloid-biased hematopoietic stem cells have extensive self-renewal capacity but generate diminished lymphoid progeny with impaired IL-7 responsiveness. Blood 103(11):4111–4118

Nijnik A, Woodbine L, Marchetti C, Dawson S, Lambe T, Liu C et al (2007) DNA repair is limiting for haematopoietic stem cells during ageing. Nature 447(7145):686–690

Noda S, Ichikawa H, Miyoshi H (2009) Hematopoietic stem cell aging is associated with functional decline and delayed cell cycle progression. Biochem Biophys Res Commun 383(2):210–215

Norddahl GL, Pronk CJ, Wahlestedt M, Sten G, Nygren JM, Ugale A et al (2011) Accumulating mitochondrial DNA mutations drive premature hematopoietic aging phenotypes distinct from physiological stem cell aging. Cell Stem Cell 8(5):499–510

Orford KW, Scadden DT (2008) Deconstructing stem cell self-renewal: genetic insights into cell-cycle regulation. Nat Rev Genet 9(2):115–128

Parmar K, Mauch P, Vergilio JA, Sackstein R, Down JD (2007) Distribution of hematopoietic stem cells in the bone marrow according to regional hypoxia. Proc Natl Acad Sci U S A 104(13):5431–5436

Passegue E, Wagers AJ, Giuriato S, Anderson WC, Weissman IL (2005) Global analysis of proliferation and cell cycle gene expression in the regulation of hematopoietic stem and progenitor cell fates. J Exp Med 202(11):1599–1611

Ramshaw HS, Crittenden RB, Dooner M, Peters SO, Rao SS, Quesenberry PJ (1995) High levels of engraftment with a single infusion of bone marrow cells into normal unprepared mice. Biol Blood Marrow Transplant 1(2):74–80

Rossi DJ, Bryder D, Zahn JM, Ahlenius H, Sonu R, Wagers AJ et al (2005) Cell intrinsic alterations underlie hematopoietic stem cell aging. Proc Natl Acad Sci U S A 102(26):9194–9199

Rossi DJ, Bryder D, Seita J, Nussenzweig A, Hoeijmakers J, Weissman IL (2007) Deficiencies in DNA damage repair limit the function of haematopoietic stem cells with age. Nature 447(7145):725–729

Salih DA, Brunet A (2008) FoxO transcription factors in the maintenance of cellular homeostasis during aging. Curr Opin Cell Biol 20(2):126–136

Simsek T, Kocabas F, Zheng J, Deberardinis RJ, Mahmoud AI, Olson EN et al (2010) The distinct metabolic profile of hematopoietic stem cells reflects their location in a hypoxic niche. Cell Stem Cell 7(3):380–390

Sudo K, Ema H, Morita Y, Nakauchi H (2000) Age-associated characteristics of murine hematopoietic stem cells. J Exp Med 192(9):1273–1280

Sun D, Luo M, Jeong M, Rodriguez B, Xia Z et al (2014) Epigenomic profiling of young and aged HSCs reveals concerted changes that reinforces self-renewal. Cell Stem Cell 14(5):673–688

Takizawa H, Regoes RR, Boddupalli CS, Bonhoeffer S, Manz MG (2011) Dynamic variation in cycling of hematopoietic stem cells in steady state and inflammation. J Exp Med 208(2): 273–284

Takubo K, Goda N, Yamada W, Iriuchishima H, Ikeda E, Kubota Y et al (2010) Regulation of the HIF-1alpha level is essential for hematopoietic stem cells. Cell Stem Cell 7(3):391–402

Tothova Z, Kollipara R, Huntly BJ, Lee BH, Castrillon DH, Cullen DE et al (2007) FoxOs are critical mediators of hematopoietic stem cell resistance to physiologic oxidative stress. Cell 128(2):325–339

van Os R, Ausema A, Dontje B, van Riezen M, van Dam G, de Haan G (2010) Engraftment of syngeneic bone marrow is not more efficient after intrafemoral transplantation than after traditional intravenous administration. Exp Hematol 38(11):1115–1123

Verovskaya EV, de Haan G (2011) The power of diversity: hematopoietic stem cell heterogeneity and its clinical relevance. Hematol Educ 5(1):132–139

Verovskaya E, Broekhuis MJC, Zwart E, Ritsema M, van Os R et al (2013) Heterogeneity of young and aged murine hematopoietic stem cells revealed by quantitative clonal analysis using cellular barcoding. Blood 122(4):523–532

Verovskaya E, Broekhuis MJC, Zwart E, Weersing E, Ritsema M, Bosman LJ et al (2014) Asymmetry in skeletal distribution of mouse hematopoietic stem cell clones and their equilibration by mobilizing cytokines. J Exp Med 211(3):487–497

Wang J, Sun Q, Morita Y, Jiang H, Gross A, Lechel A et al (2012) A differentiation checkpoint limits hematopoietic stem cell self-renewal in response to DNA damage. Cell 148(5): 1001–1014

Warr MR, Binnewies M, Flach J, Reynaud D, Garg T et al (2013) FOXO3A directs a protective autophagy in hematopoietic stem cells. Nature 494(7437):323–327

Wilson A, Laurenti E, Oser G, van der Wath RC, Blanco-Bose W, Jaworski M et al (2008) Hematopoietic stem cells reversibly switch from dormancy to self-renewal during homeostasis and repair. Cell 135(6):1118–1129

Wolf NS, Kone A, Priestley GV, Bartelmez SH (1993) In vivo and in vitro characterization of long- term repopulating primitive hematopoietic cells isolated by sequential Hoechst 33342-rhodamine123 FACS selection. Exp Hematol 21(5):614–622

Wright DE, Wagers AJ, Gulati AP, Johnson FL, Weissman IL (2001) Physiological migration of hematopoietic stem and progenitor cells. Science 294(5548):1933–1936

Xing Z, Ryan MA, Daria D, Nattamai KJ, Van Zant G, Wang L et al (2006) Increased hematopoietic stem cell mobilization in aged mice. Blood 108(7):2190–2197

Yilmaz OH, Kiel MJ, Morrison SJ (2006) SLAM family markers are conserved among hematopoietic stem cells from old and reconstituted mice and markedly increase their purity. Blood 107(3):924–930

Zhou S, Schuetz JD, Bunting KD, Colapietro AM, Sampath J, Morris JJ et al (2001) The ABC transporter Bcrp1/ABCG2 is expressed in a wide variety of stem cells and is a molecular determinant of the side-population phenotype. Nat Med 7(9):1028–1034

7 Aging of Human Haematopoietic Stem Cells

David Garrick, Dounia Djeghloul, Klaudia Kuranda, and Michele Goodhardt

Abstract

Haematopoietic aging is associated with a decline in immune function as well as an increased incidence of malignant transformation and thus has important ramifications for the health of the elderly. There is increasing evidence that changes occur within the human haematopoietic stem cell (HSC) compartment with age, including a reduced repopulating activity and perturbed differentiation potential, and that these changes contribute to haematopoietic aging and related pathologies. Understanding the mechanisms driving these age-associated changes in HSC function could potentially translate into significant public health benefit in our aging population, by reducing the risk of hospitalisation and dependency of the elderly. In this chapter we review the current state of knowledge of the causes and consequences of HSC aging in humans. Where appropriate we have highlighted the similarities and differences that have been observed between humans and mice. Although comparatively unexplored, the field of HSC aging in humans has been advanced by some recent key observations, which will stimulate and direct future research efforts in the field.

D. Garrick • D. Djeghloul • K. Kuranda • M. Goodhardt (✉)
INSERM U1126, Institut Universitaire d'Hématologie, Hôpital Saint-Louis, 1, av. Claude Vellefaux, Paris 75010, France
e-mail: michele.goodhardt@univ-paris-diderot.fr

H. Geiger et al. (eds.), *Stem Cell Aging: Mechanisms, Consequences, Rejuvenation*, DOI 10.1007/978-3-7091-1232-8_7

7.1 Introduction

Accumulating evidence indicates that aging of tissue-specific stem cells is central to the decline in tissue function and structure observed in the elderly, and the haematopoietic system is no exception. Human haematopoietic stem cells (HSC) are highly specialised multipotent cells residing in the adult bone marrow that are responsible for generating all of the blood cell lineages. Blood is one of the most highly regenerative tissues, with approximately 10^{12} cells produced daily in the bone marrow. Mature blood cells have a finite life span, ranging from 1 to 120 days in humans, so their maintenance under both steady-state conditions or in response to haematopoietic stress requires the lifelong activity of HSCs. HSCs were the first tissue-specific stem cells to be isolated, and some aspects of HSC biology have proved paradigmatic for other stem cell populations. Although much of our early knowledge came from studies of mouse HSCs, the development of xenotransplantation techniques, in which human bone marrow cells are injected into immunodeficient mice, allowed the experimental identification of a human stem cell population capable of long-term reconstitution of the haematopoietic system in radioablated recipients. Recent advances in the xenograft assay and the identification of novel cell surface markers have led to the identification of clonally pure populations of human HSCs and early progenitors (Doulatov et al. 2010; Notta et al. 2011; Goyama et al. 2015) paving the way for more detailed research on the biology and aging of human HSCs.

It is becoming clear that haematopoietic stem cells are not constant throughout life and that changes with age occur in both HSC number and function. Evidence for HSC aging was first established in mice, and the most extensive molecular characterisation of this phenomenon has been carried out in murine models (see Chap. 6). Despite similarities in many aspects of basic biology and haematology between mouse and man, there are also important species-specific differences in terms of life span, evolutionary constraints, body size and hence basal metabolic rates as well as age-associated pathologies, making it difficult to simply extrapolate from mouse models to the human system in the study of aging (Demetrius et al. 2005; Demetrius 2006). Until relatively recently, there was only indirect evidence for an age-associated decline in HSC function in humans. However, several recent key studies have clearly established that as in the mouse, age-associated changes occur also within the human HSC compartment and are likely to exert significant ramifications for the health of the aging population. In this chapter we will review the evidence and current state of knowledge of the molecular mechanisms responsible for HSC aging in humans, the clinical implications and consequences for the health of the elderly and the prospects for clinical intervention.

7.2 Haematopoietic Stem Cells and the Haematopoietic Hierarchy

Mature blood lineage cells are generated throughout life from rare HSCs that are located in the bone marrow cavity of the femur and humerus, as well as in the vertebrae, sternum and pelvis of young adults. Their location is progressively restricted

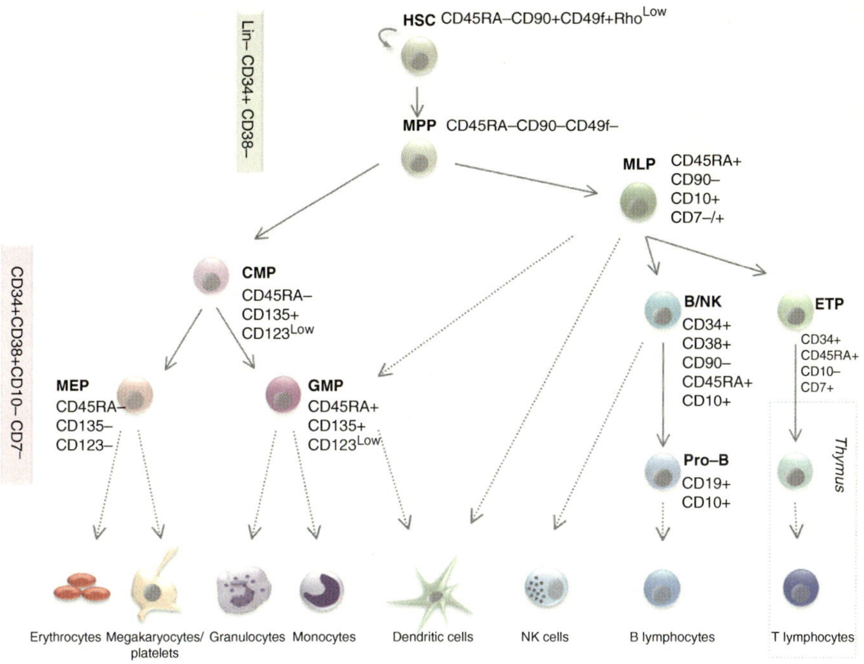

Fig. 7.1 Scheme of the human haematopoietic hierarchy and relevant cell surface markers. All blood lineage cells are derived from HSC. Differentiation of HSC gives rise to MPP that have lost self-renewal capacity. MPP in turn generate progressively more committed progenitors and eventually terminally differentiated blood cells. Phenotypic cell surface markers are indicated next to each population based on data from (Doulatov et al. 2010 #230; Notta et al. 2011 #231; Majeti et al. 2007 #257; Manz et al. 2002 #289). The *yellow box* denotes the common phenotype ($Lin^{-}CD34^{+}CD38^{-}$) shared by HSC and MPP. The *pink box* denotes the common phenotype ($CD34^{+}C38^{+}CD10^{-}CD7^{-}$) shared by myeloid progenitors. Abbreviations: *HSC* haematopoietic stem cells, *MPP* multipotent progenitors, *MLP* multilymphoid progenitors, *CMP* common myeloid progenitors, *GMP* granulocyte-macrophage progenitors, *MEP* megakaryocyte-erythrocyte progenitors, *ETP* early thymic progenitors, *NK* natural killer. Lin, a collection of surface markers derived from all terminally differentiated populations

with age. Human HSCs are also found in relatively high numbers in umbilical cord blood. In response to appropriate signals, HSCs differentiate to give rise initially to progenitor populations, which have retained multilineage potential but have limited self-renewal capacity, and then to committed progenitors whose potential is progressively restricted to one of the alternative blood cell lineages (Fig. 7.1).

A number of in vivo and in vitro experimental approaches have been developed to identify human HSC and to study human haematopoiesis (for reviews, see (Goyama et al. 2015; Doulatov et al. 2012)). These include colony-forming assays in semisolid plates in the presence of a cocktail of cytokines to give colonies of myeloid and erythroid cells, as well as in vitro culture on a variety of murine stromal cell lines capable of supporting the differentiation of primitive human progenitors to the different blood cell lineages (La Motte-Mohs et al. 2005; Baum et al. 1992; Berardi et al. 1997; Miller et al. 1999; Reynaud et al. 2003). In vivo assays, based on the capacity of the stem cells to repopulate the entire haematopoietic system after transplantation into conditioned recipients, have also been developed using a number of animal models. The most widely used in the field have been immunodeficient *Rag1*-, *Rag2*-null or *Scid* mice (for reviews, see (Shultz et al. 2007; Rongvaux et al. 2013)). The xenograft model has been refined by targeting mutation of the IL-2 receptor common gamma chain gene and crossing onto the non-obese diabetic (NOD) background to give mice (NOD-Scid $IL2rg^{-/-}$, NSG) that have impaired adaptive and innate immunity and so are more efficient in engrafting human haematopoietic progenitors (Traggiai et al. 2004; Pearson et al. 2008; Ito et al. 2002; Shultz et al. 2005). Further improvement of the humanised mice models are being generated by the expression of human cytokines and HLA genes (reviewed in (Brehm et al. 2013)). Overall, these *in vivo* animal models have been used to test the self-renewal, long-term reconstitution and differentiation potential of human HSCs and have been fundamental to characterise the age-associated changes in HSC phenotype.

Human HSCs have been isolated based on the expression of a number of cell surface markers, the first being the sialomucin protein CD34 (Civin et al. 1984). However, the $CD34^+$ population consists of both multipotent and lineage-committed progenitors, and HSC can be further purified by the exclusion of mature lineage (Lin) and CD38 markers as $Lin^-CD34^+CD38^-$ cells (Bhatia et al. 1997; Hogan et al. 2002). Subsequent studies have shown that human HSCs express the CD90 (Thy1) antigen and identified the stem cells as $Lin^-CD34^+CD38^-CD90^+CD45RA^-$, while multipotent progenitors (MPPs) are found in the $Lin^-CD34^+CD38^-CD90^-CD45RA^-$ fraction ((Doulatov et al. 2010; Majeti et al. 2007) see Fig. 7.1). Human HSCs, like murine HSCs, also have the capacity to efflux vital dyes due to the expression of the P-glycoprotein pump and can be isolated as rhodamine (Rho) dull cells (McKenzie et al. 2007). More recently, using a combination of all these markers, together with the expression of the α6 integrin CD49f, Notta et al. have isolated $Lin^-CD34^+CD38^-CD45RA^-CD90^+Rho^{lo}CD49f^+$ cells from human cord blood, which are capable of lymphoid and myeloid repopulation at the single cell level and hence represent the purest population of long-term repopulating human HSCs to date (Notta et al. 2011).

7.3 Age-Associated Changes in the Human HSC Compartment

In humans, advancing age is accompanied by significant changes in the function and composition of mature blood cells. These observations, together with the observed decrease in activity of bone marrow with age and results from HSC transplantation showing that donor age is a strong negative risk factor (Kollman et al. 2001), suggest that the proliferative and regenerative capacity of human HSC diminishes with age. As reviewed in Chap. 6, changes in the HSC compartment with age have been well documented in mice. These age-related changes are primarily characterised by an increase in the number of HSCs but a decline in their function, as indicated by decreased repopulation capacity in competitive transplantation assays, as well as a reduction in the capacity to generate lymphoid lineages. However, differences in HSC aging have been observed between different mouse strains indicating that genetic variation plays an important role in HSC aging (Chen et al. 2000; de Haan and Van Zant 2009), making it difficult to directly extrapolate from these murine studies as to the situation in humans.

Until recently, the effects of age on HSCs in humans have been surprisingly poorly characterised. Early studies were based on indirect observations of committed progenitor or differentiated populations rather than the stem cells themselves (Marley et al. 1999; Gale et al. 1997). More recent studies, including work from our group, have sought to directly assess the frequency and in vitro and in vivo differentiation potential of HSC populations isolated from the bone marrow of young and elderly individuals (Taraldsrud et al. 2009; Beerman et al. 2010; Kuranda et al. 2011; Pang et al. 2011). These studies consistently found an increase in the frequency of immunophenotypic HSCs with age, assessed as either $CD34^{+}CD38^{-}$ or more stringently as $CD34^{+}CD38^{-}CD90^{+}CD45RA^{-}$ cell populations (Table 7.1). They further suggest that MPPs may also increase in elderly bone marrow, although to a lesser extent than primitive HSCs. In contrast, the pool of $CD34^{+}$ cells, which includes both multipotent and lineage-committed progenitors, remains unchanged (Table 7.1). Functional analysis did not reveal a significant difference in the engraftment efficiency of elderly compared to young $CD34^{+}$ cells in xenotransplantation experiments in NSG mice (Kuranda et al. 2011). However, since the frequency of $CD34^{+}CD38^{-}$ HSCs is increased in the $CD34^{+}$ population of the elderly, this suggests that there is a per cell decrease in stem cell repopulating activity with age. This was indeed found to be the case when limiting numbers of more purified $CD34^{+}CD38^{-}CD90^{+}CD45RA^{-}$ HSCs were transplanted in NSG mice (Pang et al. 2011), showing that in humans, as in mice, aging is associated with a decrease in HSC function.

In addition to decreased repopulating activity, $CD34^{+}CD38^{-}CD90^{+}CD45RA^{-}$ HSCs from elderly individuals appeared to be less efficient at reconstituting lymphoid cells and are more myeloid biased than young HSC in engraftment assays (Pang et al. 2011). Consistent with a loss of lymphoid potential, the frequency of early lymphoid progenitors (CLP and B/NK progenitors) and committed B cell precursors is reduced in human bone marrow with age, while myeloid lineage progenitors (early myeloid, CMP, GMP, MEP) and committed myeloid cells remain stable (Table 7.1). Furthermore, elderly HSCs exhibit a diminished capacity to generate B cells in vitro in differentiation assays on stromal cells (Kuranda et al. 2011; Pang et al. 2011). Age-related changes in the differentiation potential of elderly HSCs are associated with increased expression of genes specifying myeloid fate and function and downregulation of genes involved in lymphopoiesis (Pang et al. 2011). Altogether these studies clearly show that aging in humans is associated with an increase in phenotypic HSCs, but that the stem cells that accumulate with age are functionally deficient, both in terms of repopulating activity and differentiation potential (summarised in Fig. 7.2). In this respect, human HSC aging resembles the C57BL6 mouse model of aging (see Chap. 6).

The molecular mechanisms responsible for the changes in human HSC function with age are still being elucidated. In mice it has been demonstrated that HSCs are

Table 7.1 Summary of age-related changes in the frequency of haematopoietic progenitors in human bone morrow (found in the literature)

Cell type	Phenotype	% MNC		Fold change	BS	References
		Young	Old			
HSC	$CD34^+CD38^-$	0.1[a]	0.7[a]	7.0 ↑	$P < 0.01$	Taraldsrud et al. (2009)
	$CD34^+CD38^-$	0.11	0.45	4.1↑	$P < 0.005$	Kuranda et al. (2011)
	$Lin\text{-}CD34^+CD38^-CD90^+$	0.04	0.1	2.5↑	$P < 0.0002$	Beerman et al. (2010)
	$Lin\text{-}CD34^+CD38^-CD90^+CD45RA^-$	0.08	0.25	3.1↑	$P < 0.002$	Pang et al. (2011)
MPP	$Lin\text{-}CD34^+CD38^-CD90^-CD45RA^-$	0.18	0.23	1.3	ns	Pang et al. (2011)
	$CD34^+CD38^-CD10^-CD90^-CD45RA^-$	0.11	0.22	2.0	ns	Kuranda et al. (2011)
MLP	$CD34^+CD38^-CD10^+CD90^-CD45RA^+$	0.01	0.02	2.0↑	$P < 0.005$	Kuranda et al. (2011)
CLP	$Lin\text{-}CD34^+CD38^+CD127^+$	0.08	0.04	2.0↓	*P < 0.05*	Pang et al. (2011)
B/NK	$CD34^+CD38^+CD90^-CD45RA^+CD10^+$	0.44[a]	0.08[a]	5.5↓	*P < 0.02*	Kuranda et al. (2011)
B committed	$CD34^+CD19^+$	0.44[a]	0.22[a]	2.0↓	*P < 0.01*	Kuranda et al. (2011)
B cells	$CD19^+$	3.2	2.5	1.3	ns	Kuranda et al. (2011)
CMP	$Lin\text{-}CD34^+CD38^+CD123^+CD45RA^-$	0.045	0.045	1.0	ns	Pang et al. (2011)
GMP	$Lin\text{-}CD34^+CD38^+CD123^+CD45RA^+$	0.04	0.04	1.0	ns	Pang et al. (2011)
MEP	$Lin\text{-}CD34^+CD38^+CD123^-CD45RA^-$	0.04	0.038	1.1	ns	Pang et al. (2011)
Early myeloid	$CD34^+CD38^+CD90^-CD45RA^{+/-}CD10^-$	0.8[a]	0.7[a]	1.1	ns	Kuranda et al. (2011)
Myeloid committed	$CD34^+CD33^+$	0.9[a]	1.2[a]	1.3	ns	Kuranda et al. (2011)
$CD34^+$progeniters	$CD34^+$	−2	−2	1.0	ns	Kuranda et al. (2011) and Pang et al. (2011)
$CD34^+$progeniters	$CD34^+$	3.5	3.6	1.0	ns	Taraldsrud et al. (2009)

Note: Percentage of cells obtained from graphs of cited papers. Statistics are stated in the cited paper. Different statistics tests were used

Abbreviations: *HSC* hematopoietic stem cell, *MPP* multipotent progenitors, *MLP* multilymphoid progenitors, *CLP* common lymphoid progenitors, *CMP* common myeloid progenitors, *GMP* granulocyte-macrophage progenitors, *MEP* megakaryocyte-erythrocyte progenitors, *MNC* mononuclear cells, *SS* statistical significance, *ns* not significant

[a]Percentage of MNC recalculated from % if $CD34^+$cells

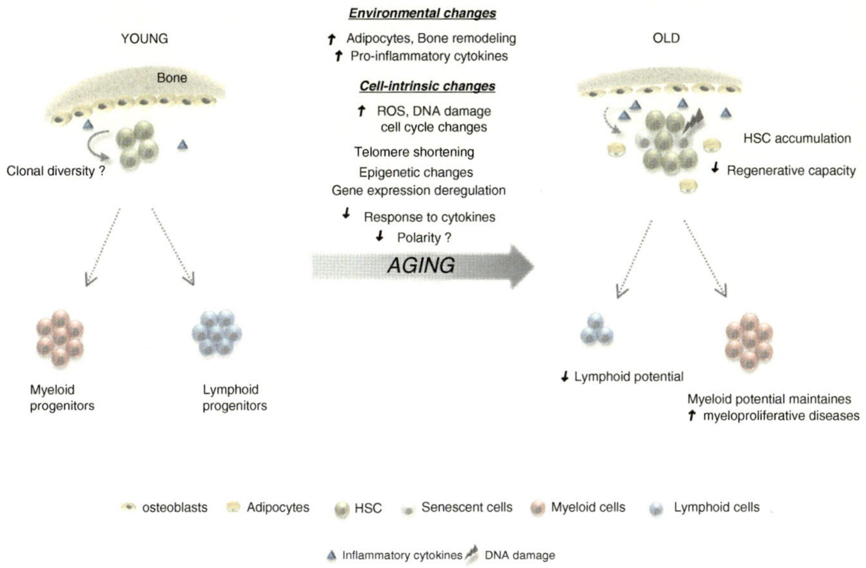

Fig. 7.2 Overview of human HSC aging. Although phenotypic HSC increase with age, both regenerative capacity and lymphoid potential of HSC is diminished. In contrast, myeloid potential is stable or even increased, leading to a propensity towards myeloproliferative disease. Potential mechanisms leading to the decline of HSC functions are indicated above the arrow and include both intrinsic modifications of HSC and environmental changes within the bone marrow milieu. Decreased cell polarity with age has not yet been demonstrated in human HSC

not homogeneous and that they can be further stratified into subpopulations which have a myeloid-biased or lymphoid-biased/balanced differentiation potential (Muller-Sieburg et al. 2002; Challen et al. 2010; Dykstra et al. 2007). Continuing from this observation, it was shown that the age-associated myeloid skewing in output observed in mice reflects a selective expansion of myeloid-biased stem cells in the aging HSC pool (see Chap. 6). At present, the existence of distinct lineage-biased subpopulations has not been demonstrated directly in human HSCs. It therefore cannot yet be stated conclusively whether the age-dependent increase in myeloid output seen in humans represents a relative shift in distinct clonal HSC subpopulations as in the mouse (the so-called clonal-shift model of HSC aging) or is the result of uniform age-associated changes in all stem cells.

7.4 Molecular Mechanisms of HSC Aging

7.4.1 Accumulation of DNA Damage

DNA damage is recognised as one of the major mechanisms driving age-associated decline in cell function and thereby leading to eventual tissue and organismal failure (reviewed in (Lombard et al. 2005)). Both human and mouse aging are associated with a decline in the efficiency of DNA repair pathways (Gorbunova et al. 2007) and an accumulation of various types of DNA damage in somatic tissues (Sedelnikova et al. 2004, 2008; Wang et al. 2009). A key role for DNA damage in aging is evidenced by the observation that many human premature aging syndromes are associated with defects in the normal mechanisms of genome surveillance and repair (Best 2009). DNA damage within stem cell pools is likely to be particularly damaging to organismal viability, as it has the potential to not only impact upon the size of the stem cell population itself but also upon the function and integrity of progenitor lineages.

Two studies have investigated DNA damage in aging human HSCs by comparing the number of foci of phosphorylated histone H2AX (γH2AX), an early response to DNA double-stranded breaks (DSB), in immunopurified haematopoietic stem populations from humans of different ages (Rube et al. 2011; Yahata et al. 2011). Both studies demonstrated that aging was associated with a significant increase in the frequency of γH2AX foci in $CD34^{+}CD38^{-}$ HSCs. The study by Yahata et al. (2011) demonstrated that this age-associated increase in DSB was correlated with an increase in the intracellular concentration of reactive oxygen species (ROS), suggesting that oxidative stress is an important driver of DNA damage in aging HSCs. Further, they demonstrated that experimentally inducing ROS and DNA damage in HSCs led to a decrease in HSC reconstitution potential. As ROS are a normal cell-intrinsic by-product of oxidative metabolism and cell cycling, it is possible that this increase in ROS and DNA damage reflects an increase in the portion of HSCs that have entered cycle with age. Similarly, increased ROS and DNA damage is also observed when HSCs are forced into cycle by transplantation, and this is also correlated with loss of HSC repopulation capacity (Yahata et al. 2011). Overall, this study demonstrates that DNA damage due to oxidative stress does increase in human HSCs with age and that this damage impacts upon normal HSC function, in terms of reconstitution in engraftment assays. Consistent with these findings, gene expression profiling in human HSCs indicated that aging is associated with the upregulation of genes involved in DNA damage repair, hinting at the possibility that aged HSCs have activated repair pathways (Pang et al. 2011). As discussed above, the number of human HSCs is increased with age, suggesting that the effects of DNA damage are not simply due to cytotoxic and cytostatic depletion of the stem cell pool and are more likely to reflect functional changes. It will be interesting to determine whether the effects of increasing intracellular ROS and DNA damage also induce other aspects of HSC aging, such as changes in differentiation potential and an increased incidence of mutations and chromosomal translocations observed in haematological malignancy.

Surprisingly, several studies have revealed important differences between how mouse and human HSCs respond to DNA damage. In response to low levels of ionising radiation, normally quiescent mouse haematopoietic stem and progenitor cells ($Lin^{-}c\text{-}Kit^{+}Sca\text{-}1^{+}Flk2^{-}$) appear relatively resistant to apoptosis and instead survive the damage by undergoing non-homologous end joining, which is an innately error-prone DNA repair pathway (Mohrin et al. 2010). In contrast, human HSCs ($Lin^{-}CD34^{+}CD38^{-}$) appear to adopt apoptosis as the default programme in response to radiation-induced damage (Milyavsky et al. 2010). Thus, these two studies suggest that humans and mice have adopted different approaches to survive genotoxic damage within the HSC compartment, with mice favouring immediate survival of the stem cell pool, but at the risk to long-term fitness in the form of mutagenic change during repair, and humans favouring removal of the damaged cells. It remains to be seen whether this surprising observation reflects technical differences between these studies or is indicative of an evolutionary response to the different life spans of these two mammals (Lane and Scadden 2010).

7.4.2 Telomere Shortening

As well as DSB, another form of DNA damage that has been characterised with age in human HSCs is telomere shortening. In most human somatic cell types, the telomere loses some 50–200 bp of terminal repeat sequence with every round of cell division as a result of incomplete end replication (reviewed in (Greenwood and Lansdorp 2003)). When telomeres reach a critically short threshold, cells are triggered to either enter a senescent state or to undergo apoptosis (Bodnar et al. 1998). In some cases, and particularly in the germ line, telomeres are protected against shortening by the activity of telomerase (Morin 1989). While telomerase is not expressed in most somatic tissues (Kim et al. 1994), telomerase activity is readily detectable in both human ($CD34^{+}CD38^{-}$) and mouse ($Thy\text{-}1.1^{lo}Sca\text{-}1^{+}Lin^{-}Mac\text{-}1^{-}CD4^{-}c\text{-}kit^{+}$) HSCs and haematopoietic progenitors, albeit at low levels (Yui et al. 1998; Morrison et al. 1996; Broccoli et al. 1995). Despite this telomerase competence, telomere shortening has been observed with advancing age in human HSCs and progenitor populations (Vaziri et al. 1994) suggesting that these low levels of telomerase activity are not sufficient to completely protect against telomere shortening in the HSCs of aged individuals. That telomere maintenance is critical for normal HSC function throughout life is suggested by the observation of acute bone marrow failure in dyskeratosis congenita and acquired aplastic anaemia, diseases associated with deficiencies in telomerase core components (reviewed in (Townsley et al. 2014)). Further, telomerase knockout mice exhibit impaired HSC function including a myeloid-lymphoid skewing (Ju et al. 2007) typical of normal age-associated changes in both mouse and human HSCs. However, this skewed phenotype is mainly due to changes extrinsic to the HSC themselves, including effects of telomere shortening on the stem cell niche (Ju et al. 2007) and the systemic environment (Song et al. 2010). Thus, the intrinsic effects of telomere shortening on HSC function are currently unclear.

7.4.3 Cell Cycle Changes and Senescence

HSCs are predominantly a quiescent population. In humans it has been shown that only a small proportion (~10 %) of HSCs are in cycle at any time (Pang et al. 2011; Laurenti et al. 2015). This primarily quiescent state appears important to minimise replication-associated mutations which could otherwise accumulate in this long-lived cell type, and in mice it has been shown that disruption of HSC quiescence leads to the accumulation of DNA damage and eventual stem cell exhaustion (Cheng et al. 2000; Walter et al. 2015). In a recent study, cell cycle profiling of human HSCs ($Lin^{-}CD34^{+}CD38^{-}CD90^{+}CD45RA^{-}$) isolated from the bone marrows of young (20–35 years) and old (>65 years) individuals revealed that aging was associated with a significant increase in the proportion of cells which had exited the quiescent (G_0) state and entered G_1 (Pang et al. 2011). Consistent with this observation, this study also found that among the genes upregulated with aging, there was significant enrichment for genes involved in cell cycle progression.

The relationship between quiescence and DNA damage is complex. While ostensibly protecting against replication-driven DNA damage, it appears that this quiescent state may also contribute to the accrual of long-term DNA damage with age, due to the attenuation of checkpoint control and DNA damage response pathways in quiescent cells (Rossi et al. 2007; Beerman et al. 2014). Consistent with this, it appears that the cells emerging from quiescence with age are functionally attenuated and may be undergoing cell cycle arrest in G_1 due to the activation of cell cycle checkpoints in response to DNA damage, leading to an eventual senescence-like state. In support of this, the age-associated increase in the number of HSCs in G_1 does not translate into expansion of downstream progenitors ((Kuranda et al. 2011; Pang et al. 2011) and Table 7.1). Furthermore, it has been reported that expression of the G_1 checkpoint protein $p16^{INK4a}$ increases with age in the HSC compartment and that $p16^{INK4a}$ deletion improves HSC function in old animals (Janzen et al. 2006). It will be interesting in the future to isolate live HSCs at different stages of the cell cycle, so that the properties of this emerging G_1 population can be investigated directly at the molecular and functional levels.

7.4.4 Epigenetic Changes

Epigenetics refers to forms of information that regulate gene expression and thereby influence cellular phenotype, but that does not involve modifications of the DNA sequence itself. Epigenetic signals can be extremely plastic, making them ideal for controlling dynamic changes in gene expression profiles that drive cell fate determination, lineage commitment and differentiation. The most important forms of epigenetic information known at present are changes in chromatin structure due to chemical modifications of DNA (DNA methylation and hydroxymethylation), covalent modifications of the histone proteins and remodelling of nucleosomes, as well as the subnuclear localisation of genes. Widespread changes in epigenetic chromatin structure and nuclear organisation have been observed during aging in a number of organisms, including yeast, *C. elegans*, *Drosophila* and mammals (reviewed in (O'Sullivan and Karlseder 2012)).

It is now clear from studies in human and mouse that epigenetic signals are critically important in maintaining HSC differentiation plasticity and multipotentiality by priming cells for the expression of different lineage-associated transcription programmes and that this permissive epigenetic programme becomes progressively restricted following lineage choice and commitment (Maes et al. 2008; Weishaupt et al. 2010). Thus, perturbation of the normal epigenetic state of HSCs could potentially have significant effects on phenotypic plasticity and differentiation output, similar to those observed with aging. Indeed, recently published high-throughput epigenomic profiling studies in mice have demonstrated that aging is associated with genome-wide changes in the epigenetic state of mouse HSCs, contributing to gene expression changes that underlie some of the observed functional alterations, including perturbed self-renewal and differentiation potential (Beerman et al. 2013; Taiwo et al. 2013; Sun et al. 2014).

While comparable epigenomic studies are still awaited in humans, emerging lines of evidence suggest that, as in the mouse, epigenetic change is also likely to be an important driver of the functional changes observed with age in human HSCs. Perhaps the epigenetic signal, which has been best characterised during HSC aging in humans, is DNA methylation. DNA methylation is the addition of a methyl group to the pyrimidine ring of cytosine residues, which are usually located within the symmetrical dinucleotide CpG, to form 5-methylcytosine (5-meC). Methylation of genomic DNA is associated with transcriptional repression and is a key regulatory mark in both normal development and in disease (Schubeler 2015). The genome-wide levels of promoter DNA methylation have been compared between human $CD34^+$ haematopoietic progenitors from cord blood and adult bone marrow (Bocker et al. 2011). This study found significant age-associated changes in DNA methylation, which exhibited a bimodal distribution; ~60 % of changed genes showed depleted methylation with age, and ~40 % had gained methylation. Interestingly, there was a significant overlap between genes which became hypomethylated in human HSPC with age and those that become hypomethylated during myeloid differentiation, leading the authors to speculate that these age-associated methylation changes may explain the myeloid-biased output observed in aged HSCs, by allowing the premature activation of the myeloid programme and loss of stem cell plasticity (Bocker et al. 2011). These findings now need to be confirmed in more purified populations of human HSCs and by comparing adult HSCs from different ages, but are highly consistent with the bimodal pattern of DNA methylation changes with age which have been reported in adult mouse HSCs (Taiwo et al. 2013; Sun et al. 2014) as well as the functional changes observed in mice lacking DNA methyltransferases (Challen et al. 2014). Together these studies suggest that DNA methylation could play a key role in maintaining HSC plasticity and suggest that age-associated changes in DNA methylation could underlie important aspects of the aging HSC phenotype.

Covalent histone modifications have been shown to play a key role in stem cell plasticity and lineage choice (Bernstein et al. 2006; Mikkelsen et al. 2007). The study of histone modifications in human HSCs has shown that lymphoid and myeloid loci that are not expressed in HSCs, but that will be expressed in downstream committed progenitors, are marked by high levels of active histone modifications (H3K4me2 and H4ac) in HSCs (Maes et al. 2008). These studies support the idea that priming of lineage-specific loci in HSC precedes lineage commitment and that these epigenetic modifications contribute to the ability of the stem cell to generate both lymphoid and myeloid lineages. Further, it was shown that these histone modifications precede changes in subnuclear localisation during activation of haematopoietic loci (Guillemin et al. 2009). Given the importance of histone modifications in establishing permissive chromatin states in HSCs, age-associated changes in these histone marks could affect the differentiation potential of the stem cell. Indeed, it has recently been shown in mice that HSC aging is associated with genome-wide changes in both activating (H3K4me3) and repressive (H3K27me3) histone modifications. In particular, with age there is an increase in the number of peaks of H3K4me3, as well as an expansion and strengthening of existing peaks, changes which positively correlate with increased gene expression (Sun et al. 2014). It will now be important to carry out similar genome-wide studies in human HSCs and to characterise the functional significance of these epigenetic changes on the HSC phenotype. Moreover, there is a need to extend these analyses to include other histone modifications and to profile aspects of chromatin compaction and heterochromatin status. Intriguingly, loss of heterochromatin structure and its associated histone mark H3K9me3 has been observed during aging in a number of organisms, leading to the proposal that a decline in heterochromatin integrity may be a common feature of aging (Tsurumi and Li 2012). That age-associated loss of heterochromatin is likely to be functionally important in driving the aging phenotype is strongly supported by the observation that similar widespread changes are observed in human syndromes of premature aging such as the Hutchinson-Gilford progeria syndrome (HGPS) (Shumaker et al. 2006). Extensive changes in nuclear architecture are also observed during cellular senescence, which is closely linked to tissue and organismal aging (reviewed in (Chandra and Narita 2013)), again strengthening the idea that the regulation of heterochromatin is likely to play an important role in the process of aging.

Another dimension of epigenetic information that has emerged relatively recently is noncoding and regulatory RNAs, which are highly conserved through evolution and appear to be important regulators of gene expression. miRNAs are a class of small noncoding RNAs of approximately 22 nucleotides in length that mediate post-transcriptional downregulation of gene expression either by translational inhibition or mRNA degradation (Wu and Belasco 2008; Filipowicz et al. 2008). There is substantial evidence that the expression of miRNAs is altered during aging and cellular senescence (Smith-Vikos and Slack 2012). While there have been no studies on the role of miRNAs in aging in the HSC compartment to date, studies in both mice and humans have identified miRNAs which regulate aspects of HSC biology that are perturbed with age, including cell cycle control (Lechman et al. 2012), stem cell pool size (Guo et al. 2010), engraftment capacity (O'Connell et al. 2010; Ooi et al. 2010), and lymphoid vs. myeloid differentiation potential (Ooi et al. 2010; Gerrits et al. 2012). It will now be of interest to compare the expression of these and other miRNAs between HSC isolated from young and old individuals. Another class of noncoding RNA, the long noncoding RNAs (lncRNA), can also act to regulate expression of downstream target genes, frequently by modifying their epigenetic state and chromatin structure (reviewed in (Batista and Chang 2013; Yang et al.

2014)). Recent studies in mice have identified lncRNAs with HSC- or HSC/MPP-restricted expression profiles (Cabezas-Wallscheid et al. 2014; Luo et al. 2015), and several of these have been shown to regulate important aspects of HSC biology, including self-renewal and lineage commitment. Intriguingly, a significant proportion of these molecules showed altered expression with age (Luo et al. 2015). However, at present the impact of lncRNAs on human HSC biology, and their involvement in the aging phenotype, has not yet been thoroughly explored.

7.4.5 Responses to Bone Marrow Environment and Inflammation

The above sections have discussed molecular pathways driving HSC aging, which are intrinsic to the stem cells themselves. It is increasingly clear that cell extrinsic factors associated with an aging bone marrow microenvironment also influence the changes in HSC function observed with age. In humans, as in mice, aging is associated with changes in bone density and bone marrow cellular composition (Syed and Hoey 2010). Furthermore, there is an age-related increase in adipogenesis in the bone marrow, which has been shown to exert a negative effect on HSC localisation and function (Naveiras et al. 2009; Tuljapurkar et al. 2011). In the mouse model, it has been demonstrated that an aged microenvironment alone is sufficient to induce changes in the properties of HSPC from young animals towards an aging phenotype, including a decline in clonality and myeloid skewing (Vas et al. 2012). This reduced clonality is in turn believed to underlie the emergence with age of clones bearing oncogenic-predisposing mutations and the age-associated increase in myeloid leukaemias (see below). Similarly, our group has recently demonstrated that changes in the bone microstructure induced in a mouse model of osteoporosis result in a reduction of B lymphoid output, similar to that which is observed with age (Lescale et al. 2015). Furthermore, aging is associated with reduced adherence of HSPC to bone marrow stromal cells, which has been attributed to elevated levels of the Rho GTPase Cdc42 in HSCs from old mice (Xing et al. 2006; Kohler et al. 2009). Strikingly, this increase of Cdc42 leads to loss of stem cell polarity and has been shown to play a critical role in the aging phenotype (Florian et al. 2012). The age-associated upregulation of Cdc42 is driven by a shift towards non-canonical Wnt signalling in HSCs from old mice (Florian et al. 2013). While the contribution of Cdc42 to HSC aging has not yet been investigated in humans, recent evidence suggests that Wnt signalling may also be perturbed with age in human HSCs (Khoo et al. 2014).

As well as changes in cellularity, aging in humans is also associated with an increase in the serum concentration of inflammatory cytokines even in the absence of acute infection or physiological stress (Singh and Newman 2011). This age-associated low-grade chronic inflammation, known as "inflammaging," is believed to have an impact upon the aging bone marrow microenvironment (Franceschi et al. 2007; Abdelmagid et al. 2015). Indeed, elevated levels of the inflammatory cytokines Rantes/Ccl5 and Mig (monokine induced by IFNγ) are observed in the bone marrow of old mice (Ergen et al. 2012), and HSCs showed an upregulation with age of genes involved in the inflammatory response, consistent with the idea that these cells are exposed to an inflammatory microenvironment (Chambers et al. 2007). There is accumulating evidence that HSCs are very sensitive to diverse inflammatory cytokines and stress signals. In particular, the major pro-inflammatory cytokine TNFα, which stimulates downstream activation of NFkB, results in an impaired maintenance of human HSCs, a myeloid-skewed differentiation output and overall depletion of reconstitution potential in xenograft models (Dybedal et al. 2001). Excessive signalling by TNFα and IFNγ has been associated with human HSC disorders, including some cases of bone marrow failure and myelodysplastic syndrome (Kitagawa et al. 1997). Further, HSCs also express Toll-like receptors (TLRs) through which they respond directly to ligands from pathogenic microbes, and it has been shown that TLR stimulation by microbial ligands biases the differentiation potential of human HSCs in favour of myeloid commitment at the expense of lymphoid output, a mechanism which is thought to be important for the rapid production of effector cells of innate immunity (De Luca et al. 2009). Overall, given that the primary responses of HSCs to inflammatory cytokines and infectious signals (impaired reconstitution potential and myeloid-biased differentiation output) are very similar to the changes observed in HSC function with age, it is likely that pro-inflammatory environment of the aging bone marrow is an important driver of human HSC aging (reviewed in (Woolthuis et al. 2011)). As inflammatory cytokines such as TNFα and IFNγ also perturb the differentiation potential of MSCs in favour of adipogenesis at the expense of osteoblastogenesis (Laschober et al. 2011), the inflammatory milieu of the aged bone marrow also contributes to the altered cellularity of the bone marrow niche described above, which in turn impacts upon HSC function.

7.5 Clinical Impact of Human HSC Aging and Future Prospects

Possibly the most important effect of HSC aging in terms of health, longevity and quality of life in the aged result from its contribution to the phenomenon of immunosenescence, the general decline or dysregulation of immune function which is observed with age. While immunosenescence is a catch-all phrase encompassing a wide range of immune-associated abnormalities, probably the most clinically important aspect of immunosenescence is a decline in adaptive immunity (Hakim and Gress 2007). The elderly exhibit a reduced capacity to mount effective immune responses to newly encountered pathogens, which contributes to an increased frequency and severity of bacterial and viral infections (Thompson et al. 2003; Janssens and Krause 2004). Further, this enhanced susceptibility to opportunistic infection is exacerbated by an impaired response to prophylactic vaccines in terms of both the quantity of antibody produced (Bouree 2003; Sasaki et al. 2011) and cell-mediated immune responses (Kang et al. 2004). While this decline in adaptive immunity is also caused by qualitative changes in mature T and B cell populations (Dorshkind and Swain 2009; Gruver et al. 2007), it is believed that functional changes in the aging HSC compartment, in particular the age-associated decline in lymphoid differentiation potential, contributes to this immune senescence.

As in most tissues, aging is associated with an almost exponential increase in the incidence of haematological malignancy (Edwards et al. 2002). In contrast to the short-lived committed progenitors and differentiated effector blood cells, it is believed that the long-lived HSCs are the likely target for the multiple genetic and/or epigenetic hits leading to oncogenic transformation in several malignancies (Kikushige and Miyamoto 2014). Among the genes which are upregulated with age in human HSCs, there is a significant enrichment for genes involved in age-associated haematological disease and in particular genes implicated in myeloid malignancies and acute myeloid leukaemia signalling, including *AURKA*, *HOXA9* and *MYC* (Pang et al. 2011). The general reduction in stem cell fitness and concomitant reduced competition within the stem cell niche is likely to contribute to the emergence with age of stem clones bearing adaptive oncogenic mutations (Marusyk and DeGregori 2008). Indeed recent reports have demonstrated that clonal haematopoiesis and the subsequent emergence of pre-leukaemic clones bearing oncogenic mutations (particularly in epigenetic modifiers such as DNMT3A, TET2 and ASXL1) is an important aspect of aging (Jaiswal et al. 2014; Xie et al. 2014). Further, as well as effects on haematological malignancy, an indirect effect of the age-associated decline in immune function could be deficient immune surveillance and a decline in immune response to non-haematological neoplasms, for which there is evidence in aging mice (Gruver et al. 2007).

HSC aging can also impact upon the health of the elderly in a number of other ways. The enhanced myeloid output of HSC observed with age is likely to contribute to the increased tendency towards inflammatory disease, which is observed in the elderly. Changes in HSC function could also affect the recovery of homeostasis after stress or injury in the elderly, and HSC aging has been suggested as the cause of the higher rate of unexplained anaemia observed with age (Guralnik et al. 2004). This is consistent with the reduced repopulating activity of HSC from aged donors observed in xenografts (Pang et al. 2011). Finally another very important clinical corollary of HSC aging arises in the context of HSC transplantations, where it is known that increased donor age is a significant prognostic indicator of poor outcome (Kollman et al. 2001). The shorter telomeres in HSCs may contribute to the reduced success of HSC transplants involving elderly donors, as the transplant process induces significant further shortening in engrafted cells (Notaro et al. 1997). As the transplant process also causes an increase in ROS and DNA damage in transplanted HSCs (Yahata et al. 2011), the higher baseline levels of accumulated damage in HSCs from elderly donors could also impact upon engraftment success.

Elucidating the mechanisms of HSC aging promises to reveal opportunities for potential therapeutic intervention to improve the health of the elderly. The relative accessibility of HSCs, together with the existence of robust protocols for HSC transplantation opens the door for the transplant of autologous HSCs that have been genetically or pharmacologically treated ex vivo. But how real are the prospects for making old HSCs young again? Recently, this field has been stimulated by studies of Florian et al. demonstrating reversal of the aging phenotype in murine HSCs by inhibition of the Rho GTPase Cdc42 and non-canonical Wnt signalling (Florian et al. 2012, 2013). Furthermore the strong implication of epigenetic pathways in the aging HSC phenotype presents particularly exciting prospects for clinical intervention. Unlike genetic events, epigenetic changes are plastic and potentially reversible by pharmacological intervention. Epigenetic therapeutics represents a major area of research interest, and several molecules targeting epigenetic processes are currently being tested at both preclinical and clinical stages (Tan et al. 2007; Creasy et al. 2012). However, as epigenetic modifiers are ubiquitously expressed, one of the disadvantages of therapeutically targeting the epigenetic processes themselves is that there is a high potential for nonspecificity and widespread deleterious side effects. For this reason, the potential implication of noncoding RNAs in the process of HSC aging is particularly interesting, because such regulatory RNAs could present therapeutic targets that offer a higher degree of specificity than the chromatin modifiers themselves. Taking an alternative approach, it was recently demonstrated that the epigenetic state could be reprogrammed by dedifferentiating aged HSCs to form induced pluripotent stem (iPS) cells (Wahlestedt et al. 2013). HSCs derived from these iPS cells exhibited a repopulating potential in competitive transplant assays, which was similar to young HSCs. The increase in ROS and DNA damage which is observed in HSCs following transplants (Yahata et al. 2011) has suggested that clinical transplant protocols may benefit from the addition of antioxidant therapy to reduce stem cell damage and improve engraftment (Hao et al. 2011). Indeed in mice it was shown that feeding transplant recipients on a diet supplemented with the antioxidant N-acetyl-L-cysteine during the reconstitution period led to a significant decrease in DNA damage in incoming HSCs and a concomitant improvement in quantity and quality of engraftment (Yahata et al. 2011).

7.6 Concluding Remarks

Although still in its infancy, the study of HSC aging in humans has progressed in recent years, and the causes and consequences of HSC aging are beginning to emerge (summarised in Fig. 7.2). Guided by observations from mice and exploiting improvements in isolation procedures, functional models and recent technological advances which have rendered large-scale genomic analyses accessible to the rare HSC population, the coming years promise significant further advance (Doulatov et al. 2012). In particular, it will now be important to carry out genome-scale methylome and chromatin profiling analyses on HSCs from different ages and to investigate the potential role played by more recently discovered epigenetic signals such as ncRNAs. As haematological decline has important consequences for the health of the elderly, there is a compelling clinical need for further rapid progress, and this promises to be a stimulating and dynamic field in the years to come.

References

Abdelmagid SM, Barbe MF, Safadi FF (2015) Role of inflammation in the aging bones. Life Sci 123C:25–34

Batista PJ, Chang HY (2013) Long noncoding RNAs: cellular address codes in development and disease. Cell 152(6):1298–1307

Baum CM et al (1992) Isolation of a candidate human hematopoietic stem-cell population. Proc Natl Acad Sci U S A 89(7):2804–2808

Beerman I et al (2010) Stem cells and the aging hematopoietic system. Curr Opin Immunol 22(4):500–506

Beerman I et al (2013) Proliferation-dependent alterations of the DNA methylation landscape underlie hematopoietic stem cell aging. Cell Stem Cell 12(4):413–425

Beerman I et al (2014) Quiescent hematopoietic stem cells accumulate DNA damage during aging that is repaired upon entry into cell cycle. Cell Stem Cell 15(1):37–50

Berardi AC et al (1997) Individual CD34+CD38lowCD19-CD10- progenitor cells from human cord blood generate B lymphocytes and granulocytes. Blood 89(10):3554–3564

Bernstein BE et al (2006) A bivalent chromatin structure marks key developmental genes in embryonic stem cells. Cell 125(2):315–326

Best BP (2009) Nuclear DNA damage as a direct cause of aging. Rejuvenation Res 12(3):199–208

Bhatia M et al (1997) Purification of primitive human hematopoietic cells capable of repopulating immune-deficient mice. Proc Natl Acad Sci U S A 94(10):5320–5325

Bocker MT et al (2011) Genome-wide promoter DNA methylation dynamics of human hematopoietic progenitor cells during differentiation and aging. Blood 117(19):e182–e189

Bodnar AG et al (1998) Extension of life-span by introduction of telomerase into normal human cells. Science 279(5349):349–352

Bouree P (2003) Immunity and immunization in elderly. Pathol Biol (Paris) 51(10):581–585

Brehm MA et al (2013) Overcoming current limitations in humanized mouse research. J Infect Dis 208(Suppl 2):S125–S130

Broccoli D, Young JW, de Lange T (1995) Telomerase activity in normal and malignant hematopoietic cells. Proc Natl Acad Sci U S A 92(20):9082–9086

Cabezas-Wallscheid N et al (2014) Identification of regulatory networks in HSCs and their immediate progeny via integrated proteome, transcriptome, and DNA Methylome analysis. Cell Stem Cell 15(4):507–522

Challen GA et al (2010) Distinct hematopoietic stem cell subtypes are differentially regulated by TGF-beta1. Cell Stem Cell 6(3):265–278

Challen GA et al (2014) Dnmt3a and Dnmt3b have overlapping and distinct functions in hematopoietic stem cells. Cell Stem Cell 15(3):350–364

Chambers SM et al (2007) Aging hematopoietic stem cells decline in function and exhibit epigenetic dysregulation. PLoS Biol 5(8), e201

Chandra T, Narita M (2013) High-order chromatin structure and the epigenome in SAHFs. Nucleus 4(1):23–28

Chen J, Astle CM, Harrison DE (2000) Genetic regulation of primitive hematopoietic stem cell senescence. Exp Hematol 28(4):442–450

Cheng T et al (2000) Hematopoietic stem cell quiescence maintained by p21cip1/waf1. Science 287(5459):1804–1808

Civin CI et al (1984) Antigenic analysis of hematopoiesis. III. A hematopoietic progenitor cell surface antigen defined by a monoclonal antibody raised against KG-1a cells. J Immunol 133(1):157–165

Creasy CL et al (2012) A novel selective EZH2 inhibitor exhibits anti-tumor activity in lymphoma with EZH2 activating mutations. In: 103rd annual meeting of the American Association for Cancer Research. AACR, Chicago

de Haan G, Van Zant G (2009) Dynamic changes in mouse hematopoietic stem cell numbers during aging. Blood 93(10):3294–3301

De Luca K et al (2009) The TLR1/2 agonist PAM(3)CSK(4) instructs commitment of human hematopoietic stem cells to a myeloid cell fate. Leukemia 23(11):2063–2074
Demetrius L (2005) Of mice and men. When it comes to studying ageing and the means to slow it down, mice are not just small humans. EMBO Rep 6 Spec No:S39–44
Demetrius L (2006) Aging in mouse and human systems: a comparative study. Ann N Y Acad Sci 1067:66–82
Dorshkind K, Swain S (2009) Age-associated declines in immune system development and function: causes, consequences, and reversal. Curr Opin Immunol 21(4):404–407
Doulatov S et al (2010) Revised map of the human progenitor hierarchy shows the origin of macrophages and dendritic cells in early lymphoid development. Nat Immunol 11(7):585–593
Doulatov S et al (2012) Hematopoiesis: a human perspective. Cell Stem Cell 10(2):120–136
Dybedal I et al (2001) Tumor necrosis factor (TNF)-mediated activation of the p55 TNF receptor negatively regulates maintenance of cycling reconstituting human hematopoietic stem cells. Blood 98(6):1782–1791
Dykstra B et al (2007) Long-term propagation of distinct hematopoietic differentiation programs in vivo. Cell Stem Cell 1(2):218–229
Edwards BK et al (2002) Annual report to the nation on the status of cancer, 1973–1999, featuring implications of age and aging on U.S. cancer burden. Cancer 94(10):2766–2792
Ergen AV, Boles NC, Goodell MA (2012) Rantes/Ccl5 influences hematopoietic stem cell subtypes and causes myeloid skewing. Blood 119(11):2500–2509
Filipowicz W, Bhattacharyya SN, Sonenberg N (2008) Mechanisms of post-transcriptional regulation by microRNAs: are the answers in sight? Nat Rev Genet 9(2):102–114
Florian MC et al (2012) Cdc42 activity regulates hematopoietic stem cell aging and rejuvenation. Cell Stem Cell 10(5):520–530
Florian MC et al (2013) A canonical to non-canonical Wnt signalling switch in hematopoietic stem-cell ageing. Nature 503(7476):392–396
Franceschi C et al (2007) Inflammaging and anti-inflammaging: a systemic perspective on aging and longevity emerged from studies in humans. Mech Ageing Dev 128(1):92–105
Gale RE et al (1997) Acquired skewing of X-chromosome inactivation patterns in myeloid cells of the elderly suggests stochastic clonal loss with age. Br J Haematol 98(3):512–519
Gerrits A et al (2012) Genetic screen identifies microRNA cluster 99b/let-7e/125a as a regulator of primitive hematopoietic cells. Blood 119(2):377–387
Gorbunova V et al (2007) Changes in DNA repair during aging. Nucleic Acids Res 35(22): 7466–7474
Goyama S, Wunderlich M, Mulloy JC (2015) Xenograft models for normal and malignant stem cells. Blood 125:2630–2640
Greenwood MJ, Lansdorp PM (2003) Telomeres, telomerase, and hematopoietic stem cell biology. Arch Med Res 34(6):489–495
Gruver AL, Hudson LL, Sempowski GD (2007) Immunosenescence of ageing. J Pathol 211(2):144–156
Guillemin C et al (2009) Chromatin modifications in hematopoietic multipotent and committed progenitors are independent of gene subnuclear positioning relative to repressive compartments. Stem Cells 27(1):108–115
Guo S et al (2010) MicroRNA miR-125a controls hematopoietic stem cell number. Proc Natl Acad Sci U S A 107(32):14229–14234
Guralnik JM et al (2004) Prevalence of anemia in persons 65 years and older in the United States: evidence for a high rate of unexplained anemia. Blood 104(8):2263–2268
Hakim FT, Gress RE (2007) Immunosenescence: deficits in adaptive immunity in the elderly. Tissue Antigens 70(3):179–189
Hao Y et al (2011) Antioxidant intervention: a new method for improving hematopoietic reconstitution capacity of peripheral blood stem cells. Med Hypotheses 76(3):421–423
Hogan CJ, Shpall EJ, Keller G (2002) Differential long-term and multilineage engraftment potential from subfractions of human CD34+ cord blood cells transplanted into NOD/SCID mice. Proc Natl Acad Sci U S A 99(1):413–418

Ito M et al (2002) NOD/SCID/gamma(c)(null) mouse: an excellent recipient mouse model for engraftment of human cells. Blood 100(9):3175–3182

Jaiswal S et al (2014) Age-related clonal hematopoiesis associated with adverse outcomes. N Engl J Med 371(26):2488–2498

Janssens JP, Krause KH (2004) Pneumonia in the very old. Lancet Infect Dis 4(2):112–124

Janzen V et al (2006) Stem-cell ageing modified by the cyclin-dependent kinase inhibitor p16INK4a. Nature 443(7110):421–426

Ju Z et al (2007) Telomere dysfunction induces environmental alterations limiting hematopoietic stem cell function and engraftment. Nat Med 13(6):742–747

Kang I et al (2004) Age-associated change in the frequency of memory CD4+ T cells impairs long term CD4+ T cell responses to influenza vaccine. J Immunol 173(1):673–681

Khoo ML et al (2014) Gene profiling reveals association between altered Wnt signaling and loss of T-cell potential with age in human hematopoietic stem cells. Aging Cell 13(4):744–754

Kikushige Y, Miyamoto T (2014) Hematopoietic stem cell aging and chronic lymphocytic leukemia pathogenesis. Int J Hematol 100(4):335–340

Kim NW et al (1994) Specific association of human telomerase activity with immortal cells and cancer. Science 266(5193):2011–2015

Kitagawa M et al (1997) Overexpression of tumor necrosis factor (TNF)-alpha and interferon (IFN)-gamma by bone marrow cells from patients with myelodysplastic syndromes. Leukemia 11(12):2049–2054

Kohler A et al (2009) Altered cellular dynamics and endosteal location of aged early hematopoietic progenitor cells revealed by time-lapse intravital imaging in long bones. Blood 114(2):290–298

Kollman C et al (2001) Donor characteristics as risk factors in recipients after transplantation of bone marrow from unrelated donors: the effect of donor age. Blood 98(7):2043–2051

Kuranda K et al (2011) Age-related changes in human hematopoietic stem/progenitor cells. Aging Cell 10(3):542–546

La Motte-Mohs RN, Herer E, Zuniga-Pflucker JC (2005) Induction of T-cell development from human cord blood hematopoietic stem cells by Delta-like 1 in vitro. Blood 105(4): 1431–1439

Lane AA, Scadden DT (2010) Stem cells and DNA damage: persist or perish? Cell 142(3): 360–362

Laschober GT et al (2011) Age-specific changes of mesenchymal stem cells are paralleled by upregulation of CD106 expression as a response to an inflammatory environment. Rejuvenation Res 14(2):119–131

Laurenti E et al (2015) CDK6 levels regulate quiescence exit in human hematopoietic stem cells. Cell Stem Cell 16(3):302–313

Lechman ER et al (2012) Attenuation of miR-126 activity expands HSC in vivo without exhaustion. Cell Stem Cell 11(6):799–811

Lescale C et al (2015) Hind limb unloading, a model of spaceflight conditions, leads to decreased B lymphopoiesis similar to aging. FASEB J 29(2):455–463

Lombard DB et al (2005) DNA repair, genome stability, and aging. Cell 120(4):497–512

Luo M et al (2015) Long non-coding RNAs control hematopoietic stem cell function. Cell Stem Cell 16:1–13

Maes J et al (2008) Lymphoid-affiliated genes are associated with active histone modifications in human hematopoietic stem cells. Blood 112(7):2722–2729

Majeti R, Park CY, Weissman IL (2007) Identification of a hierarchy of multipotent hematopoietic progenitors in human cord blood. Cell Stem Cell 1(6):635–645

Manz MG et al (2002) Prospective isolation of human clonogenic common myeloid progenitors. Proc Natl Acad Sci U S A 99(18):11872–11877

Marley SB et al (1999) Evidence for a continuous decline in haemopoietic cell function from birth: application to evaluating bone marrow failure in children. Br J Haematol 106(1):162–166

Marusyk A, DeGregori J (2008) Declining cellular fitness with age promotes cancer initiation by selecting for adaptive oncogenic mutations. Biochim Biophys Acta 1785(1):1–11

McKenzie JL et al (2007) Low rhodamine 123 retention identifies long-term human hematopoietic stem cells within the Lin-CD34+CD38- population. Blood 109(2):543–545
Mikkelsen TS et al (2007) Genome-wide maps of chromatin state in pluripotent and lineage-committed cells. Nature 448(7153):553–560
Miller JS et al (1999) Single adult human CD34(+)/Lin-/CD38(−) progenitors give rise to natural killer cells, B-lineage cells, dendritic cells, and myeloid cells. Blood 93(1):96–106
Milyavsky M et al (2010) A distinctive DNA damage response in human hematopoietic stem cells reveals an apoptosis-independent role for p53 in self-renewal. Cell Stem Cell 7(2): 186–197
Mohrin M et al (2010) Hematopoietic stem cell quiescence promotes error-prone DNA repair and mutagenesis. Cell Stem Cell 7(2):174–185
Morin GB (1989) The human telomere terminal transferase enzyme is a ribonucleoprotein that synthesizes TTAGGG repeats. Cell 59(3):521–529
Morrison SJ et al (1996) Telomerase activity in hematopoietic cells is associated with self-renewal potential. Immunity 5(3):207–216
Muller-Sieburg CE et al (2002) Deterministic regulation of hematopoietic stem cell self-renewal and differentiation. Blood 100(4):1302–1309
Naveiras O et al (2009) Bone-marrow adipocytes as negative regulators of the hematopoietic microenvironment. Nature 460(7252):259–263
Notaro R et al (1997) In vivo telomere dynamics of human hematopoietic stem cells. Proc Natl Acad Sci U S A 94(25):13782–13785
Notta F et al (2011) Isolation of single human hematopoietic stem cells capable of long-term multilineage engraftment. Science 333(6039):218–221
O'Connell RM et al (2010) MicroRNAs enriched in hematopoietic stem cells differentially regulate long-term hematopoietic output. Proc Natl Acad Sci U S A 107(32):14235–14240
O'Sullivan RJ, Karlseder J (2012) The great unravelling: chromatin as a modulator of the aging process. Trends Biochem Sci 37(11):466–476
Ooi AG et al (2010) MicroRNA-125b expands hematopoietic stem cells and enriches for the lymphoid-balanced and lymphoid-biased subsets. Proc Natl Acad Sci U S A 107(50): 21505–21510
Pang WW et al (2011) Human bone marrow hematopoietic stem cells are increased in frequency and myeloid-biased with age. Proc Natl Acad Sci U S A 108(50):20012–20017
Pearson T et al (2008) Non-obese diabetic-recombination activating gene-1 (NOD-Rag1 null) interleukin (IL)-2 receptor common gamma chain (IL2r gamma null) null mice: a radioresistant model for human lymphohematopoietic engraftment. Clin Exp Immunol 154(2):270–284
Reynaud D et al (2003) In vitro identification of human pro-B cells that give rise to macrophages, natural killer cells, and T cells. Blood 101(11):4313–4321
Rongvaux A et al (2013) Human hemato-lymphoid system mice: current use and future potential for medicine. Annu Rev Immunol 31:635–674
Rossi DJ et al (2007) Hematopoietic stem cell quiescence attenuates DNA damage response and permits DNA damage accumulation during aging. Cell Cycle 6(19):2371–2376
Rube CE et al (2011) Accumulation of DNA damage in hematopoietic stem and progenitor cells during human aging. PLoS One 6(3), e17487
Sasaki S et al (2011) Limited efficacy of inactivated influenza vaccine in elderly individuals is associated with decreased production of vaccine-specific antibodies. J Clin Invest 121(8): 3109–3119
Schubeler D (2015) Function and information content of DNA methylation. Nature 517(7534): 321–326
Sedelnikova OA et al (2004) Senescing human cells and ageing mice accumulate DNA lesions with unrepairable double-strand breaks. Nat Cell Biol 6(2):168–170
Sedelnikova OA et al (2008) Delayed kinetics of DNA double-strand break processing in normal and pathological aging. Aging Cell 7(1):89–100
Shultz LD et al (2005) Human lymphoid and myeloid cell development in NOD/LtSz-scid IL2R gamma null mice engrafted with mobilized human hemopoietic stem cells. J Immunol 174(10): 6477–6489

Shultz LD, Ishikawa F, Greiner DL (2007) Humanized mice in translational biomedical research. Nat Rev Immunol 7(2):118–130
Shumaker DK et al (2006) Mutant nuclear lamin A leads to progressive alterations of epigenetic control in premature aging. Proc Natl Acad Sci U S A 103(23):8703–8708
Singh T, Newman AB (2011) Inflammatory markers in population studies of aging. Ageing Res Rev 10(3):319–329
Smith-Vikos T, Slack FJ (2012) MicroRNAs and their roles in aging. J Cell Sci 125(Pt 1):7–17
Song Z et al (2010) Alterations of the systemic environment are the primary cause of impaired B and T lymphopoiesis in telomere-dysfunctional mice. Blood 115(8):1481–1489
Sun D et al (2014) Epigenomic profiling of young and aged HSCs reveals concerted changes during aging that reinforce self-renewal. Cell Stem Cell 14(5):673–688
Syed FA, Hoey KA (2010) Integrative physiology of the aging bone: insights from animal and cellular models. Ann N Y Acad Sci 1211:95–106
Taiwo O et al (2013) DNA methylation analysis of murine hematopoietic side population cells during aging. Epigenetics 8(10):1114–1122
Tan J et al (2007) Pharmacologic disruption of Polycomb-repressive complex 2-mediated gene repression selectively induces apoptosis in cancer cells. Genes Dev 21(9):1050–1063
Taraldsrud E et al (2009) Age and stress related phenotypical changes in bone marrow CD34+ cells. Scand J Clin Lab Invest 69(1):79–84
Thompson WW et al (2003) Mortality associated with influenza and respiratory syncytial virus in the United States. JAMA 289(2):179–186
Townsley DM, Dumitriu B, Young NS (2014) Bone marrow failure and the telomeropathies. Blood 124(18):2775–2783
Traggiai E et al (2004) Development of a human adaptive immune system in cord blood cell-transplanted mice. Science 304(5667):104–107
Tsurumi A, Li WX (2012) Global heterochromatin loss: a unifying theory of aging? Epigenetics 7(7):680–688
Tuljapurkar SR et al (2011) Changes in human bone marrow fat content associated with changes in hematopoietic stem cell numbers and cytokine levels with aging. J Anat 219(5):574–581
Vas V et al (2012) Aging of the microenvironment influences clonality in hematopoiesis. PLoS One 7(8), e42080
Vaziri H et al (1994) Evidence for a mitotic clock in human hematopoietic stem cells: loss of telomeric DNA with age. Proc Natl Acad Sci U S A 91(21):9857–9860
Wahlestedt M et al (2013) An epigenetic component of hematopoietic stem cell aging amenable to reprogramming into a young state. Blood 121:4257–4264
Walter D et al (2015) Exit from dormancy provokes DNA-damage-induced attrition in hematopoietic stem cells. Nature 520:549–552
Wang C et al (2009) DNA damage response and cellular senescence in tissues of aging mice. Aging Cell 8(3):311–323
Weishaupt H, Sigvardsson M, Attema JL (2010) Epigenetic chromatin states uniquely define the developmental plasticity of murine hematopoietic stem cells. Blood 115(2):247–256
Woolthuis CM, de Haan G, Huls G (2011) Aging of hematopoietic stem cells: Intrinsic changes or micro-environmental effects? Curr Opin Immunol 23(4):512–517
Wu L, Belasco JG (2008) Let me count the ways: mechanisms of gene regulation by miRNAs and siRNAs. Mol Cell 29(1):1–7
Xie M et al (2014) Age-related mutations associated with clonal hematopoietic expansion and malignancies. Nat Med 20(12):1472–1478
Xing Z et al (2006) Increased hematopoietic stem cell mobilization in aged mice. Blood 108(7):2190–2197
Yahata T et al (2011) Accumulation of oxidative DNA damage restricts the self-renewal capacity of human hematopoietic stem cells. Blood 118(11):2941–2950
Yang L, Froberg JE, Lee JT (2014) Long noncoding RNAs: fresh perspectives into the RNA world. Trends Biochem Sci 39(1):35–43
Yui J, Chiu CP, Lansdorp PM (1998) Telomerase activity in candidate stem cells from fetal liver and adult bone marrow. Blood 91(9):3255–3262

Aging of Stem Cells in Skin: What Is Driving the Aging Process?

8

Sabrina Gundermann, Hans-Jürgen Stark,
and Petra Boukamp

Abstract

Skin is a complex organ consisting of two very different tissue compartments, the epidermis and the dermis, which comprise cells of diverse origin and function. Establishing and maintaining the functionality of skin requires not only tightly regulated processes of maturation and differentiation of the individual cellular components but also extensive and well-coordinated interactions between those different compartments. As yet, this intricate interdependence is far from being disclosed to a satisfactory extent. Accordingly, our understanding for dysregulated conditions like skin aging is still largely insufficient. Despite the fact that skin aging is readily visible and morphologically well defined, the underlying molecular mechanisms driving that process are still a matter of debate as is the role of stem cells therein. Furthermore, the aging of skin is peculiar as it is not only driven endogenously but is largely accelerated and aggravated by external influences, mainly, UV radiation, what is reflected by the term "photoaging." By foregrounding human skin, this chapter aims at compiling the present concepts of both intrinsic and extrinsic skin aging.

8.1 The Human Skin

With about 15 % of body weight, the skin is the largest organ of humans. Covering the entire body surface, it provides a physical barrier between organism and environment, thus safeguarding us against aggressive chemicals and microbes as well as physical impacts, e.g., by irradiation. Simultaneously, the inside-out barrier

S. Gundermann • H.-J. Stark • P. Boukamp (✉)
Genetics of Skin Carcinogenesis, German Cancer Research Center,
Im Neuenheimer Feld 280, Heidelberg 69120, Germany
e-mail: p.boukamp@dkfz-heidelberg.de; p.boukamp@dkfz.de

H. Geiger et al. (eds.), *Stem Cell Aging: Mechanisms, Consequences, Rejuvenation*,
DOI 10.1007/978-3-7091-1232-8_8

protects the organism against loss of water and electrolytes. Skin also participates in the body's thermoregulation by control of blood perfusion and sweat gland activity. Furthermore, by innate and adaptive mechanisms, skin is intensively involved in immunologic defense, and it keeps us informed about our surroundings as it comprises all the sensory elements for the perception of tactile stimuli, temperature, pain, and pressure.

8.2 The Tissue Architecture of Human Skin

This highly diverse functionality results from a considerable complexity in the histological organization of the skin. In general, the skin is divided in three compartments: the stratified epithelium of the epidermis, the dermis that is the connective tissue responsible for the mechanical properties, and the subcutaneous tissue containing pads of adipocytes (panniculus adiposus).

The epidermis is a multilayered cornifying epithelium firmly anchored to the underlying dermis by a basement membrane. The vascularized dermis is rich in extracellular matrix (ECM) proteins, contains nerves and the variety of sensorial receptors, and includes – besides vascular and neural cells – fibroblasts as major cell type. Below begins without sharply defined transition the subcutaneous tissue that covers and upholsters the underlying muscles, tendons, and bones.

Further components of the skin are specialized appendages such as skin glands, hair follicles (HFs), and nails. In particular, HFs had become important objects for stem cell research in recent years.

Here, we will concentrate on the epidermis and dermis, being well aware of the fact that we dismiss many important aspects, i.e., innervation by neuronal cells; support by blood vessels; immune surveillance and response by inflammatory cells, including mast cells; and all contributions by the *panniculus adiposus*.

8.2.1 Epidermis and Keratinocytes

The ectodermally derived epidermis is a stratified epithelium consisting to 90–95 % of keratinocytes. Its external surface is the impermeable *stratum corneum*. The epidermal thickness is about 100 μm, but varies considerably according to the regional properties of the skin (Kanitakis 2002). The continuous desquamation is compensated by permanent proliferation of the keratinocytes in the basal epidermal cell layer. Those newly generated cells fuel the stratification of the epidermis and undergo the process of keratinocyte differentiation on their way upward finally ending as terminally differentiated horn squames (corneocytes). Thereby, the morphologically definable layers of the epidermis, *stratum basale, stratum spinosum, stratum granulosum*, and *stratum corneum*, all represent distinct and sequential steps of differentiation. Examples of molecular markers indicating differentiated stages are cytokeratins, cytoskeletal elements of the class of intermediate filaments that are abundant in epidermal keratinocytes, and further structural proteins involved in cellular adhesion and in formation of the cornified envelope during terminal differentiation (Alberts et al. 2004; Löffler et al. 2007).

While the robustness of the epidermal sheet relies on intense cohesion of the keratinocytes via adherent junctions and desmosomes, its firm connection to the basement membrane is brought about by specific adhesive structures called hemidesmosomes. This complex assembly of adhesion proteins and basement membrane constituents, termed dermoepidermal junction (DEJ), has been shown to crucially regulate growth, regeneration, and tissue homeostasis of the epidermis.

The *stratum basale* is the layer comprising the undifferentiated keratinocytes with mitotic potential: here also the epidermal stem cells are located in association to their stem cell niche that is at least in part defined by a particular arrangement of ECM (Muffler et al. 2008). In the suprabasal zone, the expression of differentiation-specific genes is induced which culminates in the *stratum corneum* in cornification, i.e., formation of anucleated horn squames.

Besides keratinocytes, the epidermis contains the pigment producing melanocytes and Langerhans cells (LC), antigen-presenting dendritic cells serving as outpost sentinels of the skin's immune system.

8.2.2 The Dermis and the Fibroblasts

Nutritive and mechanical support of the epidermis is provided by the dermis. This fibroelastic tissue mainly consists of dense ECM of high stability and elasticity, whereas the density of interspersed cells is rather poor. Major dermal components are collagens and elastic fibers. Collagen fibrils account for the tensile strength of the skin, and the elastic fibers are responsible for the regaining of shape after mechanical impact and deformation. These fibrous elements are embedded in a ground substance with high water-binding capacity consisting of glycoproteins and glycosaminoglycans (GAGs). The predominant cells in the dermis are fibroblasts, spindle-shaped cells, which are the major producers of the components of the ECM.

The dermis is divided into two morphologically distinct zones: the subepidermal papillary dermis (*stratum papillare*) and the deeper reticular dermis (*stratum reticulare*). The highly vascularized and innervated *stratum papillare* forms indentations with the epidermis, thus increasing the contact between the two compartments and the strength of adhesion. The *stratum reticulare* manifests a high stability and toughness due to its increased content of condensed fibrous material.

Collagen fibers are the most prominent structural proteins in skin largely contributing to its stability. This is reflected by their extraordinary high amount of 70 % of the skin's dry weight (Gniadecka et al. 1998). Also in other tissues (tendons, bones, cartilage, etc.), collagens represent the predominant proteins crucial for rigidity. At present, 27 different types of collagens have been identified, most of which form aggregates of α-polypeptide chains in triple-helical conformation. About ten collagen types can be found in dermis and most of them participate in fiber formation. Collagen type IV is one of the exceptions in that it aggregates to a planar meshwork that is the core structure of the basement membrane. The typical collagen fibrils in dermis consist of more than 80 % of collagen type I and of 10–15 % of collagen type III (Garrone et al. 1997).

Elasticity is imparted to the fibrous assembly of the dermis by an additional system of elastic fibers; nevertheless, their content is rather low with about 2 % of dermal dry weight (Uitto et al. 1983). Mature elastic fibers contain 90 % of cross-linked elastin that is associated to fibrillin microfibrils and a variety of additional glycoproteins. The dermis exhibits a high degree of organization of the elastic fiber system with thin microfibrils attaching to the basement membrane in the DEJ. Deeper down in the *stratum papillare*, these microfibrils merge and by association with elastin become thicker. In the *stratum reticulare*, those bundles combine to thick mainly horizontally oriented elastin fibers.

The extrafibrillar, amorphous matrix comprises in part GAGs and proteoglycans. These molecules are strongly hydrophilic with high binding capacities for water as well as ions, thus having properties of hydrogels. GAGs are large acidic mucopolysaccarides of repeating disaccharide units, and according to their sulfate group content and sugar composition, they are distinguished into four classes: hyaluronan, chondroitin sulfate and dermatan sulfate, heparin sulfate, and keratin sulfate. Proteoglycans are covalent complexes of GAG chains linked to a core protein and can be highly diverse and functionally specialized. They can bind and release growth factors and proteases thereby modulating signaling events and ECM remodeling. Similar functions are exerted by other adhesive glycoproteins found in dermis such as thrombospondins and tenascin (Bornstein and Sage 2002). Another important representative is cellular fibronectin, a glycoprotein secreted by fibroblasts as a primordial ECM component in regenerative processes. Fibronectin fibers formed in the fibrin clot of healing wounds guide the migration of fibroblasts and serve as templates for the deposition of collagen fibrils (Velling et al. 2002).

It is obvious that the complex composition of the dermal ECM correlates with the vast multiplicity of functional demands the skin is confronted with. Therefore, age-dependent declines in appearance and physiology of the skin have to be viewed in the context of alterations – mostly disturbances – of dermal quality.

8.3 The Aged Skin

Because skin mirrors already early signs of aging, improvement of skin quality receives more and more attention. Macroscopically, aged skin differs from young skin with regards to wrinkles, elasticity, and pigmentation. Further signs are dryness, fragility, impaired wound healing, and pathologic changes such as the development of benign neoplasms and carcinogenesis. These are the functional consequences of aberrances affecting moisture regulation, immunological responses, sweat production, thermoregulation, skin barrier sensory functions, proliferation, and DNA repair (Fritsch 2009; Giacomoni et al. 2000; Schneider et al. 2003; Zouboulis 2003). Reduction of these skin functions is often associated with skin diseases known to increase with age (Zouboulis 2003).

Skin aging can be divided into an intrinsic and an extrinsic process. Intrinsic skin aging describes a time-dependent process that is also called chronologic, natural, or cutaneous aging. Being the mechanical border between the internal organs and the environment, the skin is also exposed to exogenous noxae. Therefore, extrinsic skin aging describes the changes which are due to cumulative influences of the environment mainly ultraviolet (UV) radiation, thus coined with the term photoaging (Kohl et al. 2009; Krutmann et al. 2008). Correspondingly, it is a combination of many possible factors that causes those alterations in the aging skin. While gene mutations and changes in cell metabolism as well as hormonal balance are examples for intrinsic factors, exposure to radiation (UV and infrared, IR); chemicals; toxins, in particular cigarette smoke; as well as pathogens and mechanical stress belongs to the extrinsic factors (Makrantonaki and Zouboulis 2007, 2011; Zouboulis and Makrantonaki 2011).

8.3.1 Intrinsically Aged Skin

Intrinsic skin aging affects the skin by a slow irreversible degeneration of the tissue (Uitto 1997). The changes start in the twenties; however, aging remains hidden over decades. Finally, intrinsically aged skin has a delicate appearance called "cigarette paper" skin, as it is thin and transparent and has fine wrinkles. It shows less elasticity, increased fragility, and dryness. The skin loses the underlying fat and muscle tissue and due to bone loss, skin shrinks away from the bone causing sagging of the skin (Sjerobabski-Masnec and Situm 2010; Langton et al. 2010).

While young skin is characterized by an extensively undulated dermoepidermal junction (DEJ) forming the so-called rete ridges, one commonly cited characteristic of skin aging is the flattening of this DEJ. Thereby, rete ridge structures decrease and the contact area between epidermis and dermis reduces by about 35 % (Moragas et al. 1993). It is suggested that this flattening contributes to the increased fragility of aged skin and may also lead to reduced nutrient transfer between epidermis and dermis (Langton et al. 2010; Lavker 1979). With flattening of the DEJ, also epidermal thickness is supposed to decline. This, however, is still a matter of some debate. While some studies reported no changes in epidermal thickness (Whitton and Everall 1973; El-Domyati et al. 2002), others reported on epidermal atrophy (Lavker 1979; Branchet et al. 1990; Wu et al. 2012; Moragas et al. 1993; Lock-Andersen et al. 1997). The observed epidermal atrophy seems to affect mainly the *stratum spinosum*, while the *stratum granulosum* und *stratum corneum* remain unchanged (Lavker 1979). Epidermal atrophy goes along with a decrease in mitotic activity of the basal keratinocytes causing a decline in epidermal cell renewal (Engelke et al. 1997; Grove and Kligman 1983).

Aging affects also other cells of the epidermis. Beginning at the age of 30, the amount of melanocytes decrease by up to 20 % per decade (Gilchrest et al. 1979) and the number of LCs shows a decline from 1,200 to 800/mm^2. In addition, LCs are described to exhibit morphological changes and impaired function (Grewe 2001; Bhushan et al. 2002) resulting in impaired immune response.

Parallel to the epidermis also the dermis shows signs of atrophy. Because of alterations in the matrix, dermal thickness decreases by about 20 % (Shuster et al. 1975; Baumann 2007). Whether this is due to a loss in collagen is still a matter of debate (Lovell et al. 1987; Shuster et al. 1975). There is, however, no doubt that changes in structure and distribution of collagen type I occur, making the collagen bundles less organized and thinner with age (Fenske and Lober 1986; Waller and Maibach 2006). In parallel, the degree of cross-linking increases (Paul and Bailey 1996; Bailey 2001; Yamauchi et al. 1988). In addition, the amount of GAGs, the water-holding filling material between the collagen bundles, decreases with age (Ghersetich et al. 1994; Bernstein et al. 1996b; Fleischmajer et al. 1972), explaining the reduced hydration of the skin and possibly the impairment of cell movement and nutrition. Furthermore, the elastic fiber system in the dermis gets considerably reduced over the decades of life. At the age of 80 years, the loss of elastic fibers is as high as 50 %, which is mainly due to the progressive thinning of the skin and results in dramatic loss of elastic properties (Frances et al. 1990). All this correlates with a reduced number of dermal fibroblasts (Gunin et al. 2011a; West 1994) and mast cells (West 1994).

8.3.2 Extrinsically Aged Skin

Due to its induction by environmental noxae, extrinsic skin aging is restricted to sites such as face, décolleté, forearm, or hand (Berneburg et al. 2000). In particular, UV radiation is supposed to be responsible for 80 % of the alterations involved (Fisher et al. 1996). In affected skin, normal architecture is disrupted leading to deep wrinkles, leathery appearance, less elasticity, diffuse hyperpigmentation, and lentigines (Kohl et al. 2009; Berneburg et al. 2000). As a result of excessive UV exposure, these changes can be observed in part already early in life as "premature" aging. Nevertheless, these changes are not completely independent of the alterations caused by intrinsic aging, and it is often difficult to classify the observations as intrinsic or extrinsic skin aging since the morphological shifts of the extrinsic process often superimpose the signs of intrinsic skin aging (Krutmann et al. 2008).

In contrast to intrinsically aged skin, which shows a rather atrophic phenotype, extrinsic aging is often correlated with thickening of almost all layers of the skin (Krutmann et al. 2008). For epidermal thickening, controversial results are described. The epidermis is thereby described as acanthotic, atrophic, or even unchanged (Hughes et al. 2011; Krutmann et al. 2008; Berneburg et al. 2000; Yaar and Gilchrest 2001; Sandby-Moller et al. 2003). Furthermore, this is accompanied by aberrant epidermal differentiation and adhesion, leading to loss of polarity. Here, the reduction of β1 integrin which conveys adhesion to the DEJ seems to be instrumental (Yaar and Gilchrest 2001; Bosset et al. 2003). The DEJ appears even more flattened as described for intrinsically aged skin (Fig. 8.1a, b) and shows a reduced amount of collagen type VII, the main component of the anchoring fibrils, further destabilizing the DEJ (Contet-Audonneau et al. 1999; Craven et al. 1997).

In sun-damaged skin, also melanocytes are adversely affected. They are irregularly arranged throughout the basal epidermal layer and show strong variation in size, morphology, and melanin content. This explains the appearance of hyperpigmentation despite the fact that the number of melanocytes declines (Breathnach and Wyllie 1964; Gilchrest et al. 1979).

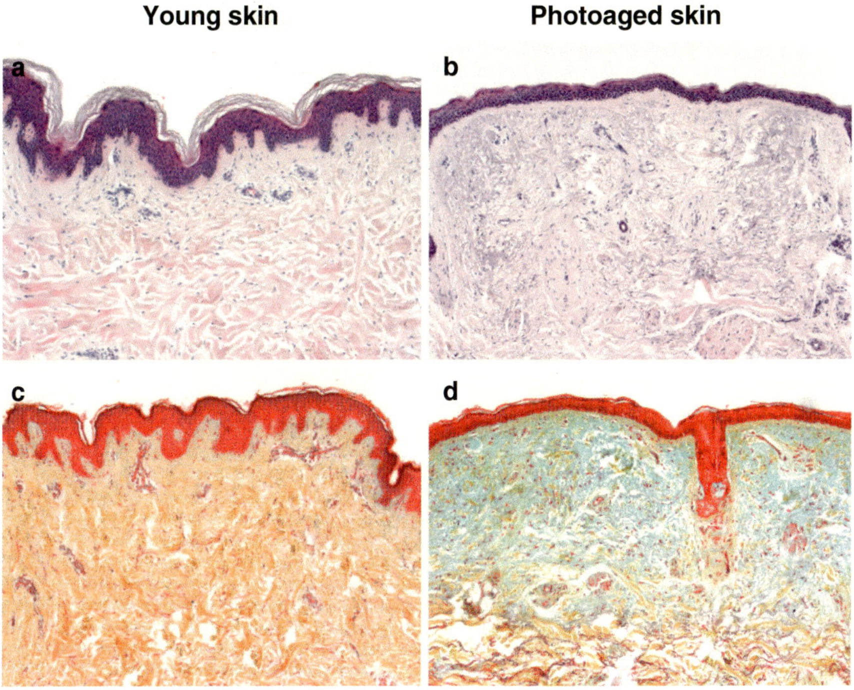

Fig. 8.1 The dermal texture of human skin changes substantially with age. Young (buttock of a 27-year-old male) (**a**, **c**, **e**, **g**) versus photoaged skin (cheek of a 75-year-old female) (**b**, **d**, **f**, **h**) processed for different histological and immunofluorescence stainings, respectively: (**a**, **b**) hematoxylin and eosin staining for morphological survey; (**c**, **d**) Movat-pentachrome staining for the simultaneous demonstration of different tissue components (cell nuclei in *dark blue*, cytoplasm in *red*, collagens in *yellow*, GAGs in *light blue*); (**e**, **f**) resorcin fuchsin with thiazine-picric acid staining for the visualization of elastic fibers in *black-violet*, collagen fibers in *red*, and nuclei in *dark brown*; and (**g**, **h**) immunofluorescence analyses for the localization of elastin in *red* and fibrillin in *green with blue* counterstain for cell nuclei. The intense indentation of the DEJ zone of young skin has largely disappeared in old skin (**a**, **b**). Collagen fibers are homogenously distributed and enriched in the dermal *str. reticulare* of young skin but diminished in the photoaged skin where few condensed fiber bundles persist (**c**, **e** vs. **d**, **f**). Oppositely, GAGs are enhanced in the dermis of photoaged in contrast to young skin (**c**, **d**) as is the staining for elastin, which, however, appears aggregated in the photoaged dermis (**e**, **f**). The organization of elastin fibers has completely changed towards amorphous aggregates with increased elastin content in photoaged versus young skin (**g**, **h**). The loss of elastin fibers and the accumulation of elastotic material (*red*) is particularly obvious in (h), whereas fibrillin-containing microfibrillar assembly (*green*) seems, at least in part, to be maintained. *Bars* in (f) (for **a**–**f**) and in (h) (for **g**–**h**) = 100 μm

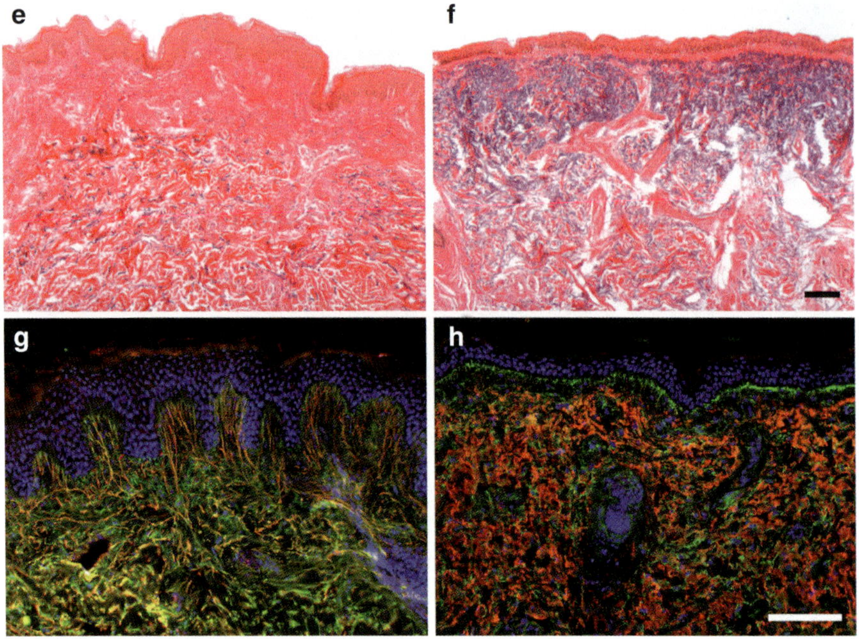

Fig. 8.1 (continued)

The most prominent changes in extrinsically aged skin are, however, seen in the dermis where the severity of changes correlates with the degree of UV damage. Solar elastosis is the most obvious and well-described alteration of the ECM (Kligman 1969). It describes an accumulation of fragmented elastic fibers (Mitchell 1967). Although the exact composition is not yet known, elastin and fibrillin, the main components of the elastic fibers, as well as fibronectin and GAGs could be detected histochemically (Chen et al. 1986; Montagna et al. 1989; Mera et al. 1987; Sams and Smith 1961), as also illustrated in Fig. 8.1. It is suggested that the fragmentation of GAGs in areas of solar elastosis contributes to loss of their hydrating properties (Bernstein et al. 1996b; Waller and Maibach 2006; Tzellos et al. 2009). In photoaged dermis, the content of collagen types I and III is generally decreased (Griffiths et al. 1993; Warren et al. 1991; Bernstein et al. 1996a), and it is discussed that it is replaced by the elastotic material (Fig. 8.1c–f). Accordingly, mature collagen fibers can no longer be detected (Bernstein et al. 1996a; El-Domyati et al. 2002; Schwartz et al. 1993) or are very thin and partly fragmented (Scharffetter-Kochanek et al. 2000). Different from intrinsically aged skin, the number of fibroblasts increases, and they adopt a stellate phenotype with a highly activated rough endoplasmic reticulum, indicative for increased metabolic activity (Uitto 1986). In addition, mast cells and mononuclear cells are attracted in extrinsically aged skin (Lavker and Kligman 1988; Krutmann 2003; Gunin et al. 2011b).

8.4 Mechanism of Aging

What are the molecular mechanisms underlying the aging process of the skin? A detailed review on mechanisms causing fibroblast aging was recently provided (Tigges et al. 2014). As we hypothesize that the major target of skin aging is the postmitotic tissue of the dermis rather than the constantly renewing epidermis, we here largely concentrate on mechanisms contributed to dermal aging.

8.4.1 Signaling Cascades and MMPs

Many studies showed that intrinsic and extrinsic skin aging have some common features, like the engagement of the same signaling pathways (Fisher et al. 2002). For example, in both cases, the MAPK pathway is dysregulated and so are the MAP kinases JNK and p38 age-dependently overexpressed (Chung et al. 2000; Fisher et al. 1998). The activation of these kinases upregulates the transcription factor AP-1 (Sardy 2009; Fisher et al. 2002), leading to an increased expression of the matrix-metalloproteinases MMP-1, MMP-3, and MMP-9, which are instrumental in the degradation of a lot of ECM components and matrix remodeling (Fisher et al. 1996, 2002; Angel et al. 2001). Furthermore, the effect of matrix degradation can be intensified by the potential of AP-1 to also downregulate the synthesis of dermal collagen by blocking the effects of TGFβ (Fisher et al. 2002).

8.4.2 Theory of Free Radicals/ROS

Free reactive oxygen species (ROS) are endogenously formed or exogenously induced and play a major role in extrinsic and intrinsic skin aging. It was shown that the ROS level in skin rises with age, while the cellular mechanisms to reduce the ROS level decline (Makrantonaki and Zouboulis 2007; Rhie et al. 2001).

High ROS level leads to the activation of the MAPK signaling pathway by activating some surface receptors like for EGF, TNF-α, insulin, and IL-1 and in turn causing upregulation of the MMPs, as it was described before (Klotz et al. 1999). Thus, a high ROS level could trigger MAPK signaling, AP-1 activation, and MMP expression thus being causal for the high amount of degraded collagen in aged skin. The main source for ROS in intrinsically aged skin is the mitochondria. About 1–5 % of the oxygen input is converted into ROS under normal physiological conditions, and it was shown that with age defects in the mitochondrial respiratory chain lead to a further increase of the ROS level (Trounce et al. 1989; Yen et al. 1989). In UV-exposed skin, additional ROS is produced by energy transfer from UV-absorbing chromophores, such as $NADH^-$/NADPH, tryptophan, riboflavin, or *trans*-urocanic acid (Hanson and Simon 1998), thereby explaining the even higher amounts of MMPs and consequently the increase in matrix degradation.

One important aspect in this context is the mutation rate of mitochondrial DNA (mtDNA). Due to its unprotected state (histone-free), mitochondrial DNA is especially susceptible to oxidative damage. As the mtDNA mainly codes for proteins being part of the oxidative phosphorylation, mutations of the mtDNA lead to defects and further increase the production of ROS; a vicious circle is established (Camougrand and Rigoulet 2001; Ozawa 1995).

It is suggested that mtDNA mutations play a role in intrinsic as well as in extrinsic skin aging. However, in extrinsically aged skin, clearly higher levels of mtDNA mutations are achieved (Berneburg et al. 1997). The causal relation between mt mutations and extrinsic aging could be confirmed by in vitro studies using fibroblasts bearing constitutive mtDNA mutations that impaired the function of the respiratory chain. Impressively, those cells, when cultured in skin equivalents, mimicked the abovementioned features of extrinsic aging, including MMP overexpression, and reduced support of keratinocyte proliferation (Krutmann 2011; Majora et al. 2009).

8.4.3 Cellular Aging, Replicative Senescence, and the Role of Telomeres

Even under optimal conditions, fibroblasts in cell culture have a life span limited to 50–70 cell divisions having resulted in the concept of "replicative senescence" (Hayflick and Moorhead 1961; Hayflick 1965). One potential mechanism for this "Hayflick limit" came about with the description of the end replication problem (Olovnikov 1973; Watson 1972) that during every round of replication leads to loss of 50–200 nucleotides from the 5'-ends of linear chromosomes, the telomeres (Harley et al. 1990; Levy et al. 1992). Nevertheless, these telomeres provide only a temporary protection as, when they progressively shorten under a critical length, the cells suffer from induction of apoptosis or senescence (Lombard et al. 2005; d'Adda di Fagagna et al. 2003; Herbig et al. 2004). Thus, critical telomere shortening is supposed to be responsible for the accumulation of senescent and functionally impaired fibroblasts in aging skin (Campisi 1996). However, this hypothesis is merely based on results with cultivated cells in vitro, and it is still controversially discussed if it is also valid for the situation in the skin in situ (Cristofalo et al. 2004; Rubin 2002; Maier and Westendorp 2009). Senescence, as commonly assessed by the expression of a specific β-galactosidase, the senescence-associated β-gal pH6.0, was originally described for aged skin (Dimri et al. 1995). However, verification still needs to be awaited. Similarly, the role of telomere loss is a matter of debate. Investigating telomere lengths in different-aged skin, an age-dependent reduction was described; however, the degree of telomere length reduction was generally very limited (Sugimoto et al. 2006; Lindsey et al. 1991). This may be attributed to the fact that the constantly regenerating epidermis expresses telomerase, the enzyme able to counteract telomere loss, and that fibroblasts only proliferate rarely in vivo, thereby bypassing proliferation-dependent telomere loss. This does not exclude that upon stress, e.g., UV radiation, telomeres of individual cells get damaged and may suffer from accelerated telomere loss (Leufke et al. 2013). However, it strongly suggests that telomere loss is not the driving force in the process of skin aging but that it is rather a consequence of damage and insufficient telomerase activity, respectively. One major factor in this damage-dependent scenario is ROS (Ayouaz et al. 2008), thus further strengthening its pivotal role in skin aging.

8.4.4 Molecular Mechanisms of Matrix Alterations

Aging of skin is accompanied by a large variety of structural changes. Because of their high half time, many matrix components are predisposed for accumulating damages and modifications leading to molecular aging. So, the half-life of fibrillar collagen is 15–95 years and for proteoglycans 11–23 years, and elastin and fibrillin persist in the tissue from embryogenesis on for the whole life of the organisms (Naylor et al. 2011).

In aged skin, proteolytic enzymes, such as MMP-1, MMP-2, MMP-3, and MMP-9, cause fragmentation of the collagens, proteoglycans, and elastic fibers (Ashworth et al. 1999; Birkedal-Hansen et al. 1993; Berton et al. 2000; Murphy and Docherty 1992). Additionally, glycation of the matrix components causes the formation of intermolecular cross-links which makes the molecules less soluble and, even more important, blocks further degradation of the fragments (Rock and Fischer 2011; Bailey et al. 1998). Finally, an increase in protein oxidation is seen, especially in UV-exposed skin where protein oxidation is induced by high ROS levels. This as well provokes the formation of protein aggregates, which resist more strongly to degradation (Sander et al. 2002). Therefore, accumulating alterations in the matrix disturb the mechanical and structural integrity of the skin and thus are most likely the molecular basis for the deteriorated appearance of aged skin.

8.4.5 Development of Solar Elastosis in Extrinsically Aged Skin

Solar elastosis, as described above, is most likely not restricted to destructive processes, i.e., degradation of preexisting elastic fibers, but it includes as well a dysregulated neosynthesis of elastin and fibrillin without proper macromolecular fiber assembly thus contributing to the accumulation of elastotic material (Jenkins 2002). The expression of the mainly responsible proteolytic enzymes like elastase or MMPs has been found to be enhanced by NF-κB, which gets activated by UV (Kohl et al. 2011). An additional consequence of dermis-specific NF-κB activation is an enhanced release of cytokines by the epidermal keratinocytes. Among these cytokines are IL1-α and GM-CSF that induce the expression of elastase in the fibroblasts (Nakajima et al. 2012; Imokawa 2008) and pro-inflammatory cytokines like IL-6, IL-8, VEGF, and TNF-α which are capable of attracting inflammatory cells (Yaar and Gilchrest 2007; Yamamoto and Gaynor 2001). These are not only per se producers of proteolytic enzymes, but also they produce and deliver ROS into their surroundings thereby extending the damaging effect on the dermal matrix. Therefore, the UV-provoked inflammatory reaction is legitimately suspected to play an important role in matrix degradation associated with aged skin (Rijken and Bruijnzeel-Koomen 2011; Rijken and Bruijnzeel 2009).

8.4.6 Loss of Mechanical Stimulation

Physiology and synthetic potential of the fibroblasts in the dermis are dependent on the tensile stress the cells are subjected to (Ingber et al. 2014). The cells sense these mechanical cues via their integrin receptors by which they are attached to the surrounding ECM, e.g., to the collagen fibrils. When they come under tensile forces, either external ones or generated by cellular contractibility, the cells adopt a stretched shape. Modification and degradation of the ECM components cause loss of adhesion points for the cells' integrins, cellular detachment, and deprivation of the positive mechanical input that, under normal conditions, ensures the synthetic phenotype of fibroblasts (Kessler et al. 2001). In consequence, a vicious circle is started that leads to increased ROS, proteolysis and a reduction of collagen, as well as ECM neosynthesis (Fisher et al. 2008, 2009). As this process is self-energizing, the final result is a dramatic loss of ECM content and structural integrity of the dermis.

8.4.7 Estrogen Depletion

Intrinsic skin aging is fundamentally influenced by hormonal changes. In particular in females in postmenopausal age, the blood concentrations of estrogen and progesterone drop considerably, and this is accompanied by the typical features of skin aging, i.e., dryness, epidermal atrophy, diminution of collagen, loss of elasticity, wrinkle formation, and impaired wound healing (Bolognia et al. 1989; Brincat 2000; Rittie et al. 2008; Sumino et al. 2004). Most importantly, hormone replacement therapies could remarkably alleviate these problems (Rittie et al. 2008; Zouboulis and Makrantonaki 2011). Although this knowledge is still mainly empirical, several indications of underlying molecular mechanisms exist, e.g., of a specific activation of the MAPK- and insulin-signaling pathways as well as of an upregulation of protective antioxidative enzymes by the ligand-stimulated estrogen receptor (Yan et al. 2011; Jackson et al. 2011). Estrogen stimulation of old fibroblasts, on the other hand, did not abrogate expression of genes correlated with extrinsic skin aging (S. Gundermann and P. Boukamp, unpublished results), suggesting that estrogen deficiency in all its complexity stands for a major pathway in the regulation of intrinsic skin aging.

8.5 Stem Cells in Skin and Aging

Can all this be related to stem cell aging? As described, skin represents a highly complex tissue and accordingly contains a broad stem cell repertoire, including resident epidermal and mesenchymal stem cells (MSCs), migratory melanocytic stem cells derived from the embryonic neural crest (NC), and dendritic precursor cells differentiating into Langerhans cells.

8.5.1 Evidence for Epidermal Stem Cell Aging

One hypothesis of skin aging is that stem cells of the epidermis decrease in number and change their function with age, leading to the characteristic phenotype commonly described for aging skin (Weinberg 2006). However, and still largely underappreciated, most of the presently available data point to a different direction. Comparison of epidermal stem cells from young and old mouse skin neither identified significant differences in stem cell number and gene expression profile nor in telomere length (Giangreco et al. 2008; Stern and Bickenbach 2007). Bickenbach and coworkers further investigated old mouse epidermal stem cells for their ability to contribute to different tissues during development. Using a blastocyst implantation assay, they unequivocally demonstrated that neither the proliferative potential nor multipotency of the epidermal stem cells did decrease with aging (Liang et al. 2004).

Alternatively, it was recently reported that the frequency and cell cycle kinetics of the transit-amplifying (TA) cells, the proliferative active descendants of the stem cells, were altered in aged mouse epidermis. Performing long-term repopulation in vivo and colony formation in vitro, Charruyer et al. reported that TA cell frequency was increased with aging and that the aged TA cells persisted longer. This argues for age-dependent alterations in TA cell kinetics with an increase in the proportion of cycling keratinocytes, as well as an increase in cell cycle duration (Charruyer et al. 2009). Due to the experimental setup, however, it has to remain elusive whether this change in cell kinetics is intrinsically regulated in the aging TA population or whether environmental influences may be the reason for this phenotypic shift.

Using well-established murine HF stem cell models, Doles and Keyes (2013) came to a quite different conclusion. Keratin 15 (K15)-positive cells are one of the best characterized stem cell populations in the HF. Utilizing a K15 promoter-reporter mouse model, they showed that the K15 stem cell population increased with age. However, their clonogenic capacity was strongly diminished. They also showed functional impairment when measuring the response to ionizing radiation and demonstrated rapid development of skin lesions upon phorbol ester treatment. Global analysis of transcript expression demonstrated that in addition to the stem-like signature, the old stem cells were markedly altered in critical signal transduction cascades. In particular, Jak-Stat signaling was shown to be responsible for inhibiting stem cell function. In their study, Doles and Keyes for the first time provided evidence that HF stem cells are susceptible to age-associated changes and may contribute to aging phenotypes (Doles and Keyes 2013).

Unfortunately, the markers identified for murine HF stem cells do not account for human HF populations. Even more so, the identity of the stem cells of the interfollicular epidermis (IFE), which predominates in human skin, is largely elusive as recently reviewed (Boehnke et al. 2012). On the other hand, there is functional evidence that also human IFE contains a stem cell population which consists of slowly cycling cells and which comprises less than 1 % of the basal keratinocytes (Muffler et al. 2008). When activated, these cells express telomerase activity, which allow them to counteract proliferation-dependent telomere loss. Accordingly, age-dependent telomere loss appears marginal (Harle-Bachor and Boukamp 1996; Krunic et al. 2009), and therefore, it is unlikely to be a driving force of epidermal stem cell aging.

Based on early studies which demonstrated different colony-forming ability for stem cells, TA cells, and differentiating keratinocytes, it was suggested that aged human keratinocyte stem cells exhibit decreased colony-forming ability (Barrandon and Green 1987). A first study on human HF stem cells also reported on a decrease in stem cell-related holoclones with age (Lecardonnel et al. 2013). Due to the small sample size, the authors were not able to sort for potential stem cell populations and instead investigated rather undefined cell populations. Although not studied extensively, our studies do not confirm this notion. We have, as yet, no indication that keratinocytes from older donors exhibit a different growth profile as determined by their proliferation potential in 2D cultures nor that they display reduced engraftment and long-term regeneration in 3D organotypic cultures (Boukamp and Stark, unpublished). It is noteworthy that also skin grafts from older individuals can essentially outlive their donors (Gallico et al. 1984; Pellegrini et al. 1999). Thus, maintaining an appropriate functional physiology into old age seems essential for survival of epidermal stem cells. This further supports the notion that they may be different from stem cells from other tissues (Winter and Bickenbach 2009).

Taken together, the concept of inhibiting epidermal stem cell function and/or number, as a major driving force of skin aging is intriguing. However, studies addressing this issue are still limited and the findings are controversial. In consequence, further detailed studies are needed to thoroughly investigate the aging processes of mouse but particularly also of human skin.

8.5.2 Aging of the Epidermal Melanocyte and Langerhans Cells

Two other types of cells resident in the epidermis are essential protagonists in skin physiology. These are the Langerhans cells (LCs) and the melanocytes.

LCs represent the dendritic cell subset which provides the first immune barrier for invading pathogens and in addition have been implicated in tolerance induction (for review see (McKenzie et al. 2006)). They originate from bone marrow precursors and were described as a nonlymphoid tissue contingent of dendritic cells in the skin (Schuler and Steinman 1985). LCs are unique in their development compared to other dendritic cells in that they are long-lived and exhibit self-renewal capacity in the epidermis. For this reason, it is presently still unclear whether self-renewal is a trait of all LCs or whether a population of LC precursor cells exists in the epidermis that are responsible for the observed stem cell characteristic (Lako et al. 2002; Ghigo et al. 2013). It is further suggested that a local pool of proliferating hematopoietic precursor cells exist that populate the skin during embryonic development (Hoetzenecker et al. 2005; Gago et al. 2009; Schuster et al. 2009) or, respectively, that fetal liver-derived LC precursors with myelo-monocytic phenotype, similar to primitive yolk sac macrophages, contribute substantially

(Hoeffel et al. 2012). Furthermore, recently two types of bone marrow-derived LCs were described, long-term and short-term LCs. Long-term LCs are transcriptionally regulated by the inhibitor of DNA binding, ID2, a TGFβ1 target gene, thus suggesting that TGFβ signaling is not only required for the development of LCs (Hacker et al. 2003) but also absolutely essential for LC homeostasis. Accordingly, TGFβ-deficient mice lack LCs completely (Borkowski et al. 1996), and TGFβ receptor-deficient mice show reduced numbers of them (Slominski and Paus 1993; Kel et al. 2010). An excellent review on the regulatory networks controlling LCs in skin was recently provided by (Hieronymus et al. 2014). Addressing LC development and function during aging of mice, no differences in the life span of LCs and apoptosis rate in aged versus young mice were found; nevertheless, their frequency and maturation were reduced in a stepwise fashion starting at 12 months of age (Xu et al. 2012). In agreement with these observations, Sprecher et al. proposed that the age-dependent decrease in LC density could result from a deficiency in local progenitors (Sprecher et al. 1990). Xu et al. also reported on a distinct aging-specific miRNA gene expression profile which might interfere with LC-related signaling pathways and elicit age-related developmental and functional defects in the LCs. In general, aging of LCs can be considered a combination of both, loss of LC number and function (Xu et al. 2012).

Similar as LCs, also melanocytes represent an important cellular fraction of the epidermis. They are specialized neural crest-derived cells that synthesize and distribute the pigment melanin packed in melanosomes to their neighboring keratinocytes. Located in the basal layer of the IFE, they spread their dendrites and contact up to 38 basal keratinocytes, thereby achieving transfer of melanosomes over a defined distance. In humans, melanocytes are located throughout all regions of the skin, i.e., the IFE, dermis, and HF, while in mouse skin, melanocytes are largely restricted to HFs with the exception of hairless regions such as ear, tail, and ventral paws. In mice, melanocyte precursor cells, the melanoblasts, are primarily formed in the neural crest during embryogenesis and then migrate through the dermis and epidermis into the bulge region of the developing HF. This process is dependent on c-kit and its ligand, the stem cell factor SCF (reviewed in (Sarin and Artandi 2007). During hair cycle, the quiescent bulge melanocyte stem cells proliferate and move into the outer root sheath of the HF (Nishimura et al. 2005). Upon differentiation and movement into the lowest part of the HF, the bulb, they gain the ability for production of melanin, which then is transferred as melanosomes into adjacent keratinocytes, thereby giving rise to the pigmentation of the growing hair.

Hair graying is a prominent feature of aging, but also in the IFE pigmentation is generally decreasing with age, as excellently reviewed by Sarin and Artandi (2007). This process is associated with loss of melanocytes and melanocyte stem cells (Steingrimsson et al. 2005). Actually, it was shown that the number of unpigmented melanocytes decreases with age rendering elderly skin (age >70 years) almost free of them (Commo et al. 2004). This implies that hair graying is due to age-dependent depletion rather than dysfunction of the melanocytes. The molecular mechanism for melanocyte depletion is still elusive. It is assumed that oxidative damage as a result of melanin synthesis may be involved, an argument supported by the fact that no melanocyte loss was seen in albino mice (Johnson and Jackson 1992). Another potential mechanism is age-dependent telomere loss, which may, upon a critical length, drive the cells into senescence or apoptosis. However, when we surveyed human skin from different-age donors for melanocytes, we did not detect that they declined from the IFE with age (Krunic et al. 2009). This questions the hypothesis that loss of melanocytes is a general consequence of skin aging. It is also of interest to note that hair graying can already occur at early age in healthy humans and may long precede the aging process of other tissues thus making these individuals an excellent cohort for studying the mechanisms behind melanocyte aging.

8.5.3 Stem Cell Aging in the Dermis

Last but not least, dermal stem cells are supposed to contribute to skin aging. Unfortunately, stem cells in the dermis have been characterized poorly, as yet, although the dermis represents a larger reservoir for adult stem cells than epidermis and hair follicles together (Shi et al. 2006).

A minor subpopulation of stem cells in the dermis fulfills the common criteria of mesenchymal stem cells (MSCs). In general, these MSCs are considered to be a cell pool with high regenerative capacities to ensure maintenance of tissue function and homeostasis throughout the life span.

In 2001, Toma et al. were the first to isolate MSCs from the murine dermis, and they called them skin-derived precursor cells (Toma et al. 2001). Some years later, multipotent stem cells were also isolated from the human dermis (Chen et al. 2007; Lorenz et al. 2008; Lysy et al. 2007; Toma et al. 2005). Besides differentiating into cells of the mesenchymal lineage, cutaneous MSCs are also able to differentiate in vitro into insulin-producing pancreatic cells (Shi and Cheng 2004), hepatocytes (Lysy et al. 2007), and keratinocytes (Crigler et al. 2007; Medina et al. 2006). Furthermore, MSCs were shown to trans-differentiate in vitro into neurons, Schwann cells, and astrocytes (Toma et al. 2001, 2005; Dyce et al. 2004; Gingras et al. 2007; Shih et al. 2005; Fernandes et al. 2006; McKenzie et al. 2006).

In skin, MSCs have been localized in the connective tissue sheath surrounding the HF, the follicular papilla, and in the adipose tissue of the hypodermis (Jahoda et al. 2003; Lako et al. 2002; Hoogduijn et al. 2006).

The mechanisms of MSC aging in general are described in detail in Chap. 11, so that we here only refer to dermally derived MSCs. Accordingly, Gago et al. reported on an age-dependent functional depletion of the stem cell pool in the human skin (Gago et al. 2009). In their study, they analyzed skin-derived precursors isolated from more than 100 individuals from 8 months to 85 years and finally showed that the aging process diminished their cell pool and/or their differentiation potential (Gago et al. 2009). Although the underlying mechanism is unclear, this age-dependent decline could contribute to an impaired tissue homeostasis.

The "classical" bone marrow-derived MSCs are defined by their potential to differentiate into three linages, i.e., bone, cartilage, and fat (da Silva Meirelles et al. 2008; Chen et al. 2007). Unfortunately, molecular markers differentiating dermal MSCs from fibroblasts are rare. A recent study comparing the expression profile of

MSCs from bone marrow, skin, and adipose tissue with dermal fibroblasts demonstrated that fibroblasts express the same immune-phenotypic markers and the same gene transcripts that are known to be expressed in stem cells (Brohem et al. 2013). Since selection for MSCs is presently still restricted to selective media (MSC basal medium), a certain bias remains. It is also noteworthy that also the crude mass population of dermal fibroblasts is able to differentiate into the three abovementioned lineages (Junker et al. 2010; Brohem et al. 2013). This finding strongly argues for the high intrinsic plasticity of fibroblasts and casts major doubts on our present understanding of dermal MSCs. Correspondingly, we have substantial evidence for dermal fibroblast to promote the aging process (Tigges et al. 2014). In particular, we can now show that extrinsic skin aging induces a phenotype that is best described by programming of the dermal fibroblasts to a chondrocyte transdifferentiation phenotype (Gundermann et al. 2015). As this specific phenotypic switch relies on a rise in TGFβ, the age-dependent upregulation of TGFβ is apparently an important molecular trigger in the process of extrinsic and potentially also intrinsic skin aging. Moreover, we found that fibroblasts having undergone this phenotypic switch drastically lose their potential to support epidermal keratinocytes in 3D organotypic cultures, which prompted us to hypothesize that the dermal fibroblasts are in the driver's seat at the process of skin aging.

Conclusions

Skin aging is a multiplex process as it encompasses a multitude of endogenous changes still poorly understood as well as numerous extrinsic noxae attacking the tissue, inducing molecular damage, and triggering cellular responses, which become superimposed on the intrinsic changes. We propose that physical impacts mainly hit the dermal fibroblasts in skin that are then reprogrammed to become drivers in the aging process while the epidermal keratinocytes are merely responders being less susceptible to damage than the fibroblasts.

Whether there is a particular role for stem cell aging remains a matter of debate. There is still an urgent need of improved understanding of the different stem cell types that requires reliable means to precisely define, isolate, and maintain pure stem cell populations. Indeed, neither MSCs nor epidermal stem cells are at present sufficiently well defined for human skin (reviewed in Boehnke et al. 2012), thus making all statements about stem cell aging in human skin still highly biased. We even would go further and challenge the concept that aging must affect stem cells directly. Protection and support of the dermal fibroblasts may be sufficient to prevent many of the signs commonly related to skin aging. Clearly, this cannot provide a total prevention of aging skin. The fact that the menopausal decline in skin texture caused by estrogen depletion can be largely reverted by hormone replacement therapy (for review see (Thornton 2013)) suggests that certain aspects of the aging process can be modulated – presumably by amelioration of the cellular microenvironment. In line with this, skin aging is definitively a more severe process in females than in males, thus questioning the general approach in searching for global molecular programs in aging of the different tissues without the gender aspect being accounted precisely and carefully.

Acknowledgment The authors are grateful to Monika Bock for her help in editing the manuscript. This work was supported by grants from the German Research Foundation (SFB873) and the BMBF GerontoSys Stromal Aging (031.5576B) and UVA Kompetenzverbund (02NUK003A).

References

Alberts B, Johnson A, Lewis J, Raff M, Roberts K, Water P (2004) Molekularbiologie der Zelle, vol 4. Wiley-VCH Verlag GmbH, Weinheim

Angel P, Szabowski A, Schorpp-Kistner M (2001) Function and regulation of AP-1 subunits in skin physiology and pathology. Oncogene 20(19):2413–2423. doi:10.1038/sj.onc.1204380

Ashworth JL, Murphy G, Rock MJ, Sherratt MJ, Shapiro SD, Shuttleworth CA, Kielty CM (1999) Fibrillin degradation by matrix metalloproteinases: implications for connective tissue remodelling. Biochem J 340(Pt 1):171–181

Ayouaz A, Raynaud C, Heride C, Revaud D, Sabatier L (2008) Telomeres: hallmarks of radiosensitivity. Biochimie 90(1):60–72. doi:10.1016/j.biochi.2007.09.011

Bailey AJ (2001) Molecular mechanisms of ageing in connective tissues. Mech Ageing Dev 122(7):735–755

Bailey AJ, Paul RG, Knott L (1998) Mechanisms of maturation and ageing of collagen. Mech Ageing Dev 106(1–2):1–56

Barrandon Y, Green H (1987) Three clonal types of keratinocyte with different capacities for multiplication. Proc Natl Acad Sci U S A 84(8):2302–2306

Baumann L (2007) Skin ageing and its treatment. J Pathol 211(2):241–251. doi:10.1002/path.2098

Berneburg M, Gattermann N, Stege H, Grewe M, Vogelsang K, Ruzicka T, Krutmann J (1997) Chronically ultraviolet-exposed human skin shows a higher mutation frequency of mitochondrial DNA as compared to unexposed skin and the hematopoietic system. Photochem Photobiol 66(2):271–275

Berneburg M, Plettenberg H, Krutmann J (2000) Photoaging of human skin. Photodermatol Photoimmunol Photomed 16(6):239–244

Bernstein EF, Chen YQ, Kopp JB, Fisher L, Brown DB, Hahn PJ, Robey FA, Lakkakorpi J, Uitto J (1996a) Long-term sun exposure alters the collagen of the papillary dermis. Comparison of sun-protected and photoaged skin by northern analysis, immunohistochemical staining, and confocal laser scanning microscopy. J Am Acad Dermatol 34(2 Pt 1):209–218

Bernstein EF, Underhill CB, Hahn PJ, Brown DB, Uitto J (1996b) Chronic sun exposure alters both the content and distribution of dermal glycosaminoglycans. Br J Dermatol 135(2):255–262

Berton A, Godeau G, Emonard H, Baba K, Bellon P, Hornebeck W, Bellon G (2000) Analysis of the ex vivo specificity of human gelatinases A and B towards skin collagen and elastic fibers by computerized morphometry. Matrix Biol 19(2):139–148

Bhushan M, Cumberbatch M, Dearman RJ, Andrew SM, Kimber I, Griffiths CE (2002) Tumour necrosis factor-alpha-induced migration of human Langerhans cells: the influence of ageing. Br J Dermatol 146(1):32–40

Birkedal-Hansen H, Moore WG, Bodden MK, Windsor LJ, Birkedal-Hansen B, DeCarlo A, Engler JA (1993) Matrix metalloproteinases: a review. Crit Rev Oral Biol Med 4(2):197–250

Boehnke K, Falkowska-Hansen B, Stark HJ, Boukamp P (2012) Stem cells of the human epidermis and their niche: composition and function in epidermal regeneration and carcinogenesis. Carcinogenesis 33(7):1247–1258. doi:10.1093/carcin/bgs136

Bolognia JL, Braverman IM, Rousseau ME, Sarrel PM (1989) Skin changes in menopause. Maturitas 11(4):295–304

Borkowski TA, Letterio JJ, Farr AG, Udey MC (1996) A role for endogenous transforming growth factor beta 1 in Langerhans cell biology: the skin of transforming growth factor beta 1 null mice is devoid of epidermal Langerhans cells. J Exp Med 184(6):2417–2422

Bornstein P, Sage EH (2002) Matricellular proteins: extracellular modulators of cell function. Curr Opin Cell Biol 14(5):608–616

Bosset S, Bonnet-Duquennoy M, Barre P, Chalon A, Lazou K, Kurfurst R, Bonte F, Schnebert S, Disant F, Le Varlet B, Nicolas JF (2003) Decreased expression of keratinocyte beta1 integrins in chronically sun-exposed skin in vivo. Br J Dermatol 148(4):770–778

Branchet MC, Boisnic S, Frances C, Robert AM (1990) Skin thickness changes in normal aging skin. Gerontology 36(1):28–35

Breathnach AS, Wyllie LM (1964) Electron microscopy of melanocytes and melanosomes in freckled human epidermis. J Invest Dermatol 42:389–394

Brincat MP (2000) Hormone replacement therapy and the skin. Maturitas 35(2):107–117

Brohem CA, de Carvalho CM, Radoski CL, Santi FC, Baptista MC, Swinka BB, de A Urban C, de Araujo LR, Graf RM, Feferman IH, Lorencini M (2013) Comparison between fibroblasts and mesenchymal stem cells derived from dermal and adipose tissue. Int J Cosmet Sci 35(5): 448–457. doi:10.1111/ics.12064

Camougrand N, Rigoulet M (2001) Aging and oxidative stress: studies of some genes involved both in aging and in response to oxidative stress. Respir Physiol 128(3):393–401

Campisi J (1996) Replicative senescence: an old lives' tale? Cell 84(4):497–500

Charruyer A, Barland CO, Yue L, Wessendorf HB, Lu Y, Lawrence HJ, Mancianti ML, Ghadially R (2009) Transit-amplifying cell frequency and cell cycle kinetics are altered in aged epidermis. J Invest Dermatol 129(11):2574–2583. doi:10.1038/jid.2009.127

Chen VL, Fleischmajer R, Schwartz E, Palaia M, Timpl R (1986) Immunochemistry of elastotic material in sun-damaged skin. J Invest Dermatol 87(3):334–337

Chen FG, Zhang WJ, Bi D, Liu W, Wei X, Chen FF, Zhu L, Cui L, Cao Y (2007) Clonal analysis of nestin(−) vimentin(+) multipotent fibroblasts isolated from human dermis. J Cell Sci 120(Pt 16):2875–2883. doi:10.1242/jcs.03478

Chung JH, Kang S, Varani J, Lin J, Fisher GJ, Voorhees JJ (2000) Decreased extracellular-signal-regulated kinase and increased stress-activated MAP kinase activities in aged human skin in vivo. J Invest Dermatol 115(2):177–182. doi:10.1046/j.1523-1747.2000.00009.x

Commo S, Gaillard O, Bernard BA (2004) Human hair greying is linked to a specific depletion of hair follicle melanocytes affecting both the bulb and the outer root sheath. Br J Dermatol 150(3):435–443. doi:10.1046/j.1365-2133.2004.05787.x

Contet-Audonneau JL, Jeanmaire C, Pauly G (1999) A histological study of human wrinkle structures: comparison between sun-exposed areas of the face, with or without wrinkles, and sun-protected areas. Br J Dermatol 140(6):1038–1047

Craven NM, Watson RE, Jones CJ, Shuttleworth CA, Kielty CM, Griffiths CE (1997) Clinical features of photodamaged human skin are associated with a reduction in collagen VII. Br J Dermatol 137(3):344–350

Crigler L, Kazhanie A, Yoon TJ, Zakhari J, Anders J, Taylor B, Virador VM (2007) Isolation of a mesenchymal cell population from murine dermis that contains progenitors of multiple cell lineages. Faseb J 21(9):2050–2063. doi:10.1096/fj.06-5880com

Cristofalo VJ, Lorenzini A, Allen RG, Torres C, Tresini M (2004) Replicative senescence: a critical review. Mech Ageing Dev 125(10–11):827–848. doi:10.1016/j.mad.2004.07.010

d'Adda di Fagagna F, Reaper PM, Clay-Farrace L, Fiegler H, Carr P, Von Zglinicki T, Saretzki G, Carter NP, Jackson SP (2003) A DNA damage checkpoint response in telomere-initiated senescence. Nature 426(6963):194–198. doi:10.1038/nature02118

da Silva Meirelles L, Caplan AI, Nardi NB (2008) In search of the in vivo identity of mesenchymal stem cells. Stem Cells 26(9):2287–2299. doi:10.1634/stemcells.2007-1122

Dimri GP, Lee X, Basile G, Acosta M, Scott G, Roskelley C, Medrano EE, Linskens M, Rubelj I, Pereira-Smith O et al (1995) A biomarker that identifies senescent human cells in culture and in aging skin in vivo. Proc Natl Acad Sci U S A 92(20):9363–9367

Doles J, Keyes WM (2013) Epidermal stem cells undergo age-associated changes. Aging (Milano) 5(1):1–2

Dyce PW, Zhu H, Craig J, Li J (2004) Stem cells with multilineage potential derived from porcine skin. Biochem Biophys Res Commun 316(3):651–658. doi:10.1016/j.bbrc.2004.02.093

El-Domyati M, Attia S, Saleh F, Brown D, Birk DE, Gasparro F, Ahmad H, Uitto J (2002) Intrinsic aging vs. photoaging: a comparative histopathological, immunohistochemical, and ultrastructural study of skin. Exp Dermatol 11(5):398–405

Engelke M, Jensen JM, Ekanayake-Mudiyanselage S, Proksch E (1997) Effects of xerosis and ageing on epidermal proliferation and differentiation. Br J Dermatol 137(2):219–225

Fenske NA, Lober CW (1986) Structural and functional changes of normal aging skin. J Am Acad Dermatol 15(4 Pt 1):571–585

Fernandes KJ, Kobayashi NR, Gallagher CJ, Barnabe-Heider F, Aumont A, Kaplan DR, Miller FD (2006) Analysis of the neurogenic potential of multipotent skin-derived precursors. Exp Neurol 201(1):32–48. doi:10.1016/j.expneurol.2006.03.018

Fisher GJ, Datta SC, Talwar HS, Wang ZQ, Varani J, Kang S, Voorhees JJ (1996) Molecular basis of sun-induced premature skin ageing and retinoid antagonism. Nature 379(6563):335–339. doi:10.1038/379335a0

Fisher GJ, Talwar HS, Lin J, Lin P, McPhillips F, Wang Z, Li X, Wan Y, Kang S, Voorhees JJ (1998) Retinoic acid inhibits induction of c-Jun protein by ultraviolet radiation that occurs subsequent to activation of mitogen-activated protein kinase pathways in human skin in vivo. J Clin Invest 101(6):1432–1440. doi:10.1172/JCI2153

Fisher GJ, Kang S, Varani J, Bata-Csorgo Z, Wan Y, Datta S, Voorhees JJ (2002) Mechanisms of photoaging and chronological skin aging. Arch Dermatol 138(11):1462–1470

Fisher GJ, Varani J, Voorhees JJ (2008) Looking older: fibroblast collapse and therapeutic implications. Arch Dermatol 144(5):666–672. doi:10.1001/archderm.144.5.666

Fisher GJ, Quan T, Purohit T, Shao Y, Cho MK, He T, Varani J, Kang S, Voorhees JJ (2009) Collagen fragmentation promotes oxidative stress and elevates matrix metalloproteinase-1 in fibroblasts in aged human skin. Am J Pathol 174(1):101–114. doi:10.2353/ajpath.2009.080599

Fleischmajer R, Perlish JS, Bashey RI (1972) Human dermal glycosaminoglycans and aging. Biochim Biophys Acta 279(2):265–275

Frances C, Branchet MC, Boisnic S, Lesty CL, Robert L (1990) Elastic fibers in normal human skin. Variations with age: a morphometric analysis. Arch Gerontol Geriatr 10(1):57–67

Fritsch P (2009) Dermatologie und Venerologie für das Studium. Springer Medizin Verlag, Heidelberg

Gago N, Perez-Lopez V, Sanz-Jaka JP, Cormenzana P, Eizaguirre I, Bernad A, Izeta A (2009) Age-dependent depletion of human skin-derived progenitor cells. Stem Cells 27(5):1164–1172. doi:10.1002/stem.27

Gallico GG 3rd, O'Connor NE, Compton CC, Kehinde O, Green H (1984) Permanent coverage of large burn wounds with autologous cultured human epithelium. N Engl J Med 311(7):448–451. doi:10.1056/NEJM198408163110706

Garrone R, Lethias C, Le Guellec D (1997) Distribution of minor collagens during skin development. Microsc Res Tech 38(4):407–412. doi:10.1002/(SICI)1097-0029(19970815)38:4<407::AID-JEMT8>3.0.CO;2-F

Ghersetich I, Lotti T, Campanile G, Grappone C, Dini G (1994) Hyaluronic acid in cutaneous intrinsic aging. Int J Dermatol 33(2):119–122

Ghigo C, Mondor I, Jorquera A, Nowak J, Wienert S, Zahner SP, Clausen BE, Luche H, Malissen B, Klauschen F, Bajenoff M (2013) Multicolor fate mapping of Langerhans cell homeostasis. J Exp Med 210(9):1657–1664. doi:10.1084/jem.20130403

Giacomoni PU, Declercq L, Hellemans L, Maes D (2000) Aging of human skin: review of a mechanistic model and first experimental data. IUBMB Life 49(4):259–263. doi:10.1080/15216540050033104

Giangreco A, Qin M, Pintar JE, Watt FM (2008) Epidermal stem cells are retained in vivo throughout skin aging. Aging Cell 7(2):250–259. doi:10.1111/j.1474-9726.2008.00372.x

Gilchrest BA, Blog FB, Szabo G (1979) Effects of aging and chronic sun exposure on melanocytes in human skin. J Invest Dermatol 73(2):141–143

Gingras M, Champigny MF, Berthod F (2007) Differentiation of human adult skin-derived neuronal precursors into mature neurons. J Cell Physiol 210(2):498–506. doi:10.1002/jcp.20889

Gniadecka M, Nielsen OF, Wessel S, Heidenheim M, Christensen DH, Wulf HC (1998) Water and protein structure in photoaged and chronically aged skin. J Invest Dermatol 111(6):1129–1133. doi:10.1046/j.1523-1747.1998.00430.x

Grewe M (2001) Chronological ageing and photoageing of dendritic cells. Clin Exp Dermatol 26(7):608–612

Griffiths CE, Russman AN, Majmudar G, Singer RS, Hamilton TA, Voorhees JJ (1993) Restoration of collagen formation in photodamaged human skin by tretinoin (retinoic acid). N Engl J Med 329(8):530–535. doi:10.1056/NEJM199308193290803

Grove GL, Kligman AM (1983) Age-associated changes in human epidermal cell renewal. J Gerontol 38(2):137–142

Gundermann S, Stark HJ, Kollar J, Busch H, Börries M, Weber S, Bickenbach JR, Boukamp P (2015) TGFβ-dependent transdifferentiation of human dermal fibroblasts towards a chondrocyte phenotype is causally related to photoaging. Submitted

Gunin AG, Kornilova NK, Petrov VV, Vasil'eva OV (2011a) Age-related changes in the number and proliferation of fibroblasts in the human skin. Adv Gerontol 24(1):43–47

Gunin AG, Kornilova NK, Vasilieva OV, Petrov VV (2011b) Age-related changes in proliferation, the numbers of mast cells, eosinophils, and cd45-positive cells in human dermis. J Gerontol A Biol Sci Med Sci 66(4):385–392. doi:10.1093/gerona/glq205

Hacker C, Kirsch RD, Ju XS, Hieronymus T, Gust TC, Kuhl C, Jorgas T, Kurz SM, Rose-John S, Yokota Y, Zenke M (2003) Transcriptional profiling identifies Id2 function in dendritic cell development. Nat Immunol 4(4):380–386. doi:10.1038/ni903

Hanson KM, Simon JD (1998) Epidermal trans-urocanic acid and the UV-A-induced photoaging of the skin. Proc Natl Acad Sci U S A 95(18):10576–10578

Harle-Bachor C, Boukamp P (1996) Telomerase activity in the regenerative basal layer of the epidermis inhuman skin and in immortal and carcinoma-derived skin keratinocytes. Proc Natl Acad Sci U S A 93(13):6476–6481

Harley CB, Futcher AB, Greider CW (1990) Telomeres shorten during ageing of human fibroblasts. Nature 345(6274):458–460. doi:10.1038/345458a0

Hayflick L (1965) The limited in vitro lifetime of human diploid cell strains. Exp Cell Res 37:614–636

Hayflick L, Moorhead PS (1961) The serial cultivation of human diploid cell strains. Exp Cell Res 25:585–621

Herbig U, Jobling WA, Chen BP, Chen DJ, Sedivy JM (2004) Telomere shortening triggers senescence of human cells through a pathway involving ATM, p53, and p21(CIP1), but not p16(INK4a). Mol Cell 14(4):501–513

Hieronymus T, Zenke M, Baek JH, Seré K (2014) The clash of Langerhans cell homeostasis in skin: Should I stay or should I go? Semin Cell Dev Biol. pii: S1084-9521(14)00021-4. doi:10.1016/j.semcdb.2014.02.009. [Epub ahead of print] Review.

Hoeffel G, Wang Y, Greter M, See P, Teo P, Malleret B, Leboeuf M, Low D, Oller G, Almeida F, Choy SH, Grisotto M, Renia L, Conway SJ, Stanley ER, Chan JK, Ng LG, Samokhvalov IM, Merad M, Ginhoux F (2012) Adult Langerhans cells derive predominantly from embryonic fetal liver monocytes with a minor contribution of yolk sac-derived macrophages. J Exp Med 209(6):1167–1181. doi:10.1084/jem.20120340

Hoetzenecker W, Ecker R, Kopp T, Stuetz A, Stingl G, Elbe-Burger A (2005) Pimecrolimus leads to an apoptosis-induced depletion of T cells but not Langerhans cells in patients with atopic dermatitis. J Allergy Clin Immunol 115(6):1276–1283. doi:10.1016/j.jaci.2005.02.011

Hoogduijn MJ, Gorjup E, Genever PG (2006) Comparative characterization of hair follicle dermal stem cells and bone marrow mesenchymal stem cells. Stem Cells Dev 15(1):49–60. doi:10.1089/scd.2006.15.49

Hughes MC, Bredoux C, Salas F, Lombard D, Strutton GM, Fourtanier A, Green AC (2011) Comparison of histological measures of skin photoaging. Dermatology 223(2):140–151. doi:10.1159/000332425

Imokawa G (2008) Recent advances in characterizing biological mechanisms underlying UV-induced wrinkles: a pivotal role of fibroblast-derived elastase. Arch Dermatol Res 300(Suppl 1):S7–S20. doi:10.1007/s00403-007-0798-x

Ingber DE, Wang N, Stamenovic D (2014) Tensegrity, cellular biophysics, and the mechanics of living systems. Rep Prog Phys 77(4):046603. doi:10.1088/0034-4885/77/4/046603

Jackson RL, Greiwe JS, Schwen RJ (2011) Ageing skin: oestrogen receptor beta agonists offer an approach to change the outcome. Exp Dermatol 20(11):879–882. doi:10.1111/j.1600-0625.2011.01362.x

Jahoda CA, Whitehouse J, Reynolds AJ, Hole N (2003) Hair follicle dermal cells differentiate into adipogenic and osteogenic lineages. Exp Dermatol 12(6):849–859

Jenkins G (2002) Molecular mechanisms of skin ageing. Mech Ageing Dev 123(7):801–810

Johnson R, Jackson IJ (1992) Light is a dominant mouse mutation resulting in premature cell death. Nat Genet 1(3):226–229. doi:10.1038/ng0692-226

Junker JP, Sommar P, Skog M, Johnson H, Kratz G (2010) Adipogenic, chondrogenic and osteogenic differentiation of clonally derived human dermal fibroblasts. Cells Tissues Organs 191(2):105–118. doi:10.1159/000232157

Kanitakis J (2002) Anatomy, histology and immunohistochemistry of normal human skin. Eur J Dermatol 12(4):390–399; quiz 400–401

Kel JM, Girard-Madoux MJ, Reizis B, Clausen BE (2010) TGF-beta is required to maintain the pool of immature Langerhans cells in the epidermis. J Immunol 185(6):3248–3255. doi:10.4049/jimmunol.1000981

Kessler D, Dethlefsen S, Haase I, Plomann M, Hirche F, Krieg T, Eckes B (2001) Fibroblasts in mechanically stressed collagen lattices assume a "synthetic" phenotype. J Biol Chem 276(39):36575–36585. doi:10.1074/jbc.M101602200

Kligman AM (1969) Early destructive effect of sunlight on human skin. JAMA 210(13): 2377–2380

Klotz LO, Pellieux C, Briviba K, Pierlot C, Aubry JM, Sies H (1999) Mitogen-activated protein kinase (p38-, JNK-, ERK-) activation pattern induced by extracellular and intracellular singlet oxygen and UVA. Eur J Biochem 260(3):917–922

Kohl E, Landthaler M, Szeimies RM (2009) Skin aging. Hautarzt 60(11):917–933. doi:10.1007/s00105-009-1790-5; quiz 934

Kohl E, Steinbauer J, Landthaler M, Szeimies RM (2011) Skin ageing. J Eur Acad Dermatol Venereol 25(8):873–884. doi:10.1111/j.1468-3083.2010.03963.x

Krunic D, Moshir S, Greulich-Bode KM, Figueroa R, Cerezo A, Stammer H, Stark HJ, Gray SG, Nielsen KV, Hartschuh W, Boukamp P (2009) Tissue context-activated telomerase in human epidermis correlates with little age-dependent telomere loss. Biochim Biophys Acta 1792(4):297–308. doi:10.1016/j.bbadis.2009.02.005

Krutmann J (2003) Premature skin aging by ultraviolet radiation and other environmental hazards. The molecular basis. Hautarzt 54(9):809–817. doi:10.1007/s00105-003-0575-5

Krutmann J (2011) How the sun ages our skin. The dermis as the driving force. Hautarzt 62(8): 588–590. doi:10.1007/s00105-011-2132-y

Krutmann J, Diepgen T, Billmann-Krutmann C (2008) Hautalterung, vol 2. Springer Medizin Verlag, Heidelberg

Lako M, Armstrong L, Cairns PM, Harris S, Hole N, Jahoda CA (2002) Hair follicle dermal cells repopulate the mouse haematopoietic system. J Cell Sci 115(Pt 20):3967–3974

Langton AK, Sherratt MJ, Griffiths CE, Watson RE (2010) A new wrinkle on old skin: the role of elastic fibres in skin ageing. Int J Cosmet Sci. doi:10.1111/j.1468-2494.2010.00574.x

Lavker RM (1979) Structural alterations in exposed and unexposed aged skin. J Invest Dermatol 73(1):59–66

Lavker RM, Kligman AM (1988) Chronic heliodermatitis: a morphologic evaluation of chronic actinic dermal damage with emphasis on the role of mast cells. J Invest Dermatol 90(3):325–330

Lecardonnel J, Deshayes N, Genty G, Parent N, Bernard BA, Rathman-Josserand M, Paris M (2013) Ageing and colony-forming efficiency of human hair follicle keratinocytes. Exp Dermatol 22(9):604–606. doi:10.1111/exd.12204

Leufke C, Leykauf J, Krunic D, Jauch A, Holtgreve-Grez H, Bohm-Steuer B, Brocker EB, Mauch C, Utikal J, Hartschuh W, Purdie KJ, Boukamp P (2013) The telomere profile distinguishes two classes of genetically distinct cutaneous squamous cell carcinomas. Oncogene. doi:10.1038/onc.2013.323

Levy MZ, Allsopp RC, Futcher AB, Greider CW, Harley CB (1992) Telomere end-replication problem and cell aging. J Mol Biol 225(4):951–960

Liang L, Chinnathambi S, Stern M, Tomanek-Chalkley A, Manuel TD, Bickenbach JR (2004) As epidermal stem cells age they do not substantially change their characteristics. J Investig Dermatol Symp Proc 9(3):229–237. doi:10.1111/j.1087-0024.2004.09309.x

Lindsey J, McGill NI, Lindsey LA, Green DK, Cooke HJ (1991) In vivo loss of telomeric repeats with age in humans. Mutat Res 256(1):45–48

Lock-Andersen J, Therkildsen P, de Fine Olivarius F, Gniadecka M, Dahlstrom K, Poulsen T, Wulf HC (1997) Epidermal thickness, skin pigmentation and constitutive photosensitivity. Photodermatol Photoimmunol Photomed 13(4):153–158

Löffler G, Petrides PE, Heinrich PC (2007) Binde- und Stützgewebe. In: Deutzmann R, Bruckner-Tuderman L, Bruckner P (eds) Biochemie und Pathobiochemie, vol 8. Springer Medizin Verlag, Heidelberg, pp 461–469

Lombard DB, Chua KF, Mostoslavsky R, Franco S, Gostissa M, Alt FW (2005) DNA repair, genome stability, and aging. Cell 120(4):497–512. doi:10.1016/j.cell.2005.01.028

Lorenz K, Sicker M, Schmelzer E, Rupf T, Salvetter J, Schulz-Siegmund M, Bader A (2008) Multilineage differentiation potential of human dermal skin-derived fibroblasts. Exp Dermatol 17(11):925–932. doi:10.1111/j.1600-0625.2008.00724.x

Lovell CR, Smolenski KA, Duance VC, Light ND, Young S, Dyson M (1987) Type I and III collagen content and fibre distribution in normal human skin during ageing. Br J Dermatol 117(4):419–428

Lysy PA, Smets F, Sibille C, Najimi M, Sokal EM (2007) Human skin fibroblasts: From mesodermal to hepatocyte-like differentiation. Hepatology 46(5):1574–1585. doi:10.1002/hep.28139

Maier AB, Westendorp RG (2009) Relation between replicative senescence of human fibroblasts and life history characteristics. Ageing Res Rev 8(3):237–243. doi:10.1016/j.arr.2009.01.004

Majora M, Wittkampf T, Schuermann B, Schneider M, Franke S, Grether-Beck S, Wilichowski E, Bernerd F, Schroeder P, Krutmann J (2009) Functional consequences of mitochondrial DNA deletions in human skin fibroblasts: increased contractile strength in collagen lattices is due to oxidative stress-induced lysyl oxidase activity. Am J Pathol 175(3):1019–1029. doi:10.2353/ajpath.2009.080832

Makrantonaki E, Zouboulis CC (2007) William J. Cunliffe Scientific Awards. Characteristics and pathomechanisms of endogenously aged skin. Dermatology 214(4):352–360. doi:10.1159/000100890

Makrantonaki E, Zouboulis CC (2011) Molecular etiology of skin aging. How important is the genetic make-up? Hautarzt 62(8):582–587. doi:10.1007/s00105-011-2136-7

McKenzie IA, Biernaskie J, Toma JG, Midha R, Miller FD (2006) Skin-derived precursors generate myelinating Schwann cells for the injured and dysmyelinated nervous system. J Neurosci 26(24):6651–6660. doi:10.1523/JNEUROSCI.1007-06.2006

Medina RJ, Kataoka K, Takaishi M, Miyazaki M, Huh NH (2006) Isolation of epithelial stem cells from dermis by a three-dimensional culture system. J Cell Biochem 98(1):174–184. doi:10.1002/jcb.20757

Mera SL, Lovell CR, Jones RR, Davies JD (1987) Elastic fibres in normal and sun-damaged skin: an immunohistochemical study. Br J Dermatol 117(1):21–27

Mitchell RE (1967) Chronic solar dermatosis: a light and electron microscopic study of the dermis. J Invest Dermatol 48(3):203–220

Montagna W, Kirchner S, Carlisle K (1989) Histology of sun-damaged human skin. J Am Acad Dermatol 21(5 Pt 1):907–918

Moragas A, Castells C, Sans M (1993) Mathematical morphologic analysis of aging-related epidermal changes. Anal Quant Cytol Histol 15(2):75–82

Muffler S, Stark HJ, Amoros M, Falkowska-Hansen B, Boehnke K, Buhring HJ, Marme A, Bickenbach JR, Boukamp P (2008) A stable niche supports long-term maintenance of human epidermal stem cells in organotypic cultures. Stem Cells 26(10):2506–2515. doi:10.1634/stemcells.2007-0991

Murphy G, Docherty AJ (1992) The matrix metalloproteinases and their inhibitors. Am J Respir Cell Mol Biol 7(2):120–125

Nakajima H, Ezaki Y, Nagai T, Yoshioka R, Imokawa G (2012) Epithelial-mesenchymal interaction during UVB-induced up-regulation of neutral endopeptidase. Biochem J 443(1):297–305. doi:10.1042/BJ20111876

Naylor EC, Watson RE, Sherratt MJ (2011) Molecular aspects of skin ageing. Maturitas 69(3): 249–256. doi:10.1016/j.maturitas.2011.04.011

Nishimura EK, Granter SR, Fisher DE (2005) Mechanisms of hair graying: incomplete melanocyte stem cell maintenance in the niche. Science 307(5710):720–724. doi:10.1126/science.1099593

Olovnikov AM (1973) A theory of marginotomy. The incomplete copying of template margin in enzymic synthesis of polynucleotides and biological significance of the phenomenon. J Theor Biol 41(1):181–190

Ozawa T (1995) Mechanism of somatic mitochondrial DNA mutations associated with age and diseases. Biochim Biophys Acta 1271(1):177–189

Paul RG, Bailey AJ (1996) Glycation of collagen: the basis of its central role in the late complications of ageing and diabetes. Int J Biochem Cell Biol 28(12):1297–1310

Pellegrini G, Ranno R, Stracuzzi G, Bondanza S, Guerra L, Zambruno G, Micali G, De Luca M (1999) The control of epidermal stem cells (holoclones) in the treatment of massive full-thickness burns with autologous keratinocytes cultured on fibrin. Transplantation 68(6):868–879

Rhie G, Shin MH, Seo JY, Choi WW, Cho KH, Kim KH, Park KC, Eun HC, Chung JH (2001) Aging- and photoaging-dependent changes of enzymic and nonenzymic antioxidants in the epidermis and dermis of human skin in vivo. J Invest Dermatol 117(5):1212–1217. doi:10.1046/j.0022-202x.2001.01469.x

Rijken F, Bruijnzeel PL (2009) The pathogenesis of photoaging: the role of neutrophils and neutrophil-derived enzymes. J Investig Dermatol Symp Proc 14(1):67–72. doi:10.1038/jidsymp.2009.15

Rijken F, Bruijnzeel-Koomen CA (2011) Photoaged skin: the role of neutrophils, preventive measures, and potential pharmacological targets. Clin Pharmacol Ther 89(1):120–124. doi:10.1038/clpt.2010.221

Rittie L, Kang S, Voorhees JJ, Fisher GJ (2008) Induction of collagen by estradiol: difference between sun-protected and photodamaged human skin in vivo. Arch Dermatol 144(9): 1129–1140. doi:10.1001/archderm.144.9.1129

Rock K, Fischer JW (2011) Role of the extracellular matrix in extrinsic skin aging. Hautarzt 62(8):591–597. doi:10.1007/s00105-011-2133-x

Rubin H (2002) The disparity between human cell senescence in vitro and lifelong replication in vivo. Nat Biotechnol 20(7):675–681. doi:10.1038/nbt0702-675

Sams WM Jr, Smith JG Jr (1961) The histochemistry of chronically sun-damaged skin. An investigation of mucopolysaccharides and basophilia in actinically damaged skin using alcian blue, Mowry's, and Hicks-Matthaei stains, methylation, and saponification. J Invest Dermatol 37:447–453

Sandby-Moller J, Poulsen T, Wulf HC (2003) Epidermal thickness at different body sites: relationship to age, gender, pigmentation, blood content, skin type and smoking habits. Acta Derm Venereol 83(6):410–413. doi:10.1080/00015550310015419

Sander CS, Chang H, Salzmann S, Muller CS, Ekanayake-Mudiyanselage S, Elsner P, Thiele JJ (2002) Photoaging is associated with protein oxidation in human skin in vivo. J Invest Dermatol 118(4):618–625. doi:10.1046/j.1523-1747.2002.01708.x

Sardy M (2009) Role of matrix metalloproteinases in skin ageing. Connect Tissue Res 50(2): 132–138. doi:10.1080/03008200802585622

Sarin KY, Artandi SE (2007) Aging, graying and loss of melanocyte stem cells. Stem Cell Rev 3(3):212–217

Scharffetter-Kochanek K, Brenneisen P, Wenk J, Herrmann G, Ma W, Kuhr L, Meewes C, Wlaschek M (2000) Photoaging of the skin from phenotype to mechanisms. Exp Gerontol 35(3):307–316

Schneider LA, Wlaschek M, Scharffetter-Kochanek K (2003) Skin aging–clinical aspects and pathogenesis. J Dtsch Dermatol Ges 1(3):223–232; quiz 233–234

Schuler G, Steinman RM (1985) Murine epidermal Langerhans cells mature into potent immunostimulatory dendritic cells in vitro. J Exp Med 161(3):526–546

Schuster C, Vaculik C, Fiala C, Meindl S, Brandt O, Imhof M, Stingl G, Eppel W, Elbe-Burger A (2009) HLA-DR+ leukocytes acquire CD1 antigens in embryonic and fetal human skin and contain functional antigen-presenting cells. J Exp Med 206(1):169–181. doi:10.1084/jem.20081747

Schwartz E, Cruickshank FA, Christensen CC, Perlish JS, Lebwohl M (1993) Collagen alterations in chronically sun-damaged human skin. Photochem Photobiol 58(6):841–844

Shi CM, Cheng TM (2004) Differentiation of dermis-derived multipotent cells into insulin-producing pancreatic cells in vitro. World J Gastroenterol 10(17):2550–2552

Shi C, Zhu Y, Su Y, Cheng T (2006) Stem cells and their applications in skin-cell therapy. Trends Biotechnol 24(1):48–52. doi:10.1016/j.tibtech.2005.11.003

Shih DT, Lee DC, Chen SC, Tsai RY, Huang CT, Tsai CC, Shen EY, Chiu WT (2005) Isolation and characterization of neurogenic mesenchymal stem cells in human scalp tissue. Stem Cells 23(7):1012–1020. doi:10.1634/stemcells.2004-0125

Shuster S, Black MM, McVitie E (1975) The influence of age and sex on skin thickness, skin collagen and density. Br J Dermatol 93(6):639–643

Sjerobabski-Masnec I, Situm M (2010) Skin aging. Acta Clin Croat 49(4):515–518

Slominski A, Paus R (1993) Melanogenesis is coupled to murine anagen: toward new concepts for the role of melanocytes and the regulation of melanogenesis in hair growth. J Invest Dermatol 101(1 Suppl):90S–97S

Sprecher E, Becker Y, Kraal G, Hall E, Harrison D, Shultz LD (1990) Effect of aging on epidermal dendritic cell populations in C57BL/6 J mice. J Invest Dermatol 94(2):247–253

Steingrimsson E, Copeland NG, Jenkins NA (2005) Melanocyte stem cell maintenance and hair graying. Cell 121(1):9–12. doi:10.1016/j.cell.2005.03.021

Stern MM, Bickenbach JR (2007) Epidermal stem cells are resistant to cellular aging. Aging Cell 6(4):439–452. doi:10.1111/j.1474-9726.2007.00318.x

Sugimoto M, Yamashita R, Ueda M (2006) Telomere length of the skin in association with chronological aging and photoaging. J Dermatol Sci 43(1):43–47. doi:10.1016/j.jdermsci.2006.02.004

Sumino H, Ichikawa S, Abe M, Endo Y, Nakajima Y, Minegishi T, Ishikawa O, Kurabayashi M (2004) Effects of aging and postmenopausal hypoestrogenism on skin elasticity and bone mineral density in Japanese women. Endocr J 51(2):159–164

Thornton MJ (2013) Estrogens and aging skin. Dermato-Endocrinology 5(2):264–270. doi:10.4161/derm.23872

Tigges J, Krutmann J, Fritsche E, Haendeler J, Schaal H, Fischer JW, Kalfalah F, Reinke H, Reifenberger G, Stuhler K, Ventura N, Gundermann S, Boukamp P, Boege F (2014) The hallmarks of fibroblast ageing. Mech Ageing Dev. doi:10.1016/j.mad.2014.03.004

Toma JG, Akhavan M, Fernandes KJ, Barnabe-Heider F, Sadikot A, Kaplan DR, Miller FD (2001) Isolation of multipotent adult stem cells from the dermis of mammalian skin. Nat Cell Biol 3(9):778–784. doi:10.1038/ncb0901-778

Toma JG, McKenzie IA, Bagli D, Miller FD (2005) Isolation and characterization of multipotent skin-derived precursors from human skin. Stem Cells 23(6):727–737. doi:10.1634/stemcells.2004-0134

Trounce I, Byrne E, Marzuki S (1989) Decline in skeletal muscle mitochondrial respiratory chain function: possible factor in ageing. Lancet 1(8639):637–639

Tzellos TG, Klagas I, Vahtsevanos K, Triaridis S, Printza A, Kyrgidis A, Karakiulakis G, Zouboulis CC, Papakonstantinou E (2009) Extrinsic ageing in the human skin is associated with alterations in the expression of hyaluronic acid and its metabolizing enzymes. Exp Dermatol 18(12):1028–1035. doi:10.1111/j.1600-0625.2009.00889.x

Uitto J (1986) Connective tissue biochemistry of the aging dermis. Age-related alterations in collagen and elastin. Dermatol Clin 4(3):433–446

Uitto J (1997) Understanding premature skin aging. N Engl J Med 337(20):1463–1465. doi:10.1056/NEJM199711133372011

Uitto J, Paul JL, Brockley K, Pearce RH, Clark JG (1983) Elastic fibers in human skin: quantitation of elastic fibers by computerized digital image analyses and determination of elastin by radioimmunoassay of desmosine. Lab Invest 49(4):499–505

Velling T, Risteli J, Wennerberg K, Mosher DF, Johansson S (2002) Polymerization of type I and III collagens is dependent on fibronectin and enhanced by integrins alpha 11beta 1 and alpha 2beta 1. J Biol Chem 277(40):37377–37381. doi:10.1074/jbc.M206286200

Waller JM, Maibach HI (2006) Age and skin structure and function, a quantitative approach (II): protein, glycosaminoglycan, water, and lipid content and structure. Skin Res Technol 12(3):145–154. doi:10.1111/j.0909-752X.2006.00146.x

Warren R, Gartstein V, Kligman AM, Montagna W, Allendorf RA, Ridder GM (1991) Age, sunlight, and facial skin: a histologic and quantitative study. J Am Acad Dermatol 25(5 Pt 1):751–760

Watson JD (1972) Origin of concatemeric T7 DNA. Nat New Biol 239(94):197–201

Weinberg RA (2006) The biology of cancer. Science, Garland

West MD (1994) The cellular and molecular biology of skin aging. Arch Dermatol 130(1):87–95

Whitton JT, Everall JD (1973) The thickness of the epidermis. Br J Dermatol 89(5):467–476

Winter MC, Bickenbach JR (2009) Aging epidermis is maintained by changes in transit-amplifying cell kinetics, not stem cell kinetics. J Invest Dermatol 129(11):2541–2543. doi:10.1038/jid.2009.236

Wu S, Li H, Zhang X, Li Z (2012) Optical features for chronological aging and photoaging skin by optical coherence tomography. Lasers Med Sci. doi:10.1007/s10103-012-1069-4

Xu YP, Qi RQ, Chen W, Shi Y, Cui ZZ, Gao XH, Chen HD, Zhou L, Mi QS (2012) Aging affects epidermal Langerhans cell development and function and alters their miRNA gene expression profile. Aging (Albany N Y) 4(11):742–754

Yaar M, Gilchrest BA (2001) Ageing and photoageing of keratinocytes and melanocytes. Clin Exp Dermatol 26(7):583–591

Yaar M, Gilchrest BA (2007) Photoageing: mechanism, prevention and therapy. Br J Dermatol 157(5):874–887. doi:10.1111/j.1365-2133.2007.08108.x

Yamamoto Y, Gaynor RB (2001) Therapeutic potential of inhibition of the NF-kappaB pathway in the treatment of inflammation and cancer. J Clin Invest 107(2):135–142. doi:10.1172/JCI11914

Yamauchi M, Woodley DT, Mechanic GL (1988) Aging and cross-linking of skin collagen. Biochem Biophys Res Commun 152(2):898–903

Yan W, Zhao Z, Zhang L, Wang D, Yan L, Yin N, Wu D, Zhang F (2011) Identification of estrogen-associated intrinsic aging genes in Chinese Han female skin by cDNA microarray technology. Biomed Environ Sci 24(4):364–373. doi:10.3967/0895-3988.2011.04.007

Yen TC, Chen YS, King KL, Yeh SH, Wei YH (1989) Liver mitochondrial respiratory functions decline with age. Biochem Biophys Res Commun 165(3):944–1003

Zouboulis CC (2003) Intrinsic skin aging. A critical appraisal of the role of hormones. Hautarzt 54(9):825–832. doi:10.1007/s00105-003-0581-7

Zouboulis CC, Makrantonaki E (2011) Clinical aspects and molecular diagnostics of skin aging. Clin Dermatol 29(1):3–14. doi:10.1016/j.clindermatol.2010.07.001

Aging of Mammalian Intestinal Stem Cells

9

Tobias Sperka and Lenhard Rudolph

Abstract

The intestinal epithelium forms the internal barrier towards all components of nutriment and a multitude of bacteria. It is the most rapidly renewing tissue in mammals and keeps up basic homeostatic function over most of the organism's lifespan. Its stress capacitance is lowered with progressive age, though. This remarkably high capacity of regeneration requires the maintenance of functional stem cells. DNA damage accumulation has been implicated in limiting organ maintenance; thus, checkpoint and repair mechanisms have been devoted to keep up genomic integrity of stem and progenitor cells. Checkpoints engaged include p53, as well as upstream regulators (Exo1, ATM, ATR, Chk1) and downstream targets (p21, PUMA). Repair or clearance of damaged cells blocks cancer development but may dampen organ maintenance thus speeding up aging. Depending on the context, modulation of checkpoint responses allows accelerating or slowing down aging and age-related malignancies. The result depends on the cells affected, the level and kind of DNA damage accumulation and the clearance of damaged cells.

9.1 Introduction

Aging is considered a multifactorial process affecting different molecular systems (Finkel et al. 2007), whose individual contributions remain to be specified. According to one hypothesis, aging-associated tissue deterioration is supported by a functional decline of somatic stem cells (Sperka et al. 2012b). The different types of tissues found in mammals depend to various degrees on stem cell-based self-renewal for

T. Sperka • L. Rudolph (✉)
Division of Stem cell aging, Leibniz Institute for Age Research, Fritz Lipmann Institute e.V. (FLI), Beutenbergstr. 11, Jena 07745, Germany
e-mail: tsperka@fli-leibniz.de; klrudolph@fli-leibniz.de

H. Geiger et al. (eds.), *Stem Cell Aging: Mechanisms, Consequences, Rejuvenation*,
DOI 10.1007/978-3-7091-1232-8_9

homeostatic maintenance or stress-induced regeneration. High-turnover organs like the haematopoietic system or the intestine are strictly dependent on stem cell-mediated self-renewal since ablation of those causes organ failure in a matter of weeks (Lorenz et al. 1951) or days (Sangiorgi and Capecchi 2008), respectively. The intestine may thus suffer aging-related declines in stem cell functionality to various degrees dependent on the situation or the external insult.

We will put a focus on molecular checkpoint mechanisms that have evolved in order to preserve homeostasis in the colon and small intestine through stem and progenitor cells and delineate potential mechanisms contributing to aging-associated deficits. An incorrect cellular quality control, inaccurate repair or improper fail-safe mechanisms may lead to the loss of healthy cells or survival of damaged cells thus causing aging-associated loss of organ maintenance, reduced stress tolerance or cancer formation.

9.2 Aging and Aging-Related Pathologies of the Intestine

The intestine is subject to aging and the development of aging-associated pathologies, some of which are linked to stem cells. The primary task of the gastrointestinal tract is to assure proper nutrition: the mammalian intestine seems to provide enough functional reserve due to its size such that aging individuals will not suffer dramatic nutritional problems under non-stressed conditions. However, in the elderly, intestinal motor dysfunction or a lowered absorptive capacity may complicate digestive function (Salles 2007), yet a more prevalent deficiency seems to develop in the form of an inability to cope with acute or chronic stress situations. Severe dietary restriction in aged rats elicits much more pronounced weight loss than in young rats, demonstrating the reduced adaptive response at old age (Chambon-Savanovitch et al. 1999). Likewise, humans tend to become undernourished after periods of severe stress, making intensive nutritional care obligate (Salles 2007; Roberts et al. 1994). The underlying mechanisms are likely to be complex but may involve absorptive epithelia of lowered functionality (Woudstra and Thomson 2002).

Ischemic colitis is caused by a partial or complete block of the intestinal blood supply inflicting ischemia/reperfusion damage to the tissue, which in most cases is completely repaired by enhanced epithelial proliferation and differentiation (El-Assal and Besner 2004; Franzin et al. 1983; Itoh et al. 2000). The elderly suffer complications that may even pose severe health threats to certain individuals, supporting the idea of reduced stress tolerance at old age (Higgins et al. 2004; Stamatakos et al. 2009).

Ionizing irradiation used during cancer treatment is another source of stress affecting high-turnover organs like the intestinal epithelium. Experiments comparing young and older mice showed a delay in organ regeneration in the older group after gamma irradiation (Hamilton and Franks 1980; Martin et al. 1998b), and clinical practice ascertains a trend towards complications in patients beyond 60 years of age after receiving pelvic irradiation, which likely affects the intestine as well (Corazza et al. 1998; Corn et al. 1994).

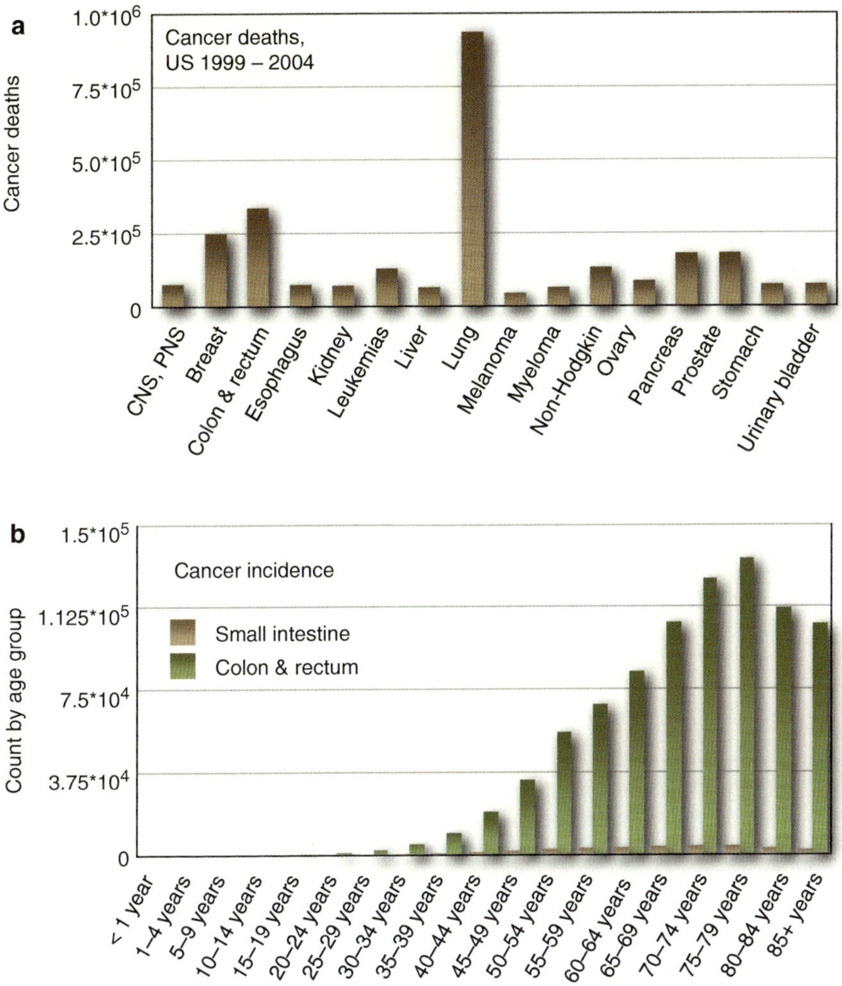

Fig. 9.1 Cancer incidence in the small intestine, colon and rectum increases with age. (**a**) Intestinal cancer is the second biggest cancer threat for human beings after lung cancer. (**b**) The colon and rectum are much more prone to cancer development, whereas the small intestine seems to be more protected (The data is retrieved from the Center for Disease Control WONDER database (http://wonder.cdc.gov) and covers the United States cancer statistics from 1999 until 2004)

Most problematic with advancing age is the increasing rate of tumour development affecting the intestine inflicting the second most cancer deaths after lung cancer (Fig. 9.1). It has recently been shown in several mouse models that intestinal stem cells are the cell type of origin for adenoma development (Barker et al. 2009; Powell et al. 2012; Sangiorgi and Capecchi 2008).

The examples above indicate that a young, healthy organism is able to respond to a variety of external stresses with different magnitudes. The young are able to balance dietary starvation, ischemia/reperfusion damage or irradiation within limits and bring about equilibrium, whereas the old encounter complications much earlier. All processes mentioned are influenced to various degrees by the intestinal epithelium and its renewal capacity suggesting that the dynamic range of appropriate stress responses becomes narrower upon old age. The reasons for these age-associated changes can be speculated upon and may partially be explained by stem cell dysfunction due to damage accumulation.

9.3 Intestinal Stem Cells

Epithelial stem cells in the small intestine reside at the bottom of narrow invaginations coined as the crypts of Lieberkühn. Dividing stem cells give rise to highly proliferative progenitor cells, so-called transit amplifying cells, which finally start to differentiate, producing absorptive and secretory cells (van der Flier and Clevers 2009). Differentiated cells are mainly located in epithelial extensions called villi where absorptive enterocytes support further breakdown of the food with subsequent nutrient uptake. Secretory goblet cells produce mucous, Paneth cells are involved in host defence, and enteroendocrine cells secrete hormones regulating digestion, blood flow and intestinal motility. Proliferation at the crypt base and apoptosis at the villus tip in combination with a cellular upward flow create a conveyor belt along the crypt-villus axis, which allows epithelial renewal in a matter of around 5 days (Fig. 9.2). The hierarchical organization of the colon is very similar to the small intestine with the exception that neither villi nor Paneth cells are generated.

Recent years gained further insight into stem cell biology of the gastrointestinal tract harnessing mouse models, which unambiguously allowed defining stem cells based on genetic marker expression and lineage-tracing experiments. Lineage tracing employs restricted Cre recombinase expression in the cell type in question and fate mapping of all genetically labelled cellular progeny produced by this cell: epithelial intestinal stem cells by definition are able to produce all differentiated cell types and persistently keep the label due to self-renewal divisions (Snippert and Clevers 2011). One of the markers for intestinal stem cells identified by these means is the seven-transmembrane receptor Lgr5 (leucine-rich repeat-containing G-protein-coupled receptor 5), which is specifically expressed in highly proliferative crypt base columnar (CBC) stem cells in the small intestine and the colon. Other experiments functionally demonstrated Bmi1- (Sangiorgi and Capecchi 2008), Tert- (Montgomery et al. 2011), and Hopx-expressing cells (Takeda et al. 2011) to be intestinal stem cells residing at the crypt position '+4', i.e. directly above Lgr5-expressing CBC stem cells. The '+4' cells are more quiescent than the proliferative CBC cells and elegant experiments demonstrated an interconversion between the cycling CBC state and the more quiescent '+4' state (Takeda et al. 2011; Montgomery et al. 2011; Tian et al. 2011; Yan et al. 2012). It has to be noted though, that the quiescent cells originally proposed to exclusively occupy crypt position '+4' may be more variable with regard to their location (Yan et al. 2012; Itzkovitz et al. 2012). These data in combination with the notion that Bmi1 is restricted to the small intestine prompted the search for markers of quiescent intestinal stem cells in the colon and led to the discovery of a subset of crypt cells

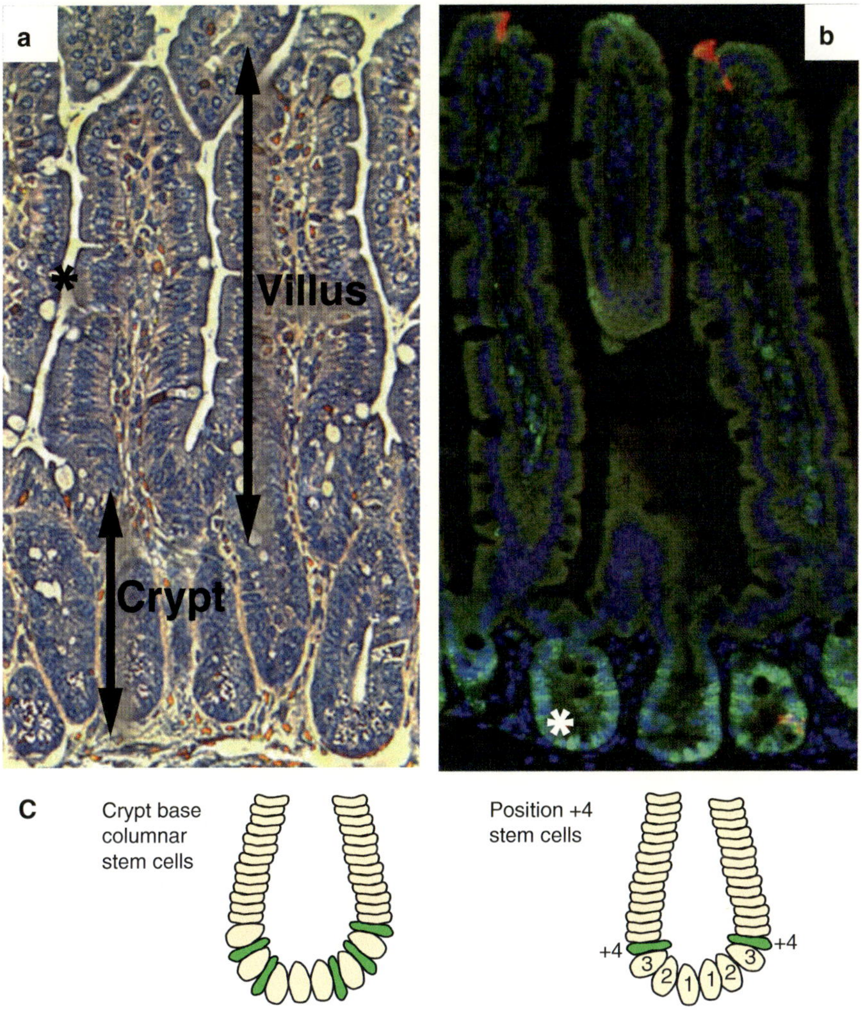

Fig. 9.2 Architecture of the small intestine. (**a**) Haematoxylin- and eosin-stained small intestine section. The small intestine is composed of luminal villi and basal crypts. The outer villus epithelium is composed of differentiated enterocytes, mucous secreting goblet cells (*asterisk*) and few enteroendocrine cells. Each villus is furthermore connected to several crypts containing proliferating cells and granular Paneth cells at the crypt bottom. (**b**) Small intestine section labelled against PCNA (*green*), cleaved caspase 3 (*red*) and DNA (*blue*). Basal crypts of Lieberkühn contain dividing stem and transit-amplifying cells and stain positive for the proliferation marker PCNA. The crypt base columnar stem cells (*asterisk*) are readily visible as cone-shaped cells at the base of the crypt interspersed with nondividing Paneth cells (*label only blue*). Shedding of exhausted cells at the tip of the villus is executed by apoptosis indicated by cleavage of caspase 3 (*red*). (**c**) Location of stem cells according to the stem cell zone (*left*) and the position +4 model (*right*)

expressing the EGF receptor inhibitor Lrig1 (leucine-rich repeats and immunoglobulin-like domains 1) (Powell et al. 2012; Wong et al. 2012). Lrig1 is expressed in the colon and small intestine and labels quiescent stem cells that can be activated to proliferate upon challenge by stress like irradiation (Powell et al. 2012).

In summary, the data suggest the existence of several stem cell states, which can be assigned in order to react to changing conditions or upcoming challenges.

Within the proliferating cell fraction, intestinal stem cells are more sensitive to ionizing irradiation than the transit-amplifying cells (Barker et al. 2007; Hua et al. 2012; Potten 1977). But it is questionable whether increased damage sensitivity would really be advantageous at the tissue level since total stem cell depletion would ultimately lead to a loss of the critical cell compartment that is required for tissue repair. In fact, recent experiments revealed an exception by demonstrating a rapid loss of Lgr5-expressing crypt base columnar stem cells upon irradiation, which is followed by a brief proliferative expansion of the Bmi1 '+4' cells. The activated Bmi1 population was suggested to subsequently give rise to new cycling Lgr5-positive crypt base columnar stem cells and afterwards re-enter the quiescent state (Yan et al. 2012). In line with this data was the demonstration that Dll1 (delta-like 1)-expressing secretory progenitor cells are also able to revert to cycling Lgr5-expressing stem cells after irradiation with ionizing radiation (van Es et al. 2012). These and other analyses reveal a cellular reserve that can be called upon under conditions of stress in order to preserve organ maintenance (Tian et al. 2011; Itzkovitz et al. 2012). The evolving concept encompasses irradiation-sensitive cycling intestinal stem cells and quiescent irradiation-resistant stem cells, which become temporarily activated in order to regenerate an irradiation injury. It will be interesting to determine if the stem cell composition is altered during the aging process and if such alteration may be involved in the changes of stress response evident at old age.

9.4 Stem Cell Protection and DNA Damage Accumulation

Different types of molecular damage affecting proteins, RNA, DNA, mitochondria and other structures accumulate during aging. There is evidence that DNA damage contributes to stem cell and organismal aging: (1) aging-dependent accumulation of DNA damage is detected in various tissues of mammals including the intestine (Dimri et al. 1995; Herbig et al. 2006; Hasty et al. 2003; Lombard et al. 2005; Jiang et al. 2007; Dolle et al. 2000). The age-dependent accumulation of DNA damage also affects stem and progenitor cells in the intestine (Rossi et al. 2007; Rube et al. 2011; Hewitt et al. 2012; Wang et al. 2009). (2) Enforced regeneration during chronic disease is associated with the accumulation of telomere dysfunction and DNA damage, leading to regenerative exhaustion, disease progression and organ failure, for example, in ulcerative colitis (O'Sullivan et al. 2002). DNA damage is not the primary cause of these diseases, but it accumulates as a consequence of elevated rates of cell turnover, and the accumulation of DNA damage leads to disease progression. (3) Intestinal stem cells express telomerase and suffer telomere length shortening with increasing age (Schepers et al. 2011). (4) Studies in genetically engineered mice with defective telomere maintenance and DNA repair capacity provided functional proof for the causal relation between DNA damage accrual and intestinal dysfunction, leading to accelerated aging (Rudolph et al. 1999; Murga et al. 2009; de Boer et al. 2002; Niedernhofer et al. 2006).

Given the highly proliferative renewal of the intestinal epithelium in the face of potentially noxious and pathogenic species (Garrett et al. 2010), the maintenance of functional stem cells appears to be of utmost importance to ensure functional integrity of the intestine during aging. Several mechanisms could be involved in protecting intestinal stem cells from damage accumulation: (1) intestinal stem cells reside in an anatomically protected site, the crypt, which is even further occluded by copious amounts of mucous, greatly reducing exposure to potentially toxic luminal content. Paneth cells at the crypt bottom additionally secrete bactericidal components (Bevins and Salzman 2011). (2) The ABC transporter P-glycoprotein/MDR1 is transiently upregulated in the crypt and villus during regeneration after ischemia-reperfusion damage (Omae et al. 2005) though it is not vital for basic function (Schinkel et al. 1997). The activity of a multidrug-resistant protein like MDR1 is able to generate a so-called side population in fluorescence-activated cell sorting. Putative stem cells from the small intestine and colon have been isolated as a side population, but their stem cell identity was not unambiguously proven at the time (Dekaney et al. 2005; Samuel et al. 2009). Stem cells in various organ systems and cancer stem cells have been shown to express high levels of multi-drug-resistant proteins (i.e. ATP-binding cassette – ABC transporters), likely increasing the resistance of stem cells to toxic substances and preventing damage accumulation (Dean et al. 2005; Visvader and Lindeman 2008). (3) There is evidence that an asymmetric separation of damaged proteins occurs during stem cell division, keeping intestinal stem cells free of damaged proteins (Rujano et al. 2006). Such mechanism of asymmetric segregation of proteins has also been described in yeast cell division in a way that the mother will keep damaged proteins away from the daughter cell (Shcheprova et al. 2008). It is tempting to speculate that this mechanism is also active in the intestine to prevent damage accumulation in long-living stem cells. (4) Asymmetric DNA strand segregation was furthermore suggested to support genomic integrity of long-lived stem cells by maintaining an immortal, unreplicated DNA strand in stem cells to protect these against replication-induced DNA mutations (Cairns 1975; Potten et al. 1978, 2002; Quyn et al. 2010). This concept has recently been challenged by several studies arguing for a random strand separation (Escobar et al. 2011; Schepers et al. 2011; Steinhauser et al. 2012) thus leaving it open if stem cells are able to modify DNA strand separation under certain conditions. (5) In line with the latter data indicating a symmetric distribution of DNA templates are findings revealing a symmetric stem cell division in the intestinal crypt (Lopez-Garcia et al. 2010; Snippert et al. 2010). Furthermore, the crypt is populated by equipotent stem cells that neutrally compete for space at the crypt base. Cell division and cell loss support a clonal drift in each crypt eventually giving rise to a clone. Stem cell populations with reduced fitness could thus easily be replaced by more healthy stem cells supporting organ maintenance during periods of stress and aging. On the other hand, damaged cells with a growth advantage could also take over and substitute the undamaged stem cell population leading to organ dysfunction and aging or to the formation of malignancies like cancer. In this light, it is clearly necessary to prevent damage accumulation via checkpoint responses in order to preserve proper organ maintenance and function.

9.4.1 Checkpoint Responses in Intestinal Stem Cells

Studies on genetically engineered irradiated rodents have shed light on the mechanisms protecting the intestine from DNA damage accumulation and organ failure. The acute survival of mice after increasing doses of gamma irradiation is determined by the function of either the haematopoietic system or the gastrointestinal tract. Doses of around 12 Gy cause death in a matter of weeks due to failure of the haematopoietic system (Lorenz et al. 1951), whereas higher doses of around 15–18 Gy lead to death after some days due to intestinal failure (Sangiorgi and Capecchi 2008; Hua et al. 2012) suggesting some kind of protection of the intestinal stem cells up to a certain irradiation threshold.

This protection is partially mediated by a tight regulation of apoptosis and proliferation at stem- and progenitor level. On the one hand, stem and progenitor cells are protected from immediate obliteration by irradiation-induced DNA damage by expression of anti-apoptotic genes Bcl2 and Bcl-W (Bcl2l2). Mice knockout for either of the two factors suffer substantially more irradiation-induced apoptosis (Merritt et al. 1995; Pritchard et al. 2000; Pritchard et al. 1999). Interestingly, expression of Bcl2 is higher in the colon than in the small intestine, and the levels of spontaneous apoptosis rise only in the colon but not in the small intestine of Bcl2 knockout mice (Merritt et al. 1995). These findings are in line with the reduced apoptotic response in the colon compared to the small intestine (Wilson et al. 1998) indicating a higher damage tolerance or improved repair mechanisms.

On the other hand, the proliferating stem and progenitor cells display a higher sensitivity towards ionizing radiation than differentiated cells (Potten 1977; Wilson et al. 1998; Hua et al. 2012) and commence apoptosis induced by the pro-apoptotic p53 target gene *PUMA* (p53-upregulated modulator of apoptosis) (Qiu et al. 2008). The transcription factor p53 has previously been implicated to mediate the apoptotic response upon irradiation since p53 knockout mice were partially protected from irradiation-induced cell death (Merritt et al. 1994, 1997). This data suggests that organ maintenance is preserved by apoptotic removal of very damaged cells and demands that a surviving fraction of stem cells must be sustained as well.

Proliferation control via cell cycle arrest or senescence induction is another way of preserving organ homeostasis to allow DNA repair or permanently suppress cell division of a damaged cell. Key components of the DNA repair machinery are the protein kinases ATM and ATR, which are activated by DNA double- and single-strand breaks, respectively (Shiloh 2003; Cimprich and Cortez 2008). The protein kinases Chk1 and Chk2 are ATM/ATR targets, which slow down cell cycle progression by reducing cyclin-dependent kinase (CDK) activity mediated by activation of the transcription factor p53 and its targets, e.g. the CDK inhibitor p21 (Kastan and Bartek 2004). Checkpoint induction may increase the time available for proper DNA repair before replication or mitosis. Deletion of ATR or Chk1 leads to rapid stem cell depletion, which is probably linked to the essential role of these kinases in S phase progression of replicating cells (Cimprich 2007; Ruzankina et al. 2007; Greenow et al. 2009). However, it is currently not known whether ATR and Chk1 contribute to the regulation of stem cell self-renewal and quiescence. In case the damage cannot be repaired, chronic checkpoint signalling triggers either cell death or permanent cell cycle arrest, i.e. senescence (Campisi and d'Adda di Fagagna 2007).

Critically short telomeres tend to fuse, induce fusion-bridge-breakage cycles and cause chronic DNA damage thus inducing permanent cell cycle arrest (O'Sullivan and Karlseder 2010; Shay et al. 1991). Generally, it has been demonstrated that intestinal stem cells are relatively resistant to radiation by quickly engaging homologous recombination and non-homologous end-joining DNA repair machineries thus restoring genomic integrity faster than transit-amplifying cells (Hua et al. 2012).

There is some evidence on CDK inhibitors and components of the DNA damage response to regulate intestinal stem and progenitor cells. Deletion of the senescence inducer p21 allows more stem and progenitor cells to proliferate under steady-state conditions and improves survival after irradiation (George et al. 2009). In aging mice with dysfunctional telomeres, p21 has been shown to limit self-renewal of intestinal stem cells, and co-deletion of p21 extends organ maintenance (Choudhury et al. 2007). The CDK inhibitors p27 and p57 have also been implicated to regulate intestinal stem and progenitor cell behaviour by regulating cell cycle progression and affecting differentiation. Upregulation of both factors upon deletion of Notch receptors causes loss of proliferating cells and a dramatic increase in differentiation into goblet cells (Riccio et al. 2008). Age-related changes in CDK inhibitor expression may thus impact on organ homeostasis through control of stem cell self-renewal via proliferation and differentiation surveillance.

Studies on senescent rodents and elderly humans have revealed changes in basal and stress-induced levels of apoptosis and proliferation. Elderly humans display with increased basal levels of apoptosis and proliferation in the small intestine (Ciccocioppo et al. 2002; Corazza et al. 1998). Starvation and re-feeding of senescent rats are associated with increased proliferation and decreased apoptosis compared to young animals (Holt et al. 1988; Xiao et al. 2001). Elevated Akt signalling has been implicated to mediate this effect by inhibiting pro-apoptotic Bad and releasing anti-apoptotic Bcl-XL (Bcl2l1) in aging rat colon (Majumdar and Du 2006). Moreover, irradiation of senescent mice revealed a stronger apoptotic response and a growth delay compared to younger mice (Martin et al. 1998a, b). These data clearly demonstrate age-related changes in cellular stress response pathways and may contribute to the lowered stress response range in older individuals. The underlying mechanistic changes and the cell populations affected are far from being understood, though.

9.5 Checkpoints Lead to Depletion of Damaged Stem Cells

As previously outlined, checkpoint genes affect intestinal stem cells under conditions of stress and during normal homeostasis. Because DNA damage contributes to limitations in stem cell functionality and tissue homeostasis during aging, it is of special interest to study the influence of checkpoint genes in this context. Given the aging-associated decline of telomere length and telomere-associated DNA damage in intestinal stem cells (Schepers et al. 2011; Hewitt et al. 2012), *Terc*

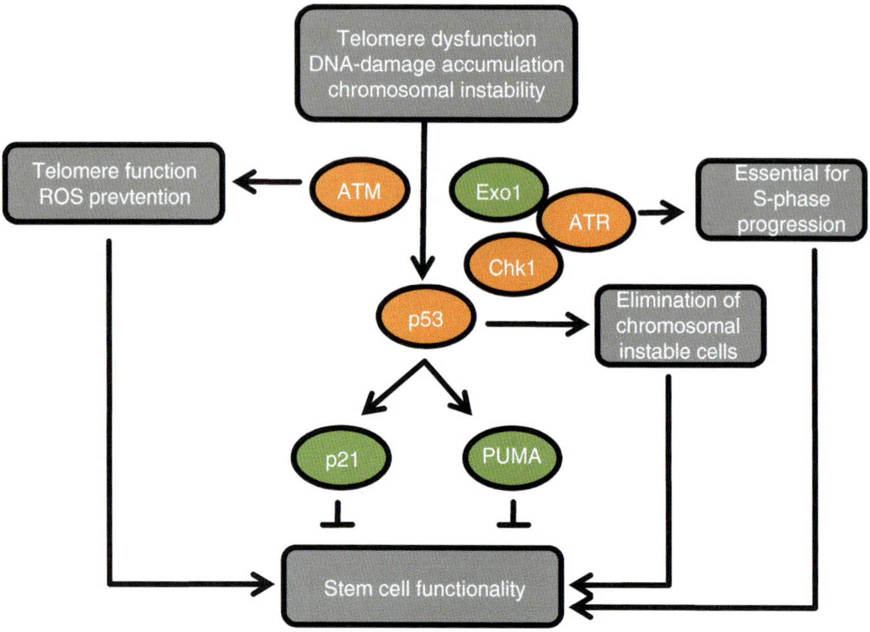

Fig. 9.3 Deletions of checkpoint genes influence DNA damage-induced impairment of somatic stem cells. Telomere dysfunction and other types of DNA damage lead to activation of the p53 pathway. In this context, deletions of checkpoint genes have positive (*green*) or negative (*red*) effects on stem cell functionality. Loss of transcriptional targets of p53 (such as p21 or PUMA (p53-upregulated modulator of apoptosis)) increases functionality and tissue maintenance. By contrast, the deletion of p53 itself leads to chromosomal instability of stem cells and accelerated tissue aging. The p53-dependent depletion of genetically instable stem cells is tissue protective. Upstream of p53 and deletion of ataxia telangiectasia mutated (*ATM*) accelerate stem cell dysfunction by increasing telomere dysfunction. Impairment of exonuclease 1 (*Exo1*)-mediated activation of ATR inhibits p53-dependent DNA damage responses and prolongs stem cell function without inducing chromosomal instability at the stem cell level. Although not directly tested in telomere-dysfunctional mice, deletion of ATR or its downstream kinase checkpoint kinase 1 (*Chk1*) is expected to have immediate adverse effects owing to its essential role in S phase progression

knockout mice with dysfunctional telomeres represent a valuable tool to study the role of checkpoint genes in DNA damage-driven aging (Rudolph et al. 1999). Studies in these aging telomere dysfunctional mice revealed some insight on the function of checkpoint genes influencing premature aging (Fig. 9.3). Interestingly, the reduced organ maintenance in mice with short telomeres is the result of impairments in stem cells, e.g. in the colon or small intestine (Rudolph et al. 1999).

9.5.1 Effects of p53 Deletion in Telomere-Dysfunctional Mice

p53 deletion in telomere-dysfunctional intestinal epithelium results in improved maintenance but chromosomal instability of intestinal stem cells, leading to defects in stem cell differentiation and accelerated atrophy of the intestinal epithelium (Begus-Nahrmann et al. 2009). These results indicate that p53 has a tissue-protective effect by removing chromosomally instable stem cells from aging somatic tissues. Decreased responsiveness of ATM–p53-dependent checkpoints occurs in aging mouse tissues (Feng et al. 2007). It is possible that this age-dependent loss of p53 checkpoint function contributes to the acceleration of tissue atrophy during aging.

9.5.2 Effects of Altered Expression of p53 Target Genes

Telomere dysfunction and irradiation-induced DNA damage also lead to increased expression of p53 target genes in somatic stem cells (Choudhury et al. 2007; Sperka et al. 2012a). In the context of telomere dysfunction, upregulation of p21 expression is associated with decreased self-renewal and functionality of stem cells, and deletion of p21 significantly rescues both stem cell self-renewal and their ability to maintain tissue integrity in the intestinal epithelium (Choudhury et al. 2007).

Telomere dysfunction also leads to an upregulation of PUMA (p53-upregulated modulator of apoptosis), a downstream target of p53 that is required for p53-dependent depletion of adult stem cells (Liu et al. 2010). Similarly to p53 deletion, PUMA deletion rescued the maintenance of intestinal stem cells in aging telomere-dysfunctional mice (Sperka et al. 2012a). However, in contrast to p53 deletion, PUMA deletion did not result in chromosomal instability of intestinal stem cells but prolonged the maintenance of the intestinal epithelium of telomere-dysfunctional mice (Sperka et al. 2012a). Cell culture studies on freshly isolated intestinal stem cells revealed that PUMA-dependent apoptosis and p21-dependent cell cycle arrest cooperate in limiting self-renewal of adult stem cells in response to telomere dysfunction. Deletion of either one of the checkpoint branches results in prolonged self-renewal of intestinal stem cells without inducing chromosomal instability. Simultaneous impairment of both checkpoint branches leads to a further increase in the self-renewal capacity of cultured intestinal stem cells with dysfunctional telomeres, but this enhancement results in chromosomal instability (Sperka et al. 2012a). Thus, each branch of the p53 checkpoint can compensate for a loss of the other branch, hence maintaining the capacity to deplete genetically instable stem cells (Sperka et al. 2012a). It is possible that chromosomal instability of stem cells contributes to the increased cancer risk that is observed in telomere-dysfunctional mice in the context of p53 deletion (Artandi et al. 2000).

9.5.3 Effects of Altering Upstream Checkpoint Components

Altering the expression of p53 inducers also affects telomere-driven aging. Loss of ATM aggravates DNA damage accumulation and accelerates tissue aging in telomere-dysfunctional mice (Wong et al. 2003). The molecular mechanisms remain to be determined but could involve the role of ATM in oxidative stress responses (Ito et al. 2004; Maryanovich et al. 2012), telomere maintenance (Smilenov et al. 1997) or replication fork stalling (Stiff et al. 2006). The first two processes have already been shown to influence stem cell maintenance, but an effect of replication fork stalling in stem cell aging remains to be investigated.

In contrast to data on ATM, deletion of exonuclease 1 (*Exo1*) improves stem cell maintenance and tissue aging in telomere-dysfunctional mice (Schaetzlein et al. 2007). Exo1 has 5'–3' exonuclease activity and is involved in DNA mismatch repair (Wei et al. 2003). Exo1 was first shown in yeast to induce DNA damage signals at dysfunctional telomeres by DNA end resection, which leads to the generation of single-stranded DNA and checkpoint induction via ATR and Chk1 (Maringele and Lydall 2002). Deletion of Exo1 impairs the processing of DNA breaks in mammalian cells with dysfunctional telomeres, which results in decreased generation of single-stranded DNA at breaks and impaired recruitment of the DNA repair factor replication protein A (RPA) (Schaetzlein et al. 2007). In telomere-dysfunctional mice in vivo, Exo1 deletion reduces the processing of DNA damage foci and blunts the induction of ATR and p53 in intestinal stem and progenitor cells. This results in decreased induction of cell cycle arrest and apoptosis, prolonged stem cell and tissue maintenance and lifespan extension (Schaetzlein et al. 2007). Loss of Exo1-dependent DNA damage signalling in telomere-dysfunctional stem and progenitor cells does not lead to an increase in chromosomal instability but impairs the formation of chromosomal fusions (Schaetzlein et al. 2007).

The studies on Exo1 deletion in telomere-dysfunctional mice reveal that Exo1-mediated processing of dysfunctional telomeres contributes to fusion formation, underlining the fatal consequences of fusions for organ maintenance (Rudolph et al. 2001; Schaetzlein et al. 2007). These studies furthermore suggest that naturally shortened telomeres may be processed in different ways dependent on the repair machinery available producing different outcomes for the organism. Intestinal stem cells are able to engage DNA repair machineries involving non-homologous end joining (NHEJ) (Hua et al. 2012). NHEJ had been implicated in telomere processing (Verdun and Karlseder 2007), and deletion of key NHEJ components DNA-PKcs (DNA protein kinase catalytic subunit) or Ku86 caused further lifespan reduction in aging telomere-dysfunctional mice (Wong et al. 2007; Maser et al. 2007; Espejel et al. 2004). DNA-PKcs or Ku86 deletion led to enforced intestinal atrophy due to additional impairments in stem and progenitor cell proliferation in mice with short telomeres compared to mice with wild-type levels of DNA-PKcs or Ku86 (Espejel et al. 2004). Levels of chromosomal fusions were unchanged in the intestinal epithelium of Terc and DNA-PKcs double knockout mice (Maser et al. 2007) indicating that the alternative processing of dysfunctional telomeres in the absence of DNA-PKcs was disadvantageous causing enforced intestinal atrophy.

Deletion of DNA mismatch repair factors, Msh2 or Pms2, both reduced intestinal atrophy in aging telomere-dysfunctional mice (Martinez et al. 2009; Siegl-Cachedenier et al. 2007). Neither telomere length nor signs of chromosomal instability were changed significantly, but checkpoint induction via p53 was reduced. The mechanisms involved are not fully clarified but implicated reduced p21 induction and rescued proliferation in intestinal crypts (Martinez et al. 2009; Siegl-Cachedenier et al. 2007). Interestingly, lifespan in the double knockout mice is not rescued since both knockouts produce an increased tumour load in mice with long telomere reserves, which is not suppressed in late-generation telomerase knockout mice (Martinez et al. 2009; Siegl-Cachedenier et al. 2007).

The examples above demonstrate the altered physiological effects of the DNA damage pathways engaged in the context of dysfunctional telomeres. Careful modulation of those may allow extending organ maintenance without increasing the danger of enhanced tumour formation.

9.6 Cancer

Recently, intestinal stem cells have been identified to be the cells of origin for intestinal adenoma formation (Barker et al. 2007; Sangiorgi and Capecchi 2008). p53-dependent checkpoint induction prevents DNA damage and mutation accrual in stem cells thus preventing tumour development at the cost of reduced organ maintenance. On the other hand, cancer incidence is increasing with progressive age (Fig. 9.1). It has already been shown that amplitudes of checkpoint responses decline with age potentially supporting accumulation of DNA mutations (Feng et al. 2007). Additionally, higher expression levels of Bcl2 in the colon compared to the small intestine (Merritt et al. 1995) may dispose the colonic stem and progenitor cells to higher damage tolerance and support the elevated neoplastic transformation.

The mutation spectrum accumulating with increasing age suggests additionally epigenetic mechanisms to play a role: in the small intestine only point mutations accumulate with progressive age, whereas the heart in comparison accumulates large genomic rearrangements (Dolle et al. 2000). The accumulation of DNA point mutations suggests insufficiencies in mismatch repair pathways, which can be caused by silencing of the mismatch repair gene *Mlh1* (Issa 2004). Methylation of genomic CpG islands has been shown to increase with progressive age (Issa et al. 1994) supporting inactivation of Mlh1 (Weisenberger et al. 2006). It will be interesting to analyse if age-associated increasing methylation of tumour suppressor genes like mismatch repair genes supports tumorigenic transformation and tumour development (Yatabe et al. 2001).

Conclusions

DNA damage checkpoints influence basic mechanisms of stem cell physiology and also promote stem cell depletion in response to DNA damage to prevent the development of cancer. These responses are clearly beneficial at a young age. However, the age-associated accumulation of DNA damage in a growing percentage of tissue stem cells can turn these protective responses into damaging responses, leading to stem cell exhaustion, tissue dysfunction and the selection of malignant stem cell clones. Alternatively, attenuation of checkpoint responses cooperating with epigenetic silencing of DNA repair mechanisms may contribute to damage accumulation, both supporting the outgrowth of dysfunctional clones. From a translational point of view, it will be important to determine the accumulation of DNA damage during natural aging. Furthermore, it is necessary to figure out the point at which checkpoint activity surpasses its beneficial effects but promotes organ dysfunction. Defined context-dependent knowledge on checkpoints may allow blocking pathways contributing to aging inducing organ dysfunction but still sustaining tumour suppression permitting healthy aging.

References

Artandi SE, Chang S, Lee SL, Alson S, Gottlieb GJ, Chin L, DePinho RA (2000) Telomere dysfunction promotes non-reciprocal translocations and epithelial cancers in mice. Nature 406(6796):641–645

Barker N, van Es JH, Kuipers J, Kujala P, van den Born M, Cozijnsen M, Haegebarth A, Korving J, Begthel H, Peters PJ, Clevers H (2007) Identification of stem cells in small intestine and colon by marker gene Lgr5. Nature 449(7165):1003–1007

Barker N, Ridgway RA, van Es JH, van de Wetering M, Begthel H, van den Born M, Danenberg E, Clarke AR, Sansom OJ, Clevers H (2009) Crypt stem cells as the cells-of-origin of intestinal cancer. Nature 457(7229):608–611

Begus-Nahrmann Y, Lechel A, Obenauf AC, Nalapareddy K, Peit E, Hoffmann E, Schlaudraff F, Liss B, Schirmacher P, Kestler H, Danenberg E, Barker N, Clevers H, Speicher MR, Rudolph KL (2009) p53 deletion impairs clearance of chromosomal-instable stem cells in aging telomere-dysfunctional mice. Nat Genet 41(10):1138–1143

Bevins CL, Salzman NH (2011) Paneth cells, antimicrobial peptides and maintenance of intestinal homeostasis. Nat Rev Microbiol 9(5):356–368. doi:10.1038/nrmicro2546

Cairns J (1975) Mutation selection and the natural history of cancer. Nature 255(5505):197–200

Campisi J, d'Adda di Fagagna F (2007) Cellular senescence: when bad things happen to good cells. Nat Rev 8(9):729–740

Chambon-Savanovitch C, Felgines C, Farges MC, Pernet P, Cezard J, Raul F, Cynober L, Vasson MP (1999) Severe dietary restriction initiated in aged rats: evidence for poor adaptation in terms of protein metabolism and intestinal functions. Eur J Clin Invest 29(6):504–511

Choudhury AR, Ju Z, Djojosubroto MW, Schienke A, Lechel A, Schaetzlein S, Jiang H, Stepczynska A, Wang C, Buer J, Lee HW, von Zglinicki T, Ganser A, Schirmacher P, Nakauchi H, Rudolph KL (2007) Cdkn1a deletion improves stem cell function and lifespan of mice with dysfunctional telomeres without accelerating cancer formation. Nat Genet 39(1):99–105

Ciccocioppo R, Di Sabatino A, Luinetti O, Rossi M, Cifone MG, Corazza GR (2002) Small bowel enterocyte apoptosis and proliferation are increased in the elderly. Gerontology 48(4): 204–208

Cimprich KA (2007) Probing ATR activation with model DNA templates. Cell Cycle (Georgetown Tex) 6(19):2348–2354

Cimprich KA, Cortez D (2008) ATR: an essential regulator of genome integrity. Nat Rev Mol Cell Biol 9(8):616–627. doi:10.1038/nrm2450

Corazza GR, Ginaldi L, Quaglione G, Ponzielli F, Vecchio L, Biagi F, Quaglino D (1998) Proliferating cell nuclear antigen expression is increased in small bowel epithelium in the elderly. Mech Ageing Dev 104(1):1–9

Corn BW, Lanciano RM, Greven KM, Noumoff J, Schultz D, Hanks GE, Fowble BL (1994) Impact of improved irradiation technique, age, and lymph node sampling on the severe complication rate of surgically staged endometrial cancer patients: a multivariate analysis. J Clin Oncol 12(3):510–515

de Boer J, Andressoo JO, de Wit J, Huijmans J, Beems RB, van Steeg H, Weeda G, van der Horst GT, van Leeuwen W, Themmen AP, Meradji M, Hoeijmakers JH (2002) Premature aging in mice deficient in DNA repair and transcription. Science (New York, NY) 296(5571): 1276–1279

Dean M, Fojo T, Bates S (2005) Tumour stem cells and drug resistance. Nat Rev Cancer 5(4): 275–284. doi:10.1038/nrc1590

Dekaney CM, Rodriguez JM, Graul MC, Henning SJ (2005) Isolation and characterization of a putative intestinal stem cell fraction from mouse jejunum. Gastroenterology 129(5): 1567–1580. doi:10.1053/j.gastro.2005.08.011

Dimri GP, Lee X, Basile G, Acosta M, Scott G, Roskelley C, Medrano EE, Linskens M, Rubelj I, Pereira-Smith O et al (1995) A biomarker that identifies senescent human cells in culture and in aging skin in vivo. Proc Natl Acad Sci U S A 92(20):9363–9367

Dolle ME, Snyder WK, Gossen JA, Lohman PH, Vijg J (2000) Distinct spectra of somatic mutations accumulated with age in mouse heart and small intestine. Proc Natl Acad Sci U S A 97(15):8403–8408

El-Assal ON, Besner GE (2004) Heparin-binding epidermal growth factor-like growth factor and intestinal ischemia-reperfusion injury. Semin Pediatr Surg 13(1):2–10

Escobar M, Nicolas P, Sangar F, Laurent-Chabalier S, Clair P, Joubert D, Jay P, Legraverend C (2011) Intestinal epithelial stem cells do not protect their genome by asymmetric chromosome segregation. Nat Commun 2:258. doi:10.1038/ncomms1260

Espejel S, Klatt P, Menissier-de Murcia J, Martin-Caballero J, Flores JM, Taccioli G, de Murcia G, Blasco MA (2004) Impact of telomerase ablation on organismal viability, aging, and tumorigenesis in mice lacking the DNA repair proteins PARP-1, Ku86, or DNA-PKcs. J Cell Biol 167(4):627–638. doi:10.1083/jcb.200407178

Feng Z, Hu W, Teresky AK, Hernando E, Cordon-Cardo C, Levine AJ (2007) Declining p53 function in the aging process: a possible mechanism for the increased tumor incidence in older populations. Proc Natl Acad Sci U S A 104(42):16633–16638. doi:10.1073/pnas.0708043104

Finkel T, Serrano M, Blasco MA (2007) The common biology of cancer and ageing. Nature 448(7155):767–774. doi:10.1038/nature05985

Franzin G, Grigioni WF, Dina R, Scarpa A, Zamboni G (1983) Mucin secretion and morphological changes of the mucosa in non-neoplastic diseases of the colon. Histopathology 7(5):707–718

Garrett WS, Gordon JI, Glimcher LH (2010) Homeostasis and inflammation in the intestine. Cell 140(6):859–870. doi:10.1016/j.cell.2010.01.023

George RJ, Sturmoski MA, May R, Sureban SM, Dieckgraefe BK, Anant S, Houchen CW (2009) Loss of p21Waf1/Cip1/Sdi1 enhances intestinal stem cell survival following radiation injury. Am J Physiol Gastrointest Liver Physiol 296(2):G245–G254. doi:10.1152/ajpgi.00021.2008

Greenow KR, Clarke AR, Jones RH (2009) Chk1 deficiency in the mouse small intestine results in p53-independent crypt death and subsequent intestinal compensation. Oncogene 28(11): 1443–1453. doi:10.1038/onc.2008.482

Hamilton E, Franks LM (1980) Cell proliferation and ageing in mouse colon. II. Late effects of repeated x-irradiation in young and old mice. Eur J Cancer 16(5):663–669

Hasty P, Campisi J, Hoeijmakers J, van Steeg H, Vijg J (2003) Aging and genome maintenance: lessons from the mouse? Science (New York, NY) 299(5611):1355–1359

Herbig U, Ferreira M, Condel L, Carey D, Sedivy JM (2006) Cellular senescence in aging primates. Science 311(5765):1257. doi:10.1126/science.1122446

Hewitt G, Jurk D, Marques FD, Correia-Melo C, Hardy T, Gackowska A, Anderson R, Taschuk M, Mann J, Passos JF (2012) Telomeres are favoured targets of a persistent DNA damage response in ageing and stress-induced senescence. Nat Commun 3:708. doi:10.1038/ncomms1708

Higgins PD, Davis KJ, Laine L (2004) Systematic review: the epidemiology of ischaemic colitis. Aliment Pharmacol Ther 19(7):729–738

Holt PR, Yeh KY, Kotler DP (1988) Altered controls of proliferation in proximal small intestine of the senescent rat. Proc Natl Acad Sci U S A 85(8):2771–2775

Hua G, Thin TH, Feldman R, Haimovitz-Friedman A, Clevers H, Fuks Z, Kolesnick R (2012) Crypt base columnar stem cells in small intestines of mice are radioresistant. Gastroenterology 143(5):1266–1276. doi:10.1053/j.gastro.2012.07.106

Issa JP (2004) CpG island methylator phenotype in cancer. Nat Rev Cancer 4(12):988–993. doi:10.1038/nrc1507

Issa JP, Ottaviano YL, Celano P, Hamilton SR, Davidson NE, Baylin SB (1994) Methylation of the oestrogen receptor CpG island links ageing and neoplasia in human colon. Nat Genet 7(4): 536–540. doi:10.1038/ng0894-536

Ito K, Hirao A, Arai F, Matsuoka S, Takubo K, Hamaguchi I, Nomiyama K, Hosokawa K, Sakurada K, Nakagata N, Ikeda Y, Mak TW, Suda T (2004) Regulation of oxidative stress by ATM is required for self-renewal of haematopoietic stem cells. Nature 431(7011):997–1002. doi:10.1038/nature02989

Itoh H, Yagi M, Fushida S, Tani T, Hashimoto T, Shimizu K, Miwa K (2000) Activation of immediate early gene, c-fos, and c-jun in the rat small intestine after ischemia/reperfusion. Transplantation 69(4):598–604

Itzkovitz S, Lyubimova A, Blat IC, Maynard M, van Es J, Lees J, Jacks T, Clevers H, van Oudenaarden A (2012) Single-molecule transcript counting of stem-cell markers in the mouse intestine. Nat Cell Biol 14(1):106–114. doi:10.1038/ncb2384

Jiang H, Ju Z, Rudolph KL (2007) Telomere shortening and ageing. Z Gerontol Geriatr 40(5):314–324

Kastan MB, Bartek J (2004) Cell-cycle checkpoints and cancer. Nature 432(7015):316–323. doi:10.1038/nature03097

Liu D, Ou L, Clemenson GD Jr, Chao C, Lutske ME, Zambetti GP, Gage FH, Xu Y (2010) Puma is required for p53-induced depletion of adult stem cells. Nat Cell Biol 12(10):993–998. doi:10.1038/ncb2100

Lombard DB, Chua KF, Mostoslavsky R, Franco S, Gostissa M, Alt FW (2005) DNA repair, genome stability, and aging. Cell 120(4):497–512. doi:10.1016/j.cell.2005.01.028

Lopez-Garcia C, Klein AM, Simons BD, Winton DJ (2010) Intestinal stem cell replacement follows a pattern of neutral drift. Science 330(6005):822–825. doi:10.1126/science.1196236

Lorenz E, Uphoff D, Reid TR, Shelton E (1951) Modification of irradiation injury in mice and guinea pigs by bone marrow injections. J Natl Cancer Inst 12(1):197–201

Majumdar AP, Du J (2006) Phosphatidylinositol 3-kinase/Akt signaling stimulates colonic mucosal cell survival during aging. Am J Physiol Gastrointest Liver Physiol 290(1):G49–G55. doi:10.1152/ajpgi.00106.2005

Maringele L, Lydall D (2002) EXO1-dependent single-stranded DNA at telomeres activates subsets of DNA damage and spindle checkpoint pathways in budding yeast yku70Delta mutants. Genes Dev 16(15):1919–1933. doi:10.1101/gad.225102

Martin K, Kirkwood TB, Potten CS (1998a) Age changes in stem cells of murine small intestinal crypts. Exp Cell Res 241(2):316–323. doi:10.1006/excr.1998.4001

Martin K, Potten CS, Roberts SA, Kirkwood TB (1998b) Altered stem cell regeneration in irradiated intestinal crypts of senescent mice. J Cell Sci 111(Pt 16):2297–2303

Martinez P, Siegl-Cachedenier I, Flores JM, Blasco MA (2009) MSH2 deficiency abolishes the anticancer and pro-aging activity of short telomeres. Aging Cell 8(1):2–17. doi:10.1111/j.1474-9726.2008.00441.x

Maryanovich M, Oberkovitz G, Niv H, Vorobiyov L, Zaltsman Y, Brenner O, Lapidot T, Jung S, Gross A (2012) The ATM-BID pathway regulates quiescence and survival of haematopoietic stem cells. Nat Cell Biol 14(5):535–541. doi:10.1038/ncb2468

Maser RS, Wong KK, Sahin E, Xia H, Naylor M, Hedberg HM, Artandi SE, DePinho RA (2007) DNA-dependent protein kinase catalytic subunit is not required for dysfunctional telomere fusion and checkpoint response in the telomerase-deficient mouse. Mol Cell Biol 27(6): 2253–2265. doi:10.1128/MCB.01354-06

Merritt AJ, Potten CS, Kemp CJ, Hickman JA, Balmain A, Lane DP, Hall PA (1994) The role of p53 in spontaneous and radiation-induced apoptosis in the gastrointestinal tract of normal and p53-deficient mice. Cancer Res 54(3):614–617

Merritt AJ, Potten CS, Watson AJ, Loh DY, Nakayama K, Hickman JA (1995) Differential expression of bcl-2 in intestinal epithelia. Correlation with attenuation of apoptosis in colonic crypts and the incidence of colonic neoplasia. J Cell Sci 108(Pt 6):2261–2271

Merritt AJ, Allen TD, Potten CS, Hickman JA (1997) Apoptosis in small intestinal epithelial from p53-null mice: evidence for a delayed, p53-independent G2/M-associated cell death after gamma-irradiation. Oncogene 14(23):2759–2766. doi:10.1038/sj.onc.1201126

Montgomery RK, Carlone DL, Richmond CA, Farilla L, Kranendonk ME, Henderson DE, Baffour-Awuah NY, Ambruzs DM, Fogli LK, Algra S, Breault DT (2011) Mouse telomerase reverse transcriptase (mTert) expression marks slowly cycling intestinal stem cells. Proc Natl Acad Sci U S A 108(1):179–184. doi:10.1073/pnas.1013004108

Murga M, Bunting S, Montana MF, Soria R, Mulero F, Canamero M, Lee Y, McKinnon PJ, Nussenzweig A, Fernandez-Capetillo O (2009) A mouse model of ATR-Seckel shows embryonic replicative stress and accelerated aging. Nat Genet 41(8):891–898. doi:10.1038/ng.420

Niedernhofer LJ, Garinis GA, Raams A, Lalai AS, Robinson AR, Appeldoorn E, Odijk H, Oostendorp R, Ahmad A, van Leeuwen W, Theil AF, Vermeulen W, van der Horst GT, Meinecke P, Kleijer WJ, Vijg J, Jaspers NG, Hoeijmakers JH (2006) A new progeroid syndrome reveals that genotoxic stress suppresses the somatotroph axis. Nature 444(7122):1038–1043. doi:10.1038/nature05456

Omae T, Goto M, Shimomura M, Masuda S, Ito K, Okuda M, Inui K (2005) Transient up-regulation of P-glycoprotein reduces tacrolimus absorption after ischemia-reperfusion injury in rat ileum. Biochem Pharmacol 69(4):561–568

O'Sullivan RJ, Karlseder J (2010) Telomeres: protecting chromosomes against genome instability. Nat Rev Mol Cell Biol 11(3):171–181. doi:10.1038/nrm2848

O'Sullivan JN, Bronner MP, Brentnall TA, Finley JC, Shen WT, Emerson S, Emond MJ, Gollahon KA, Moskovitz AH, Crispin DA, Potter JD, Rabinovitch PS (2002) Chromosomal instability in ulcerative colitis is related to telomere shortening. Nat Genet 32(2):280–284. doi:10.1038/ng989

Potten CS (1977) Extreme sensitivity of some intestinal crypt cells to X and gamma irradiation. Nature 269(5628):518–521

Potten CS, Hume WJ, Reid P, Cairns J (1978) The segregation of DNA in epithelial stem cells. Cell 15(3):899–906

Potten CS, Owen G, Booth D (2002) Intestinal stem cells protect their genome by selective segregation of template DNA strands. J Cell Sci 115(Pt 11):2381–2388

Powell AE, Wang Y, Li Y, Poulin EJ, Means AL, Washington MK, Higginbotham JN, Juchheim A, Prasad N, Levy SE, Guo Y, Shyr Y, Aronow BJ, Haigis KM, Franklin JL, Coffey RJ (2012) The pan-ErbB negative regulator Lrig1 is an intestinal stem cell marker that functions as a tumor suppressor. Cell 149(1):146–158. doi:10.1016/j.cell.2012.02.042

Pritchard DM, Potten CS, Korsmeyer SJ, Roberts S, Hickman JA (1999) Damage-induced apoptosis in intestinal epithelia from bcl-2-null and bax-null mice: investigations of the mechanistic determinants of epithelial apoptosis in vivo. Oncogene 18(51):7287–7293. doi:10.1038/sj.onc.1203150

Pritchard DM, Print C, O'Reilly L, Adams JM, Potten CS, Hickman JA (2000) Bcl-w is an important determinant of damage-induced apoptosis in epithelia of small and large intestine. Oncogene 19(34):3955–3959. doi:10.1038/sj.onc.1203729

Qiu W, Carson-Walter EB, Liu H, Epperly M, Greenberger JS, Zambetti GP, Zhang L, Yu J (2008) PUMA regulates intestinal progenitor cell radiosensitivity and gastrointestinal syndrome. Cell Stem Cell 2(6):576–583. doi:10.1016/j.stem.2008.03.009

Quyn AJ, Appleton PL, Carey FA, Steele RJ, Barker N, Clevers H, Ridgway RA, Sansom OJ, Nathke IS (2010) Spindle orientation bias in gut epithelial stem cell compartments is lost in precancerous tissue. Cell Stem Cell 6(2):175–181. doi:10.1016/j.stem.2009.12.007

Riccio O, van Gijn ME, Bezdek AC, Pellegrinet L, van Es JH, Zimber-Strobl U, Strobl LJ, Honjo T, Clevers H, Radtke F (2008) Loss of intestinal crypt progenitor cells owing to inactivation of both Notch1 and Notch2 is accompanied by derepression of CDK inhibitors p27Kip1 and p57Kip2. EMBO Rep 9(4):377–383

Roberts SB, Fuss P, Heyman MB, Evans WJ, Tsay R, Rasmussen H, Fiatarone M, Cortiella J, Dallal GE, Young VR (1994) Control of food intake in older men. JAMA 272(20):1601–1606

Rossi DJ, Bryder D, Seita J, Nussenzweig A, Hoeijmakers J, Weissman IL (2007) Deficiencies in DNA damage repair limit the function of haematopoietic stem cells with age. Nature 447(7145):725–729
Rube CE, Fricke A, Widmann TA, Furst T, Madry H, Pfreundschuh M, Rube C (2011) Accumulation of DNA damage in hematopoietic stem and progenitor cells during human aging. PLoS One 6(3), e17487. doi:10.1371/journal.pone.0017487
Rudolph KL, Chang S, Lee HW, Blasco M, Gottlieb GJ, Greider C, DePinho RA (1999) Longevity, stress response, and cancer in aging telomerase-deficient mice. Cell 96(5):701–712
Rudolph KL, Millard M, Bosenberg MW, DePinho RA (2001) Telomere dysfunction and evolution of intestinal carcinoma in mice and humans. Nat Genet 28(2):155–159
Rujano MA, Bosveld F, Salomons FA, Dijk F, van Waarde MA, van der Want JJ, de Vos RA, Brunt ER, Sibon OC, Kampinga HH (2006) Polarised asymmetric inheritance of accumulated protein damage in higher eukaryotes. PLoS Biol 4(12), e417
Ruzankina Y, Pinzon-Guzman C, Asare A, Ong T, Pontano L, Cotsarelis G, Zediak VP, Velez M, Bhandoola A, Brown EJ (2007) Deletion of the developmentally essential gene ATR in adult mice leads to age-related phenotypes and stem cell loss. Cell Stem Cell 1(1):113–126
Salles N (2007) Basic mechanisms of the aging gastrointestinal tract. Digest Dis (Basel, Switzerland) 25(2):112–117
Samuel S, Walsh R, Webb J, Robins A, Potten C, Mahida YR (2009) Characterization of putative stem cells in isolated human colonic crypt epithelial cells and their interactions with myofibroblasts. Am J Physiol Cell Physiol 296(2):C296–C305. doi:10.1152/ajpcell.00383.2008
Sangiorgi E, Capecchi MR (2008) Bmi1 is expressed in vivo in intestinal stem cells. Nat Genet 40(7):915–920
Schaetzlein S, Kodandaramireddy NR, Ju Z, Lechel A, Stepczynska A, Lilli DR, Clark AB, Rudolph C, Kuhnel F, Wei K, Schlegelberger B, Schirmacher P, Kunkel TA, Greenberg RA, Edelmann W, Rudolph KL (2007) Exonuclease-1 deletion impairs DNA damage signaling and prolongs lifespan of telomere-dysfunctional mice. Cell 130(5):863–877
Schepers AG, Vries R, van den Born M, van de Wetering M, Clevers H (2011) Lgr5 intestinal stem cells have high telomerase activity and randomly segregate their chromosomes. EMBO J 30(6):1104–1109. doi:10.1038/emboj.2011.26
Schinkel AH, Mayer U, Wagenaar E, Mol CA, van Deemter L, Smit JJ, van der Valk MA, Voordouw AC, Spits H, van Tellingen O, Zijlmans JM, Fibbe WE, Borst P (1997) Normal viability and altered pharmacokinetics in mice lacking mdr1-type (drug-transporting) P-glycoproteins. Proc Natl Acad Sci U S A 94(8):4028–4033
Shay JW, Pereira-Smith OM, Wright WE (1991) A role for both RB and p53 in the regulation of human cellular senescence. Exp Cell Res 196(1):33–39
Shcheprova Z, Baldi S, Frei SB, Gonnet G, Barral Y (2008) A mechanism for asymmetric segregation of age during yeast budding. Nature 454(7205):728–734
Shiloh Y (2003) ATM and related protein kinases: safeguarding genome integrity. Nat Rev Cancer 3(3):155–168. doi:10.1038/nrc1011
Siegl-Cachedenier I, Munoz P, Flores JM, Klatt P, Blasco MA (2007) Deficient mismatch repair improves organismal fitness and survival of mice with dysfunctional telomeres. Genes Dev 21(17):2234–2247. doi:10.1101/gad.430107
Smilenov LB, Morgan SE, Mellado W, Sawant SG, Kastan MB, Pandita TK (1997) Influence of ATM function on telomere metabolism. Oncogene 15(22):2659–2665. doi:10.1038/sj.onc.1201449
Snippert HJ, Clevers H (2011) Tracking adult stem cells. EMBO Rep 12(2):113–122. doi:10.1038/embor.2010.216
Snippert HJ, van der Flier LG, Sato T, van Es JH, van den Born M, Kroon-Veenboer C, Barker N, Klein AM, van Rheenen J, Simons BD, Clevers H (2010) Intestinal crypt homeostasis results from neutral competition between symmetrically dividing Lgr5 stem cells. Cell 143(1):134–144. doi:10.1016/j.cell.2010.09.016
Sperka T, Song Z, Morita Y, Nalapareddy K, Guachalla LM, Lechel A, Begus-Nahrmann Y, Burkhalter MD, Mach M, Schlaudraff F, Liss B, Ju Z, Speicher MR, Rudolph KL (2012a)

Puma and p21 represent cooperating checkpoints limiting self-renewal and chromosomal instability of somatic stem cells in response to telomere dysfunction. Nat Cell Biol 14(1): 73–79. doi:10.1038/ncb2388
Sperka T, Wang J, Rudolph KL (2012b) DNA damage checkpoints in stem cells, ageing and cancer. Nat Rev Mol Cell Biol 13(9):579–590. doi:10.1038/nrm3420
Stamatakos M, Douzinas E, Stefanaki C, Petropoulou C, Arampatzi H, Safioleas C, Giannopoulos G, Chatziconstantinou C, Xiromeritis C, Safioleas M (2009) Ischemic colitis: surging waves of update. Tohoku J Exp Med 218(2):83–92
Steinhauser ML, Bailey AP, Senyo SE, Guillermier C, Perlstein TS, Gould AP, Lee RT, Lechene CP (2012) Multi-isotope imaging mass spectrometry quantifies stem cell division and metabolism. Nature 481(7382):516–519. doi:10.1038/nature10734
Stiff T, Walker SA, Cerosaletti K, Goodarzi AA, Petermann E, Concannon P, O'Driscoll M, Jeggo PA (2006) ATR-dependent phosphorylation and activation of ATM in response to UV treatment or replication fork stalling. EMBO J 25(24):5775–5782. doi:10.1038/sj.emboj.7601446
Takeda N, Jain R, LeBoeuf MR, Wang Q, Lu MM, Epstein JA (2011) Interconversion between intestinal stem cell populations in distinct niches. Science (New York, NY) 334(6061): 1420–1424. doi:10.1126/science.1213214
Tian H, Biehs B, Warming S, Leong KG, Rangell L, Klein OD, de Sauvage FJ (2011) A reserve stem cell population in small intestine renders Lgr5-positive cells dispensable. Nature 478(7368):255–259. doi:10.1038/nature10408
van der Flier LG, Clevers H (2009) Stem cells, self-renewal, and differentiation in the intestinal epithelium. Annu Rev Physiol 71:241–260
van Es JH, Sato T, van de Wetering M, Lyubimova A, Nee AN, Gregorieff A, Sasaki N, Zeinstra L, van den Born M, Korving J, Martens AC, Barker N, van Oudenaarden A, Clevers H (2012) Dll1+ secretory progenitor cells revert to stem cells upon crypt damage. Nat Cell Biol 14(10):1099–1104. doi:10.1038/ncb2581
Verdun RE, Karlseder J (2007) Replication and protection of telomeres. Nature 447(7147): 924–931. doi:10.1038/nature05976
Visvader JE, Lindeman GJ (2008) Cancer stem cells in solid tumours: accumulating evidence and unresolved questions. Nat Rev Cancer 8(10):755–768. doi:10.1038/nrc2499
Wang C, Jurk D, Maddick M, Nelson G, Martin-Ruiz C, von Zglinicki T (2009) DNA damage response and cellular senescence in tissues of aging mice. Aging Cell 8(3):311–323. doi:10.1111/j.1474-9726.2009.00481.x
Wei K, Clark AB, Wong E, Kane MF, Mazur DJ, Parris T, Kolas NK, Russell R, Hou H Jr, Kneitz B, Yang G, Kunkel TA, Kolodner RD, Cohen PE, Edelmann W (2003) Inactivation of Exonuclease 1 in mice results in DNA mismatch repair defects, increased cancer susceptibility, and male and female sterility. Genes Dev 17(5):603–614. doi:10.1101/gad.1060603
Weisenberger DJ, Siegmund KD, Campan M, Young J, Long TI, Faasse MA, Kang GH, Widschwendter M, Weener D, Buchanan D, Koh H, Simms L, Barker M, Leggett B, Levine J, Kim M, French AJ, Thibodeau SN, Jass J, Haile R, Laird PW (2006) CpG island methylator phenotype underlies sporadic microsatellite instability and is tightly associated with BRAF mutation in colorectal cancer. Nat Genet 38(7):787–793. doi:10.1038/ng1834
Wilson JW, Pritchard DM, Hickman JA, Potten CS (1998) Radiation-induced p53 and p21WAF-1/CIP1 expression in the murine intestinal epithelium: apoptosis and cell cycle arrest. Am J Pathol 153(3):899–909. doi:10.1016/S0002-9440(10)65631-3
Wong KK, Maser RS, Bachoo RM, Menon J, Carrasco DR, Gu Y, Alt FW, DePinho RA (2003) Telomere dysfunction and Atm deficiency compromises organ homeostasis and accelerates ageing. Nature 421(6923):643–648
Wong KK, Maser RS, Sahin E, Bailey ST, Xia H, Ji H, McNamara K, Naylor M, Bronson RT, Ghosh S, Welsh R, DePinho RA (2007) Diminished lifespan and acute stress-induced death in DNA-PKcs-deficient mice with limiting telomeres. Oncogene 26(20):2815–2821. doi:10.1038/sj.onc.1210099
Wong VW, Stange DE, Page ME, Buczacki S, Wabik A, Itami S, van de Wetering M, Poulsom R, Wright NA, Trotter MW, Watt FM, Winton DJ, Clevers H, Jensen KB (2012) Lrig1 controls

intestinal stem-cell homeostasis by negative regulation of ErbB signalling. Nat Cell Biol 14(4):401–408. doi:10.1038/ncb2464

Woudstra T, Thomson AB (2002) Nutrient absorption and intestinal adaptation with ageing. Best Pract Res 16(1):1–15

Xiao ZQ, Moragoda L, Jaszewski R, Hatfield JA, Fligiel SE, Majumdar AP (2001) Aging is associated with increased proliferation and decreased apoptosis in the colonic mucosa. Mech Ageing Dev 122(15):1849–1864

Yan KS, Chia LA, Li X, Ootani A, Su J, Lee JY, Su N, Luo Y, Heilshorn SC, Amieva MR, Sangiorgi E, Capecchi MR, Kuo CJ (2012) The intestinal stem cell markers Bmi1 and Lgr5 identify two functionally distinct populations. Proc Natl Acad Sci U S A 109(2):466–471. doi:10.1073/pnas.1118857109

Yatabe Y, Tavare S, Shibata D (2001) Investigating stem cells in human colon by using methylation patterns. Proc Natl Acad Sci U S A 98(19):10839–10844. doi:10.1073/pnas.191225998

Aging of Muscle Stem Cells

10

Bryon R. McKay and Gianni Parise

Abstract

Skeletal Muscle has a remarkable adaptive and regenerative capacity, which is primarily due to the actions of the muscle-resident stem cell population. Aging is associated with a progressive loss in tissue function, which is associated with a decline in muscle stem cell function and total muscle stem cell pool size. Whether or not the loss in muscle stem cell function is instrumental to the progression of age-related muscle wasting (termed sarcopenia) is somewhat controversial. However, the evidence suggests that muscle stem cell dysfunction is directly related to the loss of normal muscle regeneration and response to normal physiological stimuli, which eventually leads to the loss of functional skeletal muscle. Muscle stem cell dysfunction is multifactorial and the result of alterations in local and systemic factors that we describe as niche-related or *extrinsic* factors and alterations in cell signaling and cell metabolic processes, which we describe as *intrinsic* factors. It is the combination of intrinsic and extrinsic factors that dictate the level of muscle stem cell dysfunction. Although at present it appears that the eventual decline in muscle stem cell function is inevitable, studies employing resistance exercise training regimes have shown an enhancement of the satellite

B.R. McKay, PhD (✉)
Department of Kinesiology, Faculty of Science, McMaster University, Hamilton, Ontario, ON L8S 4L8, Canada
e-mail: mckaybr@mcmaster.ca

G. Parise, PhD
Departments of Kinesiology and Medical Physics & Applied Radiation Sciences, Faculty of Science, McMaster University, Hamilton, Ontario L8S 4L8, Canada
e-mail: pariseg@mcmaster.ca

H. Geiger et al. (eds.), *Stem Cell Aging: Mechanisms, Consequences, Rejuvenation*, DOI 10.1007/978-3-7091-1232-8_10 cell pool which may mitigate or delay the decline in muscle aging. Both intrinsic and extrinsic factors must be considered in the design of novel therapeutic strategies to ensure successful and efficacious treatment of sarcopenia which is an important cause of morbidity and mortality and imposes a significant strain on the health-care system.

10.1 Overview and Importance of Muscle Stem Cells

Skeletal muscle is a multinucleated contractile tissue responsible for locomotion and plays an integral role in the maintenance of homeostasis. It is the largest organ in the body, accounting for 30–38 % of total body mass (Janssen et al. 2000). By virtue of its size relative to other organs, the skeletal muscle is the most metabolically active tissue (based on absolute energy consumption) and the principle site of protein storage in the body. Skeletal muscle is also the largest depot of glucose storage in the body and thus plays a key role in glucose homeostasis. It is a dynamic tissue capable of complete regeneration in response to overt damage (i.e., toxin-induced myotrauma) and possesses a remarkable ability for adaptation in response to physiological stimuli (i.e., exercise). For example, in response to an increase in energy demand (such as aerobic exercise), skeletal muscle is capable of upregulating oxidative machinery and inducing mitochondrial biogenesis, increasing the metabolic capacity of myofibers (Irrcher et al. 2003). Alternately, in response to heavy loads, it is capable of increasing cross-sectional area (hypertrophy) and strength (Mayhew et al. 2009). The extensive plasticity of the skeletal muscle is due, at least in part, to the presence of a tissue-specific class of stem cells that are responsible for muscle maintenance and tissue repair (Mauro 1961). Muscle-resident stem cells are indispensable for muscle regeneration, and the eventual dysfunction of these cells due to the aging process has been implicated in age-related skeletal muscle dysfunction. There are several potential mechanisms by which the aging process may adversely affect this important cell population. Although it has not been fully elucidated, the role of muscle stem cell dysfunction and importantly the loss of these progenitors in advanced age potentiate the decline in the functional capacity of skeletal muscle with age. This chapter will explore several aspects of aging, which appear to hinder muscle stem cell function and how muscle stem cell dysfunction leads to inappropriate tissue remodeling and ultimately a decline in muscle function in the later stages of life.

Although muscle stem cells have been touted as the most important progenitor for muscle growth and regeneration, there are other cell types which may contribute to myogenesis. Interstitial progenitor cells (Mitchell et al. 2010), circulating hematopoietic stem cells (LaBarge and Blau 2002), and muscle-specific stem cells known as side population cells (Gussoni et al. 1999) residing in skeletal muscle may also possess the capability to contribute to postnatal myogenesis; however, these events are rare, and the vast majority of the remarkable plasticity of skeletal muscle is due to the muscle-specific stem cell known as the satellite cell. The greater part of this chapter will focus on the satellite cell and how aging may adversely affect satellite cell function and subsequent skeletal muscle mass and function. To fully appreciate the age-related changes in satellite cell biology, it is imperative to have a basic understanding of developmental myogenesis, to which the first part of this chapter is dedicated.

10.2 Origins of Skeletal Muscle

During embryogenesis all skeletal muscles, with the exception of the muscles of the head and eyes, are derived from precursor cells from a region of paraxial mesoderm called the somites. The somite contains two sub-domains – epaxial, which mainly gives rise to the musculature of the back, and hypaxial which gives rise to the musculature of the limbs, abdomen, and thorax (Buckingham and Relaix 2007). The molecular events, which orchestrate the myogenic specification of the mesodermal precursor cells, are a combination of positive and negative signals emanating from the structures surrounding the dermomyotome, such as the neural tube and notochord (Charge and Rudnicki 2004). The main signaling molecules thought to regulate myogenesis in the embryo are Wnt proteins, Sonic hedgehog, and noggin which serve as positive growth signals and bone morphogenic proteins which serve as negative signals. These molecules regulate the expression of the myogenic regulatory factors (MRFs) and thus myogenic commitment. Following waves of migration and proliferation, precursor cells form premuscle masses in the limb buds, eventually giving rise to multinucleated primary muscle fibers (Hawke and Garry 2001). Following embryonic specification, during fetal development, the multinucleated fibers develop contractile properties (upregulation of myosin heavy chains and actin filaments) and mature into functional muscle fibers. Importantly, during the latter stages of muscle development, a distinct subpopulation of precursor cells thought to have originated from the somites (Rudnicki et al. 1993; Armand et al. 1983) resist differentiation and withdraw from the cell cycle, remaining on the periphery of the myofiber as a quiescent cell, maintaining unipotency as a progenitor cell (Kassar-Duchossoy et al. 2005; Relaix et al. 2005). Late fetal and early postnatal muscle development is dependent on these progenitor cells for myonuclear addition to support increases in cytoplasmic volume as the muscle fibers mature and hypertrophy (Hall and Ralston 1989; Petrella et al. 2008).

10.3 Muscle Stem Cells: "The Satellite Cell"

The tissue-resident stem cell known as the satellite cell is the primary driver of skeletal muscle regeneration and developmental growth. The "satellite cell" was named based on its anatomical location between the basal lamina and the plasma membrane (sarcolemma) of the muscle fiber (Mauro 1961). Although the initial study by Alexander Mauro did not classify the function of these sub-laminar cells, he speculated that satellite cells may contribute to muscle regeneration (Mauro 1961). Over the next 10 years, extensive work confirmed that satellite cells were capable of entering the cell cycle and contributing their nuclei to existing myofibers, suggesting that these cells function as resident stem cells in adult muscle (Reznik 1969; Moss and Leblond 1970). This notion gained further support from studies using isolated single fibers to illustrate that the satellite cell indeed was the source of myoblasts, which were able to proliferate and fuse together to form multinucleated myotubes (Bischoff 1975, 1986). Additionally, it was discovered that satellite cells possessed the stem cell quality of self-renewal formally classifying satellite cells as a "stem cell" population (Collins et al. 2005). Recent work (Collins et al. 2005) has demonstrated the remarkable regenerative capacity for satellite cells when as few as seven satellite cells (on an intact single fiber) were transplanted into radiation-ablated muscle and were able to give rise to a sufficient number of fusion-competent progeny to generate thousands of myonuclei and fully repopulate the satellite cell compartment. It is important to note, however, that there is evidence that atypical progenitors such as $CD45^{+}Sca1^{+}$ cells (Polesskaya et al. 2003), $PW1^{+}$ interstitial cells (PICs) (Mitchell et al. 2010), bone marrow-derived stem cells (BMDC), and hematopoietic stem cells (LaBarge and Blau 2002; Dreyfus et al. 2004) can contribute to muscle repair when satellite cells have been ablated by radiation. However, there is little evidence illustrating that any of these cell populations can contribute to postnatal muscle repair under normal physiological conditions to any significant level. It is likely that in the absence of normally functioning satellite cells or in the case of extreme muscle damage (i.e., cardiotoxin injection), these cells may contribute, in a limited capacity, to muscle regeneration.

10.4 Postnatal Myogenesis

In postnatal muscle tissue the majority of satellite cells exist in a state of quiescence between the endomysium or basal lamina and the sarcolemma of the myofiber (Mauro 1961). Figure 10.1 illustrates the anatomical location of the satellite cell (Pax7-positive cell) located between the myofiber itself and the surrounding basal lamina in a human muscle cross section. In neonatal muscle, satellite cells constitute approximately 30–50 % of all myonuclei; however, this number stabilizes in adult muscle to approximately 2–4 % of myonuclei (Hawke and Garry 2001; Kadi et al. 2005). The process orchestrating the progression of a quiescent satellite cell to a terminally differentiated muscle cell is known as myogenic determination or the myogenic program (Parise et al. 2006). A well-defined series of highly conserved transcriptional networks are responsible for the control of activation, proliferation, and differentiation and are known as the myogenic regulatory factors (MRFs). The MRFs are basic helix-loop-helix (bHLH) transcription factors, which form heterodimeric DNA-binding complexes within enhancer elements of genes that encode specific aspects of the cell cycle machinery or genes that drive terminal (myogenic) differentiation (Cao et al. 2006; Cornelison et al. 2000; Parker et al. 2006; Megeney et al. 1996; Punch et al. 2009). There are four main MRFs that control normal myogenesis: Myf5 (myogenic factor 5), MyoD (myoblast determination factor), MRF4 (muscle-specific regulatory factor 4, also known as MYF6), and myogenin. Myf5 is required for the initiation of postnatal myogenesis and is the first MRF gene that is

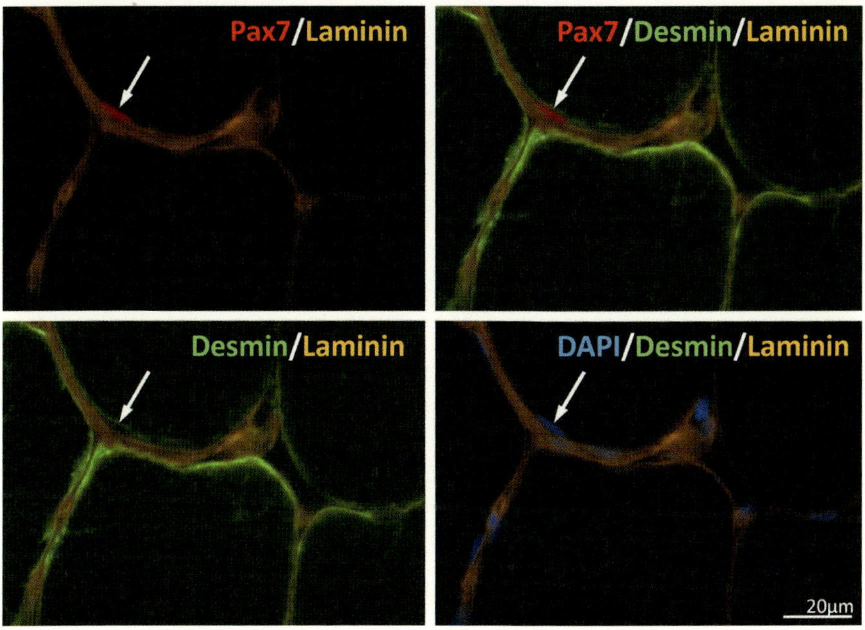

Fig. 10.1 Human muscle cross sections stained for the muscle stem cell-specific marker Pax7 (*red*), nuclei (*blue*-DAPI), basal lamina (*orange*-laminin), and the muscle cytoskeletal protein desmin (*green*). The *arrows* denote the anatomical location of the muscle stem cell between the myofiber membrane (sarcolemma) and the basal lamina. The *first panel* shows the muscle stem cell below the basal lamina, and the *second panel* shows that the muscle stem cells are between the myofiber and the basal lamina. The *third panel* verifies that the muscle stem cell is indeed between these two structures (hole where the Pax7$^+$ cell is sitting). The *last panel* confirms that the nuclear marker used to identify the muscle stem cell (Pax7) is indeed co-localizing with the nucleus of a cell

upregulated upon satellite cell activation (Cornelison and Wold 1997; Beauchamp et al. 2000). Both Myf5 and MyoD are upregulated in proliferating satellite cells (Cornelison and Wold 1997; Cooper et al. 1999), and it appears that cells that downregulate MyoD but retain Myf5 represent cells withdrawing from the myogenic program, contributing to self-renewal of the satellite cell pool (Sabourin et al. 1999; Yablonka-Reuveni et al. 1999; Zammit et al. 2004). Furthermore, it appears that MyoD is necessary for the transition from proliferation to differentiation as myoblasts from *MyoD−/−* mice exhibit impaired differentiation and fail to upregulate MRF4, but these events were reversed by reintroduction of functional MyoD (Cornelison et al. 2000).

After sufficient rounds of proliferation, myoblasts begin to downregulate Myf5 and upregulate MRF4 and myogenin. Both MRF4 and myogenin are involved in the specific events that orchestrate myogenic differentiation at both the transcriptional and epigenetic level (Punch et al. 2009; Rhodes and Konieczny 1989). Both MyoD and MRF4 act upstream of myogenin (Cao et al. 2006; Rhodes and Konieczny 1989; McKinnell et al. 2008), inducing the expression of myogenin and initiating terminal differentiation. The temporal expression of these MRFs was confirmed in culture demonstrating that MRF4 upregulation precedes myogenin and the upregulation of myogenin is immediately followed by terminal differentiation of myoblasts into nascent myotubes (Smith et al. 1994a). Thus upon activation, a quiescent satellite cell begins to express Myf5, followed by both Myf5 and MyoD. Following proliferation, the majority of proliferating satellite cells (myoblasts) downregulate Myf5 and upregulate MRF4 followed by the upregulation of myogenin leading to terminal differentiation and fusion of these cells either with the muscle fiber or by fusion with each other to form nascent myotubes. A subset of myoblasts may lose (or never upregulate) MyoD and withdraw from the cell cycle, repopulating the satellite cell pool through self-renewal.

Regulation of the postnatal myogenic program is thought to be controlled by a specific homeodomain protein Pax7. Pax7 is a member of the paired box (Pax) transcription factor family and has been shown to be important for normal maintenance of the satellite cell population (Seale et al. 2000). Currently, the precise role of Pax7 in myogenic specification remains somewhat unclear. Although some studies suggest that Pax7 is not essential for postnatal muscle growth and hypertrophy in the short term (McCarthy et al. 2011), recent data demonstrate the indispensable role of Pax7 in normal developmental myogenesis, postnatal myogenesis, and importantly the regenerative response to muscle injury (von Maltzahn et al. 2013; Lepper et al. 2011; Gunther et al. 2013). Pax7 associates with a specific histone methyltransferase complex that directs the methylation of histone H3K4 to induce chromatin modifications, which stimulates the transcription of Myf5, thus regulating entry into the myogenic program (McKinnell et al. 2008). Mice with an inducible knockdown of Pax7 display a progressive loss of satellite cells that is associated with severe muscle atrophy (von Maltzahn et al. 2013). The loss of cellular function in induced Pax7 null satellite cells is a result of cell cycle arrest and MRF dysregulation, demonstrating that Pax7 is an essential regulator of normal satellite cell function. Interestingly, mice with a complete genetic lack of the Pax7 gene (*Pax7−/−*) from conception have normal embryonic myogenesis mainly due to the increased activity of the Pax7 orthologue: Pax3 (Buckingham and Relaix 2007). However, *Pax7−/−* mice demonstrate decreased body weight and inhibited growth and die prematurely (Seale et al. 2000). Although initially these mice were thought to lack satellite cells completely (Seale et al. 2000), it was later confirmed that they do in fact possess satellite cells, albeit at low numbers (Oustanina et al. 2004), and these cells are rapidly lost as the animals age (Relaix et al. 2005). Thus, it appears that Pax7 is instrumental in the regulation of satellite cells but may not be necessary for postnatal myogenic specification (Relaix et al. 2006). In addition Pax7 appears to be necessary for satellite cell self-renewal (Olguin and Olwin 2004). In a series of studies examining the potential role of Pax7 in satellite cell self-renewal, the forced expression of Pax7 induced cells to exit the cell cycle and return to quiescence. Furthermore, Pax7 was found to be absent in myogenin-positive cells illustrating a potential reciprocal inhibition between Pax7 and myogenin, suggesting that Pax7 must be downregulated in order for differentiation to proceed (Olguin and Olwin 2004; Olguin et al. 2007). Further work by Zammit et al. (Zammit et al. 2004) demonstrated that Pax7 is co-expressed with MyoD in proliferating myoblasts and that the downregulation of Pax7 in these cells coincided with the progression of cells into terminal differentiation, while a subset of cells downregulate MyoD but retain Pax7 and withdraw from the cell cycle, thus replenishing the satellite cell pool. Collectively, these studies suggest that Pax7 is involved in the coordination of normal postnatal myogenesis. Pax7 appears to be instrumental in the initiation of the myogenic program through the transcriptional activation of Myf5 and maintenance of the satellite cell pool by increasing cell survival and preventing differentiation, thus favoring self-renewal, at least in embryonic and early postnatal myogenesis (McKinnell et al. 2008; Gunther et al. 2013; Relaix et al. 2006; Olguin and Olwin 2004).

In adult life muscle fibers are generally stable, requiring only minor myonuclear turnover due to subtle acquired structural or nuclear damage (Decary et al. 1997). Estimates for myonuclear turnover suggest that approximately 1–2 % of myonuclei are replaced on a weekly basis in rat muscle (Schmalbruch and Lewis 2000). If we extend the observation that a rat extensor digitorum longus (EDL) has approximately 2.2 million myonuclei (Viguie et al. 1997), then the average satellite cell-mediated myonuclear turnover in the EDL is ~22,000 myonuclei per week. Therefore even in the absence of myotrauma, the role of the satellite cell appears to be indispensible. Furthermore, muscle lacking functional satellite cells due to exposure to radiation is unable to undergo significant hypertrophy in response to chronic overload (Adams et al. 2002). This study demonstrates that satellite cell-mediated myonuclear addition is necessary to support significant skeletal muscle hypertrophy. However, this theory was recently refuted by data, which illustrates that satellite cells are not required for short-term muscle hypertrophy (McCarthy et al. 2011). Muscle hypertrophy was achieved in satellite cell-depleted muscle suggesting, in the short term, that intrinsic protein synthetic mechanisms are sufficient to support significant muscle hypertrophy. Skeletal muscle is an incredibly plastic tissue with significant adaptive potential, and hypertrophy is achieved without myonuclear accretion from satellite cell fusion in the short term (2 weeks). However, in the long term there is an attenuation of hypertrophy in muscle lacking satellite cells (Fry et al. 2014), an impaired regenerative capacity, as well as an increase in tissue fibrosis or extracellular matrix expansion (Fry et al. 2014, 2015). In addition, in satellite cell ablation studies, the control groups demonstrate muscle hypertrophy associated with satellite cell activation and fusion with myofibers, which suggests that the normal physiological response to overload is a combination of increased myofiber protein synthesis and satellite cell-mediated myonuclear accretion (McCarthy et al. 2011; Wang and Rudnicki 2012). Although muscle hypertrophy can occur without satellite cells in the short term, the absence of satellite cells completely prevents muscle regeneration following muscle injury. Several recent studies using different methods of satellite cell ablation have demonstrated that, in response to toxin-mediated injury and the more physiologically mediated forced exercise injury model, skeletal muscle fails to regenerate (McCarthy et al. 2011; Lepper et al. 2011; Murphy et al. 2011; Sambasivan et al. 2011). Collectively, these studies also indicate that when the satellite cell population is ablated below a threshold level, muscle regeneration is severely impaired. This lends support to the theory that when the satellite cell pool is depleted below a certain level, the in vivo regenerative capacity of the muscle is impaired. This phenomenon is observed in diseases such as

Duchenne muscular dystrophy, where the satellite cell pool becomes exhausted. Whether or not depletion of the satellite cell pool is implicated in muscle aging remains controversial. Central to our understanding of muscle stem cell function is the regulation of this cell population by extrinsic factors and is the focus of intense study in the context of normal growth and dysfunction in aging.

10.5 Regulation of Muscle Stem Cells

Mature skeletal muscle is known to produce a host of growth factors such as insulin-like growth factor-1 (IGF-1) and hepatocyte growth factor (HGF) following a physiological stimulus such as overload or myotrauma following injury. IGF-1 has been shown to activate protein synthetic machinery by its interaction with the mammalian target of rapamycin (mTOR) (Frost and Lang 2011) and is a potent mitogen believed to be involved in cell recruitment to aid in tissue repair (Mourkioti and Rosenthal 2005). The expression of the MRFs appears to be regulated, in part, by the activity of locally produced isoforms of IGF-1 (Mourkioti and Rosenthal 2005; Chakravarthy et al. 2000). IGF-1 is a growth factor found in the circulation, produced in the liver and locally by the muscle fiber itself (Mourkioti and Rosenthal 2005). Three specific isoforms of IGF-1, IGF-1Ea, IGF-1Eb, and IGF-1Ec, have been described in muscle, and each may contribute to some extent to muscle regeneration/repair (Chakravarthy et al. 2000; Rotwein et al. 1986; Yang et al. 1996). IGF-1Ea transcribes the full-length IGF-1 protein, which is identical to that produced by the liver (Yang et al. 1996). IGF-1Eb may be involved in hypertrophy or muscle regeneration; however, the physiological role of IGF-1Eb in human skeletal muscle is currently unknown. IGF-1Ec or mechano growth factor (MGF) is a splice variant of IGF-1Ea that may have both autocrine and paracrine functions, stimulating satellite cell proliferation and muscle hypertrophy following muscle stretch or muscle damage (Yang et al. 1996; Hill and Goldspink 2003).

IGF-1 has been shown to activate cultured porcine satellite cells through an mTOR-dependent mechanism (Han et al. 2008) and enhance the proliferative potential of murine satellite cells by activating the phosphatidylinositol 3′-kinase/Akt signaling pathway (Galvin et al. 2003). In response to exercise or injury, IGF-1 splice variants increase within 24 h after injury in rats (Hill and Goldspink 2003; Hill et al. 2003); however, results from human studies are conflicting.

In humans, IGF-1Ec expression following isometric knee exercise can increase in as little as 2.5 h (Greig et al. 2006), suggesting that IGF-1Ec may be important for satellite cell proliferation. Information regarding the response of IGF-1Ea to exercise is equivocal, with some animal and human work demonstrating significant increases in IGF-1Ea expression as early as 24 h post-exercise or muscle damage (Hill and Goldspink 2003; Hill et al. 2003; Greig et al. 2006; Hameed et al. 2003), while others show a marked decrease in IGF-1Ea and total IGF-1 from 4 to 12 h post-exercise that remained unchanged after 96 h (Psilander et al. 2003; Bickel et al. 2005). In humans, the IGF-1 splice variants at the mRNA level (McKay et al. 2008) and the whole muscle protein level (Philippou et al. 2009), co-localized with Pax7 (using immunofluorescent techniques) (McKay et al. 2008), appear temporally

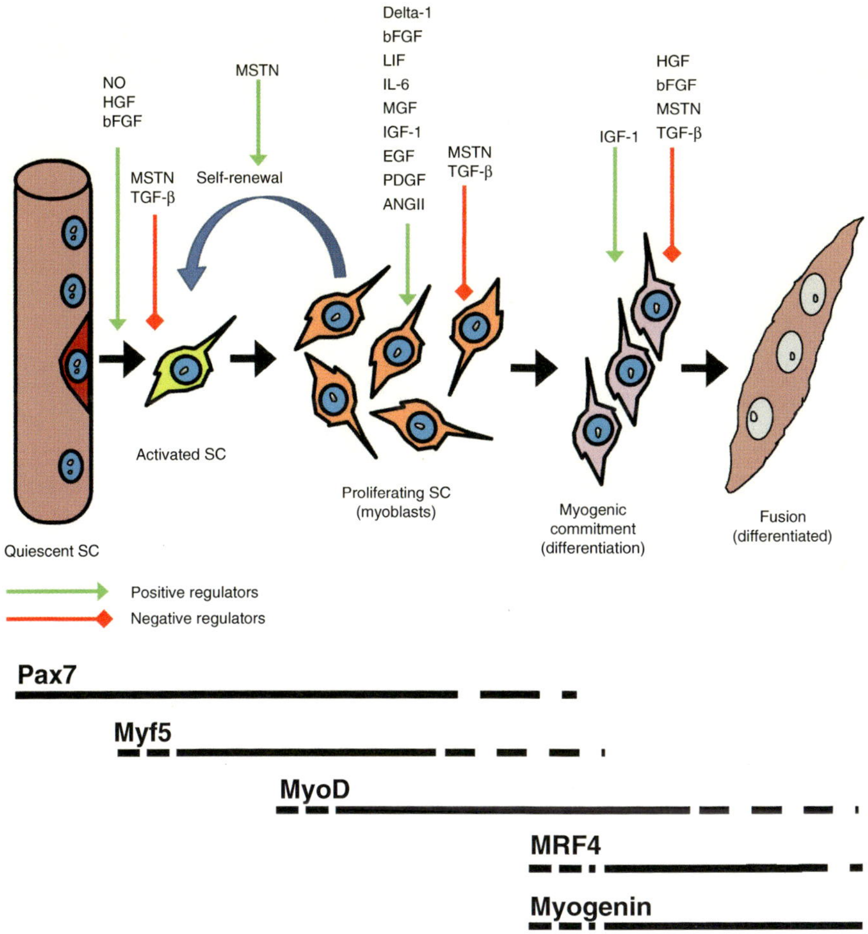

Fig. 10.2 The myogenic program from a quiescent muscle stem cell (*SC*) on the periphery of the myofiber (*dark red cell*) which then is activated (*yellow cell*) and undergoes rounds of proliferation (*orange cell*) before progressing towards terminal differentiation by either fusing together to form nascent myotubes or fusing with existing myofibers. The *green arrows* denote positive regulators of each step of the myogenic program, whereas *red arrows* denote negative regulators of each step. Below is a representation of the appearance and disappearance of Pax7 and the myogenic regulatory factors with respect to the myogenic progression of the muscle stem cells

related to MRF expression. IGF-1Ec is temporally co-expressed with Myf5, while IGF-1Ea and Eb are temporally co-expressed with MRF4, suggesting that alternate splicing of IGF-1 is associated with different phases of satellite cell progression (McKay et al. 2008; Philippou et al. 2009). Therefore, based on human and animal studies, it appears that IGF-1 may be a critical regulator of the myogenic response.

Other growth factors have been shown to influence the proliferative phase or differentiation of satellite cells in animals or in cell culture. Figure 10.2. shows a schematic representation of some of the major molecular regulators of the myogenic program and illustrates the timing of the MRFs with respect to the progression of satellite cells from quiescence to differentiation. These extracellular signaling molecules include basic fibroblast growth factor (bFGF), which has been demonstrated to induce activation/proliferation and inhibit differentiation (DiMario et al. 1989); nitric oxide (NO), which has been implicated in satellite cell activation, acting upstream of HGF to mediate the release of HGF (Tatsumi et al. 2002); Delta-1, which increases proliferation (Conboy et al. 2003); and myostatin and TGF-β, which inhibit activation and differentiation and maintain quiescence (McCroskery et al. 2003; Allen and Boxhorn 1987). Recently, angiotensin II (ANGII) has been implicated as an activator of quiescent satellite cells in vivo and in culture (Johnston et al. 2010). The actions of ANGII appear to regulate satellite cells through the angiotensin II receptor sub-type 1 (AT1), and may also play an instrumental role in satellite cell migration following muscle injury (Johnston et al. 2010). Platelet-derived growth factor (PDGF) and endothelial growth factor (EGF) also appear to increase satellite cell proliferation in culture (Hawke and Garry 2001). The stark reality is that there are many factors that influence satellite cell regulation, many of which are likely yet to be discovered. This reality underlies the complexity of understanding satellite cell regulation and emphasizes the challenge of understanding age-related satellite cell dysfunction. A growing area of interest is the role that inflammation and inflammatory markers play in regulating satellite cell function. Not surprisingly, the role of inflammatory markers on satellite cell function is not well understood.

10.6 Inflammation and Muscle Stem Cell Regulation

In young adults, myotrauma due to exercise or injury may induce damage to contractile proteins and myofibrils, or larger ultrastructural disruptions, which initiates a significant inflammatory response (Smith et al. 1994b; Beaton et al. 2002). Macrophage infiltration, which is responsible for the removal of cellular debris, is generally accompanied by lymphocytes, which coordinate the inflammatory response and importantly release factors that stimulate satellite cell proliferation (Allen et al. 1995; Cantini et al. 1995). Invading T lymphocytes can bind adhesion molecules on damaged muscle fibers and release a host of cytokines including leukemia inhibitory factor (LIF) and interleukin-6 (IL-6). In addition, damaged muscle fibers secrete several cytokines (i.e., IL-6) and growth factors (i.e., IGF-1) that may act as chemoattractants for inflammatory cells and satellite cells, augmenting the rapid repair response (Johnston et al. 2010; Pedersen and Febbraio 2008; Broholm et al. 2008; Gallucci et al. 1998; Tomiya et al. 2004). In humans, administration of nonsteroidal anti-inflammatory medication (NSAID) after damaging exercise inhibits the exercise-induced increase in satellite cells observed in controls (Mikkelsen et al. 2009; Mackey et al. 2007), illustrating that inflammation may be a key factor regulating the satellite cell response.

Cytokines released by inflammatory cells have been shown to act on satellite cells in vitro, stimulating proliferation (Cantini et al. 1995). Of the factors released, members of the interleukin-6 family of cytokines, IL-6 and LIF, appear to be instrumental in regulating satellite cell proliferation (Cantini et al. 1995; Spangenburg and Booth 2002; Austin et al. 1992). Both LIF and IL-6 bind the gp130 receptor, activating Janus-activated kinase (JAK) 2, which induces the phosphorylation of signal transducers and activators of transcription (STAT)-3 (Spangenburg and Booth 2002; Rawlings et al. 2004). Phosphorylated STAT3 (pSTAT3) forms a homodimer (or heterodimer with other STAT proteins), which translocates to the nucleus and induces the transcription of target genes such as *Cyclin D1* and *cMyc,* which are instrumental in the initiation of the cell cycle (Rawlings et al. 2004). Importantly pSTAT3 also targets the *IL-6* gene; thus, the release of IL-6 and LIF induces the production of IL-6 in target cells, creating an autocrine/paracrine loop influencing satellite cell proliferation (Rawlings et al. 2004). Until recently these findings had only been validated in cell culture models. In a landmark study by Serrano and colleagues (Serrano et al. 2008), IL-6 was shown to be an essential regulator of satellite cell function in vivo in a murine animal model. This study provided mechanistic evidence supporting the role of IL-6 as a regulator of satellite cell proliferation and that in the absence of IL-6 (*IL-6−/−* mice), satellite cell proliferation and myonuclear accretion were severely blunted (Serrano et al. 2008). Furthermore, the deficiency in proliferation observed in the IL-$6^{-/-}$ satellite cells could be rescued by the ectopic addition of IL-6 and IL-6-mediated proliferation via the IL-6/STAT3 axis (Serrano et al. 2008). This study provides strong evidence to support the role of inflammatory cytokines in regulating muscle stem cell activity.

Aging is associated with alterations in basal levels of circulating inflammatory mediators such as IL-6 and TNFα and a decrease in circulating growth factors. The changes associated with normal tissue aging confer a negative impact on the muscle stem cell compartment, which may impact the ability of the skeletal muscle to adapt and regenerate. The remainder of this chapter focuses on several aspects of the aging process in relation to the function and control of the muscle stem cells.

10.7 Muscle Stem Cells and Aging

Aging is a complex process and is associated with many physical and metabolic alterations that lead to the progressive loss of function over time (Roubenoff 2003). Sarcopenia, which is an age-related condition that includes a progressive loss of muscle mass and strength, leads to the loss of functional capacity and an increase in morbidity (Roubenoff 1999; Beccafico et al. 2007). The progressive nature of age-related muscle wasting leads to an increased incidence of injury and loss of autonomy, resulting in a significant loss in quality of life and increased risk of all-cause mortality (Roubenoff and Hughes 2000; Melton et al. 2000). Moreover, the economic impact of sarcopenia is staggering. Common age-related injuries are often associated with falls due to a loss of strength and stability. Falls represent approximately 65 % of injuries, 85 % of injury-related hospital admissions, and 58 % of injury-related deaths in individuals over the age of 65. These statistics translate into health-care costs of approximately $1 billion dollars annually in Canada and as high as $26 billion dollars annually in the United States (Janssen et al. 2004). Similar trends have been observed in Australia, Great Britain, and Europe. Thus understanding the cellular mechanisms that lead to the onset and progression of sarcopenia is necessary to propose effective countermeasures.

The normal aging process has a profound effect on the structure and function of the skeletal muscle. Aged muscle undergoes progressive and seemingly unavoidable atrophy, with a decreased cross-sectional area accompanied by a decrease in muscle fiber number and fiber size (Grounds 1998; Koopman and van Loon 2009). Importantly, muscle function or quality appears to be reduced in aged muscle, displaying a decrease in power and maximum force production even when corrected to cross-sectional area (Brooks and Faulkner 1988). In addition, aged muscle is far more susceptible to macro and micro damage compared to young muscle, and, central to this chapter, aged muscle displays a decreased capacity for growth and regeneration (Grounds 1998; Brooks and Faulkner 1996). Satellite cells have been implicated as a significant factor contributing to the inability of the muscle to repair/remodel. An age-related impairment of the myogenic program (either impaired activation of or progression through the myogenic program) is hypothesized to be a significant contributor to satellite cell dysfunction, leading to the onset or progression of sarcopenia. In adult muscle, satellite cells comprise only a small fraction of the total myonuclei (~2 %) yet possess an incredible proliferative capacity, capable of repairing extensive muscle injury. The size of the satellite cell pool is thought to remain relatively constant throughout adult life; however, it has been suggested that in the later stages of life, the size of the satellite cell pool begins to decline, impairing the regenerative capacity of muscle. Although it is unclear at present whether there is an age-related reduction in the satellite cell pool size, it is clear that aging is accompanied by an impairment in the response of satellite cells to myotrauma in humans (Dreyer et al. 2006; McKay et al. 2013; McKay et al. 2012) and animals (Conboy et al. 2003). Several studies have attributed both intrinsic factors of the satellite cells (Jejurikar et al. 2006) as well as extrinsic factors such as age-related alterations to the satellite cell niche (Conboy et al. 2003; Carlson and Conboy 2007) as major influences on the dysfunction of the satellite cell population with aging.

10.8 Muscle Stem Cell Pool Size and Aging

There is a progressive decline in the satellite cell pool after birth from ~30–50 % of total myonuclei to a relatively stable ~2–4 % of total myonuclei in adolescences and into adulthood. Whether or not aging is associated with a reduction in the total number of satellite cells has been the focus of many studies in animals and in humans and would be the simplest explanation as to why older adults suffer significant skeletal muscle loss. Although there is significant controversy in this area, there are several key studies in animals that highlight an age-related decline in the satellite cell pool. Shefer et al. (Shefer et al. 2006) compared the satellite cell pool in four groups of mice aged 3–6 months (young), 7–13 months (adult), 19–25 months (old), and 29–33 months (senile). Their data demonstrate a 60 % decline in satellite cells in the *extensor digitorum longus* (EDL) muscle between young and old mice with a further 17 % reduction between the old and senile groups (Shefer et al. 2006). Similar results were observed in the *soleus* muscle, and it was concluded that the satellite cell pool in the young was significantly larger than that of all the three other groups. Furthermore, when grown in culture, a more robust growth rate was observed in satellite cells obtained from young muscle as compared to senile muscle. Other studies in support of these findings have also reported decreased growth rates as well as increased susceptibility for satellite cells to undergo apoptosis in vitro and in vivo compared to the young (Jejurikar et al. 2006; Shefer et al. 2006; Day et al. 2010). Single muscle fiber analysis revealed that satellite cells, expressing the marker Pax7, were markedly reduced in aged muscle and had a reduced myogenic capacity in culture (Collins et al. 2007). In fact, approximately 10 % of aged myofibers did not possess any Pax7-positive cells (satellite cells). Importantly, the aged satellite cells demonstrated a marked inability to renew the progenitor pool undoubtedly contributing to the observed reduction in satellite cell number. A rather interesting observation by Collins et al. (2007) described a subpopulation of satellite cells that appeared unaffected by the aging process. Although the majority of aged satellite cells displayed altered myogenic properties, a population of relatively rare muscle stem cells was able to produce large colonies of functional satellite cells, which retained the ability to self-renew in vitro (Collins et al. 2007). This population likely represents a very small proportion of the satellite cell pool and may be a more primitive myogenic progenitor cell, which, for reasons unknown, has been able to avoid the adverse effects of aging. It may be that these cells represent a more primitive myogenic progenitor subpopulation, which retains a similar level of potency as satellite cells from young muscle. There is significant heterogeneity in the satellite cell population (Biressi and Rando 2010), and it is possible that, much like in the bone marrow, there is a small group of stem cells that rarely divide and whose progeny is mainly responsible for amplifying in response to stimuli and self-renewing to maintain the progenitor pool. It is of note, however, that the notion that aging is associated with a loss in the size of the satellite cell pool does not go unchallenged. Several studies do not report a decline in total satellite cell number in the muscle of aged mice (Conboy et al. 2003; Dreyer et al. 2006). This may be a result of different enumeration methods or molecular markers used to quantify the satellite cell pool or species of animal used. In humans, however, it has been unequivocally shown that aged muscle (>65 years) has a reduced satellite cell pool. Recently Verdijk et al. (Verdijk et al. 2007; Verdijk et al. 2009), using the cell marker N-CAM (neural cell adhesion molecule), demonstrated that the satellite cell pool in older adults was significantly lower compared to young men, and this was mainly a consequence of the loss of satellite cells associated with type II myofibers. This was subsequently confirmed using both Pax7 immunofluorescent analysis and flow cytometry in older adults (~70 years) (McKay et al. 2012, 2013). Collectively these studies demonstrate an approximate 35 % reduction in the satellite cell pool in 70-year-olds compared to 25-year-olds, the bulk of which is made up of a marked reduction in type II-associated satellite cells (McKay et al. 2012, 2013; Verdijk et al. 2007, 2009, 2012). These studies report an approximate 30 % reduction in muscle strength, 25 % reduction of type II muscle fiber cross-sectional area (CSA), and a 20 % lower type II fiber myonuclear domain compared to young lean body mass-matched controls. Importantly these studies provide evidence that the reduced type II fiber-associated SC content in the elderly may be an important factor in the etiology of type II fiber atrophy and progression of sarcopenia (Verdijk et al. 2007). Whether the absolute number of satellite cells is an important factor for muscle aging remains a somewhat contentious issue; however, the literature clearly demonstrates that satellite cells in aged muscle become dysfunctional as a consequence of the aging process. The dysfunction can be classified as *intrinsic* to the stem cells themselves or *extrinsic*, relating to factors in the stem cell niche and circulating factors known to regulate satellite cells. It appears that the combination of these *extrinsic* and *intrinsic* factors is important to the overall function or dysfunction of the muscle stem cell population.

10.9 Muscle Stem Cells and Aging: Intrinsic Cell Failure

Muscle stem cells remain in a quiescent state on the periphery of the myofiber and become activated when sufficient cellular cues from the microenvironment initiate entry into the cell cycle. A major factor influencing the success of this process is the ability of the cell to recognize these signals and activate the appropriate cell signaling pathways. Several studies have identified key factors that appear to be altered by the aging process. Furthermore, nuclei in newly fused myoblasts must function properly to allow for adequate tissue healing and to respond appropriately to physiological stimuli for the induction of adaptation. In aged muscle it appears that all of these aspects may be impaired to some degree. Several theories have been proposed to explain how quiescent cells outside the sarcolemma could acquire age-related damage. The most prominent of these theories involves oxidative damage to DNA and cellular components from years of exposure to reactive oxygen species (ROS) (Beccafico et al. 2007; Cefalu 2011; Mecocci et al. 1999). Satellite cells and myotubes (fused differentiated myoblasts) from aged humans display an age-dependent increase in lipid peroxidation and impaired calcium handling in in vitro experiments suggesting that acquired damage throughout life impairs the basic cellular processes in these cells which may contribute to the poor adaptive response of aged muscle (Beccafico et al. 2007).

10.9.1 Susceptibility to Death

Studies examining the proliferative ability of satellite cells isolated from aged muscle and assayed in culture illustrate a basic decrease in the activation and/or proliferative capacity of old cells lending support to the theory that aged satellite cells suffer from intrinsic failure (Conboy et al. 2003; McKay et al. 2012; Jejurikar et al. 2006; Collins et al. 2007). In culture, aged satellite cells and their progeny readily undergo apoptosis (Collins et al. 2007). In the presence of proapoptotic agents such as tumor necrosis factor-alpha (TNF-α), aged satellite cells demonstrate increased levels of activated caspases and seven to eightfold more fragmented DNA as well as a 40 % reduction in the antiapoptotic protein Bcl-2 compared to young controls

(Jejurikar et al. 2006). These data suggest that under stress, aged satellite cells are less capable of entering the myogenic program and fail to resist cell death, therefore impairing the response to injury. The subset of satellite cells that survive after becoming activated is able to proliferate and differentiate similarly to those of young muscle (Shefer et al. 2006; Collins et al. 2007). This suggests that if satellite cells can resist apoptosis, the intrinsic ability of some individual cells to progress through the myogenic program is retained. However, the amplification kinetics of the satellite cell pool and total yield of satellite cell progeny is significantly impaired in aged muscle (Shefer et al. 2006; Corbu et al. 2010). In addition, the ability to repopulate the reserve pool following proliferation appears to be significantly reduced in aged satellite cells (Day et al. 2010). Interestingly, in aged animals, some muscle fibers are completely devoid of satellite cells, which may indicate that some fibers have completely exhausted their satellite cell pool due to impaired self-renewal (Shefer et al. 2006; Day et al. 2010; Collins et al. 2007). The intrinsic failure of satellite cells in aged muscle may be further compounded by acquired cellular damage. Satellite cells from aged muscle have been relatively quiescent for the majority of life, which makes them susceptible to damage from genotoxic mediators such as reactive oxygen and nitrogen species from the myofiber and from the circulation (Jang et al. 2011). The accumulation of cellular damage coupled with a reduced capacity for quiescent cells to repair damage due to lower levels of DNA repair proteins and decreased response to stress (Bohr 2002) may negatively influence the ability of satellite cells to proliferate and resist apoptosis. Although the role of accumulated cellular damage in muscle stem cells has not been fully explored, it is very likely that acquired cellular damage contributes to the reduced myogenic capacity of these cells.

10.9.2 Adoption of Alternate Cell Fates

Aged satellite cells that successfully resist death and become activated in response to injury may actually activate inappropriate signaling pathways leading to inappropriate cell differentiation. In normal myogenesis there is an important balance and temporal shift from Notch signaling to Wnt signaling pathways that orchestrate proper progression down the myogenic lineage (Brack et al. 2008). Increased levels of Wnt signaling in aged animals induce satellite cells to progress away from the myogenic lineage and adopt a fibrogenic fate (Brack et al. 2007). Increased muscle fibrosis in aging is a hallmark event of the sarcopenic process and appears to be mediated, at least in part, by an increased fibrotic differentiation of satellite cells. This not only inhibits muscle repair but also decreases the quality and function of the skeletal muscle, which is particularly devastating in advanced age where there is already a loss of functional contracting tissue. To emphasize the point, myogenic-to-fibrogenic conversion of aged muscle satellite cells was substantially reduced with the injection of DKK1, a Wnt signaling inhibitor, during muscle regeneration (Brack et al. 2007). Other signaling pathways in aged muscle satellite cells involving PPARγ, and other members of the Wnt family of proteins, have also been shown to induce an adipogenic conversion of satellite cells (Taylor-Jones et al. 2002).

Although it is not fully known whether the age-related adoption of alternate fates by satellite cells actually results in greater muscle adiposity or fibrosis, this observation clearly demonstrates altered intrinsic ability of the satellite cell in advanced age. The loss of cells driven down the myogenic lineage reduces the capacity of aged muscle to respond to overload and to regenerate damaged tissue; however, intrinsic dysfunction of satellite cells represents only one potential source of muscle stem cell failure. In fact, some will argue that the intrinsic dysfunction is only a result of altered extrinsic cues driving the intrinsic response. There is certainly merit in this position as satellite cells are in constant communication with cues from the muscle fiber, surrounding cells, and systemic environment. Theoretically, any alteration in the microenvironment may have an impact on intrinsic cell function.

10.10 Muscle Stem Cells and Aging: Extrinsic Factors

The satellite cell niche and the systemic milieu have been identified as factors in age-related satellite cell dysfunction. The satellite cell niche or microenvironment that supports the satellite cell helps regulate cell activity through physical contact and direct interaction with satellite cell membrane proteins (Kuang et al. 2008; Gopinath and Rando 2008). The basal lamina is one of the main niche components, which is in constant interaction with extracellular space and circulation. It is mainly composed of type IV collagen and several structural proteins, proteoglycans, and glycoproteins such as fibronectin and laminin (Woodley et al. 1983).

10.10.1 Architectural Changes to the Niche

With aging there is a progressive thickening of the basal lamina as a result of an increased deposition of collagen (Alexakis et al. 2007), which interferes with the normal extracellular matrix (ECM) functions such as the sequestration of hepatocyte growth factor (HGF) (Tatsumi and Allen 2004) and normal diffusion of key signaling molecules. Thickening and fibrosis of the satellite cell niche may impair interactions with the systemic environment and also make it considerably more difficult for the satellite cell to degrade the ECM thus hindering satellite cell migration. The niche is also susceptible to an increase in adiposity and connective tissue deposition from an age-related increase in the fibroblast pool in the interstitial spaces of the skeletal muscle (Brack et al. 2007; Kirkland et al. 2002). This may alter the secretion of soluble factors and the sensitivity of satellite cells to systemic factors. An important component of the satellite cell niche is the presence of a blood supply in the form of capillaries associated with individual muscle fibers. Aging is associated with a reduction in the capillary network surrounding myofibers and a concomitant reduction in vascular endothelial growth factor (VEGF) expression and endothelial nitric oxide synthase (eNOS) activity, which have detrimental effects to the local vasculature and satellite cell function (Croley et al. 2005; Ryan et al. 2006). Additionally, changes to the neuronal connectivity of the skeletal muscle occur with aging (Larsson and Ansved 1995). A reduction in neuromuscular connectivity by a loss of functional motor units in aging has been shown to contribute to muscle atrophy (Larsson and Ansved 1995) which, coupled with increased fibrosis, alters the satellite cell niche architecture and may impact satellite cell function. Furthermore, a reduction in neuronal input may further impair normal satellite cell function via a reduction in electrical and neurotrophic factor stimulation (Gopinath and Rando 2008). In essence age-related changes to the architecture of the niche including thickening of the basal lamina, deposition of adipocytes, and loss of capillaries and nerves can adversely impact satellite cell function.

10.10.2 Circulating Factors

Age-related changes in the microenvironment may impair satellite cell activation (Conboy et al. 2003) and even influence muscle progenitors to adopt alternate lineages (i.e., become fibrotic (Brack et al. 2007) or adipogenic (Taylor-Jones et al. 2002)) leading to the loss of regenerative capacity of cells in the aged niche (Carlson and Conboy 2007). As mentioned above enhanced Wnt signaling in aged mice induced a shift from the normal myogenic fate to an increased propensity for fibrogenesis (Brack et al. 2007). It is entirely possible that the source of Wnt proteins in aging may originate in tissues outside of muscle circulating to the satellite cell niche. It has been shown that when young muscle satellite cells are exposed to aged serum, there is an increase in fibrogenic conversion (Brack et al. 2007). In fact, increasing a single Wnt protein, Wnt3a, in young serum is sufficient for driving a myogenic-to-fibrogenic conversion, and when administered directly into regenerating muscle, the result was increased tissue fibrosis and a reduced satellite cell proliferation (Brack et al. 2007). The most elegant of the experiments demonstrating an impaired satellite cell function as a result of an aging extrinsic environment involved parabiotic pairings of young and old mice. In experiments utilizing parabiosis, young mice and old mice are stitched together via a skin flap allowing for an anastomosis to form between the animals giving rise to essentially a single shared circulation. The idea is that this model would allow for in vivo analysis of satellite cells exposed to an aging circulation. Heterochronic (young with an old) parabiotic pairings restored the regenerative capacity of satellite cells in aged mice suggesting that age-related satellite cell dysfunction is indeed extrinsic in nature (Conboy et al. 2005). Consistent with this notion, the transmembrane protein, Notch, was implicated in age-induced satellite cell dysfunction, as it remained inactive following freeze-crush myotrauma, therefore preventing satellite cell activation (as illustrated by 75 % reduction in replicating satellite cells) (Conboy et al. 2003). The forced activation of Notch or the exposure to a young circulatory environment via heterochronic parabiotic pairing (or exposure to young serum in culture) restored the regenerative and proliferative response in aged muscle to the same levels observed to that in young animals (Conboy et al. 2003; Conboy et al. 2005). Collectively these experiments implicated Notch in age-related failure of satellite cells to activate in response to a traumatic cue. In humans, Notch signaling gene expression was impaired in both isolated satellite cell cultures and in whole muscle homogenates in muscle biopsies taken from elderly men (Carey et al. 2007). Interestingly, resistance exercise training attenuated the age-related impairment of Notch-related gene expression, suggesting that extrinsic factors can be manipulated through exercise (Carey et al. 2007). It is important to note that when human primary myoblasts were cultured with serum from either young or elderly humans, proliferation and differentiation efficiencies were not altered (George et al. 2010). Importantly, however, these experiments were carried out in young cells only, and it remains unknown whether a young circulatory environment can have a positive impact on aged satellite cells in humans.

As mentioned above altered signaling between the muscle fiber and the satellite cell, such as seen with Notch-Delta signaling, is important in dictating satellite cell function. Similarly, muscle-derived growth factors are important in normal satellite cell regulation. In the vast majority of the time, satellite cells exist in a state of quiescence maintained by the interaction of the satellite cell, muscle fiber, and other niche elements (Bischoff 1990a, b). In studies from aging mice, there appears to be an increase in muscle fiber-derived fibroblast growth factor-2 (Fgf2), which causes a subset of satellite cells to leave the quiescent state and lose the ability to self-renew (Chakkalakal et al. 2012). From this data it appears that inappropriate changes in fiber-released growth factors may be instrumental in the reduction of the satellite cell pool through inducing an inability to maintain quiescence. Although it is unclear why there is an age-related upregulation of Fgf2 from the fiber, it has been suggested that it may be an inappropriate response by the muscle fiber attempting to induce repair mechanisms (Chakkalakal et al. 2012). Interestingly, it appears that blocking Fgf2 signaling at the satellite cell may have therapeutic benefits, preventing satellite cell loss (Chakkalakal et al. 2012). Satellite cells themselves may in fact contribute to the regulation of the extracellular matrix by directly influencing fibroblast function (Fry et al. 2014). Isolated satellite cells in vitro serve to negatively regulate key extracellular matrix mRNA expression in fibroblasts, suggesting a mechanism whereby satellite cells regulate their own microenvironment to ensure appropriate extracellular architecture (Fry et al. 2014). Importantly, a reduction in satellite cell number appears to lead to increased tissue fibrosis in aging muscle (Fry et al. 2015), contributing to age-related tissue fibrosis and dysfunction.

Other key muscle-derived cytokines and growth factors such as IL-6 and IGF-1 have been shown to change as a result of the aging process. IL-6 concentration in the circulation is increased in the elderly and thought to be associated with an age-related low-grade inflammation. Interestingly, elevated levels of IL-6 are associated with a loss of muscle strength (Visser et al. 2002; Pereira et al. 2009). Prolonged elevated levels of IL-6 were shown to induce muscle atrophy (Haddad et al. 2005) and thus may also negatively impact satellite cell function. IL-6 is a pleiotropic cytokine with divergent effects on many tissues depending on spatial and temporal changes in concentration (Pedersen and Febbraio 2008; Frost and Lang 2005; Kishimoto 2005). Chronically elevated systemic IL-6 is associated with a proinflammatory environment and muscle wasting conditions such as aging, sarcopenic obesity, and cancer cachexia (Roubenoff 1999, 2003). Animal studies using direct infusion of IL-6 into the circulation report that elevated IL-6 can induce a catabolic environment, promote muscle atrophy, and impair normal muscle growth (Haddad et al. 2005; Bodell et al. 2009). Human studies also illustrate strong associations between elevated serum IL-6, lower muscle strength, and reduced muscle mass associated with the aging process (Visser et al. 2002; Pereira et al. 2009). Furthermore, it appears that systemic IL-6 can be elevated in healthy older (~70 years) individuals (Alvarez-Rodriguez et al. 2012), suggesting that IL-6 is increased as a consequence of the normal aging process. Although the precise mechanisms underpinning how chronically elevated IL-6 promotes muscle catabolism remain somewhat unclear, the role of the suppressors of cytokine signaling (SOCS) proteins has been implicated in reducing the efficacy of anabolic signaling pathways such as IGF-1 in the skeletal muscle (Al-Shanti and Stewart 2012). Recent human data demonstrated altered IL-6 signaling in elderly muscle including in satellite cells specifically (McKay et al. 2013). An age-related increase in IL-6 alters the satellite cell-specific phosphorylated signal transducer and activator of transcription-3 (STAT3) protein response in type II myofiber-associated satellite cells (McKay et al. 2013). This, coupled with an increase in muscle SOCS3 and SOCS1 proteins, appears to be inhibiting the normal temporal IL-6 response of the satellite cells, impairing proliferation (McKay et al. 2013). Low-grade chronically elevated serum IL-6 may induce an increased level of SOCS3 (and SOCS1), which acts to inhibit the IL-6/JAK/STAT3 axis on a local level. Following muscle damage, IL-6 is produced locally and by infiltrating macrophages to activate STAT3 promoting cell cycle entry and satellite cell proliferation. However, in advanced aging chronically elevated IL-6 serves to induce a feedback mechanism through SOCS3 resulting in impaired IL-6/JAK/STAT3 signaling potentially impairing the normal satellite cell response (McKay et al. 2013). Interestingly in patients with type 2 diabetes and obesity, chronically elevated IL-6 also conferred dysfunction on the muscle stem cells, impairing IL-6 receptor and STAT3/SOCS signaling (Scheele et al. 2012). Therefore it appears that altered IL-6 from low-grade chronic inflammation or more overt inflammation associated with obesity and disease processes is a major factor contributing to muscle stem cell dysfunction. However, this area of research is still in its infancy, and more research is required to fully understand the relationship between inflammation and stem cell dysfunction.

In addition to cytokine dysfunction, it appears that growth factor expression is altered with aging. Growth hormone receptor and IGF-1 protein content and circulating factors such as testosterone and IGF-1 also appear to be reduced in the elderly, which may reduce the anabolic capacity of aged muscle (Hameed et al. 2008; Musaro et al. 2001; Leger et al. 2008; Petrella et al. 2006). Since IGF-1 and testosterone are potent myogenic growth factors, a reduction of these factors in circulation will compound the dysfunction of muscle stem cells observed in aged muscle.

Although there are many local and systemic factors that may regulate muscle satellite cell function and contribute to age-related satellite cell dysfunction, no circulating factor has garnered the same level of interest as myostatin. Myostatin

(GDF-8) is a transforming growth factor-β (TGF-β) family member that plays an integral role in the growth and maintenance of muscle mass (McPherron et al. 1997). The function of myostatin is highly conserved across species (McPherron and Lee 1997), and the loss or mutation of the myostatin gene results in a "double-muscled" phenotype (McPherron et al. 1997; McPherron and Lee 1997). Muscle from myostatin null mice is typically two to three times larger than wild-type mice, which appears to be a consequence of both hypertrophy and hyperplasia of muscle fibers (McPherron et al. 1997). Postnatal blockade of myostatin signaling induces increases in muscle mass and strength, which are attributed to fiber hypertrophy in adult mice (Bogdanovich et al. 2002; Whittemore et al. 2003). Myostatin works similarly to TGF-β signaling by binding to a type II serine/threonine kinase receptor activin IIb (Lee and McPherron 2001). This activated complex phosphorylates receptor-regulated Smad proteins. Smad2 and Smad3 are the primary Smads that mediate myostatin signaling (Huang et al. 2011). This canonical myostatin signaling pathway has been shown to inhibit satellite cell proliferation by upregulating the cyclin-dependent kinase (Cdk) inhibitor p21, which hypophosphorylates phospho-retinoblastoma (pRb) and prevents progression from G_1 to DNA synthesis (S phase of the cell cycle) (Thomas et al. 2000). Satellite cells in culture, treated with exogenous myostatin, illustrate a significant increase in p21 protein, a downregulation of Cdk2, and a downregulation in hyperphosphorylated Rb with a simultaneous increase in hypophosphorylated pRb, which prevents the upregulation of cell cycle machinery and prevents the cells from passing the G_1 cell cycle checkpoint and entering S phase (McCroskery et al. 2003). Myostatin may also signal via non-Smad-mediated pathways (noncanonical), activating p38 mitogen-activated protein kinase (MAPK) via the TAK1-MKK6 cascade (Philip et al. 2005), or by induction of the extracellular signal-regulated kinase (Erk) 1/2 signaling (Yang et al. 2006), inhibiting myoblast proliferation and differentiation.

In humans there is little work demonstrating a direct relationship between myostatin and muscle wasting; however, there is some evidence to support the role of myostatin in muscle atrophy related to disease. Patients with HIV-associated muscle wasting demonstrate an elevation of serum and intramuscular concentrations of myostatin, suggesting that the upregulation of myostatin contributes to the overall loss of skeletal muscle. Most of the evidence supporting a role for myostatin in muscle wasting is derived from animal studies. In the Duchenne muscular dystrophy (*MDX* mouse model) animal model, the blockade of myostatin signaling in postnatal *MDX* mice induced an 11–30 % increase in muscle mass (Bogdanovich et al. 2002). Furthermore, the lifelong absence of myostatin (*Mstn*$^{-/-}$ mouse model) appears to reduce the aged phenotype compared to age-matched wild-type controls (Siriett et al. 2006) suggesting that myostatin is a key contributor to sarcopenia in rodents. In humans, 3 days of unloading (withdrawal of weight bearing via unilateral lower limb suspension) increased myostatin protein and mRNA compared to the control limb, suggesting that myostatin is involved in the initiation of muscle atrophy (Gustafsson et al. 2010). Interestingly, increased myostatin is accompanied by increased gene expression of atrogin-1 and MuRF-1, which are E3 ubiquitin ligases, instrumental in the regulation of muscle atrophy (Gustafsson et al. 2010; Gomes et al. 2001; Bodine et al. 2001). In mice, normal aging is accompanied by approximately doubling the concentration of MSTN in muscle lysates of old animals compared to young animals (Kovacheva et al. 2010). This is accompanied by an increase in p21 protein and muscle cell apoptosis (<1 % in young vs. 8 % apoptosis in old) (Kovacheva et al. 2010). Furthermore, excess of myostatin in circulation induced a significant reduction in muscle mass in mice (Zimmers et al. 2002). Not surprisingly in humans, elderly men have approximately twofold higher levels of myostatin protein in whole muscle homogenates compared to their young counterparts (McKay et al. 2012). Recent work in humans demonstrated that myostatin expression was reduced to a larger degree in young compared to elderly subjects (Kim et al. 2005). Interestingly, across a 2-day post-resistance exercise time course, myostatin gene expression remained over 145 % higher in older adults compared to young adults despite showing a time-dependent downregulation, which was similar to the young. Myostatin in humans and animals inhibits MyoD through increased Smad3 signaling (Langley et al. 2002; McFarlane et al. 2011). In humans, there is an age-dependent inhibition of MyoD induction in satellite cells in response to muscle damage, which coincides with increased Smad3 expression, an increased proportion of satellite cells staining positive for myostatin, and a reduced proportion of satellite cells in the cell cycle compared to young controls (McKay et al. 2012). Following muscle injury in humans, a greater proportion of satellite cells co-localize with myostatin in elderly men, and this is associated with a reduction of satellite cell activation/proliferation and a lack of progression of satellite cells into S phase of the cell cycle (McKay et al. 2012). Furthermore, the presence of myostatin was more prevalent in satellite cells of type II myofibers in the elderly, which is strongly associated with a lack of a type II satellite cell response (McKay et al. 2012). Therefore altered myostatin signaling appears to be a key contributing factor to the reduced myogenic response of older muscle to a physiological stimulus. Coupling increased myostatin with dysfunctional cytokine, Notch, and Wnt signaling, and it appears that the age-related muscle stem cell function is hampered at multiple levels of regulation and at multiple key sights of regulation (e.g., cell cycle initiation/progression, differentiation transcriptional networks, calcium homeostasis, antiapoptotic mechanisms).

10.11 Atypical Progenitors, Alternate Cell Populations, and Aging

By and large, the progenitor cell responsible for muscle regeneration is the muscle satellite cell (Relaix and Zammit 2012). However, under experimental conditions in animals, other cell populations have demonstrated myogenic potential when the satellite cell pool is compromised. Side population cells, bone marrow-derived stem cells, mesoangioblasts, pericytes, mesenchymal stem cells, and $PW1^+$ interstitial cells (PICs) have all been shown to experimentally form myotubes in vitro and/or home to the satellite cell niche (Mitchell et al. 2010; Otto et al. 2009). Importantly, in recent studies where satellite cells were ablated by various techniques, damaged muscle was unable to regenerate despite the presence of many of these other cell types in the muscle, strongly suggesting that non-satellite cell populations do not contribute significantly to muscle regeneration (McCarthy et al. 2011; Sambasivan et al. 2011). Unfortunately there is little data available that has examined the role of alternate progenitor populations in the context of aging. One would assume based on the common themes observed in tissue stem cell aging (Carlson and Conboy 2007; Brack and Rando 2012; Liu and Rando 2011) that many of the age-related changes that plague the satellite cell and the satellite cell niche would also impact the ability of non-muscle progenitors to contribute effectively to myogenesis.

10.12 Conclusions and Future Directions

Evidence from studies in animals and humans demonstrates that skeletal muscle stem cells are profoundly affected by age. Figure 10.3 illustrates some of the basic differences in muscle stem cell function between young and aged muscle. Where the young muscle stem cells are readily activated and proliferate rapidly, aged muscle stem cells show a resistance to activation and a blunted proliferative capacity. Furthermore where the majority of young muscle stem cells progress towards terminal differentiation (myogenic commitment) and the remainder reverts back to quiescence renewing the stem cell pool, aged muscle stem cells adopt multiple fates impairing efficient regeneration and repair of skeletal muscle. Activated muscle stem cells from aged muscle may begin to proliferate but become susceptible to apoptosis, while some progress down the myogenic lineage; others undergo a myogenic-to-fibrogenic or adipogenic conversion. Some aged muscle stem cells may activate and directly differentiate without proliferating, which may reduce the number for self-renewing cells and thus may decrease the absolute number of reserved cells.

Whether muscle stem cell dysfunction is a precursor to the onset of sarcopenia is currently under debate; however, it has been demonstrated that exercise training can mitigate at least some of the negative effects of aging on the skeletal muscle (Melov et al. 2007; Lanza et al. 2008; Aagaard et al. 2010) and muscle stem cell number (Verdijk et al. 2009; Leenders et al. 2013). Currently, research efforts are underway to elucidate whether exercise training actually confers a survival benefit to the muscle stem cells and whether exercise can restore and rejuvenate muscle stem cell function either through direct modification of the proteasome or through epigenetic modifications or both. There is an emerging body of literature that focuses on rejuvenating stem cell function through epigenetic reprogramming which was recently reviewed by Rando and Chang (2012). Although this area appears to be promising for the future, much work needs to be done to define how epigenetic states of young and old stem cells differ and how stem cells of different tissue types differ and importantly if epigenetic modifications are inducible in a safe and effective manner.

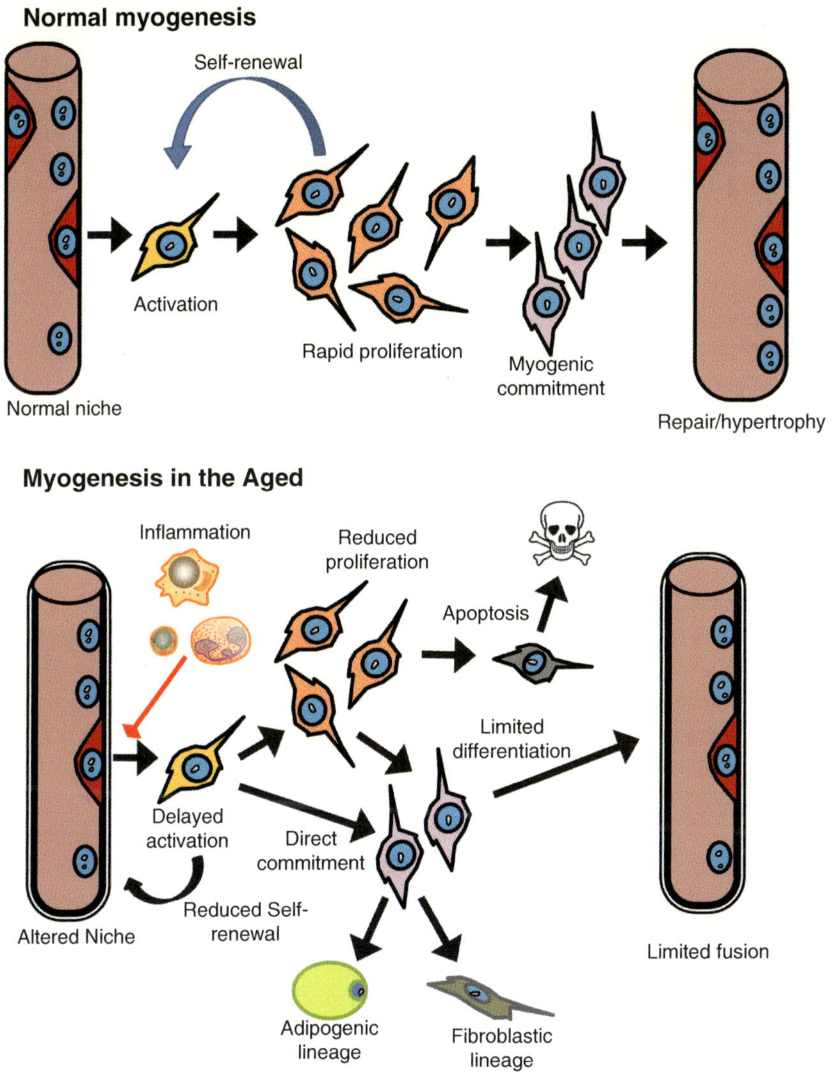

Fig. 10.3 The myogenic program as it occurs in young postnatal muscle. After activation and sufficient rounds of proliferation, muscle stem cells differentiate and fuse with existing myofibers, contributing their myonuclei. A small subset of proliferating cells withdraw from the cell cycle and revert back to quiescence thus replenishing the muscle stem cell pool. Below is a summary of the dysregulation of muscle stem cells in aged muscle. The altered stem cell niche and alteration in local and circulating factors such as chronic inflammation reduce the ability of muscle stem cells to activate and proliferate. These cells, once activated, are more susceptible to apoptosis and demonstrate a reduced ability to self-renew, which reduces the total muscle stem cell pool. This is further exacerbated by the subset of activated cells that bypass proliferation and directly differentiate, which further reduces the ability of these cells to self-renew. Furthermore, proliferating cells may adopt alternate cell fates, converting from myogenic lineage to a fibrotic and/or adipogenic lineage. Taken together these major mechanisms reduce the myogenic capacity of aged muscle stem cells, impacting the ability to repair and maintain healthy skeletal muscle which may contribute to the progression of sarcopenia

By progressively understanding how the aging process affects muscle stem cell physiology, epigenetic and proteomic regulation, and ultimately the functional manifestations of these effects, we will be able to design appropriate strategies to mitigate the effects of aging on skeletal muscle. In muscle aging, it appears that in addition to the muscle stem cell, niche elements and circulating factors must be considered in the design of therapeutic strategies to ensure successful and efficacious treatment of sarcopenia, which is a major cause of morbidity and mortality in the aging population.

References

Aagaard P, Suetta C, Caserotti P, Magnusson SP, Kjaer M (2010) Role of the nervous system in sarcopenia and muscle atrophy with aging: strength training as a countermeasure. Scand J Med Sci Sports 20:49–64

Adams GR, Caiozzo VJ, Haddad F, Baldwin KM (2002) Cellular and molecular responses to increased skeletal muscle loading after irradiation. Am J Physiol Cell Physiol 283:C1182–C1195

Alexakis C, Partridge T, Bou-Gharios G (2007) Implication of the satellite cell in dystrophic muscle fibrosis: a self-perpetuating mechanism of collagen overproduction. Am J Physiol Cell Physiol 293:C661–C669

Allen RE, Boxhorn LK (1987) Inhibition of skeletal muscle satellite cell differentiation by transforming growth factor-beta. J Cell Physiol 133:567–572

Allen RE, Sheehan SM, Taylor RG, Kendall TL, Rice GM (1995) Hepatocyte growth factor activates quiescent skeletal muscle satellite cells in vitro. J Cell Physiol 165:307–312

Al-Shanti N, Stewart CE (2012) Inhibitory effects of IL-6 on IGF-1 activity in skeletal myoblasts could be mediated by the activation of SOCS-3. J Cell Biochem 113:923–933

Alvarez-Rodriguez L, Lopez-Hoyos M, Munoz-Cacho P, Martinez-Taboada VM (2012) Aging is associated with circulating cytokine dysregulation. Cell Immunol 273:124–132

Armand O, Boutineau AM, Mauger A, Pautou MP, Kieny M (1983) Origin of satellite cells in avian skeletal muscles. Arch Anat Microsc Morphol Exp 72:163–181

Austin L, Bower J, Kurek J, Vakakis N (1992) Effects of leukaemia inhibitory factor and other cytokines on murine and human myoblast proliferation. J Neurol Sci 112:185–191

Beaton LJ, Tarnopolsky MA, Phillips SM (2002) Contraction-induced muscle damage in humans following calcium channel blocker administration. J Physiol 544:849–859

Beauchamp JR, Heslop L, Yu DS, Tajbakhsh S, Kelly RG, Wernig A, Buckingham ME, Partridge TA, Zammit PS (2000) Expression of CD34 and Myf5 defines the majority of quiescent adult skeletal muscle satellite cells. J Cell Biol 151:1221–1234

Beccafico S, Puglielli C, Pietrangelo T, Bellomo R, Fano G, Fulle S (2007) Age-dependent effects on functional aspects in human satellite cells. Ann N Y Acad Sci 1100:345–352

Bickel CS, Slade J, Mahoney E, Haddad F, Dudley GA, Adams GR (2005) Time course of molecular responses of human skeletal muscle to acute bouts of resistance exercise. J Appl Physiol 98:482–488

Biressi S, Rando TA (2010) Heterogeneity in the muscle satellite cell population. Semin Cell Dev Biol 21:845–854

Bischoff R (1975) Regeneration of single skeletal muscle fibers in vitro. Anat Rec 182:215–235

Bischoff R (1986) Proliferation of muscle satellite cells on intact myofibers in culture. Dev Biol 115:129–139

Bischoff R (1990a) Interaction between satellite cells and skeletal muscle fibers. Development 109:943–952

Bischoff R (1990b) Cell cycle commitment of rat muscle satellite cells. J Cell Biol 111:201–207

Bodell PW, Kodesh E, Haddad F, Zaldivar FP, Cooper DM, Adams GR (2009) Skeletal muscle growth in young rats is inhibited by chronic exposure to IL-6 but preserved by concurrent voluntary endurance exercise. J Appl Physiol 106:443–453

Bodine SC, Latres E, Baumhueter S, Lai VK, Nunez L, Clarke BA, Poueymirou WT, Panaro FJ, Na E, Dharmarajan K, Pan ZQ, Valenzuela DM, DeChiara TM, Stitt TN, Yancopoulos GD, Glass DJ (2001) Identification of ubiquitin ligases required for skeletal muscle atrophy. Science 294:1704–1708

Bogdanovich S, Krag TO, Barton ER, Morris LD, Whittemore LA, Ahima RS, Khurana TS (2002) Functional improvement of dystrophic muscle by myostatin blockade. Nature 420:418–421

Bohr VA (2002) Repair of oxidative DNA damage in nuclear and mitochondrial DNA, and some changes with aging in mammalian cells. Free Radic Biol Med 32:804–812

Brack AS, Rando TA (2012) Tissue-specific stem cells: lessons from the skeletal muscle satellite cell. Cell Stem Cell 10:504–514

Brack AS, Conboy MJ, Roy S, Lee M, Kuo CJ, Keller C, Rando TA (2007) Increased Wnt signaling during aging alters muscle stem cell fate and increases fibrosis. Science 317:807–810

Brack AS, Conboy IM, Conboy MJ, Shen J, Rando TA (2008) A temporal switch from notch to Wnt signaling in muscle stem cells is necessary for normal adult myogenesis. Cell Stem Cell 2:50–59

Broholm C, Mortensen OH, Nielsen S, Akerstrom T, Zankari A, Dahl B, Pedersen BK (2008) Exercise induces expression of leukaemia inhibitory factor in human skeletal muscle. J Physiol 586:2195–2201

Brooks SV, Faulkner JA (1988) Contractile properties of skeletal muscles from young, adult and aged mice. J Physiol 404:71–82

Brooks SV, Faulkner JA (1996) The magnitude of the initial injury induced by stretches of maximally activated muscle fibres of mice and rats increases in old age. J Physiol 497(Pt 2):573–580

Buckingham M, Relaix F (2007) The role of Pax genes in the development of tissues and organs: Pax3 and Pax7 regulate muscle progenitor cell functions. Annu Rev Cell Dev Biol 23:645–673

Cantini M, Massimino ML, Rapizzi E, Rossini K, Catani C, Dalla LL, Carraro U (1995) Human satellite cell proliferation in vitro is regulated by autocrine secretion of IL-6 stimulated by a soluble factor(s) released by activated monocytes. Biochem Biophys Res Commun 216:49–53

Cao Y, Kumar RM, Penn BH, Berkes CA, Kooperberg C, Boyer LA, Young RA, Tapscott SJ (2006) Global and gene-specific analyses show distinct roles for Myod and Myog at a common set of promoters. EMBO J 25:502–511

Carey KA, Farnfield MM, Tarquinio SD, Cameron-Smith D (2007) Impaired expression of Notch signaling genes in aged human skeletal muscle. J Gerontol A Biol Sci Med Sci 62:9–17

Carlson ME, Conboy IM (2007) Loss of stem cell regenerative capacity within aged niches. Aging Cell 6:371–382

Cefalu CA (2011) Theories and mechanisms of aging. Clin Geriatr Med 27:491–506

Chakkalakal JV, Jones KM, Basson MA, Brack AS (2012) The aged niche disrupts muscle stem cell quiescence. Nature 490:355–360

Chakravarthy MV, Abraha TW, Schwartz RJ, Fiorotto ML, Booth FW (2000) Insulin-like growth factor-I extends in vitro replicative life span of skeletal muscle satellite cells by enhancing G1/S cell cycle progression via the activation of phosphatidylinositol 3'-kinase/Akt signaling pathway. J Biol Chem 275:35942–35952

Charge SB, Rudnicki MA (2004) Cellular and molecular regulation of muscle regeneration. Physiol Rev 84:209–238

Collins CA, Olsen I, Zammit PS, Heslop L, Petrie A, Partridge TA, Morgan JE (2005) Stem cell function, self-renewal, and behavioral heterogeneity of cells from the adult muscle satellite cell niche. Cell 122:289–301

Collins CA, Zammit PS, Ruiz AP, Morgan JE, Partridge TA (2007) A population of myogenic stem cells that survives skeletal muscle aging. Stem Cells 25:885–894

Conboy IM, Conboy MJ, Smythe GM, Rando TA (2003) Notch-mediated restoration of regenerative potential to aged muscle. Science 302:1575–1577

Conboy IM, Conboy MJ, Wagers AJ, Girma ER, Weissman IL, Rando TA (2005) Rejuvenation of aged progenitor cells by exposure to a young systemic environment. Nature 433:760–764
Cooper RN, Tajbakhsh S, Mouly V, Cossu G, Buckingham M, Butler-Browne GS (1999) In vivo satellite cell activation via Myf5 and MyoD in regenerating mouse skeletal muscle. J Cell Sci 112(Pt 17):2895–2901
Corbu A, Scaramozza A, Badiali-DeGiorgi L, Tarantino L, Papa V, Rinaldi R, D'Alessandro R, Zavatta M, Laus M, Lattanzi G, Cenacchi G (2010) Satellite cell characterization from aging human muscle. Neurol Res 32:63–72
Cornelison DD, Wold BJ (1997) Single-cell analysis of regulatory gene expression in quiescent and activated mouse skeletal muscle satellite cells. Dev Biol 191:270–283
Cornelison DD, Olwin BB, Rudnicki MA, Wold BJ (2000) MyoD(−/−) satellite cells in single-fiber culture are differentiation defective and MRF4 deficient. Dev Biol 224:122–137
Croley AN, Zwetsloot KA, Westerkamp LM, Ryan NA, Pendergast AM, Hickner RC, Pofahl WE, Gavin TP (2005) Lower capillarization, VEGF protein, and VEGF mRNA response to acute exercise in the vastus lateralis muscle of aged vs. young women. J Appl Physiol 99:1872–1879
Day K, Shefer G, Shearer A, Yablonka-Reuveni Z (2010) The depletion of skeletal muscle satellite cells with age is concomitant with reduced capacity of single progenitors to produce reserve progeny. Dev Biol 340:330–343
Decary S, Mouly V, Hamida CB, Sautet A, Barbet JP, Butler-Browne GS (1997) Replicative potential and telomere length in human skeletal muscle: implications for satellite cell-mediated gene therapy. Hum Gene Ther 8:1429–1438
DiMario J, Buffinger N, Yamada S, Strohman RC (1989) Fibroblast growth factor in the extracellular matrix of dystrophic (mdx) mouse muscle. Science 244:688–690
Dreyer HC, Blanco CE, Sattler FR, Schroeder ET, Wiswell RA (2006) Satellite cell numbers in young and older men 24 hours after eccentric exercise. Muscle Nerve 33:242–253
Dreyfus PA, Chretien F, Chazaud B, Kirova Y, Caramelle P, Garcia L, Butler-Browne G, Gherardi RK (2004) Adult bone marrow-derived stem cells in muscle connective tissue and satellite cell niches. Am J Pathol 164:773–779
Frost RA, Lang CH (2005) Skeletal muscle cytokines: regulation by pathogen-associated molecules and catabolic hormones. Curr Opin Clin Nutr Metab Care 8:255–263
Frost RA, Lang CH (2011) mTor signaling in skeletal muscle during sepsis and inflammation: where does it all go wrong? Physiology (Bethesda) 26:83–96
Fry CS, Lee JD, Jackson JR, Kirby TJ, Stasko SA, Liu H, Dupont-Versteegden EE, McCarthy JJ, Peterson CA (2014) Regulation of the muscle fiber microenvironment by activated satellite cells during hypertrophy. FASEB J 28:1654–1665
Fry CS, Lee JD, Mula J, Kirby TJ, Jackson JR, Liu F, Yang L, Mendias CL, Dupont-Versteegden EE, McCarthy JJ, Peterson CA (2015) Inducible depletion of satellite cells in adult, sedentary mice impairs muscle regenerative capacity without affecting sarcopenia. Nat Med 21:76–80
Gallucci S, Provenzano C, Mazzarelli P, Scuderi F, Bartoccioni E (1998) Myoblasts produce IL-6 in response to inflammatory stimuli. Int Immunol 10:267–273
Galvin CD, Hardiman O, Nolan CM (2003) IGF-1 receptor mediates differentiation of primary cultures of mouse skeletal myoblasts. Mol Cell Endocrinol 200:19–29
George T, Velloso CP, Alsharidah M, Lazarus NR, Harridge SD (2010) Sera from young and older humans equally sustain proliferation and differentiation of human myoblasts. Exp Gerontol 45:875–881
Gomes MD, Lecker SH, Jagoe RT, Navon A, Goldberg AL (2001) Atrogin-1, a muscle-specific F-box protein highly expressed during muscle atrophy. Proc Natl Acad Sci U S A 98:14440–14445
Gopinath SD, Rando TA (2008) Stem cell review series: aging of the skeletal muscle stem cell niche. Aging Cell 7:590–598
Greig CA, Hameed M, Young A, Goldspink G, Noble B (2006) Skeletal muscle IGF-I isoform expression in healthy women after isometric exercise. Growth Horm IGF Res 16:373–376

Grounds MD (1998) Age-associated changes in the response of skeletal muscle cells to exercise and regeneration. Ann N Y Acad Sci 854:78–91

Gunther S, Kim J, Kostin S, Lepper C, Fan CM, Braun T (2013) Myf5-positive satellite cells contribute to Pax7-dependent long-term maintenance of adult muscle stem cells. Cell Stem Cell 13:590–601

Gussoni E, Soneoka Y, Strickland CD, Buzney EA, Khan MK, Flint AF, Kunkel LM, Mulligan RC (1999) Dystrophin expression in the mdx mouse restored by stem cell transplantation. Nature 401:390–394

Gustafsson T, Osterlund T, Flanagan JN, Von WF, Trappe TA, Linnehan RM, Tesch PA (2010) Effects of 3 days unloading on molecular regulators of muscle size in humans. J Appl Physiol 109:721–727

Haddad F, Zaldivar F, Cooper DM, Adams GR (2005) IL-6-induced skeletal muscle atrophy. J Appl Physiol 98:911–917

Hall ZW, Ralston E (1989) Nuclear domains in muscle cells. Cell 59:771–772

Hameed M, Orrell RW, Cobbold M, Goldspink G, Harridge SD (2003) Expression of IGF-I splice variants in young and old human skeletal muscle after high resistance exercise. J Physiol 547:247–254

Hameed M, Toft AD, Pedersen BK, Harridge SD, Goldspink G (2008) Effects of eccentric cycling exercise on IGF-I splice variant expression in the muscles of young and elderly people. Scand J Med Sci Sports 18(4):447–452

Han B, Tong J, Zhu MJ, Ma C, Du M (2008) Insulin-like growth factor-1 (IGF-1) and leucine activate pig myogenic satellite cells through mammalian target of rapamycin (mTOR) pathway. Mol Reprod Dev 75:810–817

Hawke TJ, Garry DJ (2001) Myogenic satellite cells: physiology to molecular biology. J Appl Physiol 91:534–551

Hill M, Goldspink G (2003) Expression and splicing of the insulin-like growth factor gene in rodent muscle is associated with muscle satellite (stem) cell activation following local tissue damage. J Physiol 549:409–418

Hill M, Wernig A, Goldspink G (2003) Muscle satellite (stem) cell activation during local tissue injury and repair. J Anat 203:89–99

Huang Z, Chen X, Chen D (2011) Myostatin: a novel insight into its role in metabolism, signal pathways, and expression regulation. Cell Signal 23:1441–1446

Irrcher I, Adhihetty PJ, Joseph AM, Ljubicic V, Hood DA (2003) Regulation of mitochondrial biogenesis in muscle by endurance exercise. Sports Med 33:783–793

Jang YC, Sinha M, Cerletti M, Dall'Osso C, Wagers AJ (2011) Skeletal muscle stem cells: effects of aging and metabolism on muscle regenerative function. Cold Spring Harb Symp Quant Biol 76:101–111

Janssen I, Heymsfield SB, Wang ZM, Ross R (2000) Skeletal muscle mass and distribution in 468 men and women aged 18–88 yr. J Appl Physiol 89:81–88

Janssen I, Shepard DS, Katzmarzyk PT, Roubenoff R (2004) The healthcare costs of sarcopenia in the United States. J Am Geriatr Soc 52:80–85

Jejurikar SS, Henkelman EA, Cederna PS, Marcelo CL, Urbanchek MG, Kuzon WM Jr (2006) Aging increases the susceptibility of skeletal muscle derived satellite cells to apoptosis. Exp Gerontol 41:828–836

Johnston AP, Baker J, Bellamy LM, McKay BR, De Lisio M, Parise G (2010) Regulation of muscle satellite cell activation and chemotaxis by angiotensin II. PLoS One 5, e15212

Kadi F, Charifi N, Denis C, Lexell J, Andersen JL, Schjerling P, Olsen S, Kjaer M (2005) The behaviour of satellite cells in response to exercise: what have we learned from human studies? Pflugers Arch 451:319–327

Kassar-Duchossoy L, Giacone E, Gayraud-Morel B, Jory A, Gomes D, Tajbakhsh S (2005) Pax3/Pax7 mark a novel population of primitive myogenic cells during development. Genes Dev 19:1426–1431

Kim JS, Cross JM, Bamman MM (2005) Impact of resistance loading on myostatin expression and cell cycle regulation in young and older men and women. Am J Physiol Endocrinol Metab 288:E1110–E1119

Kirkland JL, Tchkonia T, Pirtskhalava T, Han J, Karagiannides I (2002) Adipogenesis and aging: does aging make fat go MAD? Exp Gerontol 37:757–767

Kishimoto T (2005) Interleukin-6: from basic science to medicine–40 years in immunology. Annu Rev Immunol 23:1–21

Koopman R, van Loon LJ (2009) Aging, exercise, and muscle protein metabolism. J Appl Physiol 106:2040–2048

Kovacheva EL, Hikim AP, Shen R, Sinha I, Sinha-Hikim I (2010) Testosterone supplementation reverses sarcopenia in aging through regulation of myostatin, c-Jun NH2-terminal kinase, Notch, and Akt signaling pathways. Endocrinology 151:628–638

Kuang S, Gillespie MA, Rudnicki MA (2008) Niche regulation of muscle satellite cell self-renewal and differentiation. Cell Stem Cell 2:22–31

LaBarge MA, Blau HM (2002) Biological progression from adult bone marrow to mononucleate muscle stem cell to multinucleate muscle fiber in response to injury. Cell 111:589–601

Langley B, Thomas M, Bishop A, Sharma M, Gilmour S, Kambadur R (2002) Myostatin inhibits myoblast differentiation by down-regulating MyoD expression. J Biol Chem 277:49831–49840

Lanza IR, Short DK, Short KR, Raghavakaimal S, Basu R, Joyner MJ, McConnell JP, Nair KS (2008) Endurance exercise as a countermeasure for aging. Diabetes 57:2933–2942

Larsson L, Ansved T (1995) Effects of ageing on the motor unit. Prog Neurobiol 45:397–458

Lee SJ, McPherron AC (2001) Regulation of myostatin activity and muscle growth. Proc Natl Acad Sci U S A 98:9306–9311

Leenders M, Verdijk LB, van der Hoeven L, van Kranenburg J, Nilwik R, van Loon LJ (2013) Elderly men and women benefit equally from prolonged resistance-type exercise training. J Gerontol A Biol Sci Med Sci 68(7):769–779

Leger B, Derave W, De BK, Hespel P, Russell AP (2008) Human sarcopenia reveals an increase in SOCS-3 and myostatin and a reduced efficiency of Akt phosphorylation. Rejuvenation Res 11:163–175B

Lepper C, Partridge TA, Fan CM (2011) An absolute requirement for Pax7-positive satellite cells in acute injury-induced skeletal muscle regeneration. Development 138:3639–3646

Liu L, Rando TA (2011) Manifestations and mechanisms of stem cell aging. J Cell Biol 193:257–266

Mackey AL, Kjaer M, Dandanell S, Mikkelsen KH, Holm L, Dossing S, Kadi F, Koskinen SO, Jensen CH, Schroder HD, Langberg H (2007) The influence of anti-inflammatory medication on exercise-induced myogenic precursor cell responses in humans. J Appl Physiol 103:425–431

Mauro A (1961) Satellite cell of skeletal muscle fibers. J Biophys Biochem Cytol 9:493–495

Mayhew DL, Kim JS, Cross JM, Ferrando AA, Bamman MM (2009) Translational signaling responses preceding resistance training-mediated myofiber hypertrophy in young and old humans. J Appl Physiol 107(5):1655–1662

McCarthy JJ, Mula J, Miyazaki M, Erfani R, Garrison K, Farooqui AB, Srikuea R, Lawson BA, Grimes B, Keller C, Van Zant G, Campbell KS, Esser KA, Dupont-Versteegden EE, Peterson CA (2011) Effective fiber hypertrophy in satellite cell-depleted skeletal muscle. Development 138:3657–3666

McCroskery S, Thomas M, Maxwell L, Sharma M, Kambadur R (2003) Myostatin negatively regulates satellite cell activation and self-renewal. J Cell Biol 162:1135–1147

McFarlane C, Gu ZH, Wong ZW, Lau HY, Lokireddy S, Ge X, Mouly V, Butler-Browne G, Gluckman PD, Sharma M, Kambadur R (2011) Human myostatin negatively regulates human myoblast growth and differentiation. Am J Physiol Cell Physiol 301(1):C195–C203

McKay BR, O'Reilly CE, Phillips SM, Tarnopolsky MA, Parise G (2008) Co-expression of IGF-1 family members with myogenic regulatory factors following acute damaging muscle-lengthening contractions in humans. J Physiol 586:5549–5560

McKay BR, Ogborn DI, Bellamy LM, Tarnopolsky MA, Parise G (2012) Myostatin is associated with age-related human muscle stem cell dysfunction. FASEB J 26(6):2509–2521

McKay BR, Ogborn DI, Baker JM, Toth KG, Tarnopolsky MA, Parise G (2013) Elevated SOCS3 and altered IL-6 signaling is associated with age-related human muscle stem cell dysfunction. Am J Physiol Cell Physiol 304(8):C717–C728

McKinnell IW, Ishibashi J, Le GF, Punch VG, Addicks GC, Greenblatt JF, Dilworth FJ, Rudnicki MA (2008) Pax7 activates myogenic genes by recruitment of a histone methyltransferase complex. Nat Cell Biol 10:77–84

McPherron AC, Lee SJ (1997) Double muscling in cattle due to mutations in the myostatin gene. Proc Natl Acad Sci U S A 94:12457–12461

McPherron AC, Lawler AM, Lee SJ (1997) Regulation of skeletal muscle mass in mice by a new TGF-beta superfamily member. Nature 387:83–90

Mecocci P, Fano G, Fulle S, MacGarvey U, Shinobu L, Polidori MC, Cherubini A, Vecchiet J, Senin U, Beal MF (1999) Age-dependent increases in oxidative damage to DNA, lipids, and proteins in human skeletal muscle. Free Radic Biol Med 26:303–308

Megeney LA, Kablar B, Garrett K, Anderson JE, Rudnicki MA (1996) MyoD is required for myogenic stem cell function in adult skeletal muscle. Genes Dev 10:1173–1183

Melov S, Tarnopolsky MA, Beckman K, Felkey K, Hubbard A (2007) Resistance exercise reverses aging in human skeletal muscle. PLoS One 2, e465

Melton LJ III, Khosla S, Crowson CS, O'Connor MK, O'Fallon WM, Riggs BL (2000) Epidemiology of sarcopenia. J Am Geriatr Soc 48:625–630

Mikkelsen UR, Langberg H, Helmark IC, Skovgaard D, Andersen LL, Kjaer M, Mackey AL (2009) Local NSAID infusion inhibits satellite cell proliferation in human skeletal muscle after eccentric exercise. J Appl Physiol 107:1600–1611

Mitchell KJ, Pannerec A, Cadot B, Parlakian A, Besson V, Gomes ER, Marazzi G, Sassoon DA (2010) Identification and characterization of a non-satellite cell muscle resident progenitor during postnatal development. Nat Cell Biol 12:257–266

Moss FP, Leblond CP (1970) Nature of dividing nuclei in skeletal muscle of growing rats. J Cell Biol 44:459–462

Mourkioti F, Rosenthal N (2005) IGF-1, inflammation and stem cells: interactions during muscle regeneration. Trends Immunol 26:535–542

Murphy MM, Lawson JA, Mathew SJ, Hutcheson DA, Kardon G (2011) Satellite cells, connective tissue fibroblasts and their interactions are crucial for muscle regeneration. Development 138:3625–3637

Musaro A, McCullagh K, Paul A, Houghton L, Dobrowolny G, Molinaro M, Barton ER, Sweeney HL, Rosenthal N (2001) Localized Igf-1 transgene expression sustains hypertrophy and regeneration in senescent skeletal muscle. Nat Genet 27:195–200

Olguin HC, Olwin BB (2004) Pax-7 up-regulation inhibits myogenesis and cell cycle progression in satellite cells: a potential mechanism for self-renewal. Dev Biol 275:375–388

Olguin HC, Yang Z, Tapscott SJ, Olwin BB (2007) Reciprocal inhibition between Pax7 and muscle regulatory factors modulates myogenic cell fate determination. J Cell Biol 177:769–779

Otto A, Collins-Hooper H, Patel K (2009) The origin, molecular regulation and therapeutic potential of myogenic stem cell populations. J Anat 215(5):477–497

Oustanina S, Hause G, Braun T (2004) Pax7 directs postnatal renewal and propagation of myogenic satellite cells but not their specification. EMBO J 23:3430–3439

Parise G, O'Reilly CE, Rudnicki MA (2006) Molecular regulation of myogenic progenitor populations. Appl Physiol Nutr Metab 31:773–781

Parker MH, Perry RL, Fauteux MC, Berkes CA, Rudnicki MA (2006) MyoD synergizes with the E-protein HEB beta to induce myogenic differentiation. Mol Cell Biol 26:5771–5783

Pedersen BK, Febbraio MA (2008) Muscle as an endocrine organ: focus on muscle-derived interleukin-6. Physiol Rev 88:1379–1406

Pereira LS, Narciso FM, Oliveira DM, Coelho FM, Souza DG, Dias RC (2009) Correlation between manual muscle strength and interleukin-6 (IL-6) plasma levels in elderly community-dwelling women. Arch Gerontol Geriatr 48:313–316

Petrella JK, Kim JS, Cross JM, Kosek DJ, Bamman MM (2006) Efficacy of myonuclear addition may explain differential myofiber growth among resistance-trained young and older men and women. Am J Physiol Endocrinol Metab 291:E937–E946
Petrella JK, Kim JS, Mayhew DL, Cross JM, Bamman MM (2008) Potent myofiber hypertrophy during resistance training in humans is associated with satellite cell-mediated myonuclear addition: a cluster analysis. J Appl Physiol 104:1736–1742
Philip B, Lu Z, Gao Y (2005) Regulation of GDF-8 signaling by the p38 MAPK. Cell Signal 17:365–375
Philippou A, Papageorgiou E, Bogdanis G, Halapas A, Sourla A, Maridaki M, Pissimissis N, Koutsilieris M (2009) Expression of IGF-1 isoforms after exercise-induced muscle damage in humans: characterization of the MGF E peptide actions in vitro. In Vivo 23:567–575
Polesskaya A, Seale P, Rudnicki MA (2003) Wnt signaling induces the myogenic specification of resident CD45+ adult stem cells during muscle regeneration. Cell 113:841–852
Psilander N, Damsgaard R, Pilegaard H (2003) Resistance exercise alters MRF and IGF-I mRNA content in human skeletal muscle. J Appl Physiol 95:1038–1044
Punch VG, Jones AE, Rudnicki MA (2009) Transcriptional networks that regulate muscle stem cell function. Wiley Interdiscip Rev Syst Biol Med 1:128–140
Rando TA, Chang HY (2012) Aging, rejuvenation, and epigenetic reprogramming: resetting the aging clock. Cell 148:46–57
Rawlings JS, Rosler KM, Harrison DA (2004) The JAK/STAT signaling pathway. J Cell Sci 117:1281–1283
Relaix F, Zammit PS (2012) Satellite cells are essential for skeletal muscle regeneration: the cell on the edge returns centre stage. Development 139:2845–2856
Relaix F, Rocancourt D, Mansouri A, Buckingham M (2005) A Pax3/Pax7-dependent population of skeletal muscle progenitor cells. Nature 435:948–953
Relaix F, Montarras D, Zaffran S, Gayraud-Morel B, Rocancourt D, Tajbakhsh S, Mansouri A, Cumano A, Buckingham M (2006) Pax3 and Pax7 have distinct and overlapping functions in adult muscle progenitor cells. J Cell Biol 172:91–102
Reznik M (1969) Thymidine-3H uptake by satellite cells of regenerating skeletal muscle. J Cell Biol 40:568–571
Rhodes SJ, Konieczny SF (1989) Identification of MRF4: a new member of the muscle regulatory factor gene family. Genes Dev 3:2050–2061
Rotwein P, Pollock KM, Didier DK, Krivi GG (1986) Organization and sequence of the human insulin-like growth factor I gene. Alternative RNA processing produces two insulin-like growth factor I precursor peptides. J Biol Chem 261:4828–4832
Roubenoff R (1999) The pathophysiology of wasting in the elderly. J Nutr 129:256S–259S
Roubenoff R (2003) Catabolism of aging: is it an inflammatory process? Curr Opin Clin Nutr Metab Care 6:295–299
Roubenoff R, Hughes VA (2000) Sarcopenia: current concepts. J Gerontol A Biol Sci Med Sci 55:M716–M724
Rudnicki MA, Schnegelsberg PN, Stead RH, Braun T, Arnold HH, Jaenisch R (1993) MyoD or Myf-5 is required for the formation of skeletal muscle. Cell 75:1351–1359
Ryan NA, Zwetsloot KA, Westerkamp LM, Hickner RC, Pofahl WE, Gavin TP (2006) Lower skeletal muscle capillarization and VEGF expression in aged vs. young men. J Appl Physiol 100:178–185
Sabourin LA, Girgis-Gabardo A, Seale P, Asakura A, Rudnicki MA (1999) Reduced differentiation potential of primary MyoD–/– myogenic cells derived from adult skeletal muscle. J Cell Biol 144:631–643
Sambasivan R, Yao R, Kissenpfennig A, Van Wittenberghe L, Paldi A, Gayraud-Morel B, Guenou H, Malissen B, Tajbakhsh S, Galy A (2011) Pax7-expressing satellite cells are indispensable for adult skeletal muscle regeneration. Development 138:3647–3656
Scheele C, Nielsen S, Kelly M, Broholm C, Nielsen AR, Taudorf S, Pedersen M, Fischer CP, Pedersen BK (2012) Satellite cells derived from obese humans with type 2 diabetes and differentiated into myocytes in vitro exhibit abnormal response to IL-6. PLoS One 7, e39657

Schmalbruch H, Lewis DM (2000) Dynamics of nuclei of muscle fibers and connective tissue cells in normal and denervated rat muscles. Muscle Nerve 23:617–626

Seale P, Sabourin LA, Girgis-Gabardo A, Mansouri A, Gruss P, Rudnicki MA (2000) Pax7 is required for the specification of myogenic satellite cells. Cell 102:777–786

Serrano AL, Baeza-Raja B, Perdiguero E, Jardi M, Munoz-Canoves P (2008) Interleukin-6 is an essential regulator of satellite cell-mediated skeletal muscle hypertrophy. Cell Metab 7:33–44

Shefer G, Van de Mark DP, Richardson JB, Yablonka-Reuveni Z (2006) Satellite-cell pool size does matter: defining the myogenic potency of aging skeletal muscle. Dev Biol 294:50–66

Siriett V, Platt L, Salerno MS, Ling N, Kambadur R, Sharma M (2006) Prolonged absence of myostatin reduces sarcopenia. J Cell Physiol 209:866–873

Smith CK, Janney MJ, Allen RE (1994a) Temporal expression of myogenic regulatory genes during activation, proliferation, and differentiation of rat skeletal muscle satellite cells. J Cell Physiol 159:379–385

Smith LL, Keating MN, Holbert D, Spratt DJ, McCammon MR, Smith SS, Israel RG (1994b) The effects of athletic massage on delayed onset muscle soreness, creatine kinase, and neutrophil count: a preliminary report. J Orthop Sports Phys Ther 19:93–99

Spangenburg EE, Booth FW (2002) Multiple signaling pathways mediate LIF-induced skeletal muscle satellite cell proliferation. Am J Physiol Cell Physiol 283:C204–C211

Tatsumi R, Allen RE (2004) Active hepatocyte growth factor is present in skeletal muscle extracellular matrix. Muscle Nerve 30:654–658

Tatsumi R, Hattori A, Ikeuchi Y, Anderson JE, Allen RE (2002) Release of hepatocyte growth factor from mechanically stretched skeletal muscle satellite cells and role of pH and nitric oxide. Mol Biol Cell 13:2909–2918

Taylor-Jones JM, McGehee RE, Rando TA, Lecka-Czernik B, Lipschitz DA, Peterson CA (2002) Activation of an adipogenic program in adult myoblasts with age. Mech Ageing Dev 123:649–661

Thomas M, Langley B, Berry C, Sharma M, Kirk S, Bass J, Kambadur R (2000) Myostatin, a negative regulator of muscle growth, functions by inhibiting myoblast proliferation. J Biol Chem 275:40235–40243

Tomiya A, Aizawa T, Nagatomi R, Sensui H, Kokubun S (2004) Myofibers express IL-6 after eccentric exercise. Am J Sports Med 32:503–508

Verdijk LB, Koopman R, Schaart G, Meijer K, Savelberg HH, van Loon LJ (2007) Satellite cell content is specifically reduced in type II skeletal muscle fibers in the elderly. Am J Physiol Endocrinol Metab 292:E151–E157

Verdijk LB, Gleeson BG, Jonkers RA, Meijer K, Savelberg HH, Dendale P, van Loon LJ (2009) Skeletal muscle hypertrophy following resistance training is accompanied by a fiber type-specific increase in satellite cell content in elderly men. J Gerontol A Biol Sci Med Sci 64:332–339

Verdijk LB, Dirks ML, Snijders T, Prompers JJ, Beelen M, Jonkers RA, Thijssen DH, Hopman MT, van Loon LJ (2012) Reduced satellite cell numbers with spinal cord injury and aging in humans. Med Sci Sports Exerc 44:2322–2330

Viguie CA, Lu DX, Huang SK, Rengen H, Carlson BM (1997) Quantitative study of the effects of long-term denervation on the extensor digitorum longus muscle of the rat. Anat Rec 248:346–354

Visser M, Pahor M, Taaffe DR, Goodpaster BH, Simonsick EM, Newman AB, Nevitt M, Harris TB (2002) Relationship of interleukin-6 and tumor necrosis factor-alpha with muscle mass and muscle strength in elderly men and women: the Health ABC Study. J Gerontol A Biol Sci Med Sci 57:M326–M332

von Maltzahn MJ, Jones AE, Parks RJ, Rudnicki MA (2013) Pax7 is critical for the normal function of satellite cells in adult skeletal muscle. Proc Natl Acad Sci U S A 110:16474–16479

Wang YX, Rudnicki MA (2012) Satellite cells, the engines of muscle repair. Nat Rev Mol Cell Biol 13:127–133

Whittemore LA, Song K, Li X, Aghajanian J, Davies M, Girgenrath S, Hill JJ, Jalenak M, Kelley P, Knight A, Maylor R, O'Hara D, Pearson A, Quazi A, Ryerson S, Tan XY, Tomkinson KN,

Veldman GM, Widom A, Wright JF, Wudyka S, Zhao L, Wolfman NM (2003) Inhibition of myostatin in adult mice increases skeletal muscle mass and strength. Biochem Biophys Res Commun 300:965–971

Woodley DT, Rao CN, Hassell JR, Liotta LA, Martin GR, Kleinman HK (1983) Interactions of basement membrane components. Biochim Biophys Acta 761:278–283

Yablonka-Reuveni Z, Rudnicki MA, Rivera AJ, Primig M, Anderson JE, Natanson P (1999) The transition from proliferation to differentiation is delayed in satellite cells from mice lacking MyoD. Dev Biol 210:440–455

Yang S, Alnaqeeb M, Simpson H, Goldspink G (1996) Cloning and characterization of an IGF-1 isoform expressed in skeletal muscle subjected to stretch. J Muscle Res Cell Motil 17:487–495

Yang W, Chen Y, Zhang Y, Wang X, Yang N, Zhu D (2006) Extracellular signal-regulated kinase 1/2 mitogen-activated protein kinase pathway is involved in myostatin-regulated differentiation repression. Cancer Res 66:1320–1326

Zammit PS, Golding JP, Nagata Y, Hudon V, Partridge TA, Beauchamp JR (2004) Muscle satellite cells adopt divergent fates: a mechanism for self-renewal? J Cell Biol 166:347–357

Zimmers TA, Davies MV, Koniaris LG, Haynes P, Esquela AF, Tomkinson KN, McPherron AC, Wolfman NM, Lee SJ (2002) Induction of cachexia in mice by systemically administered myostatin. Science 296:1486–1488

Aging of Human Mesenchymal Stem Cells

11

Günter Lepperdinger and Stephan Reitinger

Abstract

Mesenchymal stem cells (MSCs) are multipotent precursors with osteogenic, chondrogenic, and adipogenic yet also anti-inflammatory and tissue-protective properties. Due to organism aging, the controlled perpetuation and continuance of developmental principles comprising cellular, interstitial, and systemic cues, stem cell proliferation, as well as progenitor differentiation appears to become constricted. Over the recent years, during which abundant information regarding MSC biology has been gathered, also a plethora of information regarding age-associated changes and molecular mechanisms has been collected, which are comprehended in this contribution, in particular regarding insights derived from both in vitro and in vivo analyses which cover alterations in DNA and chromatin modifications, as well as changes in regulatory elements such as miRNAs.

G. Lepperdinger (✉)
Department of Cell Biology and Physiology, Stem Cell and Healthy Longevity Research, University Salzburg, Hellbrunnerstr. 34, A-5020 Salzburg, Austria

Department of Extracellular Matrix Research, Institute for Biomedical Aging Research, Universität Innsbruck, Rennweg 10, A-6020 Innsbruck, Austria

Institute for Biomedical Aging Research, Austrian Academy of Sciences, Innsbruck, Austria
e-mail: guenter.lepperdinger@sbg.ac.at; guenter.lepperdinger@uibk.ac.at

S. Reitinger
Department of Extracellular Matrix Research, Institute for Biomedical Aging Research, Universität Innsbruck, Rennweg 10, A-6020 Innsbruck, Austria

Institute for Biomedical Aging Research, Austrian Academy of Sciences, Innsbruck, Austria

H. Geiger et al. (eds.), *Stem Cell Aging: Mechanisms, Consequences, Rejuvenation*,
DOI 10.1007/978-3-7091-1232-8_11

11.1 Introduction

In the bone marrow, hematopoietic cells are embedded in stroma. It became clear over the recent years that stroma should be considered more than solely being "a passive" filler: it distinctly supports hematopoietic and immunologic functions and controls proliferation and differentiation of corresponding progenitors. Besides balancing hematopoiesis, bone marrow stroma also houses specialized multipotent non-hematopoietic precursors. These are better known as mesenchymal stem cells or MSCs.

The first evidence for the presence of non-hematopoietic progenitors in the bone marrow was put forward by Julius Friedrich Cohnheim more than 140 years ago (Cohnheim 1867). About 100 years later Alexander Friedenstein rendered these particular cells more precisely (Friedenstein et al. 1970). According to these first pioneering experimental studies done in culture, this particular precursor cell type could be separated from other cells due to its very firm adherence to the surface of plastic culture dishes. Also remarkable was their nature to rapidly grow into large colonies. Based on these first findings, the concept of the MSC being a putative stromal stem cell, which is likely being involved in growth, maturation, remodeling, regeneration, and repair of many adult tissue, was put forward in the 1970s and 1980s (Luria et al. 1971; Owen and Friedenstein 1988). This notion, together with naming this cell type the MSC, was further popularized by Arnold Caplan in the late 1980s and the 1990s (Caplan 1991). Besides MSC, several other names have been concomitantly introduced by various research teams, e.g., "multipotent adult progenitors cells" (MAPCs), "marrow-isolated adult multi-lineage inducible cells" (MIAMI), or "very small embryonic-like stem cells" (VSELs) (Asahara et al. 1999; D'Ippolito et al. 2004; Jiang 2002; Kronenberg and Schipani 2009; Reyes et al. 2001). Despite this obviously confused naming, countless experimental in vitro studies helped to elucidate clearly distinct features of the MSC, many of which could only recently been corroborated in tedious in vivo studies and also in already completed clinical trials (for detailed information see www.clinicaltrials.gov). Besides many unexpected aspects such as the immune modulatory properties of MSCs (Rasmusson 2006), or their abilities to also spawn non-mesodermal lineage precursors (Woodbury et al. 2002), it became evident that MSCs are subject to aging (Fehrer and Lepperdinger 2005; Sethe et al. 2006; Stolzing and Scutt 2006).

11.2 MSC Multipotentiality

MSCs appear to be particularly interesting for further clinical applications and in tissue engineering given the fact that they can bring forth a large spectrum of mesodermal cell types as diverse as bone, cartilage, tendon, or fat precursor cells. By now it is generally accepted that the naive mesenchymal precursor cell effortlessly produces osteoblasts, prechondrocytes, and preadipocytes (Barry et al. 2001; Bruder et al. 1998; Cancedda et al. 2003; Le Blanc and Pittenger 2005; Prockop 1997). Besides these classical descendants, also terminally differentiated cell types, which are considered most widespread throughout many tissues in the body, namely, interstitial stromal cells and fibroblasts, appear to be a direct offspring from MSCs. Furthermore, it has been also reported in the literature that MSCs can be differentiated in vitro into cardiac myocytes (Wakitani et al. 1995), hepatocytes (Lee et al. 2004), or neuronal cells (Woodbury et al. 2000). Whether MSC descendants are also being recruited in vivo to contribute to the heart, liver, or neuronal tissue is controversially discussed and would need further direct experimental confirmation.

Presently, the bone marrow aspirate, subcutaneous adipose tissue, and umbilical cord blood are among the main sources for isolating MSCs. As mentioned above, the isolation procedure often relies on the MSC's firm plastic adherence. Alternatively, primary cell isolates can be affinity purified using anti-STRO1 (Williams et al. 2013) or anti-CD271 immunoglobulins (Jarocha et al. 2008) to immobilize positive cells on magnetic beads. In neither case, the respective cell culture would solely contain mesenchymal stem cells or lineage-committed mesenchymal progenitors, but is grossly contaminated with other cells (Kuznetsov et al. 1997; Klepsch et al. 2013). This limitation is mostly overcome by splitting the derived cell cultures with subsequent very low density seeding of only a few cells per cm^2 to select for the rapidly growing MSCs. It is this heterogeneity of the non-hematopoietic cell isolates which made it very difficult to precisely specify single, distinct representatives of mesenchymal precursors within the bone marrow and other stromal tissues.

Accounting for missing standards, in 2006 a group of peers in the field acting under the aegis of the "International Society for Cellular Therapy" defined minimal criteria for an isolated and culture-expanded multipotent stromal cell (Dominici et al. 2006): firstly MSCs should exhibit strong attachment to the surfaces of culture dishes (plastic adherence); secondly MSCs have to bear a set of, albeit not unique on its own but in combination sufficiently selective, surface markers, which are CD90, CD73, CD105, CD146, and CD44, and they need to lack expression of CD34, CD45, CD31, CD11b, CD14 and CD19, CD79α, as well as HLA-Class II; and thirdly, MSCs have to display tri-lineage differentiation potential into osteoblasts, preadipocytes, and chondrogenic cells. In addition to the enlisted surface markers, MSCs do also express transcription factors that are actually prominently present in embryonic stem cells such as Oct4, Nanog, and stage-specific embryonic antigen-4 (SSEA-4) (Rastegar et al. 2010).

Only recently, it could be demonstrated in murine models that MSCs can be identified in vivo using nestin expression as a distinguishing marker: nestin$^+$ MSC contains all the bone marrow colony-forming unit fibroblastic activity, and nestin$^+$ cells can be propagated as nonadherent "mesenspheres" that can self-renew and expand in serial transplantations. Also, nestin$^+$ cells appear to represent bona fide hematopoietic stem cell (HSC) niche cells, and in putting forward a unifying concept for a balanced regulation of hematopoietic and mesenchymal lineages at the stem cell level, this data is highly suggestive that in vivo HSC maintenance and MSC proliferation and differentiation are regulated in tandem (Mendez-Ferrer et al. 2010). Using nestin as a marker to specify MSCs in vivo, there are now good reasons to believe that MSCs reside in a complex three-dimensional meshwork, which comprises a plethora of other cell types such as in the case of the bone marrow, HSCs, adipocytes, and endothelial cells, altogether embedded in distinct extracellular matrix. Within this blend, MSCs guide differentiation of hematopoietic precursor cells into mature progeny, and more than that, MSCs appear to exert another pertinent function, namely, maintaining blood vessel integrity (Sacchetti et al. 2007; da Silva Meirelles et al. 2008; Crisan et al. 2008). Even though most work in this context has focused on isolated MSCs in culture, it can be envisaged that upon tissue damage and injury in vivo, MSCs are being activated and/or released from their perivascular niche, in order to support wound healing.

The current data summarizes the MSC's potential of differentiating into many different tissue-determining cell types, to name only a few bone, cartilage, muscle, tendon, heart, liver, and blood vessels and distinguish MSC not only as an important virtually ubiquitously present tissue-borne stem cells, but more than that, a powerful asset for tissue engineering strategies and clinical cell-based therapies.

11.3 MSC Aging

An obvious sign of the advancing biological age of the body is the concurrent decline of regenerative vigor, the weakened pace of wound healing and retarded repair capacities. All somatic cells are seemingly exposed to threats, such as reactive oxygen species, harmful chemical agents, or physical stressors, and hence against all expectation, even stem cells are at risk to change to the worse with age. As a last resort, cells with disproportionately accumulated hazards convert to a senescence state, which is on the long run extending the life span of the organisms, as this particular mechanism is considered to potently suppress tumor development. Since single stem cells are being activated only on rare occasions during continuously ongoing tissue remodeling, yet more distinctly immediately after traumatic stress, as in this case the stem cells are being forced to enter the cell cycle, proliferation has to be stringently controlled, in particular to avoid oncogenic growth and the deliberate emergence of cancer stem cells. Both intrinsic and extrinsic aging processes may affect stem cell properties. One possible intrinsic cellular aging mechanism is replicative or cellular senescence.

11.4 MSC Aging In Vitro

MSCs are commonly expanded in culture, and at least there, human MSCs are not immortal. On the long run in vitro cultivated MSCs display replicative senescence as independently demonstrated by a number of investigators (Fehrer et al. 2007; Stenderup et al. 2003; Stolzing et al. 2008; Estrada et al. 2013; Ho et al. 2011; Hao et al. 2013; Gruber et al. 2012; Tsai et al. 2010; Ksiazek 2009; Napolitano et al. 2007). The in vitro senescent phenotype includes the following characteristic features: (1) irreversible arrest of cell division (in contrast to quiescence, where this lock is reversible), (2) resistance to apoptotic death, and (3) the excretion of molecules normally secreted during wound repair and infection, such as inflammatory cytokines, proteases, and growth factors, the latter having detrimental consequences for the surrounding tissue (Coppe et al. 2010; Campisi et al. 2011). Most of these characteristics have also been described and studied in MSC cultures. Interestingly, the replicative capacity of cultivated MSC appears to be lower compared to that of descending fibroblasts. To that end it is however not clear whether already low levels of damage or stressful perturbations in MSC sufficiently stall the cell cycle for good. In good concordance with this assumption, MSCs do exhibit insignificant tumorigenicity in transplantation experiments.

11.4.1 Telomere Shortening

Absent telomerase activity and thus telomere shortening is distinctly associated with the in vitro senescent phenotype of MSCs. The absence of telomerase activity in MSCs during extensive replication leads to telomere attrition, which has been verified by many investigators (Pittenger et al. 1999; Zimmermann et al. 2003). Notably, constitutive expression of the catalytic subunit of telomerase gene, hTERT, prevents the onset of replicative senescence and maintains the "stemness" characteristics of MSCs (Simonsen et al. 2002). In line with this, regulated telomere maintenance by ectopic expression of hTERT sustains stem cell properties in MSC (Abdallah et al. 2005) and significantly extended their in vitro life span. Most interestingly, MSCs stalled in the cell cycle by serum deprivation transiently expressed telomerase activity during the S phase (Zhao et al. 2008). Recent work in mice suggested b-catenin as a key regulator for TERT expression in murine stem cells. Whether periodic upregulation of telomerase activity is high enough to retain telomeric capping function remains to be determined, and certainly more research needs to be performed for the detailed understanding of how MSCs actually counteract telomere shortening in vivo.

11.4.2 Chromatin Changes

Comparable to somatic cells, MSCs display variant epigenetic modes during aging. Due to their highly specialized chromatin state, MSCs bear the risk of transformation, in particular because pluripotency genes such as *SOX2* or *Nanog* are actively transcribed (Yoon et al. 2011). Epigenetic changes have also been investigated in aging MSCs; however, most of the results available to date are picturing deviations after long-term culture when cells turned replicatively senescent (Alessio et al. 2010; Jung et al. 2010; Napolitano et al. 2007; Yu and Kang 2013). For instance, an increase in EZH2 expression was noticed in adipose-derived MSCs during successive steps of long-term culture. In parallel and whether a consequence thereof is not clear, these cells showed a gradual decline of their differentiation potential. This is in due course correlated with stable histone methylation and hypo-acetylation of lineage-specific promoters (Noer et al. 2009). The global methylation pattern of MSCs from differently aged donors appears to remain unchanged with age, yet distinct exceptions manifest and stand out such as differential methylation of particular genes, like *Hoxa11*, which discriminates craniofacial MSCs from skeletal ones (Leucht et al. 2008), tumor suppressor p15 (Bork et al. 2010), or polycomb group protein target genes. Although quite evident, stemness and differentiation potential of MSCs are varying over lifetime, and the inherent dynamics are superimposed to variant epigenetic manifestations. However the detailed functions of primary epigenetic regulatory factors such as DNA methyltransferases (DNMT), histone deacetylases (HDAC), proteins of the polycomb group (PcG), and microRNAs (miRNAs) still remain to be fully elucidated. This is mainly due to the fact that their actions greatly overlap and cross-regulate each other. Similarly to other cell types, induction of cellular senescence by the downregulation of HDACs was also proposed for MSCs. Indeed, HDAC inhibitors suppress the expression of PcG genes, such as *BMI1*, *EZH2*, and *SUZ12*, while at the same time activating the jumonji domain-containing three gene, *JMJD3*. Also expression of *EZH2* and *SUZ12* is regulated by the phosphorylation status of the retinoblastoma protein, pRB, as a consequence of HDAC inhibitor treatment (Jung et al. 2010).

11.4.3 Age-Associated Expression of miRNAs

miRNAs, which are very short, noncoding RNAs, are considered very potent epigenetic regulators of gene expression. After binding to a target mRNA, miRNAs induce the degradation or a reduction in translatability of their specific target. In MSCs, miRNAs were found to target oncogenes, tumor suppressors, as well as differentiation markers, which under normal circumstances are silenced in stem and progenitor cells, in particular because self-renewal should be promoted and lineage-specific differentiation suppressed (Imam et al. 2010; Urbich et al. 2008; Wang et al. 2011). Individual miRNAs were found to be expressed that regulate epigenetic regulators such as DNMT, HDAC, high-mobility group AT-hook 2 (HMGA2), and PcG (Benetti et al. 2008; Braconi et al. 2010; Juan et al. 2009; Lee et al. 2010). Only recently, data from global analysis of miRNA expression profiles in MSCs derived from the bone marrow of old rhesus monkeys showed the downregulation of miR-let-7f, miR-23a, miR-132, miR-125b, miR-199-3p, miR-221, miR-222, miR-365, and miR-720 and the upregulation of miR-291a-5p, miR-558, miR-466, miR-466-3p, as well as miR-766 (Yu et al. 2011). To date, the activity of most of the identified miRNAs is not well comprehended. Besides miR-221 and miR-222 (Lambeth et al. 2009; Lorimer 2009), which appear to be involved in regulation of genes relevant to tumorigenesis and oncogenic activity, many of the others are clustered together in miR-17-92, a cluster which was found to be differentially regulated in many cellular aging models (Grillari et al. 2010; Hackl et al. 2010). Notably, the downregulation of miR-365 has been associated with growth arrest (Maes et al. 2009). Regulation of miRNA expression was also found to be connected to HDAC activity, as inhibition thereof was able to activate the transcription of the miRNAs let-7a1, let-7d, let-7f1, miR-23a, miR-26a, and miR-30a due to distinctly changing histone modification patterns of these genes. It could be further shown that these miRNAs once activated strongly repress the translation of HMGA2, which in turn regulates the expression of senescence genes, most importantly $p16^{INK4A}$ (Lee et al. 2010).

11.5 MSC Aging In Vivo

The presence of senescent cells in vivo, which should at least to some extent resemble cells that had been replicatively aged in culture, has by now been observed in many aged human tissues, such as the skin (Dimri et al. 1995; Ressler et al. 2006) as well as in the vascular system (Vasile et al. 2001). It is thus plausible that stem cells also reach an equivalent state of senescence in vivo. Indeed for muscle satellite cells, this could be shown in a very elegant series of experiments, which refined that this type of senescence termed "geroconversion" is dependent on the upregulation of a single cell cycle inhibitor $p16^{INK4a}$ (Sousa-Victor et al. 2014). Whether other types of stem cell pools also adopt this mechanism of geroconversion in order to control for functional fitness is still unclear. What is clear is that the maximum number of MSC population doublings and the proliferation rate of the initial passage of the primary MSCs, respectively, appear to be dependent on the age of the donors (Choudhery et al. 2014; Bajek et al. 2012; Zaim et al. 2012; Laschober et al. 2011; Kretlow et al. 2008; Baxter et al. 2004; Digirolamo et al. 1999). Yet the attenuated proliferation potential of MSCs from aged donors appears to also rely on the withdrawal of cells by cell death (Laschober et al. 2009). Conclusively MSC pools, which display slowing growth kinetics, also contain an increasing number of dying cells which augment during advancement of aging, indicative for a particular subpopulation of MSCs that fail to self-renew.

11.5.1 MSC Numbers

MSCs can be selected from mononuclear bone marrow cells by low density seeding and subsequent cultivation thereby forming fibroblastic colonies. The respective number, also called colony-forming unit fibroblastic (CFU-f), can be reliably estimated. Applying this method has been reported by several laboratories that total CFU-f decreases with age (Baxter et al. 2004; Caplan 2007; Majors et al. 1997; Muschler et al. 2001; Nishida et al. 1999; Stolzing et al. 2008). Around 30 % of the CFU-f were found to be multipotential and can thus be considered true MSCs (Kuznetsov et al. 1997). Based on this assumption, Muraglia and colleagues reported that the number of bi- or tri-potential colonies declined with age (Muraglia et al. 2000). In line with this, and in particular regarding the osteogenic capacity of clonogenic MSCs, many reports demonstrated a significant decrease in MSC numbers (Mueller and Glowacki 2001; Zhou et al. 2008). Besides the bone and bone marrow, potential sources of mesenchymal progenitor cells are the muscle (Bosch et al. 2000), vessel-associated pericytes (Collett and Canfield 2005), and blood (Eghbali-Fatourechi et al. 2005). Whether the quantity or quality of MSCs actually also varies in these tissues with advancing age remains to be determined.

11.5.2 MSC Diversity

Several functional studies have tackled with the question of whether age-associated changes other than premature or stress-induced senescence would impinge on MSC properties. Assuming that a young and fit cell population withstands accrued damage and the compounding of irreversible molecular changes longer than aged cells, a wide range of analyses have been undertaken, e.g., gross evaluations regarding the increased production of ROS, deviating SOD activity, whole genome gene expression profiles, and epigenetic signatures (Bork et al. 2010; Fehrer et al. 2007; Kemp et al. 2010; Kern et al. 2006; Laschober et al. 2011; Scaffidi and Misteli 2008; Stolzing et al. 2008; Wagner et al. 2009, 2010; Schallmoser et al. 2010). Also deviations of the age-associated MSC differentiation capacity were reported by many laboratories inasmuch as osteogenic potential of MSC isolated from aged donors appears to significantly decline, while the respective adipogenic differentiation performance remained unchanged or, worse, was found to be enhanced (Laschober et al. 2011). In the context of organism aging, several laboratories have isolated MSCs from a number of fetal tissues, e.g., human trimester fetal MSCs have been isolated from the blood, liver, bone marrow (Campagnolo et al. 2001), amniotic fluid (Tsai et al. 2004), placenta (In 't Anker et al. 2004), and cord blood (Mareschi et al. 2001). MSCs from very young organisms exhibit a higher level of naivety when compared to their adult relatives (Mirmalek-Sani et al. 2006; Gotherstrom et al. 2005). Regarding MSCs isolated form adult tissues, bone marrow-derived MSCs are certainly studied best (Pereira et al. 1995; Pittenger et al. 1999). Besides MSCs, which are juxtaposed to trabecular bone (Fehrer et al. 2007; Tuli et al. 2003), also those from adipose tissues have been investigated (Schaffler and Buchler 2007). Further sources are the periosteum (Nakahara et al. 1990), synovium (Sakaguchi et al. 2005), blood vessel (Abedin et al. 2004), tooth pulp of extracted teeth (Zhang et al. 2006; Arthur et al. 2008), and other connective tissue types, such as the dermis and muscle (Young et al. 2001). In addition, there are reports demonstrating the existence of MSCs within peripheral blood. These results have been debated though and are not always reproducible (Roufosse et al. 2004). MSCs isolated from different tissues and organs actually show a high resemblance of phenotypic characteristics. At this point, however, it appears most likely that all these cell variants are MSCs that exhibit different propensities in proliferation and differentiation capacities. For example, a recent study demonstrated that human adipose tissue contains the highest number of MSCs and umbilical cord blood the lowest. In turn, the latter come along with greatly enhanced proliferation capacity, whereas human bone marrow-derived MSCs cease growth earlier (Kern et al. 2006). Fetal MSCs also appear to be less lineage committed than MSCs from adult human individuals (Guillot et al. 2006). In consequence, it has been assumed that various vital stem cell properties of MSCs, such as their tissue regenerative capacity, are largely different. Taken together, it may further be that these limitations directly correlate with donor age.

11.6 MSC and "Inflamm-Aging"

Tissue regeneration and repair is supported by the orchestrated production and delivery of inflammatory cytokines. The controlled presentation of inflammatory stimuli supports wound and bone healing by sustained osteoblastogenesis (Duque et al. 2009). Inflammation should only be transiently turned on as high doses or chronically soaring levels favor adipogenic differentiation. Indeed bone loss, also including advancement of osteoporosis in the course of autoimmune disorders of bone, is associated with inflammatory dysregulation (Hardy and Cooper 2009; Mohanty et al. 2010). It is therefore conceivable that systemic chronic inflammation impacts on MSCs. Notably, often in older people proinflammatory cytokines are found at intermediate levels, which are considered below critical concentrations (de Gonzalo-Calvo et al. 2010). Exuberant production of proinflammatory cytokines is thought to be a consequence of lifelong antigenic burden or age-related diseases. This situation is often indicated by the term *inflamm-aging* (Salvioli et al. 2013; Biagi et al. 2011; Capri et al. 2006; Franceschi et al. 2000a, b). Moreover, chronic inflammation has been proposed a long time ago to instigate tumor formation and propagation (Dolberg et al. 1985; Dvorak 1986), and indeed there exists well-documented evidence that chronic inflammation may, for instance, cause colon cancer (Hahn et al. 2008; Khatami 2009).

MSCs treated with increasing concentrations of interferon gamma (IFNγ) or tumor necrosis factor alpha (TNFα) activate CD106/vascular cell adhesion molecule 1 (VCAM1) expression on their surface (Laschober et al. 2011). This marker was differentially expressed in primary MSCs derived from bone of differently aged donors. Regardless of CD106 expression, proliferation and self-renewal capacities were unchanged. MSCs actually exhibited no apparent difference other than being more prone to differentiate along the osteogenic lineage. Thus, dominant aberrations within the MSC microenvironment may arise from systemic chronic inflammation, which as mentioned above occurs regularly in elderly persons, or unbalanced inflammatory and anti-inflammatory networks as a consequence to lifelong antigenic burden or age-related diseases (Franceschi et al. 2007). Such cues are likely leading to a deviation from properly guiding molecular signaling mechanisms during regeneration, repair, and tissue remodeling and may thus gear a pronounced impetus for adipogenic upgrowth, to name only one prominent example, bone marrow adiposities.

Key sources of proinflammatory cytokines, such as IFNγ, are T lymphocytes and natural killer (NK) cells. In the last few years, it became clear that MSCs exhibit immune regulatory properties (Chen et al. 2006; Dazzi et al. 2006; Nauta and Fibbe 2007). It could be further shown that the suppressive activity of MSCs first requires the presence of IFNγ produced by T cells and NK cells (Krampera et al. 2006). This observation indicates that MSCs not only passively withstand inflammatory stimuli, but more than that, MSCs decisively respond to inflammatory cues, e.g., by promoting the expression of indoleamine 2,3-dioxygenase, which leads to tryptophan depletion in lymphocytes and may, by the same token, modulate lymphocyte activity and proliferation (Frumento et al. 2002; Meisel et al. 2004). Taken together, fate and function of both cell classes are tightly interconnected, and age-related changes appear to be compensated to a certain degree by complementary modes (Tokoyoda et al. 2010).

11.7 MSC and Age-Associated Disease

MSCs have the potential to ameliorate or even cure age-related degenerative diseases. However, certain limitations persist, and the health status of the donor is of particular interest. As a showcase, MSCs isolated from patients suffering from the premature aging syndrome Hutchinson-Gilford disease show altered differentiation capacities. Hence impaired or altered stem cell potency may contribute to the reduced potential of tissue homeostasis and ultimately brings about an aged phenotype in an accelerated fashion (Scaffidi and Misteli 2008). Hutchinson-Gilford disease is caused by a single factor, namely, an aberrant form of the nuclear architectural protein lamina (Eriksson et al. 2003).

The etiology of the age-associated pathology of osteoarthritis (OA) is not as well understood. It is generally accepted that besides age, multiple other factors such as obesity, history of joint trauma, and joint dysplasia are responsible. Joint resurfacing with tissue-engineered cartilage on the basis of isolated chondrocytes was shown to be greatly beneficial. However the availability of chondrocytes is a major constraint of such therapy. Yet the relatively easy availability of MSCs, which exhibit chondrogenic potential, represented an alternative cell source. There were however contradicting results; while some found no significant difference in numbers and differentiation potential irrespective of age or OA etiology when isolating MSCs from OA patients (Scharstuhl et al. 2007), others observed significant reduction of chondrogenic capacity (Murphy et al. 2002). The latter is in line with data reported earlier (Muschler et al. 2001).

In the context of osteoporosis (OP), another major age-associated health complication, in which decreased bone formation is an important pathophysiological mechanism resulting in bone loss, was investigated. Again the first results were controversial as in one study the number and the proliferative capacity of mesenchymal progenitor cells were maintained in patients with osteoporosis (Stenderup et al. 2001; Justesen et al. 2002), while results from another study demonstrated that stem cell growth, proliferative response, and osteogenic differentiation of MSCs from osteoporotic postmenopausal women were significantly affected (Rodriguez et al. 1999). Working along these lines further recent studies showed that osteogenic lineage commitment genes such as *RUNX2* or *Sp7/Osterix* were downregulated in OP patients (Dalle Carbonare et al. 2009; Rodriguez et al. 2008; Astudillo et al. 2008). It also appears likely that systemic influences dominantly impact on the osteogenic commitment of MSCs. Serum from postmenopausal OP women applied to a human osteogenic progenitor greatly inhibited osteoblastogenesis and greatly induced adipogenesis (Stringer et al. 2007; Abdallah et al. 2006). Hence during aging, distinct changes in MSCs are not only due to intrinsic modifications but also extrinsic alterations (Veronesi et al. 2012). Which bioactive factors or molecular parameters actually impinge on MSC lineage determination thereby restraining bone homeostasis and leading to bone marrow adiposities still needs to be elucidated (Rosen et al. 2009).

11.8 Outlook

It is tempting to see that over the recent years, increasingly more information became available concerning the alterations of stem cells for tissue repair as they age, and it is thus considered an emerging field of interest in the proliferating discipline of aging research. This particular theme has become increasingly important as the number of MSC therapies steadily rises, while pertinent questions about molecular mechanism driving MSC aging are still unanswered.

Acknowledgments Our work was previously funded by the Austrian Science Fund and the Jubilee Fund of the Austrian National Bank. Currently our research is supported by grants gifted by the EC (FP7 – Health VascuBone, Marie Curie Re-integration HyalStemAge), the Austrian Research Agency (Laura Bassi Centre for Excellence – DIALIFE), and the Tyrolean Future Fund (Translational Project – Smart Implants).

References

Abdallah BM, Haack-Sorensen M, Burns JS, Elsnab B, Jakob F, Hokland P, Kassem M (2005) Maintenance of differentiation potential of human bone marrow mesenchymal stem cells immortalized by human telomerase reverse transcriptase gene despite [corrected] extensive proliferation. Biochem Biophys Res Commun 326(3):527–538

Abdallah BM, Haack-Sorensen M, Fink T, Kassem M (2006) Inhibition of osteoblast differentiation but not adipocyte differentiation of mesenchymal stem cells by sera obtained from aged females. Bone 39(1):181–188

Abedin M, Tintut Y, Demer LL (2004) Mesenchymal stem cells and the artery wall. Circ Res 95(7):671–676

Alessio N, Squillaro T, Cipollaro M, Bagella L, Giordano A, Galderisi U (2010) The BRG1 ATPase of chromatin remodeling complexes is involved in modulation of mesenchymal stem cell senescence through RB-P53 pathways. Oncogene 29(40):5452–5463

Arthur A, Rychkov G, Shi S, Koblar SA, Gronthos S (2008) Adult human dental pulp stem cells differentiate towards functionally active neurons under appropriate environmental cues. Stem Cells 26(7):1787–1795

Asahara T, Masuda H, Takahashi T, Kalka C, Pastore C, Silver M, Kearne M, Magner M, Isner JM (1999) Bone marrow origin of endothelial progenitor cells responsible for postnatal vasculogenesis in physiological and pathological neovascularization. Circ Res 85:221–228

Astudillo P, Rios S, Pastenes L, Pino AM, Rodriguez JP (2008) Increased adipogenesis of osteoporotic human-mesenchymal stem cells (MSCs) characterizes by impaired leptin action. J Cell Biochem 103(4):1054–1065

Bajek A, Czerwinski M, Olkowska J, Gurtowska N, Kloskowski T, Drewa T (2012) Does aging of mesenchymal stem cells limit their potential application in clinical practice? Aging Clin Exp Res 24(5):404–411

Barry F, Boynton RE, Liu B, Murphy JM (2001) Chondrogenic differentiation of mesenchymal stem cells from bone marrow: differentiation-dependent gene expression of matrix components. Exp Cell Res 268(2):189–200

Baxter MA, Wynn RF, Jowitt SN, Wraith JE, Fairbairn LJ, Bellantuono I (2004) Study of telomere length reveals rapid aging of human marrow stromal cells following in vitro expansion. Stem Cells 22(5):675–682

Benetti R, Gonzalo S, Jaco I, Munoz P, Gonzalez S, Schoeftner S, Murchison E, Andl T, Chen T, Klatt P, Li E, Serrano M, Millar S, Hannon G, Blasco MA (2008) A mammalian microRNA cluster controls DNA methylation and telomere recombination via Rbl2-dependent regulation of DNA methyltransferases. Nat Struct Mol Biol 15(9):998

Biagi E, Candela M, Fairweather-Tait S, Franceschi C, Brigidi P (2011) Aging of the human metaorganism: the microbial counterpart. Age (Dordr) 34(1):247–267

Bork S, Pfister S, Witt H, Horn P, Korn B, Ho AD, Wagner W (2010) DNA methylation pattern changes upon long-term culture and aging of human mesenchymal stromal cells. Aging Cell 9(1):54–63

Bosch P, Musgrave DS, Lee JY, Cummins J, Shuler T, Ghivizzani TC, Evans T, Robbins TD, Huard J (2000) Osteoprogenitor cells within skeletal muscle. J Orthop Res 18(6):933–944

Braconi C, Huang N, Patel T (2010) MicroRNA-dependent regulation of DNA methyltransferase-1 and tumor suppressor gene expression by interleukin-6 in human malignant cholangiocytes. Hepatology 51(3):881–890

Bruder SP, Kraus KH, Goldberg VM, Kadiyala S (1998) The effect of implants loaded with autologous mesenchymal stem cells on the healing of canine segmental bone defects. J Bone Joint Surg Am 80(7):985–996

Campagnolo L, Russo MA, Puglianiello A, Favale A, Siracusa G (2001) Mesenchymal cell precursors of peritubular smooth muscle cells of the mouse testis can be identified by the presence of the p75 neurotrophin receptor. Biol Reprod 64(2):464–472

Campisi J, Andersen JK, Kapahi P, Melov S (2011) Cellular senescence: a link between cancer and age-related degenerative disease? Semin Cancer Biol 21(6):354–359

Cancedda R, Bianchi G, Derubeis A, Quarto R (2003) Cell therapy for bone disease: a review of current status. Stem Cells 21(5):610–619

Caplan AI (1991) Mesenchymal stem cells. J Orthop Res 9(5):641–650

Caplan AI (2007) Adult mesenchymal stem cells for tissue engineering versus regenerative medicine. J Cell Physiol 213(2):341–347

Capri M, Monti D, Salvioli S, Lescai F, Pierini M, Altilia S, Sevini F, Valensin S, Ostan R, Bucci L, Franceschi C (2006) Complexity of anti-immunosenescence strategies in humans. Artif Organs 30(10):730–742

Chen X, Armstrong MA, Li G (2006) Mesenchymal stem cells in immunoregulation. Immunol Cell Biol 84(5):413–421

Choudhery MS, Badowski M, Muise A, Pierce J, Harris DT (2014) Donor age negatively impacts adipose tissue-derived mesenchymal stem cell expansion and differentiation. J Transl Med 12:8

Cohnheim J (1867) Über Entzündung und Eiterung. J Arch Path Anat Physiol Klin Med 40:1–79

Collett GD, Canfield AE (2005) Angiogenesis and pericytes in the initiation of ectopic calcification. Circ Res 96(9):930–938

Coppe JP, Desprez PY, Krtolica A, Campisi J (2010) The senescence-associated secretory phenotype: the dark side of tumor suppression. Annu Rev Pathol 5:99–118

Crisan M, Yap S, Casteilla L, Chen CW, Corselli M, Park TS, Andriolo G, Sun B, Zheng B, Zhang L, Norotte C, Teng PN, Traas J, Schugar R, Deasy BM, Badylak S, Buhring HJ, Giacobino JP, Lazzari L, Huard J, Peault B (2008) A perivascular origin for mesenchymal stem cells in multiple human organs. Cell Stem Cell 3(3):301–313

D'Ippolito G, Diabira S, Howard G, Menei P, Ba R, Schiller PC (2004) Marrow-isolated adult multilineage inducible (MIAMI) cells, a unique population of postnatal young and old human cells with extensive expansion and differentiation potential. J Cell Sci 117:2971–2981

da Silva Meirelles L, Caplan AI, Nardi NB (2008) In search of the in vivo identity of mesenchymal stem cells. Stem Cells 26(9):2287–2299

Dalle Carbonare L, Valenti MT, Zanatta M, Donatelli L, Lo Cascio V (2009) Circulating mesenchymal stem cells with abnormal osteogenic differentiation in patients with osteoporosis. Arthritis Rheum 60(11):3356–3365

Dazzi F, Ramasamy R, Glennie S, Jones SP, Roberts I (2006) The role of mesenchymal stem cells in haemopoiesis. Blood Rev 20(3):161–171

de Gonzalo-Calvo D, Neitzert K, Fernandez M, Vega-Naredo I, Caballero B, Garcia-Macia M, Suarez FM, Rodriguez-Colunga MJ, Solano JJ, Coto-Montes A (2010) Differential inflammatory responses in aging and disease: TNF-alpha and IL-6 as possible biomarkers. Free Radic Biol Med 49(5):733–737

Digirolamo CM, Stokes D, Colter D, Phinney DG, Class R, Prockop DJ (1999) Propagation and senescence of human marrow stromal cells in culture: a simple colony-forming assay identifies samples with the greatest potential to propagate and differentiate. Br J Haematol 107(2): 275–281

Dimri GP, Lee X, Basile G, Acosta M, Scott G, Roskelley C, Medrano EE, Linskens M, Rubelj I, Pereira-Smith O et al (1995) A biomarker that identifies senescent human cells in culture and in aging skin in vivo. Proc Natl Acad Sci U S A 92(20):9363–9367

Dolberg DS, Hollingsworth R, Hertle M, Bissell MJ (1985) Wounding and its role in RSV-mediated tumor formation. Science (New York, NY) 230(4726):676–678

Dominici M, Le Blanc K, Mueller I, Slaper-Cortenbach I, Marini F, Krause D, Deans R, Keating A, Prockop D, Horwitz E (2006) Minimal criteria for defining multipotent mesenchymal stromal cells. The International Society for Cellular Therapy position statement. Cytotherapy 8(4):315–317

Duque G, Huang DC, Macoritto M, Rivas D, Yang XF, Ste-Marie LG, Kremer R (2009) Autocrine regulation of interferon gamma in mesenchymal stem cells plays a role in early osteoblastogenesis. Stem Cells 27(3):550–558

Dvorak HF (1986) Tumors: wounds that do not heal. Similarities between tumor stroma generation and wound healing. N Engl J Med 315(26):1650–1659

Eghbali-Fatourechi GZ, Lamsam J, Fraser D, Nagel D, Riggs BL, Khosla S (2005) Circulating osteoblast-lineage cells in humans. N Engl J Med 352(19):1959–1966

Eriksson M, Brown WT, Gordon LB, Glynn MW, Singer J, Scott L, Erdos MR, Robbins CM, Moses TY, Berglund P, Dutra A, Pak E, Durkin S, Csoka AB, Boehnke M, Glover TW, Collins FS (2003) Recurrent de novo point mutations in lamin A cause Hutchinson-Gilford progeria syndrome. Nature 423(6937):293–298

Estrada JC, Torres Y, Benguria A, Dopazo A, Roche E, Carrera-Quintanar L, Perez RA, Enriquez JA, Torres R, Ramirez JC, Samper E, Bernad A (2013) Human mesenchymal stem cell-replicative senescence and oxidative stress are closely linked to aneuploidy. Cell Death Dis 4, e691

Fehrer C, Lepperdinger G (2005) Mesenchymal stem cell aging. Exp Gerontol 40(12):926–930

Fehrer C, Brunauer R, Laschober G, Unterluggauer H, Reitinger S, Kloss F, Gü947lly C, Gassner R, Lepperdinger G (2007) Reduced oxygen tension attenuates differentiation capacity of human mesenchymal stem cells and prolongs their life span. Aging Cell 6:745–757

Franceschi C, Bonafe M, Valensin S, Olivieri F, De Luca M, Ottaviani E, De Benedictis G (2000a) Inflamm-aging. An evolutionary perspective on immunosenescence. Ann N Y Acad Sci 908:244–254

Franceschi C, Valensin S, Bonafe M, Paolisso G, Yashin AI, Monti D, De Benedictis G (2000b) The network and the remodeling theories of aging: historical background and new perspectives. Exp Gerontol 35(6–7):879–896

Franceschi C, Capri M, Monti D, Giunta S, Olivieri F, Sevini F, Panourgia MP, Invidia L, Celani L, Scurti M, Cevenini E, Castellani GC, Salvioli S (2007) Inflammaging and anti-inflammaging: a systemic perspective on aging and longevity emerged from studies in humans. Mech Ageing Dev 128(1):92–105

Friedenstein AJ, Chailakhjan RK, Lalykina KS (1970) The development of fibroblast colonies in monolayer cultures of guinea-pig bone marrow and spleen cells. Cell Tissue Kinet 3:393–403

Frumento G, Rotondo R, Tonetti M, Damonte G, Benatti U, Ferrara GB (2002) Tryptophan-derived catabolites are responsible for inhibition of T and natural killer cell proliferation induced by indoleamine 2,3-dioxygenase. J Exp Med 196(4):459–468

Gotherstrom C, West A, Liden J, Uzunel M, Lahesmaa R, Le Blanc K (2005) Difference in gene expression between human fetal liver and adult bone marrow mesenchymal stem cells. Haematologica 90(8):1017–1026

Grillari J, Hackl M, Grillari-Voglauer R (2010) miR-17-92 cluster: ups and downs in cancer and aging. Biogerontology 11(4):501–506

Gruber HE, Somayaji S, Riley F, Hoelscher GL, Norton HJ, Ingram J, Hanley EN Jr (2012) Human adipose-derived mesenchymal stem cells: serial passaging, doubling time and cell senescence. Biotech Histochem 87(4):303–311

Guillot PV, O'Donoghue K, Kurata H, Fisk NM (2006) Fetal stem cells: betwixt and between. Semin Reprod Med 24(5):340–347

Hackl M, Brunner S, Fortschegger K, Schreiner C, Micutkova L, Muck C, Laschober GT, Lepperdinger G, Sampson N, Berger P, Herndler-Brandstetter D, Wieser M, Kuhnel H, Strasser A, Rinnerthaler M, Breitenbach M, Mildner M, Eckhart L, Tschachler E, Trost A, Bauer JW, Papak C, Trajanoski Z, Scheideler M, Grillari-Voglauer R, Grubeck-Loebenstein B, Jansen-Durr P, Grillari J (2010) miR-17, miR-19b, miR-20a, and miR-106a are down-regulated in human aging. Aging Cell 9(2):291–296

Hahn M, Hahn T, Lee D-H, Esworthy RS, B-w K, Riggs AD, Chu F-F, Pfeifer GP (2008) Methylation of polycomb target genes in intestinal cancer is mediated by inflammation. Cancer Res 68(24):10280–10289

Hao H, Chen G, Liu J, Ti D, Zhao Y, Xu S, Fu X, Han W (2013) Culturing on Wharton's jelly extract delays mesenchymal stem cell senescence through p53 and p16INK4a/pRb pathways. PLoS One 8(3), e58314

Hardy R, Cooper MS (2009) Bone loss in inflammatory disorders. J Endocrinol 201(3):309–320

Ho JH, Chen YF, Ma WH, Tseng TC, Chen MH, Lee OK (2011) Cell contact accelerates replicative senescence of human mesenchymal stem cells independent of telomere shortening and p53 activation: roles of Ras and oxidative stress. Cell Transplant 20(8):1209–1220

Imam JS, Buddavarapu K, Lee-Chang JS, Ganapathy S, Camosy C, Chen Y, Rao MK (2010) MicroRNA-185 suppresses tumor growth and progression by targeting the Six1 oncogene in human cancers. Oncogene 29(35):4971–4979

In 't Anker PS, Scherjon SA, Kleijburg-van der Keur C, de Groot-Swings GM, Claas FH, Fibbe WE, Kanhai HH (2004) Isolation of mesenchymal stem cells of fetal or maternal origin from human placenta. Stem Cells 22(7):1338–1345

Jarocha D, Lukasiewicz E, Majka M (2008) Advantage of mesenchymal stem cells (MSC) expansion directly from purified bone marrow CD105+ and CD271+ cells. Folia Histochem Cytobiol 46(3):307–314

Jiang Y (2002) Multipotent progenitor cells can be isolated from postnatal murine bone marrow, muscle, and brain. Exp Hematol 30:896–904

Juan AH, Kumar RM, Marx JG, Young RA, Sartorelli V (2009) Mir-214-dependent regulation of the polycomb protein Ezh2 in skeletal muscle and embryonic stem cells. Mol Cell 36(1):61–74

Jung J-W, Lee S, Seo M-S, Park S-B, Kurtz A, Kang S-K, Kang K-S (2010) Histone deacetylase controls adult stem cell aging by balancing the expression of polycomb genes and jumonji domain containing 3. Cell Mol Life Sci 67(7):1165–1176

Justesen J, Stenderup K, Eriksen EF, Kassem M (2002) Maintenance of osteoblastic and adipocytic differentiation potential with age and osteoporosis in human marrow stromal cell cultures. Calcif Tissue Int 71(1):36–44

Kemp K, Gray E, Mallam E, Scolding N, Wilkins A (2010) Inflammatory cytokine induced regulation of superoxide dismutase 3 expression by human mesenchymal stem cells. Stem Cell Rev 6(4):548–559

Kern S, Eichler H, Stoeve J, Kluter H, Bieback K (2006) Comparative analysis of mesenchymal stem cells from bone marrow, umbilical cord blood, or adipose tissue. Stem Cells 24(5):1294–1301

Khatami M (2009) Inflammation, aging, and cancer: tumoricidal versus tumorigenesis of immunity: a common denominator mapping chronic diseases. Cell Biochem Biophys 55(2):55–79

Klepsch S, Jamnig A, Trimmel D, Schimke M, Kapferer W, Brunauer R, Singh S, Reitinger S, Lepperdinger G (2013) Isolation of mesenchymal stem cells from human bone and long-term cultivation under physiologic oxygen conditions. Methods Mol Biol 976:99–109

Krampera M1, Cosmi L, Angeli R, Pasini A, Liotta F, Andreini A, Santarlasci V, Mazzinghi B, Pizzolo G, Vinante F, Romagnani P, Maggi E, Romagnani S, Annunziato F (2006) Role for interferon-gamma in the immunomodulatory activity of human bone marrow mesenchymal stem cells. Stem Cells 24(2):386–398

Kretlow JD, Jin YQ, Liu W, Zhang WJ, Hong TH, Zhou G, Baggett LS, Mikos AG, Cao Y (2008) Donor age and cell passage affects differentiation potential of murine bone marrow-derived stem cells. BMC Cell Biol 9:60

Kronenberg HM, Schipani E (2009) Adult mesenchymal stem cells. In: Stembook. Stembook, (ed.). The Stem Cell Research Community, StemBook, doi:10.3824/stembook.1.38.1, http://www.stembook.org.

Ksiazek K (2009) A comprehensive review on mesenchymal stem cell growth and senescence. Rejuvenation Res 12(2):105–116

Kuznetsov SA, Krebsbach PH, Satomura K, Kerr J, Riminucci M, Benayahu D, Robey PG (1997) Single-colony derived strains of human marrow stromal fibroblasts form bone after transplantation in vivo. J Bone Miner Res 12(9):1335–1347

Lambeth LS, Yao Y, Smith LP, Zhao Y, Nair V (2009) MicroRNAs 221 and 222 target p27Kip1 in Marek's disease virus-transformed tumour cell line MSB-1. J Gen Virol 90(Pt 5):1164–1171

Laschober GT, Brunauer R, Jamnig A, Fehrer C, Greiderer B, Lepperdinger G (2009) Leptin receptor/CD295 is upregulated on primary human mesenchymal stem cells of advancing biological age and distinctly marks the subpopulation of dying cells. Exp Gerontol 44(1–2):57–62

Laschober GT, Brunauer R, Jamnig A, Singh S, Hafen U, Fehrer C, Kloss F, Gassner R, Lepperdinger G (2011) Age-specific changes of mesenchymal stem cells are paralleled by upregulation of CD106 expression as a response to an inflammatory environment. Rejuvenation Res 14(2):119–131

Le Blanc K, Pittenger M (2005) Mesenchymal stem cells: progress toward promise. Cytotherapy 7(1):36–45

Lee KD, Kuo TK, Whang-Peng J, Chung YF, Lin CT, Chou SH, Chen JR, Chen YP, Lee OK (2004) In vitro hepatic differentiation of human mesenchymal stem cells. Hepatology 40(6):1275–1284

Lee S, Jung JW, Park SB, Roh K, Lee SY, Kim JH, Kang SK, Kang KS (2010) Histone deacetylase regulates high mobility group A2-targeting microRNAs in human cord blood-derived multipotent stem cell aging. Cell Mol Life Sci 68(2):325–336

Leucht P, Kim J-B, Amasha R, James AW, Girod S, Helms JA (2008) Embryonic origin and Hox status determine progenitor cell fate during adult bone regeneration. Development 135(17):2845–2854

Lorimer IA (2009) Regulation of p27Kip1 by miRNA 221/222 in glioblastoma. Cell Cycle 8(17):2685

Luria EA, Panasyuk AF, Friedenstein AY (1971) Fibroblast colony formation from monolayer cultures of blood cells. Transfusion 11(6):345–349

Maes OC, Sarojini H, Wang E (2009) Stepwise up-regulation of MicroRNA expression levels from replicating to reversible and irreversible growth arrest states in WI-38 human fibroblasts. J Cell Physiol 221(1):109–119

Majors AK, Boehm CA, Nitto H, Midura RJ, Muschler GF (1997) Characterization of human bone marrow stromal cells with respect to osteoblastic differentiation. J Orthop Res 15(4): 546–557

Mareschi K, Biasin E, Piacibello W, Aglietta M, Madon E, Fagioli F (2001) Isolation of human mesenchymal stem cells: bone marrow versus umbilical cord blood. Haematologica 86(10):1099–1100

Meisel R, Zibert A, Laryea M, Gobel U, Daubener W, Dilloo D (2004) Human bone marrow stromal cells inhibit allogeneic T-cell responses by indoleamine 2,3-dioxygenase-mediated tryptophan degradation. Blood 103(12):4619–4621

Mendez-Ferrer S, Michurina TV, Ferraro F, Mazloom AR, Macarthur BD, Lira SA, Scadden DT, Ma'ayan A, Enikolopov GN, Frenette PS (2010) Mesenchymal and haematopoietic stem cells form a unique bone marrow niche. Nature 466(7308):829–834

Mirmalek-Sani SH, Tare RS, Morgan SM, Roach HI, Wilson DI, Hanley NA, Oreffo RO (2006) Characterization and multipotentiality of human fetal femur-derived cells: implications for skeletal tissue regeneration. Stem Cells 24(4):1042–1053

Mohanty ST, Kottam L, Gambardella A, Nicklin MJ, Coulton L, Hughes D, Wilson AG, Croucher PI, Bellantuono I (2010) Alterations in the self-renewal and differentiation ability of bone marrow mesenchymal stem cells in a mouse model of rheumatoid arthritis. Arthritis Res Ther 12(4):R149

Mueller SM, Glowacki J (2001) Age-related decline in the osteogenic potential of human bone marrow cells cultured in three-dimensional collagen sponges. J Cell Biochem 82(4): 583–590

Muraglia A, Cancedda R, Quarto R (2000) Clonal mesenchymal progenitors from human bone marrow differentiate in vitro according to a hierarchical model. J Cell Sci 113(Pt 7):1161–1166

Murphy JM, Dixon K, Beck S, Fabian D, Feldman A, Barry F (2002) Reduced chondrogenic and adipogenic activity of mesenchymal stem cells from patients with advanced osteoarthritis. Arthritis Rheum 46(3):704–713

Muschler GF, Nitto H, Boehm CA, Easley KA (2001) Age- and gender-related changes in the cellularity of human bone marrow and the prevalence of osteoblastic progenitors. J Orthop Res 19(1):117–125

Nakahara H, Bruder SP, Goldberg VM, Caplan AI (1990) In vivo osteochondrogenic potential of cultured cells derived from the periosteum. Clin Orthop Relat Res 259:223–232

Napolitano MA, Cipollaro M, Cascino A, Melone MA, Giordano A, Galderisi U (2007) Brg1 chromatin remodeling factor is involved in cell growth arrest, apoptosis and senescence of rat mesenchymal stem cells. J Cell Sci 120(Pt 16):2904–2911

Nauta AJ, Fibbe WE (2007) Immunomodulatory properties of mesenchymal stromal cells. Blood 110(10):3499–3506

Nishida S, Endo N, Yamagiwa H, Tanizawa T, Takahashi HE (1999) Number of osteoprogenitor cells in human bone marrow markedly decreases after skeletal maturation. J Bone Miner Metab 17(3):171–177

Noer A, Lindeman LC, Collas P (2009) Histone H3 modifications associated with differentiation and long-term culture of mesenchymal adipose stem cells. Stem Cells Dev 18(5): 725–736

Owen M, Friedenstein AJ (1988) Stromal stem cells: marrow-derived osteogenic precursors. Ciba Found Symp 136:42–60

Pereira RF, Halford KW, O'Hara MD, Leeper DB, Sokolov BP, Pollard MD, Bagasra O, Prockop DJ (1995) Cultured adherent cells from marrow can serve as long-lasting precursor cells for bone, cartilage, and lung in irradiated mice. Proc Natl Acad Sci U S A 92(11):4857–4861

Pittenger MF, Mackay AM, Beck SC, Jaiswal RK, Douglas R, Mosca JD, Moorman MA, Simonetti DW, Craig S, Marshak DR (1999) Multilineage potential of adult human mesenchymal stem cells. Science 284(5411):143–147

Prockop DJ (1997) Marrow stromal cells as stem cells for nonhematopoietic tissues. Science 276(5309):71–74

Rasmusson I (2006) Immune modulation by mesenchymal stem cells. Exp Cell Res 312(12):2169–2179

Rastegar F, Shenaq D, Huang J, Zhang W, Zhang B-Q, He B-C, Chen L, Zuo G-W, Luo Q, Shi Q, Wagner ER, Huang E, Gao Y, Gao J-L, Kim SH, Zhou J-Z, Bi Y, Su Y, Zhu G, Luo J, Luo X, Qin J, Reid RR, Luu HH, Haydon RC, Deng Z-L, He T-C (2010) Mesenchymal stem cells: molecular characteristics and clinical applications. World J Stem Cells 2:67–80

Ressler S, Bartkova J, Niederegger H, Bartek J, Scharffetter-Kochanek K, Jansen-Durr P, Wlaschek M (2006) p16INK4A is a robust in vivo biomarker of cellular aging in human skin. Aging Cell 5(5):379–389

Reyes M, Lund T, Lenvik T, Aguiar D, Koodie L, Verfaillie CM (2001) Purification and ex vivo expansion of postnatal human marrow mesodermal progenitor cells. Blood 98:2615–2625

Rodriguez JP, Garat S, Gajardo H, Pino AM, Seitz G (1999) Abnormal osteogenesis in osteoporotic patients is reflected by altered mesenchymal stem cells dynamics. J Cell Biochem 75(3):414–423

Rodriguez JP, Astudillo P, Rios S, Pino AM (2008) Involvement of adipogenic potential of human bone marrow mesenchymal stem cells (MSCs) in osteoporosis. Curr Stem Cell Res Ther 3(3):208–218

Rosen CJ, Ackert-Bicknell C, Rodriguez JP, Pino AM (2009) Marrow fat and the bone microenvironment: developmental, functional, and pathological implications. Crit Rev Eukaryot Gene Expr 19(2):109–124

Roufosse CA, Direkze NC, Otto WR, Wright NA (2004) Circulating mesenchymal stem cells. Int J Biochem Cell Biol 36(4):585–597

Sacchetti B, Funari A, Michienzi S, Di Cesare S, Piersanti S, Saggio I, Tagliafico E, Ferrari S, Robey PG, Riminucci M, Bianco P (2007) Self-renewing osteoprogenitors in bone marrow sinusoids can organize a hematopoietic microenvironment. Cell 31(2):324–336

Sakaguchi Y, Sekiya I, Yagishita K, Muneta T (2005) Comparison of human stem cells derived from various mesenchymal tissues: superiority of synovium as a cell source. Arthritis Rheum 52(8):2521–2529

Salvioli S, Monti D, Lanzarini C, Conte M, Pirazzini C, Bacalini MG, Garagnani P, Giuliani C, Fontanesi E, Ostan R, Bucci L, Sevini F, Yani SL, Barbieri A, Lomartire L, Borelli V, Vianello D, Bellavista E, Martucci M, Cevenini E, Pini E, Scurti M, Biondi F, Santoro A, Capri M, Franceschi C (2013) Immune system, cell senescence, aging and longevity–inflamm-aging reappraised. Curr Pharm Des 19(9):1675–1679

Scaffidi P, Misteli T (2008) Lamin A-dependent misregulation of adult stem cells associated with accelerated ageing. Nat Cell Biol 10(4):452–459

Schaffler A, Buchler C (2007) Concise review: adipose tissue-derived stromal cells–basic and clinical implications for novel cell-based therapies. Stem Cells 25(4):818–827

Schallmoser K, Bartmann C, Rohde E, Bork S, Guelly C, Obenauf AC, Reinisch A, Horn P, Ho AD, Strunk D, Wagner W (2010) Replicative senescence-associated gene expression changes in mesenchymal stromal cells are similar under different culture conditions. Haematologica 95(6):867–874

Scharstuhl A, Schewe B, Benz K, Gaissmaier C, Buhring HJ, Stoop R (2007) Chondrogenic potential of human adult mesenchymal stem cells is independent of age or osteoarthritis etiology. Stem Cells 25(12):3244–3251

Sethe S, Scutt A, Stolzing A (2006) Aging of mesenchymal stem cells. Ageing Res Rev 5(1):91–116

Simonsen JL, Rosada C, Serakinci N, Justesen J, Stenderup K, Rattan SI, Jensen TG, Kassem M (2002) Telomerase expression extends the proliferative life-span and maintains the osteogenic potential of human bone marrow stromal cells. Nat Biotechnol 20(6):592–596

Sousa-Victor P, Gutarra S, Garcia-Prat L, Rodriguez-Ubreva J, Ortet L, Ruiz-Bonilla V, Jardi M, Ballestar E, Gonzalez S, Serrano AL, Perdiguero E, Munoz-Canoves P (2014) Geriatric muscle stem cells switch reversible quiescence into senescence. Nature 506(7488):316–321

Stenderup K, Justesen J, Eriksen EF, Rattan SI, Kassem M (2001) Number and proliferative capacity of osteogenic stem cells are maintained during aging and in patients with osteoporosis. J Bone Miner Res 16(6):1120–1129

Stenderup K, Justesen J, Clausen C, Kassem M (2003) Aging is associated with decreased maximal life span and accelerated senescence of bone marrow stromal cells. Bone 33(6):919–926

Stolzing A, Scutt A (2006) Age-related impairment of mesenchymal progenitor cell function. Aging Cell 5(3):213–224

Stolzing A, Jones E, McGonagle D, Scutt A (2008) Age-related changes in human bone marrow-derived mesenchymal stem cells: consequences for cell therapies. Mech Ageing Dev 129(3):163–173

Stringer B, Waddington R, Houghton A, Stone M, Russell G, Foster G (2007) Serum from postmenopausal women directs differentiation of human clonal osteoprogenitor cells from an osteoblastic toward an adipocytic phenotype. Calcif Tissue Int 80(4):233–243

Tokoyoda K, Hauser AE, Nakayama T, Radbruch A (2010) Organization of immunological memory by bone marrow stroma. Nat Rev Immunol 10(3):193–200

Tsai MS, Lee JL, Chang YJ, Hwang SM (2004) Isolation of human multipotent mesenchymal stem cells from second-trimester amniotic fluid using a novel two-stage culture protocol. Hum Reprod 19(6):1450–1456

Tsai CC, Chen YJ, Yew TL, Chen LL, Wang JY, Chiu CH, Hung SC (2010) Hypoxia inhibits senescence and maintains mesenchymal stem cell properties through down-regulation of E2A-p21 by HIF-TWIST. Blood 117(2):459–469

Tuli R, Tuli S, Nandi S, Wang ML, Alexander PG, Haleem-Smith H, Hozack WJ, Manner PA, Danielson KG, Tuan RS (2003) Characterization of multipotential mesenchymal progenitor cells derived from human trabecular bone. Stem Cells 21(6):681–693

Urbich C, Kuehbacher A, Dimmeler S (2008) Role of microRNAs in vascular diseases, inflammation, and angiogenesis. Cardiovasc Res 79(4):581–588

Vasile E, Tomita Y, Brown LF, Kocher O, Dvorak HF (2001) Differential expression of thymosin beta-10 by early passage and senescent vascular endothelium is modulated by VPF/VEGF: evidence for senescent endothelial cells in vivo at sites of atherosclerosis. FASEB J 15(2):458–466

Veronesi F, Torricelli P, Borsari V, Tschon M, Rimondini L, Fini M (2012) Mesenchymal stem cells in the aging and osteoporotic population. Crit Rev Eukaryot Gene Expr 21(4):363–377
Wagner W, Bork S, Horn P, Krunic D, Walenda T, Diehlmann A, Benes V, Blake J, Huber F-X, Eckstein V, Boukamp P, Ho AD (2009) Aging and replicative senescence have related effects on human stem and progenitor cells. PLoS One 4(6), e5846
Wagner W, Bork S, Lepperdinger G, Joussen S, Ma N, Strunk D, Koch C (2010) How to track cellular aging of mesenchymal stromal cells? Aging (Albany NY) 2(4):224–230
Wakitani S, Saito T, Caplan AI (1995) Myogenic cells derived from rat bone marrow mesenchymal stem cells exposed to 5-azacytidine. Muscle Nerve 18(12):1417–1426
Wang HJ, Ruan HJ, He XJ, Ma YY, Jiang XT, Xia YJ, Ye ZY, Tao HQ (2011) MicroRNA-101 is down-regulated in gastric cancer and involved in cell migration and invasion. Eur J Cancer 46(12):2295–2303
Williams EL, White K, Oreffo RO (2013) Isolation and enrichment of Stro-1 immunoselected mesenchymal stem cells from adult human bone marrow. Methods Mol Biol 1035:67–73
Woodbury D, Schwarz EJ, Prockop DJ, Black IB (2000) Adult rat and human bone marrow stromal cells differentiate into neurons. J Neurosci Res 61(4):364–370
Woodbury D, Reynolds K, Black IB (2002) Adult bone marrow stromal stem cells express germline, ectodermal, endodermal, and mesodermal genes prior to neurogenesis. J Neurosci Res 69(6):908–917
Yoon DS, Kim YH, Jung HS, Paik S, Lee JW (2011) Importance of Sox2 in maintenance of cell proliferation and multipotency of mesenchymal stem cells in low-density culture. Cell Prolif 44(5):428–440
Young HE, Steele TA, Bray RA, Hudson J, Floyd JA, Hawkins K, Thomas K, Austin T, Edwards C, Cuzzourt J, Duenzl M, Lucas PA, Black AC Jr (2001) Human reserve pluripotent mesenchymal stem cells are present in the connective tissues of skeletal muscle and dermis derived from fetal, adult, and geriatric donors. Anat Rec 264(1):51–62
Yu KR, Kang KS (2013) Aging-related genes in mesenchymal stem cells: a mini-review. Gerontology 59(6):557–563
Yu JM, Wu X, Gimble JM, Guan X, Freitas MA, Bunnell BA (2011) Age-related changes in mesenchymal stem cells derived from rhesus macaque bone marrow. Aging Cell 10(1):66–79
Zaim M, Karaman S, Cetin G, Isik S (2012) Donor age and long-term culture affect differentiation and proliferation of human bone marrow mesenchymal stem cells. Ann Hematol 91(8):1175–1186
Zhang W, Walboomers XF, Shi S, Fan M, Jansen JA (2006) Multilineage differentiation potential of stem cells derived from human dental pulp after cryopreservation. Tissue Eng 12(10):2813–2823
Zhao YM, Li JY, Lan JP, Lai XY, Luo Y, Sun J, Yu J, Zhu YY, Zeng FF, Zhou Q, Huang H (2008) Cell cycle dependent telomere regulation by telomerase in human bone marrow mesenchymal stem cells. Biochem Biophys Res Commun 369(4):1114–1119
Zhou S, Greenberger JS, Epperly MW, Goff JP, Adler C, Leboff MS, Glowacki J (2008) Age-related intrinsic changes in human bone-marrow-derived mesenchymal stem cells and their differentiation to osteoblasts. Aging Cell 7(3):335–343
Zimmermann S, Voss M, Kaiser S, Kapp U, Waller CF, Martens UM (2003) Lack of telomerase activity in human mesenchymal stem cells. Leukemia 17(6):1146–1149

Aging of the Hematopoietic Stem Cell Niches

12

Jose A. Cancelas and Kyung-Hee Chang

Abstract

Adult hematopoietic stem cells (HSCs) reside in the bone marrow (BM) and provide the basis to fulfill the hematopoietic needs of an organism. Their properties of self-renewal and multilineage differentiation are controlled by direct interaction with a specific microenvironment – the so-called stem cell 'niche'. Conceptual advances in our understanding of the composition of the HSC compartment suggest that changes in the BM HSC microenvironment may reflect the aging process. The balance and extent of the effect of intrinsic versus extrinsic (environment) changes during aging on HSC are still under investigation. Growing evidence suggests that the BM HSC niche is very important in the regulation of cellular aging of HSCs. A young HSC niche would act as a protective environment, preventing HSC DNA damage, as well as replicative senescence through protection from radicals and toxic compounds, and prevention/amelioration of aging-associated signaling pathways resulting in epigenetic/genetic modifications, hematopoietic impairments, and cancer predisposition. To what extent the aging of the niche contributes to the "HSC aging" phenotype remains unknown. The analysis of the effect of aging on the activity of these specialized cell "niches" and their molecular products is the focus of this chapter.

J.A. Cancelas, MD, PhD (✉) • K.-H. Chang, PhD
Hoxworth Blood Center,
University of Cincinnati College of Medicine,
3130 Highland Ave., Cincinnati, OH 45267, USA
e-mail: jose.cancelas@cchmc.org, jose.cancelas@uc.edu

H. Geiger et al. (eds.), *Stem Cell Aging: Mechanisms, Consequences, Rejuvenation*,
DOI 10.1007/978-3-7091-1232-8_12

12.1 The Bone Marrow Hematopoietic Stem Cell Niche

In 1978, Schofield proposed a hypothesis in which stem cells are located in association with other cells which determine their behavior, the so-called stem cell 'niche' (Schofield 1978). This hypothesis has been tested in multiple models in vivo and cumulative evidence supports it. The stem cell niche represents a complex system composed of the stem cells themselves, as well as diverse cellular and architectural "matrix" components that provide a nurturing microenvironment and inputs to regulate stem cell behavior. Stem cell maintenance, survival, self-renewal, and initiation of differentiation all depend on the intimate relationship between stem cells, their niche, and other environmental, systemic factors required for homeostatic regulation of blood cell differentiation (reviewed in (Park et al. 2012; Frenette et al. 2013; Hanoun and Frenette 2013)). While this chapter is not focused on a detailed analysis of the activities of each of the components of the complex BM HSC niche, we will provide a short review of our current understanding of the BM HSC niche to allow a better understanding of the effect of aging on it.

Mammalian BM microanatomy is complex. BM is a reticular tissue containing a dense network of medullary sinuses, blood cells, and their precursors packed in the perivascular spaces between the sinusoidal capillaries. During development, migrating HSCs populate in specialized BM microenvironments where they gradually become quiescent (Wilson and Trumpp 2006) and resistant to apoptosis (Arai and Suda 2007). This migration is not static but dynamic, and HSCs remain circulating in and out of the bone and spend short intervals in the systemic circulation. HSC residence time in the circulation is very short, just a few seconds as demonstrated in parabiotic mice (Wright et al. 2001), and many HSCs transmigrate daily following circadian cycles (Mendez-Ferrer et al. 2008) controlled by neutrophil-mediated inflammatory signals (Casanova-Acebes et al. 2013).

The identification of niche cells in the BM has been complicated due to their location within the bony skeletal tissue, the difficulty of identifying specific markers that characterize subsets of stromal cells, and by the frequent infidelity of genetic markers for mesenchymal lineages. Because of these challenges, there have been conflicting reports on the cellular identity of the HSC niche. Classically, two types of niches have been described. The first one, called the endosteal niche, is located in the endosteal region, populated by osteoblasts, and believed to be a niche maintaining quiescent HSCs (Calvi et al. 2003; Li and Clevers 2010; Sugimura et al. 2012). A second niche called the vascular niche has also been described where most HSCs are found in contact or near blood vessels (Kiel and Morrison 2008; Ding et al. 2012; Ding and Morrison 2013). The difference between these two niches is probably artificial since the endosteal region is rich in blood vessels (Lichtman 1981; Nilsson et al. 2001).

The identity of candidate niche cells has been explored through the expression of specific markers or through specific of active components known to influence HSC activity. The cell components of the niche include osteoblasts (Calvi et al. 2003; Zhang et al. 2003) and osteoprogenitors (Xie et al. 2009), nestin-positive mesenchymal stem cells (MSCs) (Mendez-Ferrer et al. 2010), Schwann cells (Yamazaki et al.

2011), perivascular cells (Lichtman 1981), and endothelial cells (Ding et al. 2012). The expression of chemokines or cytokines of the stem cell niche has demonstrated the relation of the aforementioned cell types with the expression of CXCL12 (Nagasawa et al. 1996; Katayama et al. 2006; Mendez-Ferrer et al. 2008, 2010; Omatsu et al. 2010; Chow et al. 2011; Ding and Morrison 2013; Greenbaum et al. 2013), stem cell factor (SCF) (Ding et al. 2012), or transforming growth factor-beta (TGFb) (Yamazaki et al. 2009, 2011), which are important regulators of HSC migration, quiescence, and/or self-renewal.

In particular, different cell populations that express CXCL12 (Tzeng et al. 2011) or SCF (Ding et al. 2012) have been identified. In the case of CXCL12, CXCL12-abundant reticular (CAR) cells, marked by green fluorescent protein (GFP) expression inserted in the *Cxcl12* locus, have been shown to be largely perivascular cells, whereas endothelial cells and bone-lining osteoblasts express lower levels of GFP (Ara et al. 2003). Ablation of CAR cells leads to a reduction in the frequency of HSCs, as well as lymphoid and erythroid progenitors (Nagasawa et al. 2011). Deletion of *Cxcl12* in osteoblasts using *Col2.3-cre* expressing mice demonstrated no alteration in the numbers of HSCs or myeloerythroid progenitor cells (Ara et al. 2003; Nagasawa et al. 2011). However, these mice show significantly lower levels of T-cell and B-cell reconstitution and fewer early lymphoid progenitors in the BM indicating that early lymphoid progenitors accumulate in the area adjacent to the endosteum (Ara et al. 2003). While CXCL12 from mature osteoblasts, as well as osteoblast progenitors, is dispensable for HSC maintenance (Ara et al. 2003), the conditional deletion of *Cxcl12* in osteoprogenitors leads to mobilization of hematopoietic progenitors to the peripheral blood and spleen (Ara et al. 2003), suggesting differential roles for *Cxcl12* within the BM. Deletion of Cxcl12 in osteoprogenitors, but not in mature osteoblasts, reduces the number of B lymphoid progenitors, which is consistent with results following the deletion of CAR cells (Omatsu et al. 2010), and supports a role for osteoprogenitors in B lymphoid commitment.

Endothelial cell-specific *Cxcl12* deletion can be achieved using *Tie2-cre* mice, which reveal that endothelial cells synthesize a relatively modest amount of CXCL12 compared to other stromal cells. Consequently, modest defects in HSC numbers and competitive reconstitution activities are observed in these mice (Ding and Morrison 2013; Greenbaum et al. 2013). No reductions in committed myeloid or lymphoid progenitors were documented, suggesting a restricted contribution of endothelial cell-derived CXCL12 to HSC maintenance. Using similar approaches, SCF has been found to be mostly expressed by endothelial and perivascular cells with the ability to express the leptin receptor (Ding et al. 2012). Since most of the effect of SCF is associated with the expression of its transmembrane splicing form (Zsebo et al. 1990; Williams and Majumdar 1994), this data suggests that endothelial/perivascular SCF may mediate cell–cell contact-dependent activity in the HSC vascular niche. In summary, cell-specific gene deletion models for CXCL12 and SCF have shown that the endosteal niche is more likely required for lymphoid progenitor activity but not HSCs, whereas the perivascular space has been hypothesized as being the site where HSCs and other myeloid progenitors are located.

12.2 Hematopoietic Stem Cell Microenvironment Aging

Due to the complexity of the cell interactions within the stem cell niche, it is hard to pinpoint the specific and/or relevant cell types that upon aging condition the phenotype of HSCs. A number of studies have demonstrated that hematopoiesis is affected by the aging of the hematopoietic microenvironment (Larsson et al. 2008; Warren and Rossi 2009; Vas et al. 2012a, b) as would be expected from changes in the composition of the mature hematopoietic cells during their lifetime, which affects the composition of the cells in the neighborhood of the stem cells. The use of murine models of direct/reverse transplantation, and the use of transgenic animals and in vitro stem cell culture onto BM-derived stromal cell populations, have been crucial to our understanding of the role of the microenvironment on aged stem cell activity. These changes are summarized below.

A change in the cell composition of the stem cell niche(s) occurs which probably correlate with variation in the composition of the stromal cell microenvironment. It is well known that during aging, decreased bone formation, loss of bone mass, and accumulation of fat in the BM become more pronounced, leading to the altered composition of cell types and extracellular matrix in the niche(s). It has been shown that aging causes trabecular degeneration, endosteal thinning, loss of osteoblastic activity, and skewing of their differentiation toward adipocytes in the aged marrow of mice and humans (Moerman et al. 2004; Stolzing et al. 2008). It has been shown in at least two skeletal sites of the mouse that adiposity serves to inhibit hematopoiesis (Naveiras et al. 2009). The characteristic feature of age-related bone loss is the uncoupling of formation from resorption leading to a net loss of bone mass. The etiology of this uncoupling is multifactorial and includes systemic (e.g., changes in endogenous gonadal steroids), cellular (e.g., increased reactive oxygen species), and paracrine (decrease in local growth factors that promote osteoblastic differentiation factors (Meunier, Aaron *et al.* 1971)). A summary of the changes in the cell composition of the aged microenvironment can be found in Fig. 12.1. One possible mechanism of the age-associated microanatomical changes in the BM stromal composition resides in the shift of cell fate determination of MSCs toward adipogenesis (Moerman et al. 2004; Rosen and MacDougald 2006). Aging of MSCs results in impaired proliferation and differentiation and increased chromosomal instability (Wagner et al. 2010). The cellular origin of these changes can be recapitulated in vitro by activation of specific transcription factors in MSCs. One consistent finding is that PPARγ2, a critical adipocytic transcription factor, is upregulated several fold in BM from old versus young animals (Kawai et al. 2010; Kawai Paula et al. 2012). This correlates closely with the number of marrow adipocytes and inversely with bone mass. Activation of PPARγ2 in osteoblastic cells converts them to terminally differentiated adipocytes and irreversibly suppresses their phenotype, including suppression of osteoblast-specific signaling pathways (e.g., Wnt, TGFb, BMP, and IGF-1) and transcriptional factors (e.g., Dlx5, Msx2, Runx2, osterix) (Geoffroy et al. 2002; Moerman et al. 2004; Cao et al. 2007; Liu et al. 2007; Ratajczak et al. 2011).

The myeloid skewing and impaired lymphoid differentiation potential observed in aged animals are probably due to both intrinsic (Kim et al. 2003; Dykstra and de

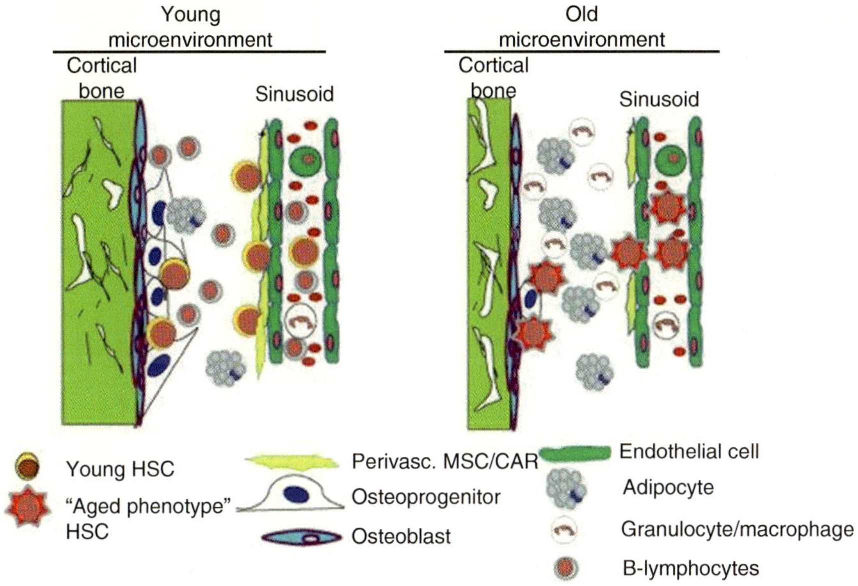

Fig. 12.1 Comparison of cell composition of the BM of the aged microenvironment versus the young microenvironment. Schema depicting the relative content of differentiated cells with recognized effect on HSC activity in young and old microenvironment (primary or transplanted) in a longitudinal section of a long bone

Haan 2008) and extrinsic signal changes (Li et al. 2001; Linton and Dorshkind 2004; Donnini et al. 2007; Zhu et al. 2007). The impaired generation of B cells in old mice reconstituted with young donor BM cells and the inefficient secretion of interleukin 7 (IL-7) by stromal cells from old mice indicate that the B lymphopoietic support potential of the hematopoietic microenvironment decreases with age (Stephan et al. 1998; Li et al. 2001).

HSCs have a reduced ability to home to their niches upon transplantation into old recipients compared to young recipients (Liang et al. 2005; Rossi et al. 2005). Modulation of HSC adhesion influences essential cellular processes required for homing and engraftment, a crucial process that allows HSCs to be transplanted in the clinical setting. The essential role of direct cell–cell contact for the regulation of self-renewal and differentiation of adult stem cells has been shown in various organisms and cell systems. Specific junctional complexes may play a similar role in the hematopoietic system. Adhesive interactions of HSCs with their niche are modulated by age-related alterations, as aged mice show an increased mobilization of HSCs in response to granulocyte colony-stimulating factor (G-CSF) (Xing et al. 2006; Ju et al. 2007). Various adhesion proteins have been implicated in this interaction such as N-cadherin (Zhang et al. 2003; Xie et al. 2009; Sugimura et al. 2012), b1/b2-integrin (Zanjani et al. 1999; Papayannopoulou et al. 2001; Priestley et al. 2006; Taniguchi Ishikawa et al. 2013), CD44 and variants (Lewinsohn et al. 1990; Dimitroff et al. 2001; Avigdor et al. 2004; Sackstein et al. 2008; Sackstein 2011), ALCAM (activated leukocyte cell adhesion molecule) (Ohneda et al. 2001; Chitteti et al. 2013), and connexin 43 (Cancelas et al. 2000; Presley et al. 2005; Schajnovitz et al. 2011; Gonzalez-Nieto et al. 2012; Ishikawa and Cancelas 2012; Taniguchi Ishikawa et al. 2012). Of these, losses of ALCAM and connexin 43 have been shown to be associated with HSC aging through microenvironment-dependent effects. Germinal loss of ALCAM results in impaired blood production; myeloid bias; upregulation of aging-associated genes including Selp (Sullivan et al. 2011), Clu (Trougakos et al. 2006), Cdc42 (Florian et al. 2012), and Foxo3 (Renault et al. 2009); and decreased serial repopulation ability of aged long-term HSCs. Connexin 43-deficiency results in an interesting mixed phenotype of bidirectional traffic defect due to microenvironment defects and loss of the replicative potential upon chemotherapy challenge resulting in senescent quiescence of HSCs due to loss of communication with the BM stromal cells (Gonzalez-Nieto et al. 2012; Taniguchi Ishikawa et al. 2012).

Stem cell mobilization is controlled by circadian rhythms (Mendez-Ferrer et al. 2009). This circadian regulation of HSC migration is likely to result in reported HSC immuno-surveillance functions, which affect the adaptive immune system (Massberg et al. 2007) and which is significantly impaired during the aging process (Plowden et al. 2004; Solana et al. 2006). Circadian rhythms of circulation of HSCs depend on central/peripheral nervous system controlled adrenergic innervation of the BM (Mendez-Ferrer et al. 2008; Scheiermann et al. 2012) and neutrophil-mediated clearance of the stem cell niches (Casanova-Acebes et al. 2013). Specifically, short-living neutrophils age in circulation, and during this aging process, they modulate their expression of adhesion and chemokine receptors L-selectin and Cxcr4. $Cd62L^{low}/Cxcr4^{high}$ neutrophils that have "aged" in the circulation infiltrate the BM and, before they die, promote reductions in the size and function of the hematopoietic niche. Modulation of the niche depends on macrophages and the activation of cholesterol-sensing nuclear receptors. Interestingly, aged BM is relatively enriched in myeloid cells, and therefore, this mechanism may play a role in the aged niche; and the expression of nuclear receptors (including PPARγ2; see above) follows circadian rhythms affecting the metabolism and is amplified by a high-fat diet (Green et al. 2008).

A change in the cytokine profile of the BM occurs during aging. The expression of TGFb in the niche decreases with age (Yamazaki et al. 2009, 2011), and age-associated telomere shortening leads to abnormal cytokine production and a functional decline in mesenchymal progenitor cells, resulting in a reduced capacity of the aged microenvironment to support HSC engraftment and differentiation (Rando 2006). The successive shortening of telomere length with every cell division is one of the best described mechanisms involved in cellular aging. Telomeres are elongated by the enzyme telomerase. Absence of telomerase results in persistent telomere shortening and finally in loss of telomere function, which is associated with replicative senescence or apoptosis. Stem cells normally express telomerase, in contrast to most somatic cells. However, telomerase activity in HSCs is low, and thus, telomere shortening still occurs during aging and limits their proliferative lifespan (Allsopp et al. 2003; Lansdorp 2005). Unfortunately, telomerase overexpression does not result in improved serial HSC engraftment (Allsopp et al. 2003).

Furthermore, telomere dysfunction and aging limited the engraftment of transplanted wild-type HSCs. Telomerase dysfunction induces alterations in the BM microenvironment by diminishing the stromal cell compartment and reducing the capacity of the BM and thymic stromal cells in their hematopoiesis-supportive activity (Ju and Rudolph 2006), and in systemic alterations that result in impaired B-and T-cell lymphopoiesis (Song et al. 2010). Terc-deficiency induced an age-dependent dysfunctional environment that correlated with progressive telomere shortening and limited the engraftment of transplanted wild-type HSC (Ju et al. 2007). These results provide clear evidence that telomere shortening is neither the unique nor an exclusively intrinsic mechanism for cellular aging in HSCs. In contrast, the loss of telomeres induces age-dependent alterations in the stem cell environment that can impair the function and engraftment of HSCs.

Aging of the microenvironment influences clonality in hematopoiesis. Clonality of identical pools of transduced HSCs exposed to a young or aged microenvironment in vivo is primarily oligoclonal within a young microenvironment, while in aged animals the transduced HSCs displayed reduced clonality (Vas et al. 2012a), providing a basis to explain a predisposition to clonal expansion in aged hematopoiesis as a putative substrate for transformation and leukemogenesis.

Aging of HSCs might also be attributed to accumulation of by-products that upon chemical modification of cellular proteins results in continuous accumulation of cellular defects. Two examples of these posttranslational modifications resulting in HSC aging have been well documented. Glycation of proteins leads to formation of cross-linked proteins (Finkel and Holbrook 2000). In HSCs, protein glycation has been demonstrated during aging and is expected to act through extrinsic and intrinsic pathways (Baraibar and Friguet 2013). Reactive oxygen species (ROS) arise as by-products of numerous metabolic processes and cause direct damage to proteins, membranes, and DNA, resulting in a vicious cycle, since accumulation of endogenous DNA mutations contributes to aging of HSCs (Kenyon and Gerson 2007). It has been shown that ROS can regulate HSC function in a concentration-dependent manner. Over time, these oxidant-damaged proteins and DNA accumulate and eventually disrupt cellular function. High levels of ROS can induce HSC senescence and apoptosis secondary to DNA damage (Ito et al. 2006). There is experimental evidence that regionally defined functional hypoxia plays a role in regulating the mesenchymal microenvironment (Fehrer et al. 2007) and cells isolated from distinct BM locations have been associated with local hypoxic gradients (Parmar et al. 2007). However, in vivo analysis of oxygen distribution in the BM has not shown distinctly hypoxic regional areas (Nombela-Arrieta et al. 2013) suggesting that HSCs are subject to similar oxygen tensions as other cells in the BM. Recently, a novel molecular pathway of hematopoietic microenvironment-dependent ROS scavenging has been reported. The integrity of this pathway is required to prevent HSC senescence. The loss connexin 43 function results in the loss of BM stromal-mediated ROS scavenging and subsequent premature senescence and/or apoptosis (Taniguchi Ishikawa et al. 2012).

Epigenetic mechanisms such as DNA methylation or histone acetylation regulate gene transcription at the chromatin level. As these modifications can be inherited by daughter cells, they have been demonstrated to play key roles in altering the long-term fate of HSCs, and hence, they might also be involved in the regulation of intrinsic HSCs (De Haan and Gerrits 2007) or microenvironment aging including MSC aging (Li et al. 2011). Gene expression analysis of HSCs from old mice compared with those from young mice revealed that genes involved in chromatin remodeling and transcriptional silencing were downregulated. Moreover, old HSCs showed transcriptional upregulation of genes involved in inflammatory response and stress, suggesting epigenetic changes, and susceptibility to modification by microenvironment signals (Chambers et al. 2007). Wagner et al. suggests that the hematopoietic microenvironment may condition multiple transcriptional signals since a large number of genes were differentially expressed in human HSC/P upon culture with stromal cells, which affected basic cell functions including cell division, reorganization of the cytoskeleton, and genetic stability (Wagner et al. 2007). The same group found a set of non-translated microRNAs implicated in the process of posttranscriptional gene expression and replicative senescence of MSCs (Wagner et al. 2008). Examples of epigenetic regulation of microenvironment aging include microRNA-dependent regulation and methyltransferase activity regulation. For instance, microRNA-34a, a known tumor suppressor that regulates silent information regulator 1 (Sirt1) expression that was shown to induce senescence in endothelial progenitors and impair angiogenesis (Zhao et al. 2010) and transcriptional expression of DNA (cytosine-5)-methyltransferase 1 (Dnmt1) isoforms, is highly upregulated upon contact between human HSC/P and BM stromal cells (Wagner et al. 2005).

Conclusion

There is clear evidence of the effects of the aging microenvironment as manifested in modified production of cytokines, chemokines, and signaling pathways that affect HSC senescence, migration, or differentiation. The remaining question is whether different HSC populations respond differently to microenvironment aging. Clonal analysis followed by expression analysis, metabolic profiling, and functional assays will provide us with the information on the crucial mechanisms responsible for microenvironment-dependent HSC aging. Results from this research are expected to provide novel therapeutic targets for intervention in the context of aging and cancer predisposition.

References

Allsopp RC, Morin GB et al (2003) Telomerase is required to slow telomere shortening and extend replicative lifespan of HSCs during serial transplantation. Blood 102(2):517–520

Ara T, Tokoyoda K et al (2003) Long-term hematopoietic stem cells require stromal cell-derived factor-1 for colonizing bone marrow during ontogeny. Immunity 19(2):257–267

Arai F, Suda T (2007) Regulation of hematopoietic stem cells in the osteoblastic niche. Adv Exp Med Biol 602:61–67

Avigdor A, Goichberg P et al (2004) CD44 and hyaluronic acid cooperate with SDF-1 in the trafficking of human CD34+ stem/progenitor cells to bone marrow. Blood 103(8):2981–2989

Baraibar MA, Friguet B (2013) Oxidative proteome modifications target specific cellular pathways during oxidative stress, cellular senescence and aging. Exp Gerontol 48(7):620–625

Calvi LM, Adams GB et al (2003) Osteoblastic cells regulate the haematopoietic stem cell niche. Nature 425(6960):841–846

Cancelas JA, Koevoet WL et al (2000) Connexin-43 gap junctions are involved in multiconnexin-expressing stromal support of hemopoietic progenitors and stem cells. Blood 96(2):498–505

Cao JJ, Kurimoto P et al (2007) Aging impairs IGF-I receptor activation and induces skeletal resistance to IGF-I. J Bone Miner Res 22(8):1271–1279

Casanova-Acebes M, Pitaval C et al (2013) Rhythmic modulation of the hematopoietic niche through neutrophil clearance. Cell 153(5):1025–1035

Chambers SM, Shaw CA et al (2007) Aging hematopoietic stem cells decline in function and exhibit epigenetic dysregulation. PLoS Biol 5(8), e201

Chitteti BR, Cheng YH et al (2013) Hierarchical organization of osteoblasts reveals the significant role of CD166 in hematopoietic stem cell maintenance and function. Bone 54(1):58–67

Chow A, Lucas D et al (2011) Bone marrow CD169+ macrophages promote the retention of hematopoietic stem and progenitor cells in the mesenchymal stem cell niche. J Exp Med 208(2):261–271

De Haan G, Gerrits A (2007) Epigenetic control of hematopoietic stem cell aging the case of Ezh2. Ann N Y Acad Sci 1106:233–239

Dimitroff CJ, Lee JY et al (2001) CD44 is a major E-selectin ligand on human hematopoietic progenitor cells. J Cell Biol 153(6):1277–1286

Ding L, Morrison SJ (2013) Haematopoietic stem cells and early lymphoid progenitors occupy distinct bone marrow niches. Nature 495(7440):231–235

Ding L, Saunders TL et al (2012) Endothelial and perivascular cells maintain haematopoietic stem cells. Nature 481(7382):457–462

Donnini A, Re F et al (2007) Intrinsic and microenvironmental defects are involved in the age-related changes of Lin - c-kit+ hematopoietic progenitor cells. Rejuvenation Res 10(4):459–472

Dykstra B, de Haan G (2008) Hematopoietic stem cell aging and self-renewal. Cell Tissue Res 331(1):91–101

Fehrer C, Brunauer R et al (2007) Reduced oxygen tension attenuates differentiation capacity of human mesenchymal stem cells and prolongs their lifespan. Aging Cell 6(6):745–757

Finkel T, Holbrook NJ (2000) Oxidants, oxidative stress and the biology of ageing. Nature 408(6809):239–247

Florian MC, Dorr K et al (2012) Cdc42 activity regulates hematopoietic stem cell aging and rejuvenation. Cell Stem Cell 10(5):520–530

Frenette PS, Pinho S et al (2013) Mesenchymal stem cell: keystone of the hematopoietic stem cell niche and a stepping-stone for regenerative medicine. Annu Rev Immunol 31:285–316

Geoffroy V, Kneissel M et al (2002) High bone resorption in adult aging transgenic mice overexpressing cbfa1/runx2 in cells of the osteoblastic lineage. Mol Cell Biol 22(17):6222–6233

Gonzalez-Nieto D, Li L et al (2012) Connexin-43 in the osteogenic BM niche regulates its cellular composition and the bidirectional traffic of hematopoietic stem cells and progenitors. Blood 119(22):5144–5154

Green CB, Takahashi JS et al (2008) The meter of metabolism. Cell 134(5):728–742

Greenbaum A, Hsu YM et al (2013) CXCL12 in early mesenchymal progenitors is required for haematopoietic stem-cell maintenance. Nature 495(7440):227–230

Hanoun M, Frenette PS (2013) This niche is a maze; an amazing niche. Cell Stem Cell 12(4):391–392

Ishikawa ET, Cancelas JA (2012) Lack of communication rusts and ages stem cells. Cell Cycle 11(17):3149–3150

Ito K, Hirao A et al (2006) Reactive oxygen species act through p38 MAPK to limit the lifespan of hematopoietic stem cells. Nat Med 12(4):446–451

Ju Z, Rudolph KL (2006) Telomeres and telomerase in stem cells during aging and disease. Genome Dyn 1:84–103

Ju Z, Jiang H et al (2007) Telomere dysfunction induces environmental alterations limiting hematopoietic stem cell function and engraftment. Nat Med 13(6):742–747
Katayama Y, Battista M et al (2006) Signals from the sympathetic nervous system regulate hematopoietic stem cell egress from bone marrow. Cell 124(2):407–421
Kawai M, Sousa KM et al (2010) The many facets of PPARgamma: novel insights for the skeleton. Am J Physiol Endocrinol Metab 299(1):E3–E9
Kawai M, de Paula FJ et al (2012) New insights into osteoporosis: the bone-fat connection. J Intern Med 272(4):317–329
Kenyon J, Gerson SL (2007) The role of DNA damage repair in aging of adult stem cells. Nucleic Acids Res 35(22):7557–7565
Kiel MJ, Morrison SJ (2008) Uncertainty in the niches that maintain haematopoietic stem cells. Nat Rev Immunol 8(4):290–301
Kim M, Moon HB et al (2003) Major age-related changes of mouse hematopoietic stem/progenitor cells. Ann N Y Acad Sci 996:195–208
Lansdorp PM (2005) Role of telomerase in hematopoietic stem cells. Ann N Y Acad Sci 1044:220–227
Larsson J, Ohishi M et al (2008) Nf2/merlin regulates hematopoietic stem cell behavior by altering microenvironmental architecture. Cell Stem Cell 3(2):221–227
Lewinsohn DM, Nagler A et al (1990) Hematopoietic progenitor cell expression of the H-CAM (CD44) homing-associated adhesion molecule. Blood 75(3):589–595
Li L, Clevers H (2010) Coexistence of quiescent and active adult stem cells in mammals. Science 327(5965):542–545
Li F, Jin F et al (2001) Impaired regeneration of the peripheral B cell repertoire from bone marrow following lymphopenia in old mice. Eur J Immunol 31(2):500–505
Li Z, Liu C et al (2011) Epigenetic dysregulation in mesenchymal stem cell aging and spontaneous differentiation. PLoS One 6(6), e20526
Liang Y, Van Zant G et al (2005) Effects of aging on the homing and engraftment of murine hematopoietic stem and progenitor cells. Blood 106(4):1479–1487
Lichtman MA (1981) The ultrastructure of the hemopoietic environment of the marrow: a review. Exp Hematol 9(4):391–410
Linton PJ, Dorshkind K (2004) Age-related changes in lymphocyte development and function. Nat Immunol 5(2):133–139
Liu H, Fergusson MM et al (2007) Augmented Wnt signaling in a mammalian model of accelerated aging. Science 317(5839):803–806
Massberg S, Schaerli P et al (2007) Immunosurveillance by hematopoietic progenitor cells trafficking through blood, lymph, and peripheral tissues. Cell 131(5):994–1008
Mendez-Ferrer S, Lucas D et al (2008) Haematopoietic stem cell release is regulated by circadian oscillations. Nature 452(7186):442–447
Mendez-Ferrer S, Chow A et al (2009) Circadian rhythms influence hematopoietic stem cells. Curr Opin Hematol 16(4):235–242
Mendez-Ferrer S, Michurina TV et al (2010) Mesenchymal and haematopoietic stem cells form a unique bone marrow niche. Nature 466(7308):829–834
Meunier P, Aaron J et al (1971) Osteoporosis and the replacement of cell populations of the marrow by adipose tissue. A quantitative study of 84 iliac bone biopsies. Clin Orthop Relat Res 80:147–154
Moerman EJ, Teng K et al (2004) Aging activates adipogenic and suppresses osteogenic programs in mesenchymal marrow stroma/stem cells: the role of PPAR-gamma2 transcription factor and TGF-beta/BMP signaling pathways. Aging Cell 3(6):379–389
Nagasawa T, Hirota S et al (1996) Defects of B-cell lymphopoiesis and bone-marrow myelopoiesis in mice lacking the CXC chemokine PBSF/SDF-1. Nature 382(6592):635–638
Nagasawa T, Omatsu Y et al (2011) Control of hematopoietic stem cells by the bone marrow stromal niche: the role of reticular cells. Trends Immunol 32(7):315–320
Naveiras O, Nardi V et al (2009) Bone-marrow adipocytes as negative regulators of the haematopoietic microenvironment. Nature 460(7252):259–263

Nilsson SK, Johnston HM et al (2001) Spatial localization of transplanted hemopoietic stem cells: inferences for the localization of stem cell niches. Blood 97(8):2293–2299
Nombela-Arrieta C, Pivarnik G et al (2013) Quantitative imaging of haematopoietic stem and progenitor cell localization and hypoxic status in the bone marrow microenvironment. Nat Cell Biol 15(5):533–543
Ohneda O, Ohneda K et al (2001) ALCAM (CD166): its role in hematopoietic and endothelial development. Blood 98(7):2134–2142
Omatsu Y, Sugiyama T et al (2010) The essential functions of adipo-osteogenic progenitors as the hematopoietic stem and progenitor cell niche. Immunity 33(3):387–399
Papayannopoulou T, Priestley GV et al (2001) Synergistic mobilization of hemopoietic progenitor cells using concurrent beta1 and beta2 integrin blockade or beta2-deficient mice. Blood 97(5):1282–1288
Park D, Sykes DB et al (2012) The hematopoietic stem cell niche. Front Biosci 17:30–39
Parmar K, Mauch P et al (2007) Distribution of hematopoietic stem cells in the bone marrow according to regional hypoxia. Proc Natl Acad Sci U S A 104(13):5431–5436
Plowden J, Renshaw-Hoelscher M et al (2004) Innate immunity in aging: impact on macrophage function. Aging Cell 3(4):161–167
Presley CA, Lee AW et al (2005) Bone marrow connexin-43 expression is critical for hematopoietic regeneration after chemotherapy. Cell Commun Adhes 12(5–6):307–317
Priestley GV, Scott LM et al (2006) Lack of alpha4 integrin expression in stem cells restricts competitive function and self-renewal activity. Blood 107(7):2959–2967
Rando TA (2006) Stem cells, ageing and the quest for immortality. Nature 441(7097): 1080–1086
Ratajczak J, Shin DM et al (2011) Higher number of stem cells in the bone marrow of circulating low Igf-1 level Laron dwarf mice–novel view on Igf-1, stem cells and aging. Leukemia 25(4):729–733
Renault VM, Rafalski VA et al (2009) FoxO3 regulates neural stem cell homeostasis. Cell Stem Cell 5(5):527–539
Rosen ED, MacDougald OA (2006) Adipocyte differentiation from the inside out. Nat Rev Mol Cell Biol 7(12):885–896
Rossi DJ, Bryder D et al (2005) Cell intrinsic alterations underlie hematopoietic stem cell aging. Proc Natl Acad Sci U S A 102(26):9194–9199
Sackstein R (2011) The biology of CD44 and HCELL in hematopoiesis: the 'step 2-bypass pathway' and other emerging perspectives. Curr Opin Hematol 18(4):239–248
Sackstein R, Merzaban JS et al (2008) Ex vivo glycan engineering of CD44 programs human multipotent mesenchymal stromal cell trafficking to bone. Nat Med 14(2):181–187
Schajnovitz A, Itkin T et al (2011) CXCL12 secretion by bone marrow stromal cells is dependent on cell contact and mediated by connexin-43 and connexin-45 gap junctions. Nat Immunol 12(5):391–398
Scheiermann C, Kunisaki Y et al (2012) Adrenergic nerves govern circadian leukocyte recruitment to tissues. Immunity 37(2):290–301
Schofield R (1978) The relationship between the spleen colony-forming cell and the haemopoietic stem cell. Blood Cells 4(1–2):7–25
Solana R, Pawelec G et al (2006) Aging and innate immunity. Immunity 24(5):491–494
Song Z, Wang J et al (2010) Alterations of the systemic environment are the primary cause of impaired B and T lymphopoiesis in telomere-dysfunctional mice. Blood 115(8):1481–1489
Stephan RP, Reilly CR et al (1998) Impaired ability of bone marrow stromal cells to support B-lymphopoiesis with age. Blood 91(1):75–88
Stolzing A, Jones E et al (2008) Age-related changes in human bone marrow-derived mesenchymal stem cells: consequences for cell therapies. Mech Ageing Dev 129(3):163–173
Sugimura R, He XC et al (2012) Noncanonical Wnt signaling maintains hematopoietic stem cells in the niche. Cell 150(2):351–365
Sullivan C, Chen Y et al (2011) Functional ramifications for the loss of P-selectin expression on hematopoietic and leukemic stem cells. PLoS One 6(10), e26246

Taniguchi Ishikawa E, Gonzalez-Nieto D et al (2012) Connexin-43 prevents hematopoietic stem cell senescence through transfer of reactive oxygen species to bone marrow stromal cells. Proc Natl Acad Sci U S A 109(23):9071–9076

Taniguchi Ishikawa E, Chang KH et al (2013) Klf5 controls bone marrow homing of stem cells and progenitors through Rab5-mediated beta1/beta2-integrin trafficking. Nat Commun 4:1660

Trougakos IP, Pawelec G et al (2006) Clusterin/Apolipoprotein J up-regulation after zinc exposure, replicative senescence or differentiation of human haematopoietic cells. Biogerontology 7(5–6):375–382

Tzeng YS, Li H et al (2011) Loss of Cxcl12/Sdf-1 in adult mice decreases the quiescent state of hematopoietic stem/progenitor cells and alters the pattern of hematopoietic regeneration after myelosuppression. Blood 117(2):429–439

Vas V, Senger K et al (2012a) Aging of the microenvironment influences clonality in hematopoiesis. PLoS One 7(8), e42080

Vas V, Wandhoff C et al (2012b) Contribution of an aged microenvironment to aging-associated myeloproliferative disease. PLoS One 7(2), e31523

Wagner W, Saffrich R et al (2005) Hematopoietic progenitor cells and cellular microenvironment: behavioral and molecular changes upon interaction. Stem Cells 23(8):1180–1191

Wagner W, Wein F et al (2007) Adhesion of hematopoietic progenitor cells to human mesenchymal stem cells as a model for cell-cell interaction. Exp Hematol 35(2):314–325

Wagner W, Wein F et al (2008) Adhesion of human hematopoietic progenitor cells to mesenchymal stromal cells involves CD44. Cells Tissues Organs 188(1–2):160–169

Wagner W, Bork S et al (2010) How to track cellular aging of mesenchymal stromal cells? Aging (Albany NY) 2(4):224–230

Warren LA, Rossi DJ (2009) Stem cells and aging in the hematopoietic system. Mech Ageing Dev 130(1–2):46–53

Williams DA, Majumdar MK (1994) Analysis of steel factor (stem cell factor) isoforms in the hematopoietic microenvironment. Stem Cells 12(Suppl 1):67–74, discussion 75–67

Wilson A, Trumpp A (2006) Bone-marrow haematopoietic-stem-cell niches. Nat Rev Immunol 6(2):93–106

Wright DE, Wagers AJ et al (2001) Physiological migration of hematopoietic stem and progenitor cells. Science 294(5548):1933–1936

Xie Y, Yin T et al (2009) Detection of functional haematopoietic stem cell niche using real-time imaging. Nature 457(7225):97–101

Xing Z, Ryan MA et al (2006) Increased hematopoietic stem cell mobilization in aged mice. Blood 108(7):2190–2197

Yamazaki S, Iwama A et al (2009) TGF-beta as a candidate bone marrow niche signal to induce hematopoietic stem cell hibernation. Blood 113(6):1250–1256

Yamazaki S, Ema H et al (2011) Nonmyelinating Schwann cells maintain hematopoietic stem cell hibernation in the bone marrow niche. Cell 147(5):1146–1158

Zanjani ED, Flake AW et al (1999) Homing of human cells in the fetal sheep model: modulation by antibodies activating or inhibiting very late activation antigen-4-dependent function. Blood 94(7):2515–2522

Zhang J, Niu C et al (2003) Identification of the haematopoietic stem cell niche and control of the niche size. Nature 425(6960):836–841

Zhao T, Li J et al (2010) MicroRNA-34a induces endothelial progenitor cell senescence and impedes its angiogenesis via suppressing silent information regulator 1. Am J Physiol Endocrinol Metab 299(1):E110–E116

Zhu X, Gui J et al (2007) Lymphohematopoietic progenitors do not have a synchronized defect with age-related thymic involution. Aging Cell 6(5):663–672

Zsebo KM, Williams DA et al (1990) Stem cell factor is encoded at the Sl locus of the mouse and is the ligand for the c-kit tyrosine kinase receptor. Cell 63(1):213–224

Part IV

Stem Cells, Diseases of Aging and Rejuvenation

Hematopoietic Stem Cell Aging and Leukemogenesis

13

Curtis J. Henry, Andrii I. Rozhok, and James DeGregori

Abstract

Age is the single most important prognostic factor associated with many cancers, including most leukemias and lymphomas. Aging is also associated with dramatic changes in hematopoiesis. As all mature hematopoietic cells are derived from multipotent hematopoietic stem cells (HSCs), age-dependent changes in these cells likely contribute to both alterations in hematopoiesis and increases in blood cancers. In this chapter, we will review the various changes in both HSCs and their microenvironments that could underlie the striking association between multiple major hematopoietic malignancies and old age. Traditionally, the association between aging and cancer has been explained by the requirement for sufficient time for the requisite number of oncogenic mutations to accumulate, as these mutations are thought to limit cancer incidence. We will describe how other aging-associated changes in hematopoietic stem/progenitor cells and their niches, including alterations in cellular fitness, inflammation, localization, and differentiation, could substantially impact age-dependent leukemogenesis. Aging of the hematopoietic system is highly complicated and multifactorial, and likewise increased leukemogenesis in the elderly will likely belie simple explanations.

C.J. Henry • A.I. Rozhok
Department of Biochemistry and Molecular Genetics,
University of Colorado Anschutz Medical Campus, Aurora, CO, USA
e-mail: Curtis.Henry@ucdenver.edu; Andrii.Rozhok@gmail.com

J. DeGregori (✉)
Department of Biochemistry and Molecular Genetics,
University of Colorado Anschutz Medical Campus, Aurora, CO, USA

Department of Pediatrics, University of Colorado Anschutz Medical Campus,
Aurora, CO, USA

Integrated Department of Immunology, National Jewish Health, Denver, CO, USA
e-mail: James.DeGregori@ucdenver.edu

H. Geiger et al. (eds.), *Stem Cell Aging: Mechanisms, Consequences, Rejuvenation*,
DOI 10.1007/978-3-7091-1232-8_13

13.1 Introduction

Aging in humans and animals involves noticeable (and for those of us over 40, regrettable) phenotypic changes. Natural selection is relatively blind to the elderly, which during most of our evolutionary history was relatively rare (Medawar 1952; Kirkwood 2005; Williams 1957). The chances of a human beyond what we today refer to as "middle aged" contributing to the gene pool of future generations were low, as an earlier demise due to disease, starvation, predators, or other causes was more likely. The decline in the soma with age has been postulated to reflect the lack of selection for somatic maintenance beyond the age when the animal is likely to contribute to population maintenance by passing its genes on to future generations.

The same reasoning should apply to cancer, as selective pressure to limit cancer in old animals should be minimal (DeGregori 2011). Thus, cancer is infrequent during the period when any given member of an animal species is likely to contribute genetically to future generations in the "wild." Although cancer does not appear to substantially limit survival past breeding age, this is not to suggest that cancer has been unimportant in vertebrate evolution. Indeed, the evolution of long-lived and large animals, such as vertebrates, necessarily entailed the acquisition of potent tumor-suppressive mechanisms (DeGregori 2011). Although there may have been some advantage for better suppression of cancer in the elderly, enhanced tumor suppression probably comes with a cost, which, if it required increased energy allocation in youth, may have been disfavored. The same logic applies to somatic maintenance (preventing diseases of the elderly) in general. Thus, evolution has in effect weighed the costs and benefits of somatic cell maintenance and tumor suppression, favoring a strategy that maximizes reproductive success.

Just like other tissues and systems of the organism, the hematopoietic stem cell (HSC) pool demonstrates clearly detectable aging-associated changes, marking the general decline of hematopoiesis at middle to advanced ages (Dykstra et al. 2011). Among the major features of this process are a notable decline in cell proliferation rates (Pietras et al. 2011; Bowie et al. 2006), decreasing lymphopoiesis due in part to a shift towards increased frequency of myeloid-biased cells (Pang et al. 2011), and dysregulation of gene expression (Chambers et al. 2007; Rossi et al. 2005). Following this functional decline in HSC pools, hematopoietic malignancies demonstrate an exponential increase in incidence (http://seer.cancer.gov/) starting in the late 40s to early 50s in humans, which is typical of most other cancers. Most research thus far has been concentrated around searching for defined sets of oncogenic mutations that lead to cancer. However, the universal pattern of the age-dependent cancer incidence curve, common even to known animal models, casts doubt as to whether it is the accumulation of mutations per se that solely drives cancer evolution in aged populations, as different cancers seem to require distinct sets of mutations, while following the same incidence statistics with age. This pattern also seems to "scale" according to life span in different animal species (DeGregori 2011), despite roughly comparable rates of mutation across mammals (Lynch 2010a). Together, this

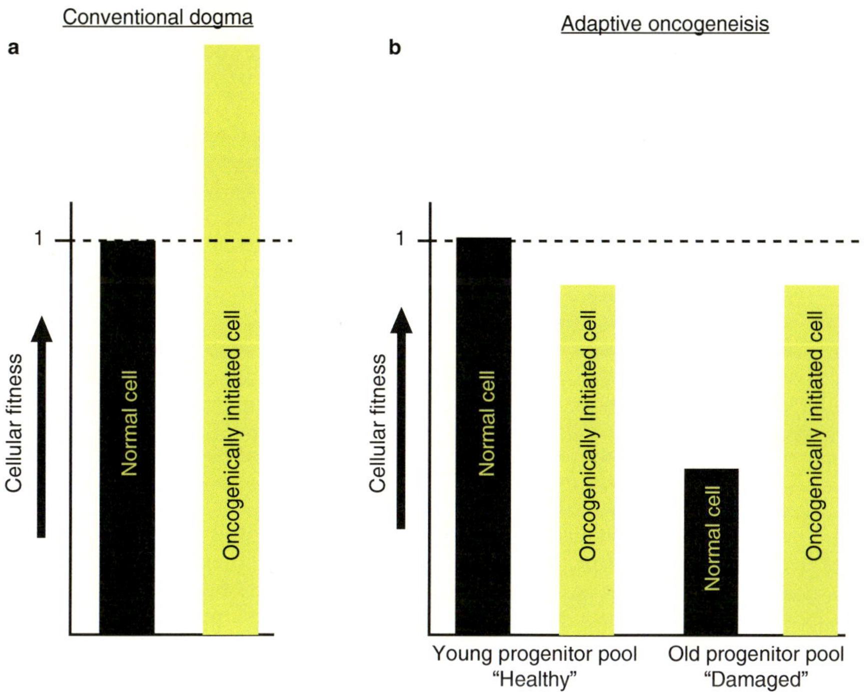

Fig. 13.1 Contrasting conventional and alternative models for why cancer incidence increases with age. (**a**) Illustrates the conventional dogma that suggests oncogenic mutations create "super cells" that have surpassed the fitness optimum leading to oncogenesis. (**b**) The Adaptive Oncogenesis Hypothesis suggests that the incidence of cancer is not just dependent on the accumulation of mutations, but the ability of these mutations to make the harboring cell more "fit" relative to other cells (competitors) in the niche. This model suggests that mutations are typically maladaptive in healthy backgrounds due to their inability to improve optimal fitness levels; however, in damaged pools (such as in aged backgrounds), these mutations improve cellular fitness leading to increased selection for oncogenically initiated cells. Fitness is defined as the ability of cells of a specific genotype or epigenotype to contribute to current and subsequent cell generations due to their phenotypic properties (proliferative capacity, survival capabilities, maintenance of stemness, etc.)

evidence suggests that additional universal age-dependent mechanism(s) may modify the fate of oncogenic mutations leading to cancer development.

Societal changes have significantly increased the life span of humans over the last century, contributing to a greater emphasis on research to understand aging-associated diseases such as cancer. Laboratory settings have also allowed for much longer life spans for mice (2–3 years) than would be experienced in nature (<1 year). It is well known that old age is the greatest risk factor for cancer, as the incidence of most cancers rises exponentially with age (Weinberg 2007; DePinho 2000). This rise has traditionally been linked to the occurrence of oncogenic events which are believed to generally improve cellular fitness (Fig. 13.1) and promote clonal selection (Weinberg 2007). Based on this paradigm, the occurrence of oncogenic events is believed to be the main time-limiting step in cancer initiation and progression (Kennedy et al. 2012; Vogelstein and Kinzler 2004; Weinberg 2007; Serrano and Blasco 2007; Hoeijmakers 2009; Peto et al. 1975). However, a number of observations are difficult to explain from the standpoint of this paradigm (see DeGregori 2013 for the full argument), suggesting a more complex relationship between oncogenic mutations and cancer. Due to much higher cell division rates during prenatal and early postnatal periods (e.g., Bowie et al. 2006), a major fraction of mutations should accumulate during ontogeny (Lynch 2010a; Frank 2010; Vijg et al. 2005), which seems inconsistent with low cancer incidence before the late 40s and early 50s in humans. Besides, oncogenic mutations are frequently detected in normal tissues (e.g., Biernaux et al. 1995; Bose et al. 1998; Matioli 2002; Crawford et al. 2004; Mutter et al. 2001) without apparently promoting cancer. Thus, the accumulation of oncogenic mutations per se appears insufficient to initiate cancer. In fact, many oncogenic mutations have been shown to contribute to loss of HSC self-renewal (reviewed in DeGregori 2013), which challenges the view that they are typically advantageous. Finally, alterations in the niche architecture and declining immune function with age, which may significantly alter the selective forces acting on both normal and oncogenically initiated cells, introduce additional complexity into the actual link between oncogenic mutations, aging and cancer.

Therefore, after summarizing what is currently known about HSC aging and malignancies associated with this process, we will discuss current attempts to explain the aging-dependent mechanism of carcinogenesis. It is clear that both aging hematopoiesis and aging-associated hematopoietic malignancies will be multifactorial. We will also describe an alternative model (adaptive oncogenesis), which suggests that fitness effects of oncogenic mutations are context dependent, generally conferring disadvantages to incipient cells in highly fit pools but gaining selective advantages within aged/damaged stem cell pools in the aged tissue microenvironment. This model predicts that aging alters the general adaptive landscape for stem cell pools, which promotes increased selection for oncogenically initiated cells.

This chapter will address a number of open questions in understanding age-dependent changes in hematopoiesis and increases in leukemia incidence, such as:

- Why does the HSC pool become more biased towards the production of myeloid cells with increasing age, and what is the relationship between this changing HSC bias and declining HSC fitness with age?
- How does this shift in HSC bias and HSC fitness with age influence leukemogenesis?
- How do other changes in hematopoiesis with age influence leukemogenesis?
- Given that leukemias associated with aging, such as acute myeloid leukemia (AML), chronic myeloid leukemia (CML), and chronic lymphocytic leukemia (CLL), are thought to have HSC origins, what changes in HSC with age could account for the dramatically increased incidence of these leukemias in old age?

13.2 Aging-Associated Changes in Hematopoietic Development and Malignancies

The proportion of elderly people is progressively rising throughout the world, with the most pronounced increase in the developed world. Elderly people (>65 years old) are expected to comprise greater than 20 % of the global population by 2050 (Dorshkind et al. 2009). This increase is posing major challenges for healthcare systems, as aging is associated with marked increases in a number of diseases, including most types of cancers (Rossi et al. 2007a; Benz and Yau 2008; Blagosklonny et al. 2010). With more than 80 % of human cancers being diagnosed after the age of 50 (DePinho 2000), aging represents the single most important prognostic factor for many cancers, including lung, colon, prostate, and certain leukemias (Benz and Yau 2008; Edwards et al. 2002; Balducci and Beghe 2001). Strong associations between aging and cancer are traditionally used to support the mutation-centric view of clonal evolution of cancers: aging leads to an accumulation of random mutations, and since (a) some of these random mutations are expected to activate cellular oncogenes or silence suppressor genes and (b) transformation is thought to require cooperation between several oncogenic events, aging should translate into increased risk of cancer initiation (Serrano and Blasco 2007). On the other hand, aging is also associated with substantial cell-autonomous and microenvironmental changes (Henry et al. 2011). In principle, these changes are likely to modify the ability of oncogenic mutations to drive clonal expansion. In fact, some of the age-related changes such as increased inflammation and decreased immune surveillance have been clearly implicated in carcinogenesis (Henry et al. 2011). In addition, intracellular changes such as telomere shortening as well as growth inhibitory changes in the microenvironment could create a scenario for a strong selection for "oncogenic resistance to growth inhibitory conditions" (Blagosklonny 2002). Thus, the effects of aging are highly complex and span organismal, systemic, and cellular levels (Anisimov 2009). Cancer development is similarly complex. Thus it would be naïve to expect a simple relationship between mutational accumulation and the increasing incidence of aging-associated cancers.

One of the more notable age-related changes in hematopoiesis is immunologic decline (Kirkwood 2005; Henry et al. 2010, 2011; Anisimov 2009; Greaves and Maley 2012; Greaves and Wiemels 2003; Naumov et al. 2006a), which impairs responses to pathogens and reduces vaccination efficacy in elderly populations. Decreased immune function is not compartmentalized; reduced immune cell function (and in some cases cell numbers) has been observed in both the myeloid and lymphoid lineages (Weiskopf et al. 2009; Weinberger et al. 2008; Linton and Dorshkind 2004). Furthermore, recent studies have indicated that these functional reductions result at least in part from aging-associated defects in HSC function, which are transferred to their lineage-committed progeny (Weiskopf et al. 2009; Linton and Dorshkind 2004). The causes of the aging-associated decline in HSC and hematopoietic cell function are still under investigation.

13.2.1 Aging and HSC Fitness

Most of our current knowledge on the impact of aging on HSC function comes from studies using mouse models (Weiskopf et al. 2009; Linton and Dorshkind 2004; Sharpless and DePinho 2007; Miller and Allman 2005). These studies implicate both cell-intrinsic and cell-extrinsic factors towards impairments of age-dependent alterations in HSC behavior, although within animals both intrinsic and extrinsic influences are tightly interwoven and are sometimes difficult to disentangle (Chambers and Goodell 2007). Extrinsic factors, including changes in niche composition and hormone production, have been postulated to play a major role in declining HSC function with age (Chambers and Goodell 2007; Yin and Li 2010). In vivo, multiphoton intravital microscopic analysis of HSC in the bone marrow of young and old mice has revealed that old HSC resides further away from the endosteum than young HSC progenitors, indicating age-related extrinsic changes in niche composition that could impact HSC function (Kohler et al. 2009). HSC normally resides in the hypoxic endosteum microenvironment (Kohler et al. 2009; Simsek et al. 2010; Takubo et al. 2010). The effects of hypoxia on HSC function are just beginning to be delineated. It has been postulated that hypoxia regulates HSC self-renewal by limiting cell cycle entry (increasing self-renewal), which would limit potential accumulation of damage (Takubo et al. 2010; Eliasson et al. 2010). Thus, residing further away from the endosteum, by increasing oxidative damage to DNA, could increase the mutational load in HSC (although this has yet to be demonstrated). In addition, it is believed that hypoxia limits the flow of extracellular fluids into the endosteum niche, which reduces HSC exposure to potentially hazardous toxins and proinflammatory cytokines, some of which are known to be tumor promoting (Takubo et al. 2010; Eliasson et al. 2010; Karin et al. 2006; Rossi et al. 2007b). Indeed, aging in mice is associated with increased detection of γH2AX foci in HSCs (Rossi et al. 2007b), which could reflect DNA damage. Notably, a recent study showed that these foci reflect increased replicative stress in old HSCs, coinciding with decreased expression of components of the replication complex (MCMs) (Flach et al. 2014). Should mutation load increase in HSCs in old age, the chances for acquisition of oncogenic mutations would be increased, which could contribute to the increased initiation and progression of cancers in the elderly (whether mouse or human). Moreover, another consequence of increasing replicative stress and mutational load in HSCs (as in other stem cell types) could be the reduction in stem cell fitness, as DNA damage is more likely to negatively impact on cellular functions than improve them. Beyond the accumulation of genetic damage, epigenetic drift with age also likely contributes to HSC fitness decline (Chambers et al. 2007; Wahlestedt et al. 2013; Bell and Spector 2011; Sun et al. 2014). Moreover, other cell-autonomous changes likely contribute. Indeed, HSCs in old mice exhibit greater dependence on autophagy, which is selectively needed for old HSCs to maintain cellular energetics and survival (Warr et al. 2013). Notably, recent computational studies of changes in HSC fitness with age indicate that cell-autonomous accumulation of genetic and epigenetic events is insufficient to account for fitness decline with age, instead revealing that microenvironmental alterations with age are required to achieve the observed severalfold reductions in HSC fitness in old age (Rozhok et al. 2014).

Aging-related, cell-autonomous changes have also been shown to contribute to altered HSC function (Chambers et al. 2007; Rossi et al. 2005; Chambers and Goodell 2007; Warren and Rossi 2009; Beerman et al. 2010a, b). When highly purified long-term HSCs (LT-HSCs) from young or old mice are used to reconstitute young irradiated recipient mice, old LT-HSCs are two- to fourfold less efficient at reconstitution per HSC, and their production is skewed in favor of myelopoiesis, with accompanying reductions in lymphopoiesis (Dykstra et al. 2011; Chambers et al. 2007; Rossi et al. 2005; Henry et al. 2010; Miller and Allman 2005; Chambers and Goodell 2007; Morrison et al. 1996). Thus, as defined above, old HSCs are less "fit," in that young competitors are better able to contribute to hematopoiesis. Aged mouse HSCs also exhibit decreased homing and engraftment posttransplantation (Morrison et al. 1996; Liang et al. 2005).

The hematopoietic system may attempt to compensate for decreased HSC function (and decreased production rates of mature progeny) by increasing the size of the HSC compartment, as indicated by the observation that the number of phenotypic HSCs in C57Bl/6 mice increases with age (Chambers et al. 2007; Rossi et al. 2005; Chambers and Goodell 2007; Morrison et al. 1996; Sudo et al. 2000; de Haan and Van Zant 1999; Noda et al. 2009). Such increased homeostatic mechanisms, both through increased cytokine levels and perhaps increased turnover rates for HSCs and other progenitors, should contribute to higher chances for mutation fixation. Larger numbers of HSCs should also increase the target size for oncogenic mutations. However, increased HSC numbers have not been observed in all mouse strains; BALB/c and DBA/2 do not show an age-dependent increase in the numbers of phenotypic HSC (Van Zant et al. 1990; Chen et al. 1999, 2000). Still, increased numbers of HSCs, at least as assessed using phenotypic markers, are observed in elderly humans (Pang et al. 2011). Although the effects of aging on human HSCs have not been as extensively studied, human HSC function also appears to be affected by aging given the observation that the proliferative potential of human HSCs declines with age (Vaziri et al. 1994). In addition, bone marrow from older humans is less efficient at reconstituting recipients when compared to the reconstitution capacity of bone marrow derived from younger patients (Kollman et al. 2001). Therefore, based on experiments in mice and more limited observations in humans, it is clear that a number of age-associated alterations in the HSC compartment, due both to extrinsic and intrinsic factors, could contribute to increased frequencies of hematopoietic malignancies in the elderly.

13.2.2 HSC Lineage Bias Changes with Age

Aging-associated changes in HSC reflect more than just a decline in fitness. In aged mice, it has been demonstrated that a developmental shift occurs, with reduced lymphopoiesis in favor of greater myelopoiesis. Microarray analyses of HSCs isolated from old mice reveal a gene expression profile consistent with this age-dependent bias towards production of myeloid cells (Chambers et al. 2007; Rossi et al. 2005; Chambers and Goodell 2007). The myeloid bias appears to be at least in part cell autonomous, as adoptive transfer of young and old HSC into young, irradiated recipients recapitulates the age-dependent bias towards myelopoiesis (Rossi et al. 2005; Dorshkind et al. 2009; Guerrettaz et al. 2008). This bias appears to be mediated by a shift towards greater frequency of myeloid-biased (mb-HSCs) relative to lymphoid-biased HSCs (lb-HSCs) (Sudo et al. 2000; Muller-Sieburg et al. 2004; Cho et al. 2008), and these differentially biased HSCs can now be prospectively isolated and shown to transfer their developmental bias to recipient mice (Beerman et al. 2010a, b; Challen et al. 2010; Kent et al. 2009). Notably, transplantation of mb-HSCs from young and old mice at limiting dilution (so that a single HSC contributes to hematopoiesis in the recipient) reveals a significant two- to threefold reduction in engraftment for each successfully engrafting old mb-HSC relative to young mb-HSC (indicating a per cell reduction in HSC fitness in old mice, even when similarly "biased" HSCs are compared) (Dykstra et al. 2011). It is important to note that while myelopoiesis becomes favored over lymphopoiesis in old mice, myeloid progenitors are not "more fit" relative to the young (Dorshkind et al. 2009; Signer et al. 2007). Moreover, elderly humans exhibit alterations in HSC representation and decreased developmental potential of both lymphoid and myeloid progenitors (Pang et al. 2011; Sharpless and DePinho 2007; Kuranda et al. 2011). Notably, human BM cells with HSC phenotypic markers also show myeloid skewing and decreased repopulation potential in immunocompromised mice, suggesting decreased fitness (Pang et al. 2011).

Recently, it has been demonstrated that increases in the activation of the noncanonical Wnt signaling pathways, mediated primarily through Wnt5a, are an important contributor to HSC aging (Florian et al. 2013). Specifically, aging promotes increased expression of Wnt5a in long-term HSCs which leads to decreased β-catenin activity in the nucleus (Florian et al. 2013). Furthermore, Wnt5a treatment of young HSC induces an aging-like phenotype, indicated by increases in Cdc42 levels, decreases in the percentage of polarized LT-HSCs, and reductions in B lymphopoiesis with accompanying increases in myelopoiesis in vivo (all of which are phenotypes observed in aged HSC (Florian et al. 2012)). Additionally, aged LT-HSCs isolated from Wnt5a-haploinsufficient mice have reduced Wnt5a and Cdc42 activity, corresponding with better maintenance of B lymphopoiesis without accompanying increases in myelopoiesis in vivo (Florian et al. 2013). Transplantation studies revealed that reducing Wnt5a activity in aged LT-HSCs reversed the aging-associated, cell-intrinsic deficiencies exhibited by a significant increase in the capacity of these cells to make B cells. As a result of the findings from this study, we now have a better understanding of the molecular processes that underlies LT-HSC aging and how this shift from canonical to noncanonical Wnt signaling pathways promotes aging hematopoiesis. Importantly, these studies suggest that at least some aspects of the aging hematopoietic phenotype are reversible by altering signaling, which could also provide insight into how particular oncogenic events may be adaptive in aged HSC pools.

13.2.3 The Impact of Aging on Innate and Adaptive Immunity

The biological consequences of aging-associated alterations in hematopoiesis and the impact on the ability of mature immune cells to control aging-associated cancers are being actively investigated. It has been extensively shown that innate and adaptive immune cells from older individuals exhibit reduced function and/or numbers, which decrease their ability to eliminate aging-associated cancers. Decreased cellular function in old age has been documented in innate cells such as dendritic cells (DC) (reduced T-cell stimulatory capacity) and natural killer (NK) cells (reduced cytolytic potential) (Weiskopf et al. 2009; Shaw et al. 2010). Age-associated impairment of the adaptive immune system, composed of T cells and B cells, is also observed in both aged mice and humans (Dorshkind et al. 2009; Miller and Allman 2005; Dorshkind and Swain 2009; Allman and Miller 2005) and could facilitate cancer development in older individuals. In mice, the production of naïve $CD4^+$ T cells (T-helper cells) declines with age (Aspinall and Andrew 2000a, b). $CD8^+$ T-cell function, as measured by decreased proliferation in response to IL-2 and reduced expansion and differentiation upon antigenic stimulation, also declines with age (Weiskopf et al. 2009). Aging B lymphopoiesis in mice is characterized by reductions in progenitor cell populations, reduced responsiveness to growth factors such as interleukin-7 (IL-7), and decreases in their ability to undergo class switching. In humans, the aging-associated decline in B-cell lymphopoiesis appears less drastic compared to mice, and it has been demonstrated that B-cell lymphopoiesis in humans is less dependent on IL-7 stimulation (Rawlings et al. 1995; Prieyl and LeBien 1996). The decline in humoral responses in elderly people is largely attributed to changes in the composition of the cells comprising the B-cell repertoire (Weiskopf et al. 2009; Weksler et al. 2002; Weksler and Szabo 2000). Although the effects of aging on B lymphopoiesis differ between mice and humans, there is significant overlap in the decline in B-cell function found in aged population from both species. While reductions in B lymphopoiesis in old age could contribute to reduced tumor immune surveillance, there has been more debate about the implications of perturbed B-cell development towards leukemogenesis (described below).

13.2.4 Inflamm-aging and the HSC Repertoire

Aging-associated changes in HSC representation and function, some of which are subsequently imprinted on mature immune cells, prompt two important questions: (1) What factor(s) induces the aging-associated alterations in HSCs? (2) What is the biological cause (and consequence) of these changes? Regarding the former, increases in aging-associated inflammation, dubbed "inflamm-aging," may impinge on stem cell populations leading to the altered stem cell dynamics discussed in the previous sections. Indeed, inflammatory cytokines have been shown to regulate HSC activity with cytokines such as TGF-β and TNF-α showing the capacity to promote quiescence by repressing HSC expansion (Pronk et al. 2011; Yamazaki et al. 2009; Mirantes et al. 2014; Pietras et al. 2014), while the cycling of HSCs can be induced through the actions of IFN-α and IFN-γ (Essers et al. 2009; Baldridge et al. 2010). Furthermore, recent evidence suggests that HSC skewing towards the myeloid lineage in old age may be in part promoted by cytokines like TGF-β1 (Chambers et al. 2007; Challen et al. 2010; Song et al. 2010) or by perturbation of the bone marrow microenvironment (Omatsu et al. 2010). "Inflamm-aging" is characterized by increased levels of tumor necrosis factor-alpha (TNF-α), interleukin-6 (IL-6), and interleukin-1β (IL-1β) in the plasma (Weiskopf et al. 2009; Franceschi et al. 2000). This increase in the circulating levels of inflammatory cytokines likely reflects cell-autonomous changes in myeloid gene expression resulting in microenvironmental alterations that could favor tumor initiation and progression (Weiskopf et al. 2009). In addition, aged HSCs themselves have been shown to exhibit proinflammatory and stress response biases in their gene expression profiles (Chambers et al. 2007). The cause of this aging-associated increase in inflammation is still under investigation. However, proposed culprits of "inflamm-aging" are chronic infection by pathogens past reproductive ages, elevated aging-associated stress responses, declines in mechanisms responsible for dealing with these stresses, and altered cellular metabolism (Chambers et al. 2007; Simsek et al. 2010; Franceschi et al. 2000; Garaycoechea et al. 2012). In fact, Franceschi et al. propose that humans (and presumably other mammals) evolved to deal with chronic infections during our "normally" short lives (at least for most of our evolutionary history) (Franceschi et al. 2000). These same mechanisms, which were beneficial for controlling infections through reproductive years, could be contributing to increased inflammation in old age (of course, natural selection acted on the former to a far greater extent than the latter).

Chronic inflammation is a well-defined contributor to tumorigenesis, with tumor-associated macrophages (TAMs) playing a key role in driving tumorigenesis (Franceschi et al. 2000). The cytokines produced by these TAMs, such as TNF-α, could influence cancer progression both by promoting proliferation of oncogenically initiated or more advanced tumor cells and by inducing death of nonmalignant stem and progenitor cells (thus stimulating compensatory proliferation) (Karin et al. 2006; Karin and Greten 2005). Inflammation may also create a hazardous microenvironment, with increased ROS and cell death, thus promoting selection for oncogenic events that confer resistance or are otherwise adaptive to this altered microenvironment.

On the HSC level, inflammation likely induces changes in the microenvironment that increases exposure of HSCs to insults that could induce damage and subsequently lead to decreased fitness (Chambers et al. 2007; King and Goodell 2011). As mentioned above, HSCs in the bone marrow of old mice reside further away from the protective hypoxic endosteum than their young counterparts (Kohler et al. 2009). This altered localization correlates with increased detection of γH2AX foci and loss of quiescence (mediated by reduced HIF-1α levels) (Takubo et al. 2010), which may result from elevated exposure to toxins and cytokines (King and Goodell 2011). Furthermore, loss of HIF-1α has been shown to alter HSC metabolism (Simsek et al. 2010), which could alter cellular fitness. Thus, aging-associated inflammation may also promote cell-intrinsic (whether genetic or epigenetic) and cell-extrinsic changes, which alter cellular fitness, subsequently leading to increased selection for oncogenically initiated cells.

Regarding the latter, *why* does the representation and fitness of HSC change with age? As discussed previously, hematopoietic aging in mice and humans is accompanied by a skewing towards myelopoiesis. This shift is attributed at least in good measure to changes in the genetic program of HSC and a reduction in the number of lymphoid-biased HSC. From an evolutionary perspective, this shift in the representation of hematopoietic progenitors appears programmed, since the majority of the lymphocyte repertoire is established during youth. Indeed, T lymphopoiesis declines starting before full maturity (Dorshkind et al. 2009; Dorshkind and Swain 2009) and thus clearly represents an evolved program (not a consequence of decreased maintenance in old age). Similarly, B lymphopoiesis declines gradually with age (Miller and Allman 2005; Milne and Paige 2006), similarly suggesting that this programmed switch in hematopoiesis (to more myeloid) provides an adaptive advantage to mammals. Thus, these changes in hematopoiesis are advantageous for organismal fitness: the lymphocytic repertoire is established relatively early in life and can be subsequently maintained at the mature T- and B-cell levels. In contrast, a constant supply of myeloid cells, which do not self-renew and represent the first line of defense against pathogens (in addition to playing many other roles in tissue maintenance) (Cain et al. 2009), is required throughout life.

Notably, not all aging-associated changes are cell intrinsic. Several reports show that microenvironmental alterations that occur during aging alter HSC function. These include the reduced ability of old bone marrow stromal cells to support hematopoietic progenitor cell expansion, reduced adhesion of aged HSC and progenitor cells to bone marrow stroma, and reduced polarity of aged HSC to BM stromal cells (Florian et al. 2012; Stephan et al. 1998; Xing et al. 2006). Moreover, inflammation has been shown to directly impair B lymphopoiesis (and thus favor myelopoiesis), by preventing B-progenitor localization to the IL-7-rich niche (Ueda et al. 2005). While inflammation is important for animal survival in youth (for eliminating infections and repairing tissues), it can become detrimental in old age in its chronic manifestation (Goto 2008). These observations suggest that microenvironmental stimuli play a role in the regulation of stem and progenitor cells and the subsequent skewing that accompanies aging. Thus, while some myeloid skewing could be the result of the programs described above, additional skewing towards myelopoiesis could result from age-associated inflammation (at ages beyond the point where natural selection has much influence).

Taken together, these data suggest that aging hematopoiesis potentially results from two phenomena: (1) aging-associated decline in lymphoid-biased HSC populations resulting from a program favored by natural selection (B- and T-cell repertoires are established early, allowing hematopoiesis to focus on myeloid cell production, as myeloid cells are the first line of defense against pathogens) and (2) aging-associated changes in HSC that result truly in old age (beyond much influence from natural selection, such as due to inflammation), which decrease the fitness of HSC and subsequently progenitor cells (Fig. 13.2).

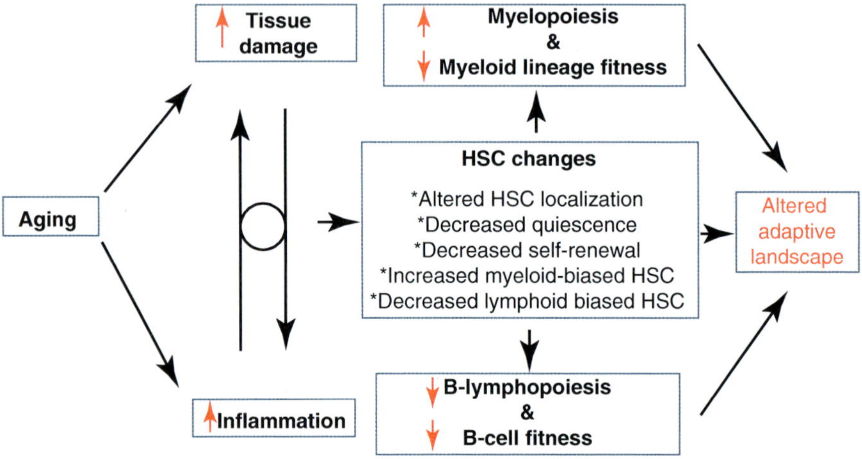

Fig. 13.2 Model for aging-associated changes in the adaptive landscape. Aging is accompanied by increasing tissue damage and elevated levels of inflammation "inflamm-aging." We postulate that these events modulate HSC function by promoting localization away from the protective hypoxic endosteal niches. Aging HSCs also exhibit cell-intrinsic alterations which lead to decreased quiescence, reduced self-renewal, and altered differentiation, favoring the expansion of myeloid-biased HSCs while suppressing the maintenance of lymphoid-biased HSCs. Overall, these age-dependent alterations of the HSC pool should result in fitness declines in stem and progenitor cells ultimately resulting in drastic changes in the adaptive landscape

13.2.5 Aging, HSC, and Leukemogenesis: A Complicated Relationship

So how might aging-associated changes in HSC contribute to increased leukemogenesis? Stochastic aging-associated changes, such as inflamm-aging, could stress HSCs in older individuals. For example, elevated levels of proinflammatory cytokines such as TNF-α (which is frequently increased in the serum of older individuals) have been shown to increase cycling of mb-HSCs (Ueda et al. 2004). Chronic inflammation has also been associated with decreased HSC self-renewal, which is attributed to persistent mitochondrial ROS levels (Chambers et al. 2007; Goto 2008). Increased cycling of HSCs together with higher ROS levels (and subsequent damage to cellular macromolecules including DNA) could thus increase the chances for fixation of oncogenic mutations.

As we argued above, reducing the fitness of a cell population should increase selection for adaptive oncogenic events, thus promoting cancer development. Thus, an important question remains: how do changes in HSCs with age influence the fitness of committed progenitor cells? As described above, aging-associated defects in B and T lymphopoiesis are particularly apparent, which in mice can be attributed to both cell-intrinsic and cell-extrinsic effects of aging (Linton and Dorshkind 2004; Guerrettaz et al. 2008; Allman and Miller 2005; Miller and Allman 2003; Lescale et al. 2010). For T lymphopoiesis, age-dependent reductions (starting in youth) in T-cell production can be partially (but not entirely) ascribed to the involution of the thymus. B-cell progenitors exhibit clear cell-autonomous functional defects, such as reduced responsiveness to IL-7 (Miller and Allman 2003; Andrew and Aspinall 2001) and reduced kinase-dependent signaling (Henry et al. 2010). These functional defects have consequences, as we have shown that defective signaling and reduced fitness of old B-progenitor pools lead to potent selection for oncogenes like BCR–ABL that correct signaling defects (Henry et al. 2010). Moreover, stochastic modeling of somatic evolution in HSC pools revealed that non-cell-autonomous influences on HSC fitness through the microenvironment were necessary to explain the age dependence of leukemias (Rozhok et al. 2014).

But why would B-cell progenitors in old mice exhibit functional defects, as these progenitors are not themselves "old" (with relatively short life spans of a few weeks), but were derived from old HSCs? Notably, Muller-Sieburg et al. have shown that even young murine mb-HSCs (which dominate the hematopoietic system of old mice and humans) make lymphoid progenitors with reduced function compared to those derived from lymphoid-biased HSCs that are predominant during youth (Muller-Sieburg et al. 2004). Why would natural selection favor a mechanism whereby functionally impaired B progenitors are produced in young mice? In one speculative scenario, we propose that fit B lymphopoiesis is indeed maintained in young mice, as most B progenitors will be derived from lb-HSCs, and the few derived from mb-HSC will be outcompeted by those derived from lb-HSC (thus maintaining overall B-progenitor pool fitness). Indeed, Challen et al. showed that in competitive repopulation experiments, myeloid progenitor pools were dominated by progeny of mb-HSCs and lymphoid progenitor pools by progeny of lb-HSCs (Challen et al. 2010). Notably, this lineage bias was amplified by competition from progeny of HSCs with the opposing bias (i.e., lymphoid progenitors from mb-HSC appear to be outcompeted by lymphoid progenitors from lb-HSC) (Challen et al. 2010). Thus, only when mb-HSCs far outnumber lb-HSCs (in old animals, beyond the period of strong natural selection) would B progenitors actually be largely derived from mb-HSCs and thus exhibit reduced fitness.

Notably, impaired B lymphopoiesis in old age appears to be more than just mediated by changes in HSC subsets. Phenotypically distinct age-associated B cells (ABCs) have been shown to accumulate with advanced age in mice (Hao et al. 2011), and these ABCs exhibit heightened Toll-like receptor (TLR) responses (Hao et al. 2011) and accumulate in response to TLR signals (Rubtsov et al. 2011, 2013). ABCs from old mice also exhibit increased expression of TNFα, which can inhibit IL-7-dependent clonogenicity of pro-B cells (Frasca et al. 2012; Ratliff et al. 2010). Importantly, the Melamed lab has shown that depletion of mature B cells in old mice leads to restored B lymphopoiesis (Keren et al. 2011a, b).

So what are the implications for leukemogenesis? Beyond the obvious effects of reduced lymphopoiesis on immunity against cancers, the impaired fitness of lymphoid progenitors could promote selection for adaptive oncogenic mutations, as we have shown (Henry et al. 2010). While such oncogenic mutations might be expected to be adaptive in mb-HSC-derived B-progenitor pools (even if derived from young mb-HSCs), in a young animal B progenitors from lb-HSCs will dominate such pools (maintaining pool fitness and thus preventing oncogenic selection). On the other hand, the *lack* of lb-HSC-derived B lymphopoiesis in old animals could be responsible for the poor fitness of B-progenitor pools and consequent increased leukemogenesis. Finally, it is also interesting to consider how alterations in B lymphopoiesis mediated by mature B cells (such as ABC) might impact oncogenic adaptation and leukemogenesis in old mice. Importantly, whether mediated by changes at the level of HSCs or committed B progenitors, it will be important to determine whether modulation of inflammation can impinge on B lymphopoiesis and the adaptive landscape for oncogenesis.

If aging-associated mechanisms such as inflammation are indeed a major contributor to aging hematopoiesis, it stands to reason that if these insults could be reduced, it would lead to improvements in cellular fitness and reduced cancer incidence. Several observations support that this may be the case. First, it is notable that polymorphisms associated with reduced inflammation are preferentially found in centenarians (Franceschi et al. 2005; Salvioli et al. 2009). Second, a history of infectious or autoimmune disease (and associated chronic immune stimulation) is a significant risk factor for the development of acute myeloid leukemia (AML) or myelodysplastic syndrome (MDS) (Kristinsson et al. 2011). Third, the extent of hematopoietic decline varies widely between aged individuals, which may partially be attributed to lifestyle choice. Indeed, smoking (Malfertheiner and Schutte 2006) and obesity (Roberts et al. 2010; Renehan et al. 2008) are both linked to higher cancer incidence, while caloric restriction in mice (Anisimov 2009) and prophylactics in humans (i.e., aspirin) (Thakkar et al. 2013; Carlson et al. 2013) have been shown to have antitumor effects. A common feature of these cancer-promoting and cancer-reducing contexts is the observation that they all modulate inflammation, with smoking and obesity promoting inflammation and caloric restriction and aspirin reducing inflammation.

13.3 Aging-Associated Hematopoietic Malignancies in Humans: Forest or the Trees?

Two categories of hematopoietic cancers that show striking increases in incidence in individuals over 65 years of age are leukemias and lymphomas (Fig. 13.3). In the following section we will review three aging-associated leukemias, each thought to initiate within HSCs: chronic lymphocytic leukemia (CLL), chronic myelogenous leukemia (CML), and acute myelogenous leukemia (AML). We will focus on the etiologies of these malignancies and how aging-associated hematopoietic changes in stem and progenitor cell populations may promote disease evolution in older populations.

13.3.1 Etiologies of Aging-Associated Cancers

CLL is the most common type of leukemia/lymphoma in adults, with 72 being the median age of diagnosis (Gaidano et al. 2012). CLL is characterized by the

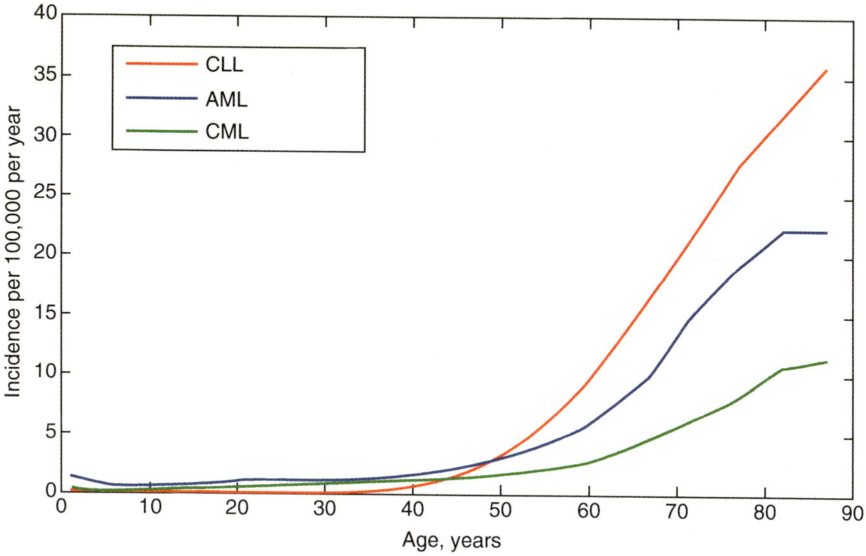

Fig. 13.3 Leukemia incidence in aging populations. The incidence of chronic lymphocytic leukemia (*CLL*), acute myelogenous leukemia (*AML*), and chronic myelogenous leukemia (*CML*) increases drastically after the age of 50 with CLL being the most common leukemia in aged populations (Data are from http://seer.cancer.gov/, for years 1992–2000, and include all sexes and races)

accumulation of antigen-stimulated mature B lymphocytes and proliferating precursor cells (Chiorazzi et al. 2005). Although chromosomal translocations and cytogenetic lesions are rare in leukemic clones at earlier points of disease progression, these do appear as the disease progresses (Gaidano et al. 2012; Chiorazzi et al. 2005). Indeed, deletions such as 13q14, 11q22-23, 17p13, 14q32, and 6p21 are found in over half of CLL reported cases (Gaidano et al. 2012). These genetic changes likely have profound biological effects given that they modulate the expression of microRNAs (13q14), p53 (17p13), and the ataxia telangiectasia mutated (ATM; 11q22-23) gene (Chiorazzi et al. 2005). In addition to genetic aberrations in malignant cells isolated from CLL patients, tumorigenic B cells also exhibit altered intracellular signaling. Clinically, CD38 (which augments BCR signaling) or ZAP-70-positive B cells are associated with more aggressive forms of the disease, while decreased activation of these molecules usually indicates a more indolent disease status (Chiorazzi et al. 2005). Recently it has been demonstrated that HSCs from CLL patients are aberrant and contribute to disease progression, which conflicts with the prevailing view that this leukemia initiates in more mature B cells (Alizadeh and Majeti 2011; Kikushige et al. 2011; Damm et al. 2014). In fact, HSCs from CLL patients exhibit cell-intrinsic lymphoid lineage bias that generates clonal CLL-like B cells (Alizadeh and Majeti 2011; Kikushige et al. 2011). Given the clonal nature of B cells that arise from HSCs from CLL patients, it is likely that a selection for oncogenically initiated HSCs occurs, predisposing patients to this malignancy resulting from additional oncogenic events at more B-lineage-committed stages. However, little is known regarding either the initiating genetic/epigenetic events or the aging-associated contexts that favor initiation and progression of this disease.

CML is a clonal disorder of HSCs that results in a malignant increase of myeloid cells, erythroid cells, and platelets in the bone marrow and the peripheral blood. The median age of presentation of CML is 53 years of age (Sawyers 1999). There is no clear genetic or hereditary predisposition for this disease (Garcia-Manero et al. 2003). The acquisition of the Philadelphia chromosome, resulting from a translocation and fusion of the breakpoint cluster region (BCR) gene on chromosome 22 with the ABL gene produces BCR–ABL, a constitutively active, pleiotropic tyrosine kinase capable of promoting cellular growth, proliferation, and survival independent of growth factor regulation (Sawyers 1999). Notably, CML in chronic phase is thought to be a simple leukemia (a myeloproliferative disorder), with Bcr–Abl as the only recurrent mutation (Mullighan et al. 2008). In contrast, the more advanced CMLs and BCR–ABL$^+$ ALLs show numerous additional genetic changes beyond the BCR–ABL translocation (Mullighan et al. 2008), typical of most hematologic malignancies and other cancers. Although the presence of BCR–ABL transcripts is relatively common in the general population, the clinical manifestation of CML only occurs in 1–2 in 100,000 individuals per year (Biernaux et al. 1995; Bose et al. 1998; Matioli 2002; Garcia-Manero et al. 2003). These observations indicate that the acquisition of the driver mutation alone is insufficient to cause disease; thus, additional factors must contribute to disease progression (or lack thereof).

AML is a common form of leukemia in adults with the median age of diagnosis being 69 (Lowenberg et al. 1999). AML refers to a heterogeneous subset of leukemia characterized by clonal expansion of myeloid progenitors (granulocytic, monocytic, erythroid, or megakaryocytic) with a reduced capacity to differentiate into mature myeloid cells. Unlike CML in chronic phase, AML is recognized as a heterogeneous disease with a vast array of driver and secondary mutations required for the development of the disease. These mutations can be broadly classified into two categories: mutations that promote proliferation and those that impair differentiation (Kelly and Gilliland 2002). For the former, activating mutations in FLT3 are the most frequently observed molecular defect in AML, occurring in approximately 30–35 % of adults and 20–25 % of children with AML (Armstrong et al. 2004; Small 2006). Mutations that reduce differentiation include those in genes encoding tet methylcytosine dioxygenase 2 (TET2) and DNA (cytosine-5)-methyltransferase 3 alpha (DNMT3A), which substantially impact on the epigenetic profiles of these cells (Sanders and Valk 2013).

13.3.2 Why Are These Leukemias Associated with Advanced Age?

The field of cancer research is dominated by the view that oncogenesis is rate limited by the incidence of oncogenic mutations and that these mutations are typically advantageous when they occur in the right cell type. Such oncogenic mutations (including activation of proto-oncogenes and deactivation of tumor suppressor genes, whether by genetic or by epigenetic mechanisms) are thought to provide cells with various "Hallmarks of Cancer," including sustained proliferative signaling and resistance to growth-suppressive and cell death signals (Hanahan and Weinberg 2011). In particular, it is widely accepted that the exponential increase of cancer incidence with age reflects the time required for cells to accumulate sufficient numbers of genetic and epigenetic mutations to confer the cancer phenotype (Kennedy et al. 2012; Vogelstein and Kinzler 2004; Weinberg 2007; Serrano and Blasco 2007; Hoeijmakers 2009; Peto et al. 1975; Nowell 1976). The logic is quite simple: aging leads to mutations (including oncogenic mutations) eventually resulting in cancers such as CLL, CML, and AML. Oncogenic mutations are clearly required for cancer evolution, and increases in genetic/epigenetic diversity in somatic cells associated with aging should contribute to cancer incidence. Increased rates of genomic instability in some cancers can also help promote tumor evolution (Crawford et al. 2004; Agrawal et al. 2012; Chang et al. 2001). However, is the axiomatic attribution of the rising incidence of cancers with age primarily due to the accumulation of oncogenic mutations sufficiently justified? Could it be so simple that the aging-associated increase in cancer incidence is solely explained by the requirement of the time it takes to increase the accumulation of oncogenic mutations?

As referred to above, many, if not most, mutations appear to accumulate during ontogeny, rather than during adulthood (Lynch 1988, 2010a, b; Frank 2010; Vijg et al. 2005). The maintenance of self-renewing adult tissues may require relatively few stem cell divisions. It is estimated that any given HSC will divide on average five to ten times through the life of an adult mouse (from maturity to 2–3 years of age) (Wilson et al. 2008), and yet one would surmise that the generation of each HSC in a young mouse required far larger numbers of cell divisions (counting from the one-celled zygote). In fact, mouse HSCs divide more than once per day through fetal development up until around 3 weeks of age and then abruptly switch to rare cell divisions (once every few months) through old age (Bowie et al. 2006). Similarly, telomere shortening in human hematopoietic cells occurs mostly during ontogeny (Rufer et al. 1999; Weng et al. 1998). Thus, it is not surprising that a substantial fraction of mutations and epigenetic changes would occur and accumulate during ontogeny (Frank 2010; Vijg et al. 2005; Horvath 2013), followed by a more modest rate of accumulation during tissue maintenance postmaturity. That the incidence of most cancers rises late in life, with kinetics that are quite disconnected from the time-dependent accumulation of mutations in the tissues from when these cancers arise, argues that age-dependent acquisition of oncogenic mutations per se is not a rate-limiting step in tumorigenesis.

13.3.3 Oncogenic Mutations Do Not Always Lead to Cancer

Many barriers to tumor progression exist, and it is well known that multiple oncogenic mutations are required for the emergence of clinically detectable cancers (Greaves and Maley 2012; Gatenby and Gillies 2008). Still, if the incidence of oncogenic mutations were a rate-limiting step in tumorigenesis, one would not expect an abundance of oncogenic mutations in the absence of tumorigenesis. Nonetheless, cell clones with mutational or epigenetic inactivation of the *PTEN* or *INK4A* tumor suppressor genes are frequently found in histologically normal endometria and breast (respectively) of cancer-free women (Crawford et al. 2004; Mutter et al. 2001), far outpacing the incidence of the corresponding cancers. Furthermore, the presence of TEL-AML1 and AML1-ETO translocations in the blood cells of newborns is approximately 100-fold greater than the risk of the associated leukemias (Greaves and Maley 2012). Perhaps most surprisingly, histologically advanced microscopic tumors are detected in many tissues of adult humans (Greaves and Maley 2012; Almog et al. 2006; Naumov et al. 2006b), but they appear to be mostly held in check by unknown mechanisms. In addition, even though it is thought that the incidence of CML is limited by the occurrence of the initiating BCR–ABL translocation (Vickers 1996), in-frame BCR–ABL fusions are detected in leukocytes of approximately one in three healthy individuals (Biernaux et al. 1995; Bose et al. 1998), and the vast majority of which will fortunately never develop this leukemia (despite persistence of the translocation in leukocytes for long enough to suggest an HSC origin (Matioli 2002)). In mouse models, the expression of BCR–ABL provides a much greater advantage to progenitor cells in an aged as compared with a young hematopoietic system, leading to increased clonal expansion and leukemogenesis (Henry et al. 2010). Thus, at least in this mouse model, the occurrence of the same oncogenic mutation, BCR–ABL, results in very different outcomes depending on the age of the target tissue.

13.3.4 Tissue Decline with Age: Tumor Suppressive or Tumor Promoting?

It is frequently argued that aging is in part the consequence of organisms attempting to repress cancer evolution (Serrano and Blasco 2007; Hoeijmakers 2009; Dorshkind et al. 2009; Campisi 2005a, b; Sahin and Depinho 2010; Sharpless and DePinho 2002, 20042007; Hinkal et al. 2009a, b). In this view, aging and tissue decline limit cancer development, through telomere shortening and tumor suppressor gene action. Thus, preventing cancer during youth (when animals are most likely to contribute to future generations) requires mechanisms, such as limited telomere maintenance, which contribute to aging phenotypes later in life. Thus, aging is thought to in part represent a cost of tumor suppression. According to this model, aging-associated events (i.e., telomere shortening) increase the activation of tumor suppressor genes, which inhibits cancer at the expense of accelerating the aging process (Dorshkind et al. 2009; Sharpless and DePinho 2007; Campisi 2005a; Sahin and Depinho 2010; Finkel et al. 2007; Jaskelioff et al. 2011). Indeed, aging-associated reductions in telomere length and function activate p53, a critical tumor suppressor (Sharpless and DePinho 2007; Wynford-Thomas et al. 1995). In addition to p53, expression of the *INK4A* locus-encoded p16 and Arf tumor suppressor genes also increases with age in certain hematopoietic compartments

(Signer et al. 2008). Moreover, increases in p16 and Arf have been shown to contribute to the aging-associated decline in lymphopoiesis (Signer et al. 2008). Decreasing the expression of these tumor suppressors in aged lymphoid progenitors reverses the senescent phenotype, resulting in an increased susceptibility to transformation (Signer et al. 2008). However, if tumor-suppressive mechanisms are elevated in aging tissues, this will be at odds with the strong association between age and the increased incidence of most cancers and with the fact that interventions that delay aging (like caloric restriction) also delay cancer occurrence (Blagosklonny 2008).

13.4 Explaining the Link Between Age and Cancer: An Alternative View

As described above, the prevalent model that the accumulation of oncogenic mutations is the rate-limiting step in the evolution of aging-associated cancers may be an oversimplification. Consideration of the extensive changes that occur during the aging process and how these changes may affect *selection* for the fittest cells will likely provide more insight into why cancer incidences increase with age (referred to as adaptive oncogenesis (Fig. 13.1)). For example, we know that aging in humans is associated with HSC fitness declines, resulting in fitness reductions in both myeloid and lymphoid progenitors (Pang et al. 2011; Sharpless and DePinho 2007; Vaziri et al. 1994; Kollman et al. 2001; Kuranda et al. 2011). Thus, oncogenic mutations in aged HSCs in an aged landscape that are adaptive (thus improving fitness) in this context should be selected for. These same mutations would be much less likely to be adaptive in a young healthy HSC pool, as these cells are near a peak of fitness (DeGregori 2013). Considering CML, we know that manifestation of this disease is dependent on the acquisition of the BCR–ABL oncoprotein (Randolph 2005). Thus, due to the properties that BCR–ABL expression confers to cells, it is easy to surmise how the expression of this oncoprotein could improve the fitness of aged HSCs, conferring a selective advantage to oncogenically initiated cells in an aged background. Indeed, we have shown that cell signaling is impaired in lymphoid progenitors from old mice, promoting selection for BCR–ABL expression (which restores signaling) (Henry et al. 2010). The evolution of AML likely follows similar rules: selection for the more fit, oncogenically initiated HSC in a pool of less fit, and less competitive aged HSCs. For example, IDH mutations (common in AMLs in the elderly) have been shown to increase the self-renewal capacity of transformed cells (Prensner and Chinnaiyan 2011) and to confer a selective advantage to cells when growth-promoting cytokines are limiting, but *not* when sufficient cytokines are provided (Losman et al. 2013). Aged HSCs localize further from protective hypoxic niches (Kohler et al. 2009). This relocation leads to decreased expression of HIF-1α, which is known to regulate cellular metabolism, cytokine production, and self-renewal (Takubo et al. 2010; Eliasson et al. 2010). Notably, mutations in IDH have been shown to increase HIF-1α expression leading to increased self-renewal capacities (Prensner and Chinnaiyan 2011). We speculate that in aged HSC pools (but not young), reductions in self-renewal and cytokine-dependent signaling will lead to selection for oncogenic mutations in genes such as IDH that promote these properties. The HSC origins of CLL are just beginning to be unraveled (Puente et al. 2011; Quesada et al. 2011), and future experiments identifying how recurrent mutations associated with malignant CLL-associated HSC might be adaptive in aged HSC pools could shed light on the etiology of this disease.

We have argued that long-lived multicellular organisms have evolved stem cell populations with high fitness, not only as a means of efficiently maintaining a tissue, but also because high fitness in a cell population will oppose somatic evolution (DeGregori 2011; Marusyk and DeGregori 2008; Fleenor et al. 2010; Casas-Selves and DeGregori 2011). Like with animal populations well adapted to their environments, stabilizing selection should limit changes that alter fitness in a population of stem cells with high fitness. In fact, this link between fitness and the extent of positive selection has also been shown for viruses and other organisms independent of their biology and may be universal (Silander et al. 2007). Thus, highly effective competition in a young healthy stem cell population should serve to maintain the status quo, preventing somatic evolution. Effective competition will also facilitate the elimination of the occasional damaged cell from the stem cell pool, helping to maintain tissue fitness to maximize the reproductive success of the animal. But in stem cell pools damaged by aging, irradiation, or other insults, the fitness landscape will be dramatically altered (Fig. 13.1; "old progenitor pool"). The fitness of the stem cell pool will be reduced, promoting selection for mutations and epigenetic events that improve cellular fitness (Henry et al. 2011; Marusyk and DeGregori 2008). Still, just as the evolution of species is driven by mutation, selection, and drift, our model does not negate the importance of genetic and epigenetic diversity, which increases with age and following carcinogenic exposure, in providing fuel for selection (Garinis et al. 2008). Moreover, it is important to reemphasize that the decline of stem cell fitness with age or other insults will not just reflect cell-intrinsic damage, but will also involve microenvironmental alterations (indirectly reducing the fitness of stem cells) (Rozhok et al. 2014; Alizadeh and Majeti 2011; Bagby and Meyers 2009; Laconi et al. 2008).

One hypothetical mechanism behind such a context-dependent effect of oncogenic mutations might reside in part in the coupling between the cell cycle and differentiation machineries, demonstrated for hematopoietic and other stem cell systems and indicating increased differentiation in cells following increased cell cycling (Hindley and Philpott 2012; Steinman 2002). The general tendency of initiating oncogenic mutations to affect the cell cycle machinery by most often increasing cell division rates is generally explained as selectively advantageous for the incipient cell in a stem cell pool. However, in the long run increased cell cycling may be disadvantageous by purging such cells from the stem cell pool via increased rates of differentiation. Indeed, many oncogenic mutations engineered within uncompromised HSC pools in young mice lead to increased HSC cycling and decreased self-renewal (reviewed in DeGregori 2013). But in damaged and aged pools, where cell cycling is decreased, oncogenic events may potentially bring the incipient cells closer to the self-renewal *versus* differentiation optimum, giving them a selective advantage over the rest of the pool.

Conclusion

In this chapter we have described how the myriad of changes in HSCs and hematopoiesis in general associated with aging could contribute to increased blood cancers in old age. Similarly to other tissues, HSCs and the hematopoietic compartment demonstrate clear signs of senescence with age, generally revealed in reduced cell signaling, decreased lymphopoiesis, and gene deregulation due to accumulation of DNA damage and other progressive insults, like chronic inflammation. This overall decline, followed by reduced immune function and other systemic alterations, coincides with an increased risk of many age-related diseases, including cancer. Hematopoietic malignancies fit well within the typical pattern of age-associated cancers, with a steep rise in incidence that begins during the late 40s to early 50s, which coincides with the onset of the mostly post-reproductive period in humans and putatively reflects relaxing selection for maintenance of tissue integrity. Given overall cancer incidence statistics, this link between age and cancer is one of the central problems in health care, as a clear understanding of what drives cancer at advanced ages would help develop better strategies of prevention and treatment.

Both aging and cancer are highly complex, and their causes are certainly multifactorial, and models to explain their clear association will also need to be multifaceted. We have reviewed the various changes within the hematopoietic system with aging that are expected to contribute to increased cancer incidence in the elderly, including accumulating mutations and epigenetic changes, increased inflammation, decreased adaptive immunity, altered numbers and fitness of progenitors, and other alterations that could change the adaptive landscape. While oncogenic mutations are undoubtedly required for the evolution of cancer, the development of these malignancies most likely heavily depends on the forces that shape the population fate of these oncogenic events within the competitive pools of stem and progenitor cells. We discuss evidence suggesting that oncogenic mutations per se, while being the necessary priming events, may be insufficient to drive the development of cancerous tissue without more systemic alterations in the adaptive hematopoietic landscape, which in turn alters the selective pressures on different mutations and/or cell phenotypes within HSC pools. The obvious link between fitness and the strength of stabilizing or directional selection clearly demonstrated from natural organismal populations suggests that clonal progression of oncogenically initiated cells (including HSCs) will be shaped by similar forces in tissues. This alternative perspective needs to be considered when explaining links between aging and cancers, including hematopoietic malignancies. Overlooking the role of these forces is perhaps the biggest problem of our current understanding of cancer development and may potentially hinder development of more efficacious strategies to deal with aging-associated malignancies.

References

Agrawal N et al (2012) Comparative genomic analysis of esophageal adenocarcinoma and squamous cell carcinoma. Cancer Discov 2(10):899–905

Alizadeh AA, Majeti R (2011) Surprise! HSC are aberrant in chronic lymphocytic leukemia. Cancer Cell 20(2):135–136

Allman D, Miller JP (2005) The aging of early B-cell precursors. Immunol Rev 205:18–29

Almog N et al (2006) Prolonged dormancy of human liposarcoma is associated with impaired tumor angiogenesis. FASEB J 20(7):947–949

Andrew D, Aspinall R (2001) Il-7 and not stem cell factor reverses both the increase in apoptosis and the decline in thymopoiesis seen in aged mice. J Immunol 166(3):1524–1530

Anisimov VN (2009) Carcinogenesis and aging 20 years after: escaping horizon. Mech Ageing Dev 130(1–2):105–121

Armstrong SA et al (2004) FLT3 mutations in childhood acute lymphoblastic leukemia. Blood 103(9):3544–3546

Aspinall R, Andrew D (2000a) Thymic atrophy in the mouse is a soluble problem of the thymic environment. Vaccine 18(16):1629–1637

Aspinall R, Andrew D (2000b) Thymic involution in aging. J Clin Immunol 20(4):250–256

Bagby GC, Meyers G (2009) Myelodysplasia and acute leukemia as late complications of marrow failure: future prospects for leukemia prevention. Hematol Oncol Clin North Am 23(2):361–376

Baldridge MT et al (2010) Quiescent haematopoietic stem cells are activated by IFN-gamma in response to chronic infection. Nature 465(7299):793–797

Balducci L, Beghe C (2001) Cancer and age in the USA. Crit Rev Oncol Hematol 37(2):137–145

Beerman I et al (2010a) Functionally distinct hematopoietic stem cells modulate hematopoietic lineage potential during aging by a mechanism of clonal expansion. Proc Natl Acad Sci U S A 107(12):5465–5470

Beerman I et al (2010b) Stem cells and the aging hematopoietic system. Curr Opin Immunol 22(4):500–506

Bell JT, Spector TD (2011) A twin approach to unraveling epigenetics. Trends Genet 27(3):116–125

Benz CC, Yau C (2008) Ageing, oxidative stress and cancer: paradigms in parallax. Nat Rev Cancer 8(11):875–879

Biernaux C et al (1995) Detection of major bcr-abl gene expression at a very low level in blood cells of some healthy individuals. Blood 86(8):3118–3122

Blagosklonny MV (2002) Oncogenic resistance to growth-limiting conditions. Nat Rev Cancer 2(3):221–225

Blagosklonny MV (2008) Prevention of cancer by inhibiting aging. Cancer Biol Ther 7(10):1520–1524

Blagosklonny MV et al (2010) Impact papers on aging in 2009. Aging (Albany NY) 2(3):111–121

Bose S et al (1998) The presence of typical and atypical BCR-ABL fusion genes in leukocytes of normal individuals: biologic significance and implications for the assessment of minimal residual disease. Blood 92(9):3362–3367

Bowie MB et al (2006) Hematopoietic stem cells proliferate until after birth and show a reversible phase-specific engraftment defect. J Clin Invest 116(10):2808–2816

Cain D et al (2009) Effects of acute and chronic inflammation on B-cell development and differentiation. J Invest Dermatol 129(2):266–277

Campisi J (2005a) Suppressing cancer: the importance of being senescent. Science 309(5736): 886–887

Campisi J (2005b) Senescent cells, tumor suppression, and organismal aging: good citizens, bad neighbors. Cell 120(4):513–522

Carlson LM et al (2013) Low-dose aspirin delays an inflammatory tumor progression in vivo in a transgenic mouse model of neuroblastoma. Carcinogenesis 34(5):1081–1088

Casas-Selves M, DeGregori J (2011) How cancer shapes evolution, and how evolution shapes cancer. Evol Educ Outreach 4:624–634
Challen GA et al (2010) Distinct hematopoietic stem cell subtypes are differentially regulated by TGF-beta1. Cell Stem Cell 6(3):265–278
Chambers SM, Goodell MA (2007) Hematopoietic stem cell aging: wrinkles in stem cell potential. Stem Cell Rev 3(3):201–211
Chambers SM et al (2007) Aging hematopoietic stem cells decline in function and exhibit epigenetic dysregulation. PLoS Biol 5(8), e201
Chang S, Khoo C, DePinho RA (2001) Modeling chromosomal instability and epithelial carcinogenesis in the telomerase-deficient mouse. Semin Cancer Biol 11(3):227–239
Chen J, Astle CM, Harrison DE (1999) Development and aging of primitive hematopoietic stem cells in BALB/cBy mice. Exp Hematol 27(5):928–935
Chen J, Astle CM, Harrison DE (2000) Genetic regulation of primitive hematopoietic stem cell senescence. Exp Hematol 28(4):442–450
Chiorazzi N, Rai KR, Ferrarini M (2005) Chronic lymphocytic leukemia. N Engl J Med 352(8):804–815
Cho RH, Sieburg HB, Muller-Sieburg CE (2008) A new mechanism for the aging of hematopoietic stem cells: aging changes the clonal composition of the stem cell compartment but not individual stem cells. Blood 111(12):5553–5561
Crawford YG et al (2004) Histologically normal human mammary epithelia with silenced p16(INK4a) overexpress COX-2, promoting a premalignant program. Cancer Cell 5(3):263–273
Damm F et al (2014) Acquired initiating mutations in early hematopoietic cells of CLL patients. Cancer Discov 4(9):1088–1101
de Haan G, Van Zant G (1999) Dynamic changes in mouse hematopoietic stem cell numbers during aging. Blood 93(10):3294–3301
DeGregori J (2011) Evolved tumor suppression: why are we so good at not getting cancer? Cancer Res 71(11):3739–3744
DeGregori J (2013) Challenging the axiom: does the occurrence of oncogenic mutations truly limit cancer development with age? Oncogene 32(15):1869–1875
DePinho RA (2000) The age of cancer. Nature 408(6809):248–254
Dorshkind K, Swain S (2009) Age-associated declines in immune system development and function: causes, consequences, and reversal. Curr Opin Immunol 21(4):404–407
Dorshkind K, Montecino-Rodriguez E, Signer RA (2009) The ageing immune system: is it ever too old to become young again? Nat Rev Immunol 9(1):57–62
Dykstra B et al (2011) Clonal analysis reveals multiple functional defects of aged murine hematopoietic stem cells. J Exp Med 208(13):2691–2703
Edwards BK et al (2002) Annual report to the nation on the status of cancer, 1973–1999, featuring implications of age and aging on U.S. cancer burden. Cancer 94(10):2766–2792
Eliasson P et al (2010) Hypoxia mediates low cell-cycle activity and increases the proportion of long-term-reconstituting hematopoietic stem cells during in vitro culture. Exp Hematol 38(4):301–310. e2
Essers MA et al (2009) IFNalpha activates dormant haematopoietic stem cells in vivo. Nature 458(7240):904–908
Finkel T, Serrano M, Blasco MA (2007) The common biology of cancer and ageing. Nature 448(7155):767–774
Flach J et al (2014) Replication stress is a potent driver of functional decline in ageing haematopoietic stem cells. Nature 512(7513):198–202, Advance online publication
Fleenor CJ, Marusyk A, DeGregori J (2010) Ionizing radiation and hematopoietic malignancies: altering the adaptive landscape. Cell Cycle 9(15):3005–3011
Florian MC et al (2012) Cdc42 activity regulates hematopoietic stem cell aging and rejuvenation. Cell Stem Cell 10(5):520–530
Florian MC et al (2013) A canonical to non-canonical Wnt signalling switch in haematopoietic stem-cell ageing. Nature 503(7476):392–396

Franceschi C et al (2000) Inflamm-aging. An evolutionary perspective on immunosenescence. Ann N Y Acad Sci 908:244–254
Franceschi C et al (2005) Genes involved in immune response/inflammation, IGF1/insulin pathway and response to oxidative stress play a major role in the genetics of human longevity: the lesson of centenarians. Mech Ageing Dev 126(2):351–361
Frank SA (2010) Evolution in health and medicine Sackler colloquium: somatic evolutionary genomics: mutations during development cause highly variable genetic mosaicism with risk of cancer and neurodegeneration. Proc Natl Acad Sci U S A 107 Suppl 1:1725–1730
Frasca D et al (2012) A molecular mechanism for TNF-alpha-mediated downregulation of B cell responses. J Immunol 188(1):279–286
Gaidano G, Foa R, Dalla-Favera R (2012) Molecular pathogenesis of chronic lymphocytic leukemia. J Clin Invest 122(10):3432–3438
Garaycoechea JI et al (2012) Genotoxic consequences of endogenous aldehydes on mouse haematopoietic stem cell function. Nature 489(7417):571–575
Garcia-Manero G et al (2003) Chronic myelogenous leukemia: a review and update of therapeutic strategies. Cancer 98(3):437–457
Garinis GA et al (2008) DNA damage and ageing: new-age ideas for an age-old problem. Nat Cell Biol 10(11):1241–1247
Gatenby RA, Gillies RJ (2008) A microenvironmental model of carcinogenesis. Nat Rev Cancer 8(1):56–61
Goto M (2008) Inflammaging (inflammation + aging): a driving force for human aging based on an evolutionarily antagonistic pleiotropy theory? Biosci Trends 2(6):218–230
Greaves M, Maley CC (2012) Clonal evolution in cancer. Nature 481(7381):306–313
Greaves MF, Wiemels J (2003) Origins of chromosome translocations in childhood leukaemia. Nat Rev Cancer 3(9):639–649
Guerrettaz LM, Johnson SA, Cambier JC (2008) Acquired hematopoietic stem cell defects determine B-cell repertoire changes associated with aging. Proc Natl Acad Sci U S A 105(33):11898–11902
Hanahan D, Weinberg RA (2011) Hallmarks of cancer: the next generation. Cell 144(5):646–674
Hao Y et al (2011) A B-cell subset uniquely responsive to innate stimuli accumulates in aged mice. Blood 118(5):1294–1304
Henry CJ et al (2010) Declining lymphoid progenitor fitness promotes aging-associated leukemogenesis. Proc Natl Acad Sci U S A 107(50):21713–21718
Henry CJ, Marusyk A, DeGregori J (2011) Aging-associated changes in hematopoiesis and leukemogenesis: what's the connection? Aging (Albany NY) 3(6):643–656
Hindley C, Philpott A (2012) Co-ordination of cell cycle and differentiation in the developing nervous system. Biochem J 444(3):375–382
Hinkal G, Parikh N, Donehower LA (2009a) Timed somatic deletion of p53 in mice reveals age-associated differences in tumor progression. PLoS One 4(8), e6654
Hinkal GW et al (2009b) Altered senescence, apoptosis, and DNA damage response in a mutant p53 model of accelerated aging. Mech Ageing Dev 130(4):262–271
Hoeijmakers JH (2009) DNA damage, aging, and cancer. N Engl J Med 361(15):1475–1485
Horvath S (2013) DNA methylation age of human tissues and cell types. Genome Biol 14(10): R115
Jaskelioff M et al (2011) Telomerase reactivation reverses tissue degeneration in aged telomerase-deficient mice. Nature 469(7328):102–106
Karin M, Greten FR (2005) NF-kappaB: linking inflammation and immunity to cancer development and progression. Nat Rev Immunol 5(10):749–759
Karin M, Lawrence T, Nizet V (2006) Innate immunity gone awry: linking microbial infections to chronic inflammation and cancer. Cell 124(4):823–835
Kelly LM, Gilliland DG (2002) Genetics of myeloid leukemias. Annu Rev Genomics Hum Genet 3:179–198
Kennedy SR, Loeb LA, Herr AJ (2012) Somatic mutations in aging, cancer and neurodegeneration. Mech Ageing Dev 133(4):118–126

Kent DG et al (2009) Prospective isolation and molecular characterization of hematopoietic stem cells with durable self-renewal potential. Blood 113(25):6342–6350
Keren Z et al (2011a) Chronic B cell deficiency from birth prevents age-related alterations in the B lineage. J Immunol 187(5):2140–2147
Keren Z et al (2011b) B-cell depletion reactivates B lymphopoiesis in the BM and rejuvenates the B lineage in aging. Blood 117(11):3104–3112
Kikushige Y et al (2011) Self-renewing hematopoietic stem cell is the primary target in pathogenesis of human chronic lymphocytic leukemia. Cancer Cell 20(2):246–259
King KY, Goodell MA (2011) Inflammatory modulation of HSCs: viewing the HSC as a foundation for the immune response. Nat Rev Immunol 11(10):685–692
Kirkwood TB (2005) Understanding the odd science of aging. Cell 120(4):437–447
Kohler A et al (2009) Altered cellular dynamics and endosteal location of aged early hematopoietic progenitor cells revealed by time-lapse intravital imaging in long bones. Blood 114(2):290–298
Kollman C et al (2001) Donor characteristics as risk factors in recipients after transplantation of bone marrow from unrelated donors: the effect of donor age. Blood 98(7):2043–2051
Kristinsson SY et al (2011) Chronic immune stimulation might act as a trigger for the development of acute myeloid leukemia or myelodysplastic syndromes. J Clin Oncol 29(21): 2897–2903
Kuranda K et al (2011) Age-related changes in human hematopoietic stem/progenitor cells. Aging Cell 10(3):542–546
Laconi E, Doratiotto S, Vineis P (2008) The microenvironments of multistage carcinogenesis. Semin Cancer Biol 18:322–329
Lescale C et al (2010) Reduced EBF expression underlies loss of B cell potential of hematopoietic progenitors with age. Aging Cell 9(3):410–419
Liang Y, Van Zant G, Szilvassy SJ (2005) Effects of aging on the homing and engraftment of murine hematopoietic stem and progenitor cells. Blood 106(4):1479–1487
Linton PJ, Dorshkind K (2004) Age-related changes in lymphocyte development and function. Nat Immunol 5(2):133–139
Losman JA et al (2013) (R)-2-hydroxyglutarate is sufficient to promote leukemogenesis and its effects are reversible. Science 339(6127):1621–1625
Lowenberg B, Downing JR, Burnett A (1999) Acute myeloid leukemia. N Engl J Med 341(14):1051–1062
Lynch M (1988) The rate of polygenic mutation. Genet Res 51(2):137–148
Lynch M (2010a) Evolution of the mutation rate. Trends Genet 26(8):345–352
Lynch M (2010b) Rate, molecular spectrum, and consequences of human mutation. Proc Natl Acad Sci U S A 107(3):961–968
Malfertheiner P, Schutte K (2006) Smoking–a trigger for chronic inflammation and cancer development in the pancreas. Am J Gastroenterol 101(1):160–162
Marusyk A, DeGregori J (2008) Declining cellular fitness with age promotes cancer initiation by selecting for adaptive oncogenic mutations. Biochim Biophys Acta 1785(1):1–11
Matioli GT (2002) BCR-ABL insufficiency for the transformation of human stem cells into CML. Med Hypotheses 59(5):588–589
Medawar P (1952) An unsolved problem of biology. H.K. Lewis, London
Miller JP, Allman D (2003) The decline in B lymphopoiesis in aged mice reflects loss of very early B-lineage precursors. J Immunol 171(5):2326–2330
Miller JP, Allman D (2005) Linking age-related defects in B lymphopoiesis to the aging of hematopoietic stem cells. Semin Immunol 17(5):321–329
Milne CD, Paige CJ (2006) IL-7: a key regulator of B lymphopoiesis. Semin Immunol 18(1):20–30
Mirantes C, Passegue E, Pietras EM (2014) Pro-inflammatory cytokines: emerging players regulating HSC function in normal and diseased hematopoiesis. Exp Cell Res 329(2): 248–254
Morrison SJ et al (1996) The aging of hematopoietic stem cells. Nat Med 2(9):1011–1016

Muller-Sieburg CE et al (2004) Myeloid-biased hematopoietic stem cells have extensive self-renewal capacity but generate diminished lymphoid progeny with impaired IL-7 responsiveness. Blood 103(11):4111–4118

Mullighan CG et al (2008) BCR-ABL1 lymphoblastic leukaemia is characterized by the deletion of Ikaros. Nature 453(7191):110–114

Mutter GL et al (2001) Molecular identification of latent precancers in histologically normal endometrium. Cancer Res 61(11):4311–4314

Naumov GN, Akslen LA, Folkman J (2006a) Role of angiogenesis in human tumor dormancy: animal models of the angiogenic switch. Cell Cycle 5(16):1779–1787

Naumov GN et al (2006b) A model of human tumor dormancy: an angiogenic switch from the nonangiogenic phenotype. J Natl Cancer Inst 98(5):316–325

Noda S, Ichikawa H, Miyoshi H (2009) Hematopoietic stem cell aging is associated with functional decline and delayed cell cycle progression. Biochem Biophys Res Commun 383(2):210–215

Nowell PC (1976) The clonal evolution of tumor cell populations. Science 194(4260):23–28

Omatsu Y et al (2010) The essential functions of adipo-osteogenic progenitors as the hematopoietic stem and progenitor cell niche. Immunity 33(3):387–399

Pang WW et al (2011) Human bone marrow hematopoietic stem cells are increased in frequency and myeloid-biased with age. Proc Natl Acad Sci U S A 108(50):20012–20017

Peto R et al (1975) Cancer and ageing in mice and men. Br J Cancer 32(4):411–426

Pietras EM, Warr MR, Passegue E (2011) Cell cycle regulation in hematopoietic stem cells. J Cell Biol 195(5):709–720

Pietras EM et al (2014) Re-entry into quiescence protects hematopoietic stem cells from the killing effect of chronic exposure to type I interferons. J Exp Med 211(2):245–262

Prensner JR, Chinnaiyan AM (2011) Metabolism unhinged: IDH mutations in cancer. Nat Med 17(3):291–293

Prieyl JA, LeBien TW (1996) Interleukin 7 independent development of human B cells. Proc Natl Acad Sci U S A 93(19):10348–10353

Pronk CJ et al (2011) Tumor necrosis factor restricts hematopoietic stem cell activity in mice: involvement of two distinct receptors. J Exp Med 208(8):1563–1570

Puente XS et al (2011) Whole-genome sequencing identifies recurrent mutations in chronic lymphocytic leukaemia. Nature 475(7354):101–105

Quesada V et al (2011) Exome sequencing identifies recurrent mutations of the splicing factor SF3B1 gene in chronic lymphocytic leukemia. Nat Genet 44(1):47–52

Randolph TR (2005) Chronic myelocytic leukemia–part I: history, clinical presentation, and molecular biology. Clin Lab Sci 18(1):38–48

Ratliff M et al (2010) In senescence, age-associated B cells secrete TNFalpha and inhibit survival of B-cell precursors. Aging Cell 12(2):303–311

Rawlings DJ et al (1995) Long-term culture system for selective growth of human B-cell progenitors. Proc Natl Acad Sci U S A 92(5):1570–1574

Renehan AG, Roberts DL, Dive C (2008) Obesity and cancer: pathophysiological and biological mechanisms. Arch Physiol Biochem 114(1):71–83

Roberts DL, Dive C, Renehan AG (2010) Biological mechanisms linking obesity and cancer risk: new perspectives. Annu Rev Med 61:301–316

Rossi DJ et al (2005) Cell intrinsic alterations underlie hematopoietic stem cell aging. Proc Natl Acad Sci U S A 102(26):9194–9199

Rossi DJ et al (2007a) Hematopoietic stem cell quiescence attenuates DNA damage response and permits DNA damage accumulation during aging. Cell Cycle 6(19):2371–2376

Rossi DJ et al (2007b) Deficiencies in DNA damage repair limit the function of haematopoietic stem cells with age. Nature 447(7145):725–729

Rozhok AI, Salstrom JL, DeGregori J (2014) Stochastic modeling indicates that aging and somatic evolution in the hematopoietic system are driven by non-cell-autonomous processes. Aging (Albany NY) 6(12):1033–1048

Rubtsov AV et al (2011) Toll-like receptor 7 (TLR7)-driven accumulation of a novel CD11c(+) B-cell population is important for the development of autoimmunity. Blood 118(5): 1305–1315
Rubtsov AV et al (2013) TLR7 drives accumulation of ABCs and autoantibody production in autoimmune-prone mice. Immunol Res 55(1–3):210–216
Rufer N et al (1999) Telomere fluorescence measurements in granulocytes and T lymphocyte subsets point to a high turnover of hematopoietic stem cells and memory T cells in early childhood. J Exp Med 190(2):157–167
Sahin E, Depinho RA (2010) Linking functional decline of telomeres, mitochondria and stem cells during ageing. Nature 464(7288):520–528
Salvioli S et al (2009) Why do centenarians escape or postpone cancer? The role of IGF-1, inflammation and p53. Cancer Immunol Immunother 58(12):1909–1917
Sanders MA, Valk PJ (2013) The evolving molecular genetic landscape in acute myeloid leukaemia. Curr Opin Hematol 20(2):79–85
Sawyers CL (1999) Chronic myeloid leukemia. N Engl J Med 340(17):1330–1340
Serrano M, Blasco MA (2007) Cancer and ageing: convergent and divergent mechanisms. Nat Rev Mol Cell Biol 8(9):715–722
Sharpless NE, DePinho RA (2002) p53: good cop/bad cop. Cell 110(1):9–12
Sharpless NE, DePinho RA (2004) Telomeres, stem cells, senescence, and cancer. J Clin Invest 113(2):160–168
Sharpless NE, DePinho RA (2007) How stem cells age and why this makes us grow old. Nat Rev Mol Cell Biol 8(9):703–713
Shaw AC et al (2010) Aging of the innate immune system. Curr Opin Immunol 22(4):507–513
Signer RA et al (2007) Age-related defects in B lymphopoiesis underlie the myeloid dominance of adult leukemia. Blood 110(6):1831–1839
Signer RA et al (2008) Aging and cancer resistance in lymphoid progenitors are linked processes conferred by p16Ink4a and Arf. Genes Dev 22(22):3115–3120
Silander OK, Tenaillon O, Chao L (2007) Understanding the evolutionary fate of finite populations: the dynamics of mutational effects. PLoS Biol 5(4), e94
Simsek T et al (2010) The distinct metabolic profile of hematopoietic stem cells reflects their location in a hypoxic niche. Cell Stem Cell 7(3):380–390
Small D (2006) FLT3 mutations: biology and treatment. Hematology Am Soc Hematol Educ Program 178–184
Song Z et al (2010) Alterations of the systemic environment are the primary cause of impaired B and T lymphopoiesis in telomere-dysfunctional mice. Blood 115(8):1481–1489
Steinman RA (2002) Cell cycle regulators and hematopoiesis. Oncogene 21(21):3403–3413
Stephan RP, Reilly CR, Witte PL (1998) Impaired ability of bone marrow stromal cells to support B-lymphopoiesis with age. Blood 91(1):75–88
Sudo K et al (2000) Age-associated characteristics of murine hematopoietic stem cells. J Exp Med 192(9):1273–1280
Sun D et al (2014) Epigenomic profiling of young and aged HSCs reveals concerted changes during aging that reinforce self-renewal. Cell Stem Cell 14(5):673–688
Takubo K et al (2010) Regulation of the HIF-1alpha level is essential for hematopoietic stem cells. Cell Stem Cell 7(3):391–402
Thakkar A et al (2013) The molecular mechanism of action of aspirin, curcumin and sulforaphane combinations in the chemoprevention of pancreatic cancer. Oncol Rep 29(4): 1671–1677
Ueda Y et al (2004) Inflammation controls B lymphopoiesis by regulating chemokine CXCL12 expression. J Exp Med 199(1):47–58
Ueda Y, Kondo M, Kelsoe G (2005) Inflammation and the reciprocal production of granulocytes and lymphocytes in bone marrow. J Exp Med 201(11):1771–1780
Van Zant G et al (1990) Genotype-restricted growth and aging patterns in hematopoietic stem cell populations of allophenic mice. J Exp Med 171(5):1547–1565

Vaziri H et al (1994) Evidence for a mitotic clock in human hematopoietic stem cells: loss of telomeric DNA with age. Proc Natl Acad Sci U S A 91(21):9857–9860
Vickers M (1996) Estimation of the number of mutations necessary to cause chronic myeloid leukaemia from epidemiological data. Br J Haematol 94(1):1–4
Vijg J et al (2005) Aging and genome maintenance. Ann N Y Acad Sci 1055:35–47
Vogelstein B, Kinzler KW (2004) Cancer genes and the pathways they control. Nat Med 10(8):789–799
Wahlestedt M et al (2013) An epigenetic component of hematopoietic stem cell aging amenable to reprogramming into a young state. Blood 121(21):4257–4264
Warr MR et al (2013) FOXO3A directs a protective autophagy program in haematopoietic stem cells. Nature 494(7437):323–327
Warren LA, Rossi DJ (2009) Stem cells and aging in the hematopoietic system. Mech Ageing Dev 130(1–2):46–53
Weinberg R (2007) The biology of cancer (Chapter 11). Garland Science, New York
Weinberger B et al (2008) Biology of immune responses to vaccines in elderly persons. Clin Infect Dis 46(7):1078–1084
Weiskopf D, Weinberger B, Grubeck-Loebenstein B (2009) The aging of the immune system. Transpl Int 22(11):1041–1050
Weksler ME, Szabo P (2000) The effect of age on the B-cell repertoire. J Clin Immunol 20(4):240–249
Weksler ME, Goodhardt M, Szabo P (2002) The effect of age on B cell development and humoral immunity. Springer Semin Immunopathol 24(1):35–52
Weng NP, Hathcock KS, Hodes RJ (1998) Regulation of telomere length and telomerase in T and B cells: a mechanism for maintaining replicative potential. Immunity 9(2):151–157
Williams GC (1957) Pleiotropy, natural selection, and the evolution of senescence. Evolution 11:398–411
Wilson A et al (2008) Hematopoietic stem cells reversibly switch from dormancy to self-renewal during homeostasis and repair. Cell 135(6):1118–1129
Wynford-Thomas D et al (1995) Does telomere shortening drive selection for p53 mutation in human cancer? Mol Carcinog 12(3):119–123
Xing Z et al (2006) Increased hematopoietic stem cell mobilization in aged mice. Blood 108(7):2190–2197
Yamazaki S et al (2009) TGF-beta as a candidate bone marrow niche signal to induce hematopoietic stem cell hibernation. Blood 113(6):1250–1256
Yin T, Li L (2010) Fountain of Youth: aged blood-forming stem cells could be rejuvenated by young microenvironment. Cell Res 20(5):504–505

Mesenchymal Stem Cells in Wound Repair, Tissue Homeostasis, and Aging

14

Dongsheng Jiang and Karin Scharffetter-Kochanek

Abstract

Wound healing and scar remodeling are complex, multicellular processes that involve coordinated regulation of many cell types and various cytokines. The repair capacity gradually decreases with aging, constituting a severe health problem that frequently affects aged individuals. The decrease in cell number and function of mesenchymal stem cells (MSCs) is most likely responsible for the decline of tissue regeneration and wound healing. MSCs are endowed with the unique capacity for self-renewal and differentiation into histogenetically distinct cell types required for tissue repair. In addition, by their potential to sense danger signals at the wound site, MSCs are able to adaptively respond to infections and unrestrained macrophage activation and thus control inflammation. These properties make them promising for the treatment of chronic nonhealing wounds in the elderly or even for rejuvenation of the skin and other organs. This review will focus on the physiological and therapeutic roles of MSCs in cutaneous wound healing in the context of age-related chronic wounds and will help to decipher how the aging process affects the overall wound repair capacity of MSCs.

D. Jiang • K. Scharffetter-Kochanek (✉)
Department of Dermatology and Allergic Diseases, University of Ulm,
Albert-Einstein Allee 23, Ulm 89091, Germany
e-mail: karin.scharffetter-kochanek@uniklinik-ulm.de

H. Geiger et al. (eds.), *Stem Cell Aging: Mechanisms, Consequences, Rejuvenation*,
DOI 10.1007/978-3-7091-1232-8_14

Abbreviations

AT-MSCs	Adipose tissue-derived MSCs
MSCs	Mesenchymal stem cells
BM-MSCs	Bone-marrow-derived MSCs
EPC	Endothelial progenitor cells
ESC	Epidermal stem cells
MHC	Major histocompatibility complex

14.1 Introduction

Normal tissue repair follows a sequence of events involving clotting, inflammation, granulation tissue formation with wound contraction and angiogenesis, scar remodeling, and reepithelialization (Fig. 14.1) (Wlaschek and Scharffetter-Kochanek 2005; Singer and Caplan 2011; Gurtner et al. 2008). However, chronic skin wounds fail to progress through the normal pattern of wound repair, but instead remain in a chronic inflammatory state with little signs of healing. The prevalence of chronic wounds is closely related to the aging population. In industrialized countries with a constantly increasing aged population, difficult-to-treat chronic wounds represent a major public health problem with high socioeconomic impact. Worldwide 200 million difficult-to-treat chronic wounds severely affect the quality of life of elderly patients. Thus, medically effective and cost-efficient treatment modalities are urgently needed, and in this regard topical application of mesenchymal stem cells (MSCs) may qualify as promising strategy to enhance healing of chronic wounds.

Aging occurs on a molecular, cellular, tissue, organ, and organismal level and involves a variety of distinct cell types, among them MSCs. Mitochondrial dysfunction with increased levels of reactive oxygen species (ROS) contributes to DNA damage accumulation, telomere dysfunction, and perturbed proteostasis observed in aged skin with the activation of distinct cellular check point responses such as apoptosis, replicative senescence, differentiation, repressed self-renewal, and the adoption of a senescence-associated secretory phenotype (SASP) eventually leading to the overall decline of tissue and organ homeostasis, repair, and regeneration. Currently, the contribution of intrinsic MSC aging to skin aging and impaired tissue repair is not well understood, and most of the studies on MSC aging are relying on in vitro evidence. This is mainly due to the lack of a MSC-specific marker, which would allow tracing individual MSCs in situ in skin and other organs. Nevertheless, the rare and specialized MSCs are supposed to be essential for the maintenance of cutaneous homeostasis and integrity (wound repair) throughout life. With advancing age, the number of MSCs and possibly their function decline with the observed overall compromised potential for regeneration and repair of damaged cells and tissues.

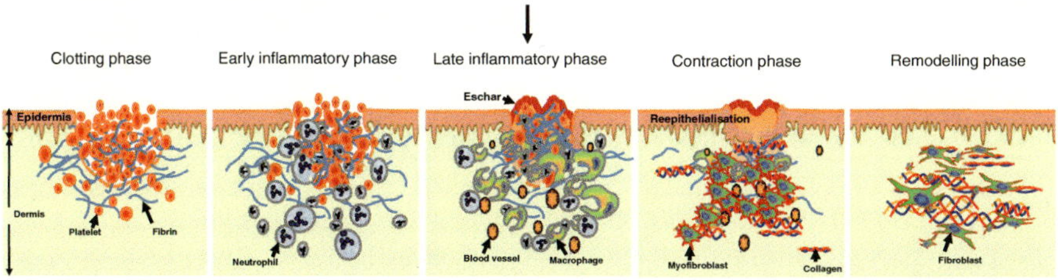

Fig. 14.1 Physiological wound healing. Wound healing involves a series of partly overlapping phases. Clot formation ensures a first sealing of the wound to prevent microbial invasion. Within a few hours after injury, first polymorphonuclear neutrophils (PMN) and later macrophages invade the wound site in the inflammatory phase of wound healing to further combat contaminating microorganisms and to phagocytose cellular debris. The late inflammatory phase with macrophages as the prevailing inflammatory cells is followed by granulation tissue formation with wound contraction and angiogenesis and the matrix remodeling phase. The arrow indicates that in chronic wounds, the inflammatory phase of wound healing is prolonged and often persists, leading to chronic wounds which fail to progress through the pattern of normal healing

In this chapter, we present an overview of the physiological and therapeutic roles of MSCs on cutaneous wound repair and attempt to highlight the impact of age-induced changes in MSCs affecting their capacity for wound healing and scar remodeling. We conclude with a perspective for clinical applications with MSC-based treatment for age-associated wound-healing defects.

14.2 Mesenchymal Stem Cells in Skin Wound Repair

The focus of dermatological stem cell research is predominantly on the epidermis and the hair follicle. In contrast, the characterization of stem cells in the mesenchymal compartments of the skin has largely escaped the attention of the dermatologic community, though the dermis represents a larger reservoir for adult stem cells than the epidermis and the hair follicle together (Sellheyer and Krahl 2010). Currently, no in situ aging data on MSC populations are available. Nevertheless, MSCs have already shown great therapeutic promise for the treatment of age-related chronic wounds in the recent decade. Therefore, this review will preferentially focus on functions and cellular mechanisms of MSCs in the repair of chronic wounds.

14.2.1 Physiological Localization of Dermal MSCs in the Skin and Their Function During Wound Healing

In 2001, MSCs residing within the dermis were first isolated. They are endowed with the capacity to differentiate into adipocytes, smooth muscle cells, osteocytes, chondrocytes, neurons and glia, and hematopoietic cells of myeloid and erythroid lineages (Toma et al. 2001). The perifollicular connective tissue sheath and the dermal papilla represent the likely anatomical niche for these multipotent dermal cells, a finding proven in animal models (Jahoda et al. 2003; Hoogduijn et al. 2006). The primary adult hair follicle dermal papilla and dermal sheath MSCs revealed a fibroblastic morphology and were $CD44^+$, $CD73^+$, and $CD90^+$. These MSC subpopulations were capable of dye exclusion suggesting a multidrug resistance property and had the capacity to differentiate into various mesenchymal lineages, such as osteoblasts, adipocytes, chondrocytes, myocytes, and expressed neuroprogenitor cell markers (Jahoda et al. 2003; Hoogduijn et al. 2006). Subsequently, MSC-like cells have been isolated from human dermis by other investigators (Bartsch et al. 2005; Lorenz et al. 2008; Vaculik et al. 2012); however, in these studies the detailed in situ localization of the reported MSC subpopulations was not illustrated.

During anagen hair development, connective tissue sheath and dermal papilla are not only remodeled but also impressively expanded (Tobin et al. 2003). This strongly suggests the presence of resident mesenchymal precursor cell populations. During each anagen hair development, angiogenesis occurs, which likely arises from mesenchymal stem cells residing in the connective tissue sheath (Jahoda et al. 2003). Moreover, the follicular connective tissue sheath serves as a crucial cell pool for the formation of granulation tissue, which is formed by fibroblasts, endothelial cells, mast cells, and macrophages, after substantial skin trauma (Jahoda and Reynolds 2001).

Besides the skin-resident dermal MSCs, the endogenous MSCs originated from other organs are reported to participate in wound healing of the skin. In fact, early studies on adult stem cells in the dermal compartment showed that progenitors from bone marrow are chemokine-dependently recruited to the site of skin injury and participate in skin restoration. These cells express mesenchymal markers (collagen I, fibronectin) and are mainly understood as progenitors of fibroblasts; therefore, they are designated "fibrocytes" (Bucala et al. 1994). Fibrocytes constitutively produce ECM components as well as ECM-modifying enzymes and can further differentiate into myofibroblasts playing a crucial role in wound healing (Abe et al. 2001; Quan et al. 2004). Another study showed that upon wounding, stimulated bone-marrow-derived stem cells traffic through both wounded and non-wounded skin and incorporate into skin compartments (Badiavas et al. 2003). This concept was strongly supported by a study of Opalenik and Davidson (2005). Using a luciferase and b-galactosidase reporter under the control of collagen I a2 chain, the authors were able to show that the transplanted donor bone-marrow-derived cells occurred in the granulation tissue of recipient mice, differentiated to fibroblasts, and contributed significantly to total collagen I production during later stages of repair (Opalenik and Davidson 2005). Very recently, with a similar experimental strategy, these bone-marrow-derived cells have convincingly been proven to be bone-marrow-derived endogenous MSCs (Seppanen et al. 2013). Notably, ceramide-1-phosphate (C1P) which is upregulated during tissue damage was identified as an important chemoattractant for the recruitment of bone-marrow-derived MSCs to the site of injury (Kim et al. 2013). All these studies suggest that endogenous MSCs are native constituents of the wound bed and play a crucial role in cutaneous wound healing.

14.2.2 Therapeutic Function and Cellular Mechanisms of Transplanted MSCs in Wound-Healing Repair and Skin Aging

Despite the lack of knowledge on physiological functions of dermal MSCs in the skin in situ, the therapeutic benefits of transplanted MSCs in accelerating wound healing have been extensively documented in humans and various animal models regardless of the source of MSCs (skin, bone marrow, adipose tissue, skeletal muscle, umbilical cord blood, gingival, dental pulp, Wharton's jelly, synovial, amniotic fluid) and donor (autograft, allograft, xenograft) (De Bari et al. 2001; Falanga 2012; Falanga et al. 2007; Heo et al. 2011; In't Anker et al. 2003; Jiang et al. 2013; Miura et al. 2005; Qi et al. 2014; Sasaki et al. 2008; Stoff et al. 2009; Tark et al. 2010; Wan et al. 2013; Williams et al. 1999; Wu et al. 2007; Xie et al. 2013; Yew et al. 2011; Zhang et al. 2010; Ksander et al. 2014). MSCs apparently also enhance the quality of adult wound healing with reduced scar formation and regeneration of skin appendages (Qi et al. 2014; Stoff et al. 2009; Wu et al. 2014; Jackson et al. 2012; Liu et al. 2014).

By the release of a variety of bioactive trophic molecules and immunosuppressive functions, MSCs distinctly contribute to establish a regenerative microenvironment, which stimulates the proliferation of lineage-specific cells such as fibroblasts, epidermal cells, and endothelial cells to repair and restore damaged tissue. The MSC-imprinted regenerative microenvironment also stimulates and stabilizes angiogenesis and vessel formation and dampens the unrestrained activation of proinflammatory M1 macrophages eventually leading to scar-reduced or even scarless healing in adults. Here we summarize the major mechanisms that MSCs participate in improving wound healing (Fig. 14.2).

14.2.3 Paracrine Trophic Effects

Topical application and local or intravenous injection of MSCs for the treatment of acute and chronic wounds result in accelerated wound closure with increased epithelialization, granulation tissue formation, and angiogenesis (Jiang et al. 2013; Falanga et al. 2007; Heo et al. 2011; Wu et al. 2007). Although there is evidence for MSC differentiation in the wound, most of the therapeutic effects are likely due to MSCs releasing soluble factors that regulate local cellular responses to cutaneous injury (Hocking and Gibran 2010). MSCs are described as a factory of cytokines, chemokines, and ECM molecules (Caplan and Dennis 2006) or even a "drugstore" during injury (Caplan and Correa 2011; Murphy et al. 2013).

Chen et al. (2008) found that compared to dermal fibroblasts, BM-MSCs released vascular endothelial growth factor (VEGF), insulin-like growth factor-1 (IGF-1), epidermal growth factor (EGF), keratinocyte growth factor (KGF), angiopoietin-1, stromal cell-derived factor-1 (SDF-1), macrophage inflammatory protein-1 (MCP-1), and erythropoietin, which are essential for physiological wound healing. Moreover, in contrast to noneffective fibroblast-conditioned medium,

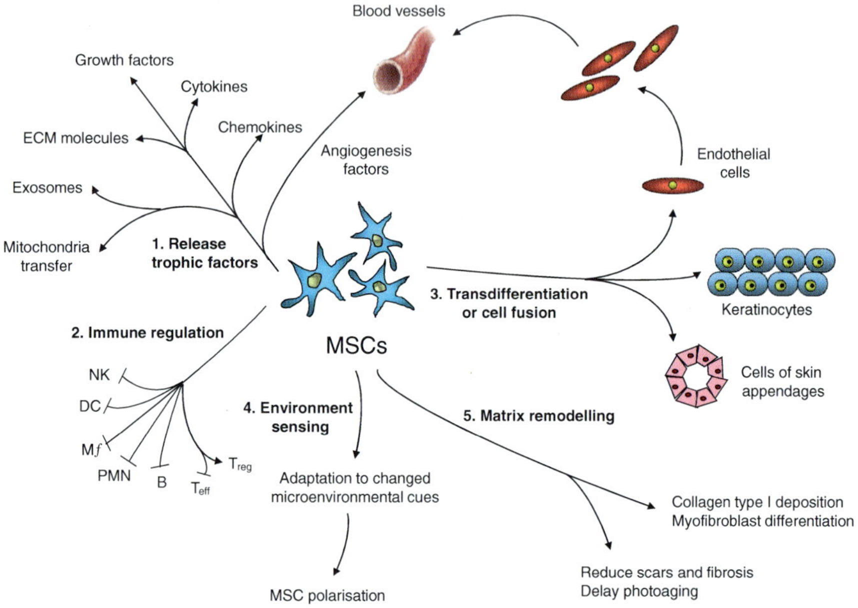

Fig. 14.2 Mechanisms of MSCs accelerated wound healing. MSCs participate in wound repair via several beneficial mechanisms, including (*1*) release of trophic factors, (*2*) suppression of overwhelming inflammation, (*3*) transdifferentiation into or fusion with specialized cells, (*4*) sensing of the microenvironment, and (*5*) remodeling extracellular matrices. *ECM* extracellular matrix, *Mf* macrophages, *PMN* polymorphonuclear neutrophil, *DC* dendritic cells, *NK* natural killer cells, *Teff* effector T cells, *Treg* regulator T cells

BM-MSC-conditioned medium significantly enhanced migration and proliferation of macrophages, keratinocytes, and endothelial cells in vitro and in vivo and enhanced wound healing in a mouse model of excisional wound (Chen et al. 2008). Injected MSCs accelerate wound healing via the production and release of transforming growth factor beta (TGF-β and VEGF which induce myofibroblast differentiation and angiogenesis (Jiang et al. 2013; Nemeth et al. 2010; Popova et al. 2010; Salvolini et al. 2010a). Coculture experiments with dermal MSCs and endothelia cells suggest that MSCs have the potential to activate endothelial cells. This activation is mediated via the release of VEGF and NO with subsequent alteration of the endothelial barrier overall facilitating the transmigration of MSCs across the endothelium to the wound site (Salvolini et al. 2010b). Injection of TNF-α-activated human AT-MSCs accelerate wound closure, angiogenesis, proliferation, and infiltration of immune cells into the cutaneous wound through paracrine mechanisms involving IL-6 and IL-8 in a rat excisional wound model (Heo et al. 2011). Conditioned medium from AT-MSCs stimulates collagen synthesis and promotes proliferation and migration of human keratinocytes and fibroblasts during the wound-healing process (Lee et al. 2012; Rose et al. 2008). Interestingly, the synthesis of therapeutically relevant trophic factors such as IL-6, SDF-1, HGF, and VEGF was induced in MSCs following activation of the toll-like receptor 3 (TLR3) (Mastri et al. 2012). The ligand of

TLR3 polyinosinic:polycytidylic acid (poly I:C), in fact, stimulates wound healing (Lin et al. 2012). These data in conjunction with the assumption that danger signals like virus-derived double-strand RNA (a TLR3 ligand) of virally induced tissue damage occur in wounds would support the notion that TRL3 activation on MSCs enhances tissue repair.

Dermal fibroblasts and myofibroblasts are mainly responsible for wound contraction, extracellular matrix deposition, and scar tissue remodeling during tissue repair (Smith et al. 2010). In response to cocultured MSCs, dermal fibroblasts upregulate integrin a_7 and downregulate ICAM1, VCAM1, and MMP11 expression suggesting that MSCs may provide important early signals to enhance dermal fibroblast responses to enhance tissue repair after cutaneous injuries (Smith et al. 2010). In this context, MSCs enhance the directed migration of dermal fibroblasts to a gradient of soluble chemotactic agents released by MSCs (Rodriguez-Menocal et al. 2012). Notably, the migration of fibroblasts was found to be dramatically reduced in "wound" scratch assays on cocultures of unperturbed MSCs with fibroblasts isolated from tissue of chronic wound patients and lowest when both MSCs and fibroblasts stem from biopsies of chronic wound patients (Smith et al. 2010). These data suggest that both fibroblasts and MSCs isolated from chronic wounds – most likely due to the hostile microenvironment of chronic wounds (Wlaschek and Scharffetter-Kochanek 2005) – reveal distinctly impaired functions critical for proper tissue repair. In addition to the paracrine-mediated enhanced fibroblast migration, MSCs from human exfoliated deciduous teeth were found to increase proliferation and collagen synthesis of dermal fibroblasts (Ueda and Nishino 2010).

Besides the conventional secretion of soluble proteins, MSC-derived exosomes have recently been identified to effectively serve as paracrine mediators in accelerating tissue repair and regeneration. This conclusion was based on the interesting observation that the active components consisting of 50–200 nm-sized complexes which were isolated from conditioned medium of human embryonic stem cells derived from MSCs or fetal MSCs have the potential to markedly reduce the infarct size in a mouse model of myocardial ischemia/reperfusion (MI/R) injury, if intravenously infused 5 min before reperfusion (Lai et al. 2010a, b). The MSC-secreted complexes were identified as exosomes, organized as bi-lipid membrane vesicles of endosomal origin containing proteins and RNA. Exosomes are secreted through fusion of multivesicular bodies with MSC cell membrane. The exosomes' cargo is efficiently protected from degradation by proteases and RNase and uptaken by recipient cells via endocytosis or membrane fusion (Lai et al. 2011).

Impaired mitochondrial function plays an essential role in aging, reduced tissue homeostasis, and repair. In 2006, it was reported that mitochondria and mtDNA can be actively transferred from human BM-MSCs to mammalian cells with nonfunctional mitochondria without cell fusion, and this transfer, in fact, rescues the aerobic respiration (Spees et al. 2006). This was confirmed in vitro by two subsequent studies showing that mitochondria were transferred from human BM-MSCs to cocultured rat cardiomyocytes (Plotnikov et al. 2008) or human osteosarcoma cells (Cho et al. 2012) without functional mitochondria. The evidence that the transfer of intact mitochondria can contribute to tissue repair in vivo was provided very recently. Islam and colleagues (2012) reported that human and mouse BM-MSCs attached to LPS-injured mouse alveolar epithelial cells transfer intact mitochondria by forming connexin-43-containing gap junctions, nanotubes (e.g., filopodia), and microvesicles (e.g., exosomes) and, thus, contribute to the repair of an injured lung in vivo (Islam et al. 2012; Prockop 2012). Whether the transfer of mitochondria from MSCs to injured or senescent cells occurs during wound healing of the skin is currently unclear and needs further investigation.

14.2.4 Angiogenesis

Angiogenesis is an essential process in tissue repair, regeneration, and reconstruction. Signals that control endothelial cell proliferation and differentiation have been studied on MSCs themselves or on their target tissues and organs. MSC differentiation into endothelial cells can be practically accomplished by exposure to angiogenic factors in vitro (Oswald et al. 2004). Exposure to VEGF evokes the differentiation of MSCs into endothelial-like cells, expressing VEGFR-1, VEGFR-2, and von Willebrand factor (vWF) as well as the formation of the characteristic capillary-like structures (Oswald et al. 2004). Forced expression of secreted Frizzled-related protein-1 enhanced MSCs surrounding neovessel in vivo and strongly increased PDGF-BB expression in MSCs and enhanced b-catenin-dependent cell-cell contacts between MSCs themselves and epithelial cells or smooth muscle cells in vitro (Dufourcq et al. 2008). After injury, the level of TGF-β_1 in the wound dramatically increases and plays a pivotal role in the regulation of angiogenesis. TGF-β_1 stimulated the synthesis of VEGF and the phosphorylation of Akt and ERK1/2 in MSCs (Wang et al. 2008). On the other hand, the paracrine factors, including TGF-β_1, VEGF, bFGF, angiogenin, procathepsin B, IL-11, and BMP2, all secreted by MSCs, affect endothelial cell migration, ECM invasion, proliferation, and survival in vitro (Potapova et al. 2007) and induce angiogenesis and enhance wound healing in vivo (Wu et al. 2007).

Recently, MSCs have been proposed to arise from perivascular stromal cells including pericytes, which are localized on the abluminal side of blood vessels, immediately opposed to endothelial cells (Crisan et al. 2008; Traktuev et al. 2008). BM-MSCs express pericyte markers such as α-SMA, desmin, calponin, smoothelin, and NG2 (Au et al. 2008) and can home to skin wound sites where they may exert pericyte functions to support and stabilize new vessel formation (Ozerdem et al. 2005). Au et al. (2008) found that BM-MSCs elongated and co-aligned in close proximity to vascular tubelike structures. MSC migration towards vascular cells depends on PDGF-BB, a growth factor that also enhances pericyte migration. In vivo data suggest that pericytes are attracted to newly formed vessels (Au et al. 2008). The preferred perivascular localization of MSCs may be related to their origin and interaction with vascular structures after homing. Pericyte coverage is important for the stability and remodeling of blood vessels in wound healing (Bergers and Song 2005).

14.2.5 Immune Regulation in the Inflammatory Phase of Tissue Repair

MSCs are low immunogenic cells. Despite constitutive expression of MHC-I and upregulation of MHC-II induced by IFN-γ, human MSCs fail to stimulate proliferation of allogeneic lymphocytes in vitro, even in the presence of CD28-stimulating antibodies or B7 costimulator (Prockop et al. 2010), which makes allogeneic transplantation of MSCs possible. MSCs have shown promising regulatory effects on both adaptive and innate immune systems in general and during tissue repair in particular. MSCs are reported to suppress proliferation and function of T-effector cells and induce their apoptosis by indoleamine 2,3-dioxygenase (IDO)-mediated tryptophan depletion (Meisel et al. 2004; Selmani et al. 2008; Akiyama et al. 2012; Ren et al. 2008); promote the generation of regulatory T cells (Selmani et al. 2008; Kavanagh and Mahon 2011; Sun et al. 2011; Ghannam et al. 2010); reduce B-cell activation and proliferation (Corcione et al. 2006; Maby-El Hajjami et al. 2009); suppress differentiation, maturation, and antigen-presenting functions of dendritic cells (Nauta et al. 2006; Aggarwal and Pittenger 2005); inhibit proliferation and cytotoxicity of natural killer cells (Selmani et al. 2008; Sotiropoulou et al. 2006; Aggarwal and Pittenger 2005); induce macrophage polarization from a proinflammatory M1 to an anti-inflammatory M2 state (Dayan et al. 2011; Jiang et al. 2013; Kim and Hematti 2009; Qi et al. 2014; Zhang et al. 2010); and inhibit neutrophil apoptosis and activation (Jiang and Scharffetter-Kochanek unpublished data) (Raffaghello et al. 2008; Tso et al. 2010). Several recent reviews provide further information (Prockop et al. 2010; Nasef et al. 2008; Atoui and Chiu 2012; Bunnell et al. 2010; Shi et al. 2010).

TNF-α released from proinflammatory M1 macrophages stimulates inducible nitric oxide synthase (iNOS) with subsequent generation of microbicidal NO• to combat invading microbes during early phases of wound healing (Daley et al. 2010). By contrast to scarless fetal wound healing with little or no TNF-α release, TNF-α concentrations in acute physiological healing result in enhanced microbicidal activity at the expense of enhanced scar formation. Also persisting overexpression of TNF-α as observed in chronic wounds results in the nonhealing state (Sindrilaru et al. 2011). A vicious cycle of persistent high concentrations of reactive oxygen species and matrix metalloproteinase-1 activation with subsequent tissue breakdown and premature senescence of resident wound fibroblasts are responsible for the delay in tissue restoration (Sindrilaru et al. 2011). In addition, high TNF-α concentrations – via an autocrine loop – maintain and amplify proinflammatory M1 macrophage activation (Sindrilaru et al. 2011; Wang et al. 2006). MSCs significantly suppressed TNF-α release from macrophages in vitro and in vivo and thus contribute to the accelerated wound healing in various animal models (Pietila et al. 2012; Qi et al. 2014; Zhang et al. 2010).

TNF-α-stimulated gene 6 protein (TSG-6) has recently been identified as a key anti-inflammatory molecule released from activated MSCs. TSG-6 released from intravenously infused human MSCs decreased inflammatory responses, reduced infarct size, and improved cardiac function in experimental myocardial infarction

(Lee et al. 2009). Moreover, through the interaction with CD44 on macrophages, MSC-derived TSG-6 attenuates zymosan-induced mouse peritonitis due to its ability to decrease the TLR2-mediated nuclear translocation of NFκB in resident macrophages (Choi et al. 2011). Recently, our group has demonstrated that TSG-6 released from MSCs improves wound healing by limiting macrophage activation, inflammation, and scar formation in a murine model of full-thickness skin wounds (Qi et al. 2014).

14.2.6 Transdifferentiation

Stem cells are believed to have tissue regeneration capacity during injury through self-renewal, engrafting, differentiation, and transdifferentiation into histogenetically distinct tissue cells. In vitro, MSCs have shown great potential to transdifferentiate into non-mesenchymal lineages such as endothelial cells (Oswald et al. 2004), keratinocytes (Fu et al. 2006; Sasaki et al. 2008), hepatocytes (Stock et al. 2010; Wu and Tao 2012; Liang et al. 2012), cardiomyocytes (Toma et al. 2002; Xu et al. 2004), and neurons (Joannides et al. 2004; Yang et al. 2004). However, the transdifferentiation capacity of MSCs in vivo is still under debate. A mesenchymal population resident in neonatal murine dermis has the potential for transdifferentiation into epidermal cells in a three-dimensional culture (Crigler et al. 2007). Approximately 10 % of ex vivo-expanded murine BM-MSCs differentiate into $CD31^+$ and $VEGF^+$ endothelial cells in vivo (Al-Khaldi et al. 2003). Fu and colleagues (2006) have demonstrated the differentiation of MSCs into vascular endothelial cells, epidermal cells, or sweat gland cells under different lineage-specific culture conditions. More importantly, a few percent of autograft MSCs transdifferentiated into vascular endothelial cells in granulation tissues, sebaceous duct cells, and epidermal cells in regenerated skin in vivo in a full-thickness wound model in minipigs (Fu et al. 2006; Li et al. 2006). These authors also showed that wound-healing quality, such as the speed of reepithelialization, the number of epidermal rete ridges, thickness of the regenerated epidermis, the morphology, the number, and the arrangement of microvasculature, fibroblasts, and collagen, were markedly enhanced (Fu et al. 2006). These authors even managed to regenerate functional sweat gland-like structures based on transplantation of ex vivo predifferentiated human BM-MSCs (Sheng et al. 2009). In addition, Sasaki et al. (2008) demonstrated that MSCs were recruited into wounded skin and contribute to wound repair by transdifferentiation into multiple skin cell type in vivo. This conclusion was based on the observation that markers for keratinocytes, endothelial cells, and pericytes were detected on donor GFP-tagged MSCs (Sasaki et al. 2008). However, whether these cells retained their mesenchymal markers and whether they were functional keratinocytes or endothelial cells was not addressed in this study.

Unfortunately, clinical observations so far have failed to demonstrate a therapeutically relevant level of long-term engraftment or transdifferentiation of transplanted MSCs. The concept of MSC transdifferentiation is still under discussion and may not reflect what is occurring during tissue repair in vivo (Sanges et al. 2011;

Zech 2005; Lee et al. 2009). For example, BM-MSCs express cardiac-specific markers, retain the stromal phenotype, and do not become functional cardiomyocytes, which generate action potentials or display ionic currents (Rose et al. 2008). Schneider and colleagues (2008) showed that multipotent human MSCs failed to transdifferentiate into E-cadherin- or cytokeratin-expressing cells under optimized organotypic culture conditions for keratinocytes but differentiated into myofibroblast-like cells with the key ability to contract and organize extracellular matrices (Schneider et al. 2008). Despite the viability and evident three-dimensional epidermis-like growth pattern, MSCs showed a persistent expression of mesenchymal, but not of epithelial markers, thus indicating a lack of epidermal differentiation (Schneider et al. 2008). Therefore, the authors argued that MSCs might rather contribute to wound-healing processes like wound contraction and reorganization of the newly deposited extracellular matrix, but do not significantly contribute to transdifferentiation into keratinocytes. Nevertheless, the partial cellular reprogramming, leading to the acquisition of at least some characteristics of the desired lineage, may possibly contribute to the tissue regeneration capacity of MSCs (Rose et al. 2008). In fact, emerging studies have suggested that the observed tissue regeneration of transplanted MSCs could be due to cell fusion of MSCs with resident cells at the wound site, thus enhancing cellular plasticity with subsequent changes in their developmental potential (Sanges et al. 2011; Zech 2005). It will be interesting for the future studies to investigate whether MSCs simply fuse with histologically distinct resident skin cells during wound healing.

14.2.7 Microenvironment Sensing

MSCs are evolutionary optimized to sense the microenvironment to protect the supportive tissue-specific niche for other stem cells and themselves (Meraviglia et al. 2014; Murphy et al. 2013). In simple organisms such as planarians, hydra, and starfish, stem cells in the mesenchyme (the blastema) are well positioned to interpret cues from the environment and to execute decisions about the direction of wound repair and tissue regeneration (reviewed in (Stappenbeck and Miyoshi 2009)). Wnt, BMP, and Hox signals, neuropeptides, and microRNAs are suggested to be important regulators in sensing the anterior-posterior axis. However, how mammalian MSCs communicate with the overlying epithelium and inflammatory cells in the mesenchyme during wound healing is not well understood.

Emerging studies suggest that pattern recognition receptors (PRRs) including pathogen-associated molecular patterns (PAMPs) and damage-associated molecular patterns (DAMPs), for example, toll-like receptors (TLRs), may qualify as environmental sensor of MSCs (Hwa Cho et al. 2006; Pevsner-Fischer et al. 2007; Raicevic et al. 2010; Mastri et al. 2012; Nurmenniemi et al. 2010), although their expression level on MSCs is very low. A recent report demonstrated that TLR4-primed MSCs were proinflammatory (MSC1), while TLR3-primed MSCs were immunosuppressive (MSC2) (Waterman et al. 2010). The authors suggested a new MSC1-MSC2 polarization paradigm, which has attracted much attention in the field (Keating 2012;

Bunnell et al. 2010). However, in apparent contradiction to the findings that TLR4-primed MSCs are proinflammatory by Waterman et al. (2010), several studies have reported beneficial effects of MSC treatment in animal models of sepsis or LPS-induced lung injury, in which MSCs were administered within 1 h following LPS challenge (Nemeth et al. 2009; Nemeth et al. 2010; Delarosa et al. 2012; Gonzalez-Rey et al. 2009). Based on the therapeutic benefit observed in these experimental models, the conclusion was drawn that high concentrations of LPS did not polarize MSCs towards a proinflammatory phenotype. In addition to these in vivo findings, upon the addition of IFN-γ and LPS (TLR4 ligand) to MSC-macrophage cocultures, MSCs exhibited an excellent suppressive effect on TNF-α release from cocultured macrophages (Jiang et al. 2013; Qi et al. 2014). It is possible that the dosage and duration of TLR ligands that MSCs are exposed to and model inherent differences are critical for MSC polarization. This also suggests that other mechanisms exist to fine-tune the environment-sensing capability of MSCs. Overall the concept of a sensing mechanism in MSCs is clinically relevant as this would imply that adaptive responses will be provided, eventually even qualifying MSCs as versatile "adaptive drugstores," which – in contrast to the application of recombinant growth factors or a designed cocktail thereof – would have the advantage to perfectly adapt to the specific needs of defined wounds.

14.2.8 Matrix Remodeling

MSCs also showed promising effects on the synthesis, deposition, and remodeling of the extracellular matrix reducing scar tissue formation and preventing skin aging. BM-MSCs were shown to attenuate bleomycin-induced skin (Wu et al. 2014) and lung fibrosis (Ortiz et al. 2003) in mice. The detailed mechanisms of how MSCs reduce scar formation during the remodeling phase of wound healing are far from being fully defined. It is most likely that besides dampening inflammation in early phases of wound repair, antifibrotic factors released by MSCs contribute to further reduce fibrogenic responses (Jackson et al. 2012). Recently, we and others reported that human bone-marrow-derived MSCs injected at the wound site released the anti-inflammatory factor TSG-6 and concomitantly modulate the TGF-β_3 to TGF-β_1 ratio, resulting in reduced scar formation in murine full-thickness wounds (Liu et al. 2014; Qi et al. 2014).

Moreover, MSCs exert beneficial effects for the prevention and reduction of skin aging. Transplantation of dermal MSCs from young BALB/c mice into aged mice resulted in a substantial increase in dermal thickness and collagen type I content in the dermis (Xue and Li 2011). Subcutaneous injection of AT-MSCs significantly stimulated collagen synthesis and angiogenesis and also increased dermal thickness, collagen density, and fibroblast numbers in the skin of hairless mice (Heo et al. 2011). These studies suggest that MSCs, especially dermal MSCs, play an important role in improving the phenotype of aged skin, which – if untreated – is characterized by severe atrophy, reduced collagen deposition, and rarefaction of elastic fibers with reduced skin resilience and enhanced wrinkle formation (Kligman and Kligman 1986; Fisher et al. 2008). In vitro data suggest that by the activation of

TGF-β/Smad and p38 MAPK signaling pathways, dermal MSCs secrete various growth factors that control and reroute senescent neighboring fibroblasts to secrete and synthesize collagen or elastin, enhance extracellular matrix deposition, and thus may contribute to improve wrinkles and strengthen overall skin elasticity (Wang et al. 2008). Conditioned medium from AT-MSCs stimulated both collagen synthesis and migration of dermal fibroblasts, which improved the wrinkling and accelerated wound healing in animal models. MSC-conditioned medium also protected dermal fibroblasts from oxidative stress induced by chemicals and UVB irradiation and even has the capacity to reverse the noxious UVB-induced suppression of fibroblast proliferation, reduced collagen type I synthesis, and enhanced proteolytic MMP-1 activity in dermal fibroblasts towards tissue homeostasis and rejuvenation (Kim and Hematti 2009).

A contradiction is currently obvious for effects of MSCs on myofibroblasts differentiation and collagen deposition. In wound healing and photoaging studies, MSCs were found to induce myofibroblasts differentiation, collagen deposition, and granulation tissue formation and enhance dermal thickness (Xue and Li 2011; Jiang et al. 2013; Wan et al. 2013), while in studies on fibrosis and scar formation, MSCs seem to have the opposite effects with reduction in collagen synthesis and reduced myofibroblast differentiation (Liu et al. 2014; Qi et al. 2014; Wu et al. 2014). It is likely that microenvironmental differences of the studied models may be responsible for the apparently opposite MSC responses, and it will be especially exciting to decipher the underlying mechanisms in future studies.

In summary, MSCs exert their effects by multiple ways that stimulate tissue recovery on many important levels, including stimulation of endogenous progenitor cells, suppression of apoptosis of vulnerable cells, remodeling of extracellular matrices, and stimulation of new blood vessel formation. However, the majority of these mentioned studies were performed in vitro, and often the MSC numbers assessed in culture were higher when compared to physiological or therapeutic situations in vivo. Therefore, the identification of MSC-specific markers with subsequent development of transgenic models to trace endogenous MSCs or the refinement of state-of-the-art technology to track exogenously added MSCs in skin and tissues is urgently required for meaningful in vivo studies to clearly dissect MSC functions in situ in physiological and pathologic conditions of tissue repair and aging.

14.3 Impact of Mesenchymal Stem Cell Aging on Wound Healing

The cellular mechanisms of aging of skin stem cells (Chap. 8) and in particular MSCs (Chap. 11) have been detailed in the previous chapters. Here we would like to emphasize the consequence of MSC aging on impaired wound healing. The most obvious and striking evidence that aging plays an important role in wound healing comes from the fact that fetal wounds can heal rapidly without scars, while in adults wound healing decreases with aging and frequently though not always reveal enhanced fibrosis and scar formation. Among other reasons, this most likely reflects changes in the functional state of mesenchymal stem cells.

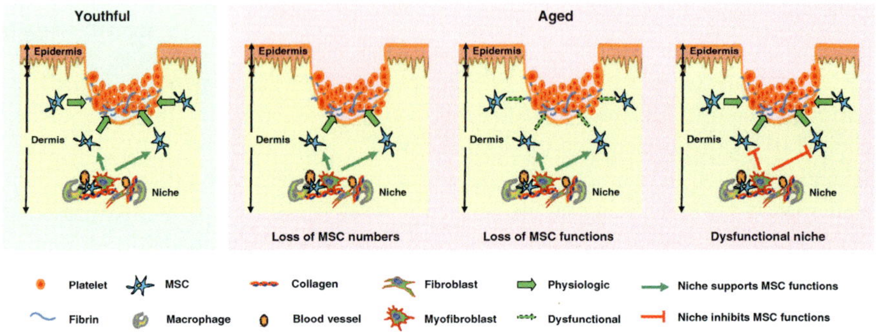

Fig. 14.3 Impact of aging on MSCs. The decline of MSC numbers and functions in wound repair with age can be assigned to age-related changes at several levels, including (1) decreased numbers of MSCs, (2) decline in MSC functions required for wound healing, and (3) persisting proinflammatory oxidative niche/microenvironment

14.3.1 Age-Induced Changes in MSCs and Impact on Wound Healing

Like other stem cells, MSCs have evolutionary developed sophisticated mechanisms to protect themselves during aging. For example, MSCs have a high resistance to oxidative-stress-induced death, which correlates with the low level of intracellular reactive species by effective ROS scavenging, constitutive expression of antioxidant enzymes required to counteract oxidative stress, and high levels of total intracellular glutathione (Valle-Prieto and Conget 2010). Human MSCs also constitutively express methionine sulfoxide reductase A at high levels, an enzyme which is crucial for the repair of oxidized proteins and for the recovery of methionine residues that efficiently serve as scavengers for oxidants (Salmon et al. 2009). However, a recent study has shown that excessive activation of Wnt/b-catenin signaling induces MSC aging through promoting the intracellular production of ROS despite higher expression of antioxidant enzymes (Zhang et al. 2013). This suggests that the aging process inevitably brings detrimental changes to MSCs. Aging affects MSC proliferation and differentiation and their cell surface marker expression profile. The declined capacity for wound repair with age may be due to age-related changes in MSCs (Fig. 14.3).

14.3.2 Decrease of MSC Numbers Across the Life Span

Like other adult stem cells, MSCs are a rare population among other cells in their resident tissues. For example, clonal analysis revealed that 6.4 % of the single-cell-derived dermal clones were tripotent and that dermal stem cells represent approximately only 0.3 % of human dermal foreskin fibroblasts (Chen et al. 2007). Unfortunately, with such a small pool, the number of MSCs declines dramatically with age. MSCs in human bone marrow diminish in number almost exponentially from birth (1 in 10,000 nucleated cells) to the eighth decade of life (1 in 2,000,000 nucleated cells) (Haybesworth et al. 1994). With increasing age, osteoprogenitor cells derived from MSCs in rat bone marrow display decreased potential for colony formation and size (Dodson et al. 1996) and a reduced ability to form bone (Quarto et al. 1995). The same conclusion was obtained in human MSCs. By comparing BM-MSCs from old and young human donors, Stenderup and colleagues (2003) found aging was associated with decreased proliferative capacity of osteoprogenitor cells, thus the decreased osteoblastic cell number (Stenderup et al. 2003). In the skin compartment, it has recently been reported that the abundance and differentiation potential of human skin-derived precursors (SKPs), a subpopulation of mesenchymal stem cells with the capacity to differentiate into neurons and smooth muscle cells, decrease sharply with age (Gago et al. 2009). The authors found that SKPs are extremely difficult to isolate, expand, and differentiate when obtained from the elderly (Gago et al. 2009). However, it is still unclear whether human fetal skin differs in MSC numbers, location, and function from adult skin and if so which mechanisms is responsible for such sharp decline in MSC numbers in adult skin of the elderly. Fetal tissue has the ability to rapidly heal skin wounds without a scar unlike adult tissue. Differences in MSCs between fetal and adult skin in the context of wound healing has not been extensively studied, but will be an important area for further investigation.

14.3.3 Loss of Function in Aged MSCs

Experts in the field argue that impaired wound healing in aged skin is rather induced by impaired stem cell mobilization or reduced ability to respond to proliferative signals than reduced stem cell numbers per se (Sharpless and DePinho 2007; Zouboulis et al. 2008), a situation similar to hematopoietic stem cells (Geiger et al. 2013) and skeletal muscle satellite cells in aged individuals (Rando 2006).

It is well known that aging perturbs cytokine and hormone networks. Similarly, aged MSCs reveal a significantly altered cytokines prolife that is less effective in tissue repair (Galderisi et al. 2009; Kretlow et al. 2008; Crisostomo et al. 2006). By contrast to young BM-MSCs, aged BM-MSCs with significantly higher $p16^{INK4a}$ expression downregulate the secretion of IGF-1, fibroblast growth factor (FGF-2), VEGF, and hepatocyte growth factor (HGF), leading to a decline in their capacity to induce angiogenesis exemplified by reduced tube-forming on Matrigel (Sen et al. 2002). Choudhery et al. (2012) also showed that compared to young MSCs, aged MSCs display significantly reduced SDF-1 and VEGF expression (Choudhery et al. 2012) with severe impact on wound repair. SDF-1 is involved in homing, engraftment, neovascularization, and cell proliferation. VEGF increases vascular permeability, promotes angiogenesis, attracts monocytes to the site of injury, and is also involved in neovascularization (Matsumoto et al. 2005; Tang et al. 2005).

Furthermore, there is an age-related decline in overall MSC "fitness" as suggested by Stolzing et al. (2008). They found human BM-MSCs from older donors to have reduced CFU-F capacity and increased oxidative damage, ROS levels, and p21 and p53 expression (Stolzing et al. 2008). In line with this, BM-MSCs from aged mice revealed reduced angiogenesis, proliferation, and antiapoptotic effects with overall reduced wound healing promoting potential (Choudhery et al. 2012). The authors compared BM-MSCs from young and aged C57BL/6 mice. By contrast to aged MSCs, they found a significantly higher expression the antiaging NAD-dependent histone deacetylase Sirt1 in young MSCs. Aged MSCs revealed a reduction in VEGF, SDF-1, and AKT and a consistently increased expression of p53, p21, Bax, and p16^{INK4a}. Tube formation, wound healing, and proliferative potential of MSCs from young mice were significantly better compared to that of MSCs from aged mice. This study suggests that the age-related increase in the expression of apoptotic and senescent genes, with the concomitantly occurring decrease in Sirt1 expression, is at least to some extent responsible for impaired MSC functions. In mammalian cells, Sirt1 is known to enhance proliferation by the suppression of p53 (Chua and Laurent 2006) and to inhibit cell apoptosis and senescence (Jung-Hynes and Ahmad 2009). Lamin A has been identified as a direct activator of Sirt1 (Liu et al. 2012) and if mutated leads to premature aging as earlier reported for Hutchinson-Gilford progeria syndrome (Liu et al. 2011). This suggests that lamin A via Sirt1 induction possibly contributes to the regenerative ability of MSCs and – if disrupted – may play a critical role in MSC aging. In line with these findings, Park et al. (2005) reported that the adipogenic differentiation potential of human MSCs sharply declines at the end of their proliferative life span. In senescent MSCs, the basal phosphorylation level of ERK was also significantly increased, a finding commonly shared by senescent cells (Park et al. 2005). A striking observation suggests that aged bone-marrow cells rather inhibit than induce skin wound vascularization. Accordingly, lineage-negative (lin^-) bone-marrow cells from young mice greatly increase vascularization and decrease wound size in healing-impaired mice. In contrast, as with lin^- cells from obese diabetic mice, lin^- cells from old mice had no significant effect on wound size, but profoundly impaired neovascularization in wounds (Schatteman and Ma 2006). These studies suggest transplantation of aged MSCs is likely to result in reduced therapeutic effects in older patients and may actually be detrimental.

Recently, miRNAs were found to be involved in replicative senescence (Li et al. 2011; Hackl et al. 2010; Scheid et al. 2002) and physiological aging of MSCs (Li et al. 2011). The role of miRNAs in the proliferation and differentiation of MSCs in the context of wound healing has also been explored (reviewed (Guo et al. 2011)). Altered expression of a panel of miRNAs with upregulation of miR-31, −21, −223, −142, −205, −203, −18b, −19a, −130b, −16, −26b, −125b, and let-7f and downregulation of miR-133a, −181, −30a-3p, −193b, −30a-5p, −204, −200b, −96, −127, −181c, −182, and -130a was detected in injured tissue at the stage of active granulation formation in murine skin excision models (Zou et al. 2010; Shilo et al. 2007). Furthermore, Zou and colleagues (2010) found that TGF-β, a key growth factor elevated at the wound site, stimulates upregulation of miR-21 in MSCs, thus promoting proliferation and differentiation of MSCs in vitro. Consistently, knockdown of miR-21 in the wound bed delayed the healing process (Zou et al. 2010). These results suggest that miR-21 regulates gene expression and, subsequently, the behavior of MSCs in wound healing. It remains to be seen whether the changes of miRNA expression pattern during aging contribute to the decline in MSC potential and impaired wound healing and tissue regeneration in elderly.

14.3.4 Dysfunctional Microenvironment in Aged Skin

Skin homeostasis is gradually lost as a consequence of the intrinsic aging process, ultraviolet light, and other environmental threats resulting in a dramatic change in the structure of dermis. The most remarkable change is referred to the switch of dermal fibroblasts from a collagen-synthesizing phenotype to a more proteolytic phenotype (Varani et al. 2004), which clinically translates to the gradual appearance of wrinkles and severe skin sagging.

Senescent cells mainly dermal fibroblasts develop during intrinsic aging, as a result of a repetitive stress response or a combination thereof. Senescent fibroblasts are characterized by their resistance to apoptosis and the manifestation of their senescence-associated secretory phenotype (SASP) with the release of a vast variety of proinflammatory factors that alter the tissue microenvironment and may affect MSC functions (Campisi 2005). In addition, age-associated immunosenescence leads to an elevated endogenous inflammatory milieu (Gruver et al. 2007; Aw et al. 2007; Peters et al. 2009). We recently reported that the persistent inflammatory microenvironment in chronic venous leg ulcers with elevated TNF-α and unrestrained release of hydroxyl radical by macrophages results in the induction of senescent wound bed fibroblasts and impaired wound healing (Sindrilaru et al. 2011). It is tempting to speculate that such a hostile microenvironment also causes senescence and functional loss of MSCs, eventually promoting the state of never-healing chronic wounds.

BM-MSCs from aged donors displayed diminished long-term proliferation potential and elevated expression of vascular cell adhesion molecule-1 (VCAM-1; CD106) as a response to the inflammatory environment (Laschober et al. 2011). The authors demonstrated that moderate levels of inflammatory stimuli are interpreted by MSCs at a young age as instructive signals for osteoblastogenesis, whereas at old age, an inflammatory microenvironment may effectively suppress bone remodeling and repair by tissue-borne MSCs, while promoting adipogenic differentiation eventually leads to an increase in adipose tissue (Laschober et al. 2011).

Notably, reduced b_1 integrin expression on MSCs as occurring during aging leads to reduced proliferation potential especially during wound healing (Piwko-Czuchra et al. 2009). Since b_1 integrins play important roles in interacting with other resident cells and ECM molecules, it is most likely that changes in b_1 integrin expression on aged MSCs disrupt downstream signaling and thus exert still not fully understood effects on MSC behavior during injury in the elderly.

Several lines of evidence suggest that defective neovascularization with aging is related to depressed signaling by hypoxia-inducible factor-1 (HIF-1), the master regulator of neovascularization, which regulates the expression of VEGF, SDF-1, and CXC chemokine receptor-4 (CXCR4). The function and mobilization of MSCs have been shown to decrease with decline in HIF-1 signaling with age, which may lead to depressed neovascularization and wound healing (Hoenig et al. 2008).

The aged MSCs themselves also have impact on other cell populations in the environment, for example, on fibroblasts. MSCs derived from healthy donors were significantly better than MSCs derived from chronic wound patients in stimulating directed fibroblast migration (Rodriguez-Menocal et al. 2012). Accordingly, the aging process is amplified by the vicious feedback loop between MSCs and their niche.

14.3.5 MSCs in Aging-Related Chronic Wounds: Chronic Wounds due to Intrinsic Aging

Optimal healing of cutaneous wounds requires a well-orchestrated regulation of complex biological and molecular events including cell migration, proliferation, extracellular matrix deposition, and scar remodeling. Cellular responses to inflammatory mediators, growth factors, and cytokines, and to mechanical forces, must be appropriate and precise. However, this orderly progression of tissue repair becomes increasingly compromised with advancing age, resulting in the increased incidence of chronic nonhealing wounds, like diabetic foot ulcers (Falanga 2005; Yao et al. 2012), pressure ulcers (Jaul 2010), and chronic venous ulcers (Raju and Neglen 2009). Lack of wound healing is life-threatening as invading bacteria very often promote infection and sepsis. Several pathogenic mechanisms that contribute to wound-healing failure, ranging from disease-specific intrinsic deficiencies in blood supply, angiogenesis, and matrix turnover to extrinsic factors like infection and persisting inflammation, have major effects on MSC functions and may even deplete skin-resident MSCs. Thus treatment of these conditions with exogenous MSCs may hold substantial promise for the benefit of our patients.

Diabetic foot ulcer represents the major complication of diabetes mellitus, occurring in 15 % of all patients with diabetes and accounting for more than 80 % of all lower-leg amputations. Preclinical research has generated exciting results using MSCs for the treatment of diabetic wounds. Systemic or local injections of both BM-MSCs and AT-MSCs, or topical application of allogeneic nondiabetic BM-MSCs seeded in a collagen scaffold, have shown promising effects on augmenting wound healing and increasing angiogenesis in diabetic animal models (Kwon et al. 2008; Kim et al. 2011; O'Loughlin et al. 2013; Wan et al. 2013). Dash et al. (2009) showed that implantation of autologous BM-MSCs in diabetic foot ulcers of lower extremity in human patients significantly accelerates the healing process and improves clinical parameters like ulcer size and pain-free walking distance (Dash et al. 2009). In a double-blind, randomized, controlled trial, BM-MSCs transplantation has shown to increase lower-limb perfusion and promote foot ulcer healing in diabetic patients (Lu et al. 2011). Interestingly, infusion of MSCs not only improves wound healing but also ameliorates hyperglycemia, improves insulin sensitivity, and promotes repair of pancreatic islets in type 2 diabetic rodents (Si et al. 2012; Lee et al. 2006).

Similarly, MSCs from skin or bone marrow showed beneficial effects in patients suffering from type IV pressure ulcers (Sarasua et al. 2011) and chronic venous ulcers (Ravari et al. 2011). In contrast, other studies did not find conclusive beneficial effects of MSC treatment in accelerating healing of pressure ulcers in rats (de la Garza-Rodea et al. 2011) or of lower-extremity chronic wounds in a preclinical study of very small sample size (Mulder et al. 2010).

14.3.6 Chronic Wounds of Other Origin with Progeroid Syndromes

Chronic wounds represent unsolved problems for patients suffering from genetic progeroid syndromes.

The Hutchinson-Gilford progeria syndrome (HGPS) is a rare and fatal human premature aging disease, characterized by premature arteriosclerosis and degeneration of vascular smooth muscle cells. HGPS is caused by a single-point mutation in the lamin A gene, resulting in the generation of progerin, a truncated splicing mutant of lamin A. Accumulation of progerin leads to various aging-associated nuclear defects including disorganization of nuclear lamina and loss of heterochromatin (Liu et al. 2011). HGPS affects mesenchymal lineages, including the dermis, the skeletal system, and the vascular smooth muscle (VSMC). Because progerin also accumulates during physiological aging, A-type lamins should be regarded as intrinsic modulators of aging within adult stem cells and their niches that are essential for survival to old age (Pekovic and Hutchison 2008). The expression of progerin in immortalized human MSCs negatively affects their molecular identity and differentiation potential by interfering with the Notch signaling pathway, which is essential in stem cell regulation and maintenance (Scaffidi and Misteli 2008). When fibroblasts derived from HGPS patients were used for the induction of pluripotent stem cells (iPSCs), Zhang et al. observed that progerin was accumulated in MSCs and VSMCs differentiated from HGPS-iPSCs and, furthermore, displayed increased DNA damage and nuclear abnormalities. In addition, the viability of HGPS-derived MSCs was compromised by stress and hypoxia in vitro and in vivo (Zhang et al. 2011). The Zmpste24-null mouse, a murine model for human HGPS, shows a defective proliferative potential of bulge cells in the hair follicle niche due to a decreased Wnt signaling pathway, which leads to a severe dysfunction of epidermal stem cell renewal (Espada et al. 2008). A further study by Liu et al. (2012) has demonstrated that lamin A interacts with and activates Sirt1, whereas in the presence of progerin or prelamin A, Sirt1 exhibits reduced association with the nuclear matrix and decreased deacetylase activity and thus eventually leads to rapid depletion of adult stem cells in Zmpste24-null mice (Liu et al. 2012). Rosengardten et al. (2011) have found that induced expression of the mutated lamin A results in a loss of adult stem cells and impaired wound healing in mice, which may be due to downregulation of p63 with concomitant activation of the DNA damage response and premature senescence programs. In addition, these mutant mice have an activated inflammatory response with upregulation of a variety of genes involved in major inflammatory pathways, which is also associated with normal aging (Rosengardten et al. 2011). These findings strongly support the hypothesis that lamin A regulates stem cell maintenance via a range of dysregulated regenerative and inflammatory signaling pathways.

Mammalian aging is accompanied by accumulation of genomic DNA damage and progressive decline in the ability of tissues to regenerate. DNA damage activates the tumor suppressor p53, which leads to cell-cycle arrest, senescence, or apoptosis. The constitutively activated p53 mutant mice have shown segmental progeria that is correlated with the depletion of adult stem cells in multiple tissues, including bone marrow, brain, and testes. Notably, deficiency of Puma, which is required for p53-dependent apoptosis after DNA damage, rescues segmental progeria and prevents the depletion of adult stem cells (Liu et al. 2010). Su and colleagues showed that the p53 family member, TAp63, is essential for maintenance of epidermal and dermal precursors. TAp63$^{-/-}$ mice age prematurely, develop blisters and skin ulcerations, lose hair follicles with age, and have an impaired wound-healing response. TAp63 regulates self-renewal of dermal precursors via p57^{Kip2}. Loss of TAp63 leads to genomic instability and causes dermal precursor cells to senesce (Su et al. 2009). These findings suggest that the apoptotic pathway mediated by p53 and its family members plays a key role in the depletion of adult stem cells after the accumulation of DNA damage during aging, which leads to a decrease in tissue regeneration.

Lack of CD18 (common b chain of the β_2 integrins) leads to leukocyte adhesion deficiency syndrome type 1 (LAD-1), a recessively inherited disorder with impaired b_2 integrin function (Kishimoto et al. 1987; Arnaout 1990). Concerned individuals reveal phenotypes reminiscent of aging such as impaired wound healing, osteoporosis, an impaired control of infections, and defective vaccine responses due to a hampered B- and T-cell activation (Fischer et al. 1986; Anderson and Smith 2001). In fact, our own studies on the CD18-deficient mouse model also revealed impaired wound healing (Peters et al. 2005; Peters et al. 2006; Sindrilaru et al. 2009; Sindrilaru et al. 2007). Remarkably, local application of human AT-MSCs significantly accelerates wound healing of CD18-deficient mice, largely due to the release of paracrine trophic factors (Jiang and Scharffetter-Kochanek, unpublished data).

14.4 Perspectives

The most prevalent difference between adult and fetal wound healing is scar formation, exclusively occurring in adult tissue repair. Relatively little is known on differences between fetal and adult dermal MSCs and the interactions between MSCs and their niche with aging. Identification and assessment of genes responsible for beneficial MSC functions should advance our understanding of MSC physiology and potentially can be translated for improved MSC-based therapy in tissue repair and aging. In addition, understanding of whether skin MSCs retain their tissue regeneration potential throughout the life span of the individual or undergo an aging process with an overall decreased stem cell pool and/or function remains an important area of investigation. The slow progress in this area is largely due to the limitations of research tools such as specific dermal MSC markers, suitable animal models, and very low frequency of MSCs in situ. The mechanisms underlying tissue regeneration and immune modulation by therapeutic MSC applications also require further elucidation, particularly the extent to which the two processes are interrelated. Therefore, the development of suitable tools for tracking primary MSCs in vivo in order to understand how to activate or stimulate regeneration of human dermal MSCs will be urgent tasks in the near future to foster dermal MSC-based therapies for medical applications.

Conclusions

Many studies suggest that dermal MSCs play important roles in tissue homeostasis and regeneration during skin injury, and MSC-based therapy holds a great potential for the development of personalized allogeneic "off-the-shelf" medicine for nonhealing wounds. For therapeutic use, ex vivo expansion of MSCs is often required. However, BM-MSCs and possibly also AT-MSCs are sensitive to expansion-induced senescence during culture. The decrease in functionality has caused increasing concerns on the efficacy of MSC-based therapies (Sensebe et al. 2010). With this caveat, adipose tissue may be the better MSC source for therapeutic purposes, as adequate MSC numbers are easily obtained for direct application without any need for expansion and manipulation in vitro. Moreover, the majority of ongoing clinical trials involving MSCs prefer autologous sources. Unfortunately, MSCs from aged donors most likely reveal diminished efficacy in the treatment of age-associated chronic wounds. Thus, transplantation efficacy of allogeneic MSCs from healthy young donors rather than autologous MSCs from elderly patients with chronic wounds needs to be taken into consideration. Alternatively, it is already possible to isolate MSCs at young age and keep them under GMP conditions frozen for future use.

Acknowledgment We thank Dr. Meinhard Wlaschek for his helpful suggestion and critical discussion for preparing the manuscript. We are grateful to John Wiley & Sons publications for the copyright permission to reproduce Fig. 14.1 in our previous publication in Wound Repair and Regeneration (Wlaschek and Scharffetter-Kochanek 2005). Our work is supported by the European Commission (CASCADE HEALTH-FP7-223236), the German Research Foundation (DFG KFO142 and DFG SFB1149), and the Baden-Württemberg Stiftung (P-BWS-ASII/15) to K.S.-K., and Bausteinförderung from the Medical Faculty, University of Ulm (LSBN.0100) to D.J.

References

Abe R, Donnelly SC, Peng T, Bucala R, Metz CN (2001) Peripheral blood fibrocytes: differentiation pathway and migration to wound sites. J Immunol 166(12):7556–7562

Aggarwal S, Pittenger MF (2005) Human mesenchymal stem cells modulate allogeneic immune cell responses. Blood 105(4):1815–1822

Akiyama K, Chen C, Wang D, Xu X, Qu C, Yamaza T, Cai T, Chen W, Sun L, Shi S (2012) Mesenchymal-stem-cell-induced immunoregulation involves FAS-ligand-/FAS-mediated T cell apoptosis. Cell Stem Cell 10(5):544–555

Al-Khaldi A, Eliopoulos N, Martineau D, Lejeune L, Lachapelle K, Galipeau J (2003) Postnatal bone marrow stromal cells elicit a potent VEGF-dependent neoangiogenic response in vivo. Gene Ther 10(8):621–629

Anderson DC, Smith CW (2001) Leukocyte adhesion deficiencies. In: The metabolic and molecular basis of inherited diseases. McGraw-Hill, New York, pp 4829–4856

Arnaout MA (1990) Structure and function of the leukocyte adhesion molecules CD11/CD18. Blood 75(5):1037–1050

Atoui R, Chiu RC (2012) Concise review: immunomodulatory properties of mesenchymal stem cells in cellular transplantation: update, controversies, and unknowns. Stem Cells Transl Med 1(3):200–205

Au P, Tam J, Fukumura D, Jain RK (2008) Bone marrow-derived mesenchymal stem cells facilitate engineering of long-lasting functional vasculature. Blood 111(9):4551–4558

Aw D, Silva AB, Palmer DB (2007) Immunosenescence: emerging challenges for an ageing population. Immunology 120(4):435–446

Badiavas EV, Abedi M, Butmarc J, Falanga V, Quesenberry P (2003) Participation of bone marrow derived cells in cutaneous wound healing. J Cell Physiol 196(2):245–250

Bartsch G, Yoo JJ, De Coppi P, Siddiqui MM, Schuch G, Pohl HG, Fuhr J, Perin L, Soker S, Atala A (2005) Propagation, expansion, and multilineage differentiation of human somatic stem cells from dermal progenitors. Stem Cells Dev 14(3):337–348

Bergers G, Song S (2005) The role of pericytes in blood-vessel formation and maintenance. Neuro Oncol 7(4):452–464

Bucala R, Spiegel LA, Chesney J, Hogan M, Cerami A (1994) Circulating fibrocytes define a new leukocyte subpopulation that mediates tissue repair. Mol Med 1(1):71–81

Bunnell BA, Betancourt AM, Sullivan DE (2010) New concepts on the immune modulation mediated by mesenchymal stem cells. Stem Cell Res Ther 1(5):34

Campisi J (2005) Senescent cells, tumor suppression, and organismal aging: good citizens, bad neighbors. Cell 120(4):513–522

Caplan AI, Correa D (2011) The MSC: an injury drugstore. Cell Stem Cell 9(1):11–15

Caplan AI, Dennis JE (2006) Mesenchymal stem cells as trophic mediators. J Cell Biochem 98(5):1076–1084

Chen FG, Zhang WJ, Bi D, Liu W, Wei X, Chen FF, Zhu L, Cui L, Cao Y (2007) Clonal analysis of nestin(−) vimentin(+) multipotent fibroblasts isolated from human dermis. J Cell Sci 120 (Pt 16):2875–2883

Chen L, Tredget EE, Wu PY, Wu Y (2008) Paracrine factors of mesenchymal stem cells recruit macrophages and endothelial lineage cells and enhance wound healing. PLoS One 3(4), e1886

Cho YM, Kim JH, Kim M, Park SJ, Koh SH, Ahn HS, Kang GH, Lee JB, Park KS, Lee HK (2012) Mesenchymal stem cells transfer mitochondria to the cells with virtually no mitochondrial function but not with pathogenic mtDNA mutations. PLoS One 7(3), e32778

Choi H, Lee RH, Bazhanov N, Oh JY, Prockop DJ (2011) Anti-inflammatory protein TSG-6 secreted by activated MSCs attenuates zymosan-induced mouse peritonitis by decreasing TLR2/NF-{kappa}B signaling in resident macrophages. Blood 118(2):330–338

Choudhery MS, Khan M, Mahmood R, Mehmood A, Khan SN, Riazuddin S (2012) Bone marrow derived mesenchymal stem cells from aged mice have reduced wound healing, angiogenesis, proliferation and anti-apoptosis capabilities. Cell Biol Int 36(8):747–753

Chua F, Laurent GJ (2006) Neutrophil elastase: mediator of extracellular matrix destruction and accumulation. Proc Am Thorac Soc 3(5):424–427

Corcione A, Benvenuto F, Ferretti E, Giunti D, Cappiello V, Cazzanti F, Risso M, Gualandi F, Mancardi GL, Pistoia V, Uccelli A (2006) Human mesenchymal stem cells modulate B-cell functions. Blood 107(1):367–372

Crigler L, Kazhanie A, Yoon TJ, Zakhari J, Anders J, Taylor B, Virador VM (2007) Isolation of a mesenchymal cell population from murine dermis that contains progenitors of multiple cell lineages. FASEB J 21(9):2050–2063

Crisan M, Yap S, Casteilla L, Chen CW, Corselli M, Park TS, Andriolo G, Sun B, Zheng B, Zhang L, Norotte C, Teng PN, Traas J, Schugar R, Deasy BM, Badylak S, Buhring HJ, Giacobino JP,

Lazzari L, Huard J, Peault B (2008) A perivascular origin for mesenchymal stem cells in multiple human organs. Cell Stem Cell 3(3):301–313

Crisostomo PR, Wang M, Wairiuko GM, Morrell ED, Terrell AM, Seshadri P, Nam UH, Meldrum DR (2006) High passage number of stem cells adversely affects stem cell activation and myocardial protection. Shock 26(6):575–580

Daley JM, Brancato SK, Thomay AA, Reichner JS, Albina JE (2010) The phenotype of murine wound macrophages. J Leukoc Biol 87(1):59–67

Dash NR, Dash SN, Routray P, Mohapatra S, Mohapatra PC (2009) Targeting nonhealing ulcers of lower extremity in human through autologous bone marrow-derived mesenchymal stem cells. Rejuvenation Res 12(5):359–366

Dayan V, Yannarelli G, Billia F, Filomeno P, Wang XH, Davies JE, Keating A (2011) Mesenchymal stromal cells mediate a switch to alternatively activated monocytes/macrophages after acute myocardial infarction. Basic Res Cardiol 106(6):1299–1310

De Bari C, Dell'Accio F, Tylzanowski P, Luyten FP (2001) Multipotent mesenchymal stem cells from adult human synovial membrane. Arthritis Rheum 44(8):1928–1942

de la Garza-Rodea AS, Knaan-Shanzer S, van Bekkum DW (2011) Pressure ulcers: description of a new model and use of mesenchymal stem cells for repair. Dermatology 223(3):266–284

Delarosa O, Dalemans W, Lombardo E (2012) Toll-like receptors as modulators of mesenchymal stem cells. Front Immunol 3:182

Dodson SA, Bernard GW, Kenney EB, Carranza FA (1996) In vitro comparison of aged and young osteogenic and hemopoietic bone marrow stem cells and their derivative colonies. J Periodontol 67(3):184–196

Dufourcq P, Descamps B, Tojais NF, Leroux L, Oses P, Daret D, Moreau C, Lamaziere JM, Couffinhal T, Duplaa C (2008) Secreted frizzled-related protein-1 enhances mesenchymal stem cell function in angiogenesis and contributes to neovessel maturation. Stem Cells 26(11):2991–3001

Espada J, Varela I, Flores I, Ugalde AP, Cadinanos J, Pendas AM, Stewart CL, Tryggvason K, Blasco MA, Freije JM, Lopez-Otin C (2008) Nuclear envelope defects cause stem cell dysfunction in premature-aging mice. J Cell Biol 181(1):27–35

Falanga V (2005) Wound healing and its impairment in the diabetic foot. Lancet 366(9498): 1736–1743

Falanga V (2012) Stem cells in tissue repair and regeneration. J Invest Dermatol 132(6): 1538–1541

Falanga V, Iwamoto S, Chartier M, Yufit T, Butmarc J, Kouttab N, Shrayer D, Carson P (2007) Autologous bone marrow-derived cultured mesenchymal stem cells delivered in a fibrin spray accelerate healing in murine and human cutaneous wounds. Tissue Eng 13(6):1299–1312

Fischer A, Durandy A, Sterkers G, Griscelli C (1986) Role of the LFA-1 molecule in cellular interactions required for antibody production in humans. J Immunol 136(9):3198–3203

Fisher GJ, Varani J, Voorhees JJ (2008) Looking older: fibroblast collapse and therapeutic implications. Arch Dermatol 144(5):666–672

Fu X, Fang L. Li X, Chang B, Sheng Z (2006) Enhanced wound-healing quality with bone marrow mesenchymal stem cells autografting after skin injury. Wound Repair Regen 14(3):325–335

Gago N, Perez-Lopez V, Sanz-Jaka JP, Cormenzana P, Eizaguirre I, Bernad A, Izeta A (2009) Age-dependent depletion of human skin-derived progenitor cells. Stem Cells 27(5):1164–1172

Galderisi U, Helmbold H, Squillaro T, Alessio N, Komm N, Khadang B, Cipollaro M, Bohn W, Giordano A (2009) In vitro senescence of rat mesenchymal stem cells is accompanied by downregulation of stemness-related and DNA damage repair genes. Stem Cells Dev 18(7):1033–1042

Geiger H, de Haan G, Florian MC (2013) The ageing haematopoietic stem cell compartment. Nat Rev Immunol 13(5):376–389

Ghannam S, Pene J, Torcy-Moquet G, Jorgensen C, Yssel H (2010) Mesenchymal stem cells inhibit human Th17 cell differentiation and function and induce a T regulatory cell phenotype. J Immunol 185(1):302–312

Gonzalez-Rey E, Anderson P, Gonzalez MA, Rico L, Buscher D, Delgado M (2009) Human adult stem cells derived from adipose tissue protect against experimental colitis and sepsis. Gut 58(7):929–939

Gruver AL, Hudson LL, Sempowski GD (2007) Immunosenescence of ageing. J Pathol 211(2):144–156

Guo L, Zhao RC, Wu Y (2011) The role of microRNAs in self-renewal and differentiation of mesenchymal stem cells. Exp Hematol 39(6):608–616

Gurtner GC, Werner S, Barrandon Y, Longaker MT (2008) Wound repair and regeneration. Nature 453(7193):314–321

Hackl M, Brunner S, Fortschegger K, Schreiner C, Micutkova L, Muck C, Laschober GT, Lepperdinger G, Sampson N, Berger P, Herndler-Brandstetter D, Wieser M, Kuhnel H, Strasser A, Rinnerthaler M, Breitenbach M, Mildner M, Eckhart L, Tschachler E, Trost A, Bauer JW, Papak C, Trajanoski Z, Scheideler M, Grillari-Voglauer R, Grubeck-Loebenstein B, Jansen-Durr P, Grillari J (2010) miR-17, miR-19b, miR-20a, and miR-106a are down-regulated in human aging. Aging Cell 9(2):291–296

Haybesworth SE, Goldberg VM, Caplan AI (1994) Diminution of the number mesenchymal stem cells as a cause for skeletal aging. In: Woo SL-Y (ed) Musculoskeletal soft-tissue aging: impact on mobility. AAOS Workshop, Rosemont, pp 79–87

Heo SC, Jeon ES, Lee IH, Kim HS, Kim MB, Kim JH (2011) Tumor necrosis factor-alpha-activated human adipose tissue-derived mesenchymal stem cells accelerate cutaneous wound healing through paracrine mechanisms. J Invest Dermatol 131(7):1559–1567

Hocking AM, Gibran NS (2010) Mesenchymal stem cells: paracrine signaling and differentiation during cutaneous wound repair. Exp Cell Res 316(14):2213–2219

Hoenig MR, Bianchi C, Rosenzweig A, Sellke FW (2008) Decreased vascular repair and neovascularization with ageing: mechanisms and clinical relevance with an emphasis on hypoxia-inducible factor-1. Curr Mol Med 8(8):754–767

Hoogduijn MJ, Gorjup E, Genever PG (2006) Comparative characterization of hair follicle dermal stem cells and bone marrow mesenchymal stem cells. Stem Cells Dev 15(1):49–60

Hwa Cho H, Bae YC, Jung JS (2006) Role of toll-like receptors on human adipose-derived stromal cells. Stem Cells 24(12):2744–2752

In't Anker PS, Scherjon SA, der Kleijburg-van KC, Noort WA, Claas FH, Willemze R, Fibbe WE, Kanhai HH (2003) Amniotic fluid as a novel source of mesenchymal stem cells for therapeutic transplantation. Blood 102(4):1548–1549

Islam MN, Das SR, Emin MT, Wei M, Sun L, Westphalen K, Rowlands DJ, Quadri SK, Bhattacharya S, Bhattacharya J (2012) Mitochondrial transfer from bone-marrow-derived stromal cells to pulmonary alveoli protects against acute lung injury. Nat Med 18(5):759–765

Jackson WM, Nesti LJ, Tuan RS (2012) Mesenchymal stem cell therapy for attenuation of scar formation during wound healing. Stem Cell Res Ther 3(3):20

Jahoda CA, Reynolds AJ (2001) Hair follicle dermal sheath cells: unsung participants in wound healing. Lancet 358(9291):1445–1448

Jahoda CA, Whitehouse J, Reynolds AJ, Hole N (2003) Hair follicle dermal cells differentiate into adipogenic and osteogenic lineages. Exp Dermatol 12(6):849–859

Jaul E (2010) Assessment and management of pressure ulcers in the elderly: current strategies. Drugs Aging 27(4):311–325

Jiang D, Qi Y, Walker NG, Sindrilaru A, Hainzl A, Wlaschek M, Macneil S, Scharffetter-Kochanek K (2013) The effect of adipose tissue derived MSCs delivered by a chemically defined carrier on full-thickness cutaneous wound healing. Biomaterials 34(10):2501–2515

Joannides A, Gaughwin P, Schwiening C, Majed H, Sterling J, Compston A, Chandran S (2004) Efficient generation of neural precursors from adult human skin: astrocytes promote neurogenesis from skin-derived stem cells. Lancet 364(9429):172–178

Jung-Hynes B, Ahmad N (2009) SIRT1 controls circadian clock circuitry and promotes cell survival: a connection with age-related neoplasms. FASEB J 23(9):2803–2809

Kavanagh H, Mahon BP (2011) Allogeneic mesenchymal stem cells prevent allergic airway inflammation by inducing murine regulatory T cells. Allergy 66(4):523–531

Keating A (2012) Mesenchymal stromal cells: new directions. Cell Stem Cell 10(6):709–716

Kim J, Hematti P (2009) Mesenchymal stem cell-educated macrophages: a novel type of alternatively activated macrophages. Exp Hematol 37(12):1445–1453

Kim EK, Li G, Lee TJ, Hong JP (2011) The effect of human adipose-derived stem cells on healing of ischemic wounds in a diabetic nude mouse model. Plast Reconstr Surg 128(2):387–394

Kim C, Schneider G, Abdel-Latif A, Mierzejewska K, Sunkara M, Borkowska S, Ratajczak J, Morris AJ, Kucia M, Ratajczak MZ (2013) Ceramide-1-phosphate regulates migration of multipotent stromal cells and endothelial progenitor cells–implications for tissue regeneration. Stem Cells 31(3):500–510

Kishimoto TK, Hollander N, Roberts TM, Anderson DC, Springer TA (1987) Heterogeneous mutations in the beta subunit common to the LFA-1, Mac-1, and p150,95 glycoproteins cause leukocyte adhesion deficiency. Cell 50(2):193–202

Kligman LH, Kligman AM (1986) The nature of photoaging: its prevention and repair. Photodermatol 3(4):215–227

Kretlow JD, Jin YQ, Liu W, Zhang WJ, Hong TH, Zhou G, Baggett LS, Mikos AG, Cao Y (2008) Donor age and cell passage affects differentiation potential of murine bone marrow-derived stem cells. BMC Cell Biol 9:60

Ksander BR, Kolovou PE, Wilson BJ, Saab KR, Guo Q, Ma J, McGuire SP, Gregory MS, Vincent WJ, Perez VL, Cruz-Guilloty F, Kao WW, Call MK, Tucker BA, Zhan Q, Murphy GF, Lathrop KL, Alt C, Mortensen LJ, Lin CP, Zieske JD, Frank MH, Frank NY (2014) ABCB5 is a limbal stem cell gene required for corneal development and repair. Nature 511(7509):353–357

Kwon DS, Gao X, Liu YB, Dulchavsky DS, Danyluk AL, Bansal M, Chopp M, McIntosh K, Arbab AS, Dulchavsky SA, Gautam SC (2008) Treatment with bone marrow-derived stromal cells accelerates wound healing in diabetic rats. Int Wound J 5(3):453–463

Lai RC, Arslan F, Lee MM, Sze NS, Choo A, Chen TS, Salto-Tellez M, Timmers L, Lee CN, El Oakley RM, Pasterkamp G, de Kleijn DP, Lim SK (2010a) Exosome secreted by MSC reduces myocardial ischemia/reperfusion injury. Stem Cell Res 4(3):214–222

Lai RC, Arslan F, Tan SS, Tan B, Choo A, Lee MM, Chen TS, Teh BJ, Eng JK, Sidik H, Tanavde V, Hwang WS, Lee CN, El Oakley RM, Pasterkamp G, de Kleijn DP, Tan KH, Lim SK (2010b) Derivation and characterization of human fetal MSCs: an alternative cell source for large-scale production of cardioprotective microparticles. J Mol Cell Cardiol 48(6):1215–1224

Lai RC, Chen TS, Lim SK (2011) Mesenchymal stem cell exosome: a novel stem cell-based therapy for cardiovascular disease. Regen Med 6(4):481–492

Laschober GT, Brunauer R, Jamnig A, Singh S, Hafen U, Fehrer C, Kloss F, Gassner R, Lepperdinger G (2011) Age-specific changes of mesenchymal stem cells are paralleled by upregulation of CD106 expression as a response to an inflammatory environment. Rejuvenation Res 14(2):119–131

Lee RH, Seo MJ, Reger RL, Spees JL, Pulin AA, Olson SD, Prockop DJ (2006) Multipotent stromal cells from human marrow home to and promote repair of pancreatic islets and renal glomeruli in diabetic NOD/scid mice. Proc Natl Acad Sci U S A 103(46): 17438–17443

Lee RH, Pulin AA, Seo MJ, Kota DJ, Ylostalo J, Larson BL, Semprun-Prieto L, Delafontaine P, Prockop DJ (2009) Intravenous hMSCs improve myocardial infarction in mice because cells embolized in lung are activated to secrete the anti-inflammatory protein TSG-6. Cell Stem Cell 5(1):54–63

Lee SH, Jin SY, Song JS, Seo KK, Cho KH (2012) Paracrine effects of adipose-derived stem cells on keratinocytes and dermal fibroblasts. Ann Dermatol 24(2):136–143

Li H, Fu X, Ouyang Y, Cai C, Wang J, Sun T (2006) Adult bone-marrow-derived mesenchymal stem cells contribute to wound healing of skin appendages. Cell Tissue Res 326(3):725–736

Li Q, Ke F, Zhang W, Shen X, Xu Q, Wang H, Yu XZ, Leng Q, Wang H (2011) Plasmin plays an essential role in amplification of psoriasiform skin inflammation in mice. PLoS One 6(2), e16483

Liang XJ, Chen XJ, Yang DH, Huang SM, Sun GD, Chen YP (2012) Differentiation of human umbilical cord mesenchymal stem cells into hepatocyte-like cells by hTERT gene transfection in vitro. Cell Biol Int 36(2):215–221

Lin Q, Wang L, Lin Y, Liu X, Ren X, Wen S, Du X, Lu T, Su SY, Yang X, Huang W, Zhou S, Wen F, Su SB (2012) Toll-like receptor 3 ligand polyinosinic:polycytidylic acid promotes wound healing in human and murine skin. J Invest Dermatol 132(8):2085–2092

Liu D, Ou L, Clemenson GD Jr, Chao C, Lutske ME, Zambetti GP, Gage FH, Xu Y (2010) Puma is required for p53-induced depletion of adult stem cells. Nat Cell Biol 12(10):993–998

Liu GH, Barkho BZ, Ruiz S, Diep D, Qu J, Yang SL, Panopoulos AD, Suzuki K, Kurian L, Walsh C, Thompson J, Boue S, Fung HL, Sancho-Martinez I, Zhang K, Yates J 3rd, Izpisua Belmonte JC (2011) Recapitulation of premature ageing with iPSCs from Hutchinson-Gilford progeria syndrome. Nature 472(7342):221–225

Liu B, Ghosh S, Yang X, Zheng H, Liu X, Wang Z, Jin G, Zheng B, Kennedy BK, Suh Y, Kaeberlein M, Tryggvason K, Zhou Z (2012) Resveratrol rescues SIRT1-dependent adult stem cell decline and alleviates progeroid features in laminopathy-based progeria. Cell Metab 16(6):738–750

Liu S, Jiang L, Li H, Shi H, Luo H, Zhang Y, Yu C, Jin Y (2014) Mesenchymal stem cells prevent hypertrophic scar formation via inflammatory regulation when undergoing apoptosis. J Invest Dermatol 134(10):2648–2657

Lorenz K, Sicker M, Schmelzer E, Rupf T, Salvetter J, Schulz-Siegmund M, Bader A (2008) Multilineage differentiation potential of human dermal skin-derived fibroblasts. Exp Dermatol 17(11):925–932

Lu D, Chen B, Liang Z, Deng W, Jiang Y, Li S, Xu J, Wu Q, Zhang Z, Xie B, Chen S (2011) Comparison of bone marrow mesenchymal stem cells with bone marrow-derived mononuclear cells for treatment of diabetic critical limb ischemia and foot ulcer: a double-blind, randomized, controlled trial. Diabetes Res Clin Pract 92(1):26–36

Maby-El Hajjami H, Ame-Thomas P, Pangault C, Tribut O, DeVos J, Jean R, Bescher N, Monvoisin C, Dulong J, Lamy T, Fest T, Tarte K (2009) Functional alteration of the lymphoma stromal cell niche by the cytokine context: role of indoleamine-2,3 dioxygenase. Cancer Res 69(7):3228–3237

Mastri M, Shah Z, McLaughlin T, Greene CJ, Baum L, Suzuki G, Lee T (2012) Activation of Toll-like receptor 3 amplifies mesenchymal stem cell trophic factors and enhances therapeutic potency. Am J Physiol Cell Physiol 303(10):C1021–C1033

Matsumoto R, Omura T, Yoshiyama M, Hayashi T, Inamoto S, Koh KR, Ohta K, Izumi Y, Nakamura Y, Akioka K, Kitaura Y, Takeuchi K, Yoshikawa J (2005) Vascular endothelial growth factor-expressing mesenchymal stem cell transplantation for the treatment of acute myocardial infarction. Arterioscler Thromb Vasc Biol 25(6):1168–1173

Meisel R, Zibert A, Laryea M, Gobel U, Daubener W, Dilloo D (2004) Human bone marrow stromal cells inhibit allogeneic T-cell responses by indoleamine 2,3-dioxygenase-mediated tryptophan degradation. Blood 103(12):4619–4621

Meraviglia V, Azzimato V, Piacentini L, Chiesa M, Kesharwani RK, Frati C, Capogrossi MC, Gaetano C, Pompilio G, Colombo GI, Rossini A (2014) Syngeneic cardiac and bone marrow stromal cells display tissue-specific microRNA signatures and microRNA subsets restricted to diverse differentiation processes. PLoS One 9(9), e107269

Miura Y, Miura M, Gronthos S, Allen MR, Cao C, Uveges TE, Bi Y, Ehirchiou D, Kortesidis A, Shi S, Zhang L (2005) Defective osteogenesis of the stromal stem cells predisposes CD18-null mice to osteoporosis. Proc Natl Acad Sci U S A 102(39):14022–14027

Mulder GD, Lee DK, Faghihnia N (2010) Autologous bone marrow-derived stem cells for chronic wounds of the lower extremity: a retrospective study. Wounds 22(9):219–225

Murphy MB, Moncivais K, Caplan AI (2013) Mesenchymal stem cells: environmentally responsive therapeutics for regenerative medicine. Exp Mol Med 45, e54

Nasef A, Ashammakhi N, Fouillard L (2008) Immunomodulatory effect of mesenchymal stromal cells: possible mechanisms. Regen Med 3(4):531–546

Nauta AJ, Kruisselbrink AB, Lurvink E, Willemze R, Fibbe WE (2006) Mesenchymal stem cells inhibit generation and function of both CD34+–derived and monocyte-derived dendritic cells. J Immunol 177(4):2080–2087

Nemeth K, Leelahavanichkul A, Yuen PS, Mayer B, Parmelee A, Doi K, Robey PG, Leelahavanichkul K, Koller BH, Brown JM, Hu X, Jelinek I, Star RA, Mezey E (2009) Bone marrow stromal cells attenuate sepsis via prostaglandin E(2)-dependent reprogramming of host macrophages to increase their interleukin-10 production. Nat Med 15(1): 42–49

Nemeth K, Keane-Myers A, Brown JM, Metcalfe DD, Gorham JD, Bundoc VG, Hodges MG, Jelinek I, Madala S, Karpati S, Mezey E (2010) Bone marrow stromal cells use TGF-beta to suppress allergic responses in a mouse model of ragweed-induced asthma. Proc Natl Acad Sci U S A 107(12):5652–5657

Nurmenniemi S, Kuvaja P, Lehtonen S, Tiuraniemi S, Alahuhta I, Mattila RK, Risteli J, Salo T, Selander KS, Nyberg P, Lehenkari P (2010) Toll-like receptor 9 ligands enhance mesenchymal stem cell invasion and expression of matrix metalloprotease-13. Exp Cell Res 316(16): 2676–2682

O'Loughlin A, Kulkarni M, Creane M, Vaughan EE, Mooney E, Shaw G, Murphy M, Dockery P, Pandit A, O'Brien T (2013) Topical administration of allogeneic mesenchymal stromal cells seeded in a collagen scaffold augments wound healing and increases angiogenesis in the diabetic rabbit ulcer. Diabetes 62(7):2588–2594

Opalenik SR, Davidson JM (2005) Fibroblast differentiation of bone marrow-derived cells during wound repair. FASEB J 19(11):1561–1563

Ortiz LA, Gambelli F, McBride C, Gaupp D, Baddoo M, Kaminski N, Phinney DG (2003) Mesenchymal stem cell engraftment in lung is enhanced in response to bleomycin exposure and ameliorates its fibrotic effects. Proc Natl Acad Sci U S A 100(14):8407–8411

Oswald J, Boxberger S, Jorgensen B, Feldmann S, Ehninger G, Bornhauser M, Werner C (2004) Mesenchymal stem cells can be differentiated into endothelial cells in vitro. Stem Cells 22(3):377–384

Ozerdem U, Alitalo K, Salven P, Li A (2005) Contribution of bone marrow-derived pericyte precursor cells to corneal vasculogenesis. Invest Ophthalmol Vis Sci 46(10): 3502–3506

Park JS, Kim HY, Kim HW, Chae GN, Oh HT, Park JY, Shim H, Seo M, Shin EY, Kim EG, Park SC, Kwak SJ (2005) Increased caveolin-1, a cause for the declined adipogenic potential of senescent human mesenchymal stem cells. Mech Ageing Dev 126(5):551–559

Pekovic V, Hutchison CJ (2008) Adult stem cell maintenance and tissue regeneration in the ageing context: the role for A-type lamins as intrinsic modulators of ageing in adult stem cells and their niches. J Anat 213(1):5–25

Peters T, Sindrilaru A, Hinz B, Hinrichs R, Menke A, Al-Azzeh EA, Holzwarth K, Oreshkova T, Wang H, Kess D, Walzog B, Sulyok S, Sunderkotter C, Friedrich W, Wlaschek M, Krieg T, Scharffetter-Kochanek K (2005) Wound-healing defect of CD18(–/–) mice due to a decrease in TGF-beta1 and myofibroblast differentiation. EMBO J 24(19):3400–3410

Peters T, Sindrilaru A, Wang H, Oreshkova T, Renkl AC, Kess D, Scharffetter-Kochanek K (2006) CD18 in monogenic and polygenic inflammatory processes of the skin. J Investig Dermatol Symp Proc 11(1):7–15

Peters T, Weiss JM, Sindrilaru A, Wang H, Oreshkova T, Wlaschek M, Maity P, Reimann J, Scharffetter-Kochanek K (2009) Reactive oxygen intermediate-induced pathomechanisms contribute to immunosenescence, chronic inflammation and autoimmunity. Mech Ageing Dev 130(9):564–587

Pevsner-Fischer M, Morad V, Cohen-Sfady M, Rousso-Noori L, Zanin-Zhorov A, Cohen S, Cohen IR, Zipori D (2007) Toll-like receptors and their ligands control mesenchymal stem cell functions. Blood 109(4):1422–1432

Pietila M, Lehtonen S, Tuovinen E, Lahteenmaki K, Laitinen S, Leskela HV, Natynki A, Pesala J, Nordstrom K, Lehenkari P (2012) CD200 positive human mesenchymal stem cells suppress TNF-alpha secretion from CD200 receptor positive macrophage-like cells. PLoS One 7(2), e31671

Piwko-Czuchra A, Koegel H, Meyer H, Bauer M, Werner S, Brakebusch C, Fassler R (2009) Beta1 integrin-mediated adhesion signalling is essential for epidermal progenitor cell expansion. PLoS One 4(5), e5488

Plotnikov EY, Khryapenkova TG, Vasileva AK, Marey MV, Galkina SI, Isaev NK, Sheval EV, Polyakov VY, Sukhikh GT, Zorov DB (2008) Cell-to-cell cross-talk between mesenchymal stem cells and cardiomyocytes in co-culture. J Cell Mol Med 12(5A):1622–1631

Popova AP, Bozyk PD, Goldsmith AM, Linn MJ, Lei J, Bentley JK, Hershenson MB (2010) Autocrine production of TGF-beta1 promotes myofibroblastic differentiation of neonatal lung mesenchymal stem cells. Am J Physiol Lung Cell Mol Physiol 298(6):L735–L743

Potapova IA, Gaudette GR, Brink PR, Robinson RB, Rosen MR, Cohen IS, Doronin SV (2007) Mesenchymal stem cells support migration, extracellular matrix invasion, proliferation, and survival of endothelial cells in vitro. Stem Cells 25(7):1761–1768

Prockop DJ (2012) Mitochondria to the rescue. Nat Med 18(5):653–654

Prockop DJ, Brenner M, Fibbe WE, Horwitz E, Le Blanc K, Phinney DG, Simmons PJ, Sensebe L, Keating A (2010) Defining the risks of mesenchymal stromal cell therapy. Cytotherapy 12(5):576–578

Qi Y, Jiang D, Sindrilaru A, Stegemann A, Schatz S, Treiber N, Rojewski M, Schrezenmeier H, Vander Beken S, Wlaschek M, Bohm M, Seitz A, Scholz N, Durselen L, Brinckmann J, Ignatius A, Scharffetter-Kochanek K (2014) TSG-6 released from intradermally injected mesenchymal stem cells accelerates wound healing and reduces tissue fibrosis in murine full-thickness skin wounds. J Invest Dermatol 134(2):526–537

Quan TE, Cowper S, Wu SP, Bockenstedt LK, Bucala R (2004) Circulating fibrocytes: collagen-secreting cells of the peripheral blood. Int J Biochem Cell Biol 36(4):598–606

Quarto R, Thomas D, Liang CT (1995) Bone progenitor cell deficits and the age-associated decline in bone repair capacity. Calcif Tissue Int 56(2):123–129

Raffaghello L, Bianchi G, Bertolotto M, Montecucco F, Busca A, Dallegri F, Ottonello L, Pistoia V (2008) Human mesenchymal stem cells inhibit neutrophil apoptosis: a model for neutrophil preservation in the bone marrow niche. Stem Cells 26(1):151–162

Raicevic G, Rouas R, Najar M, Stordeur P, Boufker HI, Bron D, Martiat P, Goldman M, Nevessignsky MT, Lagneaux L (2010) Inflammation modifies the pattern and the function of toll-like receptors expressed by human mesenchymal stromal cells. Hum Immunol 71(3):235–244

Raju S, Neglen P (2009) Clinical practice. Chronic venous insufficiency and varicose veins. N Engl J Med 360(22):2319–2327

Rando TA (2006) Stem cells, ageing and the quest for immortality. Nature 441(7097):1080–1086

Ravari H, Hamidi-Almadari D, Salimifar M, Bonakdaran S, Parizadeh MR, Koliakos G (2011) Treatment of non-healing wounds with autologous bone marrow cells, platelets, fibrin glue and collagen matrix. Cytotherapy 13(6):705–711

Ren G, Zhang L, Zhao X, Xu G, Zhang Y, Roberts AI, Zhao RC, Shi Y (2008) Mesenchymal stem cell-mediated immunosuppression occurs via concerted action of chemokines and nitric oxide. Cell Stem Cell 2(2):141–150

Rodriguez-Menocal L, Salgado M, Ford D, Van Badiavas E (2012) Stimulation of skin and wound fibroblast migration by mesenchymal stem cells derived from normal donors and chronic wound patients. Stem Cells Transl Med 1(3):221–229

Rose RA, Jiang H, Wang X, Helke S, Tsoporis JN, Gong N, Keating SC, Parker TG, Backx PH, Keating A (2008) Bone marrow-derived mesenchymal stromal cells express cardiac-specific markers, retain the stromal phenotype, and do not become functional cardiomyocytes in vitro. Stem Cells 26(11):2884–2892

Rosengardten Y, McKenna T, Grochova D, Eriksson M (2011) Stem cell depletion in Hutchinson-Gilford progeria syndrome. Aging Cell 10(6):1011–1020

Salmon AB, Perez VI, Bokov A, Jernigan A, Kim G, Zhao H, Levine RL, Richardson A (2009) Lack of methionine sulfoxide reductase a in mice increases sensitivity to oxidative stress but does not diminish life span. FASEB J 23(10):3601–3608

Salvolini E, Lucarini G, Zizzi A, Orciani M, Di Benedetto G, Di Primio R (2010a) Human skin-derived mesenchymal stem cells as a source of VEGF and nitric oxide. Arch Dermatol Res 302(5):367–374

Salvolini E, Orciani M, Vignini A, Mattioli-Belmonte M, Mazzanti L, Di Primio R (2010b) Skin-derived mesenchymal stem cells (S-MSCs) induce endothelial cell activation by paracrine mechanisms. Exp Dermatol 19(9):848–850

Sanges D, Lluis F, Cosma MP (2011) Cell-fusion-mediated reprogramming: pluripotency or transdifferentiation? Implications for regenerative medicine. Adv Exp Med Biol 713:137–159

Sarasua JG, Lopez SP, Viejo MA, Basterrechea MP, Rodriguez AF, Gutierrez AF, Gala JG, Menendez YM, Augusto DE, Arias AP, Hernandez JO (2011) Treatment of pressure ulcers with autologous bone marrow nuclear cells in patients with spinal cord injury. J Spinal Cord Med 34(3):301–307

Sasaki M, Abe R, Fujita Y, Ando S, Inokuma D, Shimizu H (2008) Mesenchymal stem cells are recruited into wounded skin and contribute to wound repair by transdifferentiation into multiple skin cell type. J Immunol 180(4):2581–2587

Scaffidi P, Misteli T (2008) Lamin A-dependent misregulation of adult stem cells associated with accelerated ageing. Nat Cell Biol 10(4):452–459

Schatteman GC, Ma N (2006) Old bone marrow cells inhibit skin wound vascularization. Stem Cells 24(3):717–721

Scheid A, Wenger RH, Schaffer L, Camenisch I, Distler O, Ferenc A, Cristina H, Ryan HE, Johnson RS, Wagner KF, Stauffer UG, Bauer C, Gassmann M, Meuli M (2002) Physiologically low oxygen concentrations in fetal skin regulate hypoxia-inducible factor 1 and transforming growth factor-beta3. FASEB J 16(3):411–413

Schneider RK, Neuss S, Stainforth R, Laddach N, Bovi M, Knuechel R, Perez-Bouza A (2008) Three-dimensional epidermis-like growth of human mesenchymal stem cells on dermal equivalents: contribution to tissue organization by adaptation of myofibroblastic phenotype and function. Differentiation 76(2):156–167

Sellheyer K, Krahl D (2010) Skin mesenchymal stem cells: prospects for clinical dermatology. J Am Acad Dermatol 63(5):859–865

Selmani Z, Naji A, Zidi I, Favier B, Gaiffe E, Obert L, Borg C, Saas P, Tiberghien P, Rouas-Freiss N, Carosella ED, Deschaseaux F (2008) Human leukocyte antigen-G5 secretion by human mesenchymal stem cells is required to suppress T lymphocyte and natural killer function and to induce CD4+CD25highFOXP3+ regulatory T cells. Stem Cells 26(1): 212–222

Sen CK, Khanna S, Babior BM, Hunt TK, Ellison EC, Roy S (2002) Oxidant-induced vascular endothelial growth factor expression in human keratinocytes and cutaneous wound healing. J Biol Chem 277(36):33284–33290

Sensebe L, Krampera M, Schrezenmeier H, Bourin P, Giordano R (2010) Mesenchymal stem cells for clinical application. Vox Sang 98(2):93–107

Seppanen E, Roy E, Ellis R, Bou-Gharios G, Fisk NM, Khosrotehrani K (2013) Distant mesenchymal progenitors contribute to skin wound healing and produce collagen: evidence from a murine fetal microchimerism model. PLoS One 8(5), e62662

Sharpless NE, DePinho RA (2007) How stem cells age and why this makes us grow old. Nat Rev Mol Cell Biol 8(9):703–713

Sheng Z, Fu X, Cai S, Lei Y, Sun T, Bai X, Chen M (2009) Regeneration of functional sweat gland-like structures by transplanted differentiated bone marrow mesenchymal stem cells. Wound Repair Regen 17(3):427–435

Shi Y, Hu G, Su J, Li W, Chen Q, Shou P, Xu C, Chen X, Huang Y, Zhu Z, Huang X, Han X, Xie N, Ren G (2010) Mesenchymal stem cells: a new strategy for immunosuppression and tissue repair. Cell Res 20(5):510–518

Shilo S, Roy S, Khanna S, Sen CK (2007) MicroRNA in cutaneous wound healing: a new paradigm. DNA Cell Biol 26(4):227–237

Si Y, Zhao Y, Hao H, Liu J, Guo Y, Mu Y, Shen J, Cheng Y, Fu X, Han W (2012) Infusion of mesenchymal stem cells ameliorates hyperglycemia in type 2 diabetic rats: identification of a novel role in improving insulin sensitivity. Diabetes 61(6):1616–1625

Sindrilaru A, Seeliger S, Ehrchen JM, Peters T, Roth J, Scharffetter-Kochanek K, Sunderkotter CH (2007) Site of blood vessel damage and relevance of CD18 in a murine model of immune complex-mediated vasculitis. J Invest Dermatol 127(2):447–454

Sindrilaru A, Peters T, Schymeinsky J, Oreshkova T, Wang H, Gompf A, Mannella F, Wlaschek M, Sunderkotter C, Rudolph KL, Walzog B, Bustelo XR, Fischer KD, Scharffetter-Kochanek K (2009) Wound healing defect of Vav3–/– mice due to impaired {beta}2-integrin-dependent macrophage phagocytosis of apoptotic neutrophils. Blood 113(21):5266–5276

Sindrilaru A, Peters T, Wieschalka S, Baican C, Baican A, Peter H, Hainzl A, Schatz S, Qi Y, Schlecht A, Weiss JM, Wlaschek M, Sunderkotter C, Scharffetter-Kochanek K (2011) An unrestrained proinflammatory M1 macrophage population induced by iron impairs wound healing in humans and mice. J Clin Invest 121(3):985–997

Singer NG, Caplan AI (2011) Mesenchymal stem cells: mechanisms of inflammation. Annu Rev Pathol 6:457–478

Smith AN, Willis E, Chan VT, Muffley LA, Isik FF, Gibran NS, Hocking AM (2010) Mesenchymal stem cells induce dermal fibroblast responses to injury. Exp Cell Res 316(1):48–54

Sotiropoulou PA, Perez SA, Gritzapis AD, Baxevanis CN, Papamichail M (2006) Interactions between human mesenchymal stem cells and natural killer cells. Stem Cells 24(1):74–85

Spees JL, Olson SD, Whitney MJ, Prockop DJ (2006) Mitochondrial transfer between cells can rescue aerobic respiration. Proc Natl Acad Sci U S A 103(5):1283–1288

Stappenbeck TS, Miyoshi H (2009) The role of stromal stem cells in tissue regeneration and wound repair. Science 324(5935):1666–1669

Stenderup K, Justesen J, Clausen C, Kassem M (2003) Aging is associated with decreased maximal life span and accelerated senescence of bone marrow stromal cells. Bone 33(6):919–926

Stock P, Bruckner S, Ebensing S, Hempel M, Dollinger MM, Christ B (2010) The generation of hepatocytes from mesenchymal stem cells and engraftment into murine liver. Nat Protoc 5(4):617–627

Stoff A, Rivera AA, Sanjib Banerjee N, Moore ST, Michael Numnum T, Espinosa-de-Los-Monteros A, Richter DF, Siegal GP, Chow LT, Feldman D, Vasconez LO, Michael Mathis J, Stoff-Khalili MA, Curiel DT (2009) Promotion of incisional wound repair by human mesenchymal stem cell transplantation. Exp Dermatol 18(4):362–369

Stolzing A, Jones E, McGonagle D, Scutt A (2008) Age-related changes in human bone marrow-derived mesenchymal stem cells: consequences for cell therapies. Mech Ageing Dev 129(3):163–173

Su X, Paris M, Gi YJ, Tsai KY, Cho MS, Lin YL, Biernaskie JA, Sinha S, Prives C, Pevny LH, Miller FD, Flores ER (2009) TAp63 prevents premature aging by promoting adult stem cell maintenance. Cell Stem Cell 5(1):64–75

Sun J, Han ZB, Liao W, Yang SG, Yang Z, Yu J, Meng L, Wu R, Han ZC (2011) Intrapulmonary delivery of human umbilical cord mesenchymal stem cells attenuates acute lung injury by expanding CD4+CD25+ Forkhead Boxp3 (FOXP3)+regulatory T cells and balancing anti- and pro-inflammatory factors. Cell Physiol Biochem 27(5):587–596

Tang YL, Zhao Q, Qin X, Shen L, Cheng L, Ge J, Phillips MI (2005) Paracrine action enhances the effects of autologous mesenchymal stem cell transplantation on vascular regeneration in rat model of myocardial infarction. Ann Thorac Surg 80(1):229–236

Tark KC, Hong JW, Kim YS, Hahn SB, Lee WJ, Lew DH (2010) Effects of human cord blood mesenchymal stem cells on cutaneous wound healing in leprdb mice. Ann Plast Surg 65(6):565–572

Tobin DJ, Gunin A, Magerl M, Handijski B, Paus R (2003) Plasticity and cytokinetic dynamics of the hair follicle mesenchyme: implications for hair growth control. J Invest Dermatol 120(6):895–904

Toma JG, Akhavan M, Fernandes KJ, Barnabe-Heider F, Sadikot A, Kaplan DR, Miller FD (2001) Isolation of multipotent adult stem cells from the dermis of mammalian skin. Nat Cell Biol 3(9):778–784

Toma C, Pittenger MF, Cahill KS, Byrne BJ, Kessler PD (2002) Human mesenchymal stem cells differentiate to a cardiomyocyte phenotype in the adult murine heart. Circulation 105(1):93–98

Traktuev DO, Merfeld-Clauss S, Li J, Kolonin M, Arap W, Pasqualini R, Johnstone BH, March KL (2008) A population of multipotent CD34-positive adipose stromal cells share pericyte and mesenchymal surface markers, reside in a periendothelial location, and stabilize endothelial networks. Circ Res 102(1):77–85

Tso GH, Law HK, Tu W, Chan GC, Lau YL (2010) Phagocytosis of apoptotic cells modulates mesenchymal stem cells osteogenic differentiation to enhance IL-17 and RANKL expression on CD4+ T cells. Stem Cells 28(5):939–954

Ueda M, Nishino Y (2010) Cell-based cytokine therapy for skin rejuvenation. J Craniofac Surg 21(6):1861–1866

Vaculik C, Schuster C, Bauer W, Iram N, Pfisterer K, Kramer G, Reinisch A, Strunk D, Elbe-Burger A (2012) Human dermis harbors distinct mesenchymal stromal cell subsets. J Invest Dermatol 132(3 Pt 1):563–574

Valle-Prieto A, Conget PA (2010) Human mesenchymal stem cells efficiently manage oxidative stress. Stem Cells Dev 19(12):1885–1893

Varani J, Schuger L, Dame MK, Leonard C, Fligiel SE, Kang S, Fisher GJ, Voorhees JJ (2004) Reduced fibroblast interaction with intact collagen as a mechanism for depressed collagen synthesis in photodamaged skin. J Invest Dermatol 122(6):1471–1479

Wan J, Xia L, Liang W, Liu Y, Cai Q (2013) Transplantation of bone marrow-derived mesenchymal stem cells promotes delayed wound healing in diabetic rats. J Diabetes Res 2013:647107

Wang H, Peters T, Kess D, Sindrilaru A, Oreshkova T, Van Rooijen N, Stratis A, Renkl AC, Sunderkotter C, Wlaschek M, Haase I, Scharffetter-Kochanek K (2006) Activated macrophages are essential in a murine model for T cell-mediated chronic psoriasiform skin inflammation. J Clin Invest 116(8):2105–2114

Wang XJ, Dong Z, Zhong XH, Shi RZ, Huang SH, Lou Y, Li QP (2008) Transforming growth factor-beta1 enhanced vascular endothelial growth factor synthesis in mesenchymal stem cells. Biochem Biophys Res Commun 365(3):548–554

Waterman RS, Tomchuck SL, Henkle SL, Betancourt AM (2010) A new mesenchymal stem cell (MSC) paradigm: polarization into a pro-inflammatory MSC1 or an Immunosuppressive MSC2 phenotype. PLoS One 5(4), e10088

Williams JT, Southerland SS, Souza J, Calcutt AF, Cartledge RG (1999) Cells isolated from adult human skeletal muscle capable of differentiating into multiple mesodermal phenotypes. Am Surg 65(1):22–26

Wlaschek M, Scharffetter-Kochanek K (2005) Oxidative stress in chronic venous leg ulcers. Wound Repair Regen 13(5):452–461

Wu XB, Tao R (2012) Hepatocyte differentiation of mesenchymal stem cells. Hepatobiliary Pancreat Dis Int 11(4):360–371

Wu Y, Chen L, Scott PG, Tredget EE (2007) Mesenchymal stem cells enhance wound healing through differentiation and angiogenesis. Stem Cells 25(10):2648–2659

Wu Y, Huang S, Enhe J, Ma K, Yang S, Sun T, Fu X (2014) Bone marrow-derived mesenchymal stem cell attenuates skin fibrosis development in mice. Int Wound J 11(6):701–710

Xie MW, Gorodetsky R, Micevicz ED, Mackenzie NC, Gaberman E, Levdansky L, McBride WH (2013) Marrow-derived stromal cell delivery on fibrin microbeads can correct radiation-induced wound-healing deficits. J Invest Dermatol 133(2):553–561

Xu B, deWaal RM, Mor-Vaknin N, Hibbard C, Markovitz DM, Kahn ML (2004) The endothelial cell-specific antibody PAL-E identifies a secreted form of vimentin in the blood vasculature. Mol Cell Biol 24(20):9198–9206

Xue S, Li L (2011) Upregulation of collagen type 1 in aged murine dermis after transplantation of dermal multipotent cells. Clin Exp Dermatol 36(7):775–781

Yang LY, Zheng JK, Liu XM, Hui GZ, Guo LH (2004) Culture of skin-derived precursors and their differentiation into neurons. Chin J Traumatol 7(2):91–95

Yao H, Ting X, Minjie W, Yemin C, Xiqiao W, Yuzhi J, Ming T, Weida W, Peifen Q, Shuliang L (2012) The investigation of demographic characteristics and the health-related quality of life in patients with diabetic foot ulcers at first presentation. Int J Low Extrem Wounds 11(3):187–193

Yew TL, Hung YT, Li HY, Chen HW, Chen LL, Tsai KS, Chiou SH, Chao KC, Huang TF, Chen HL, Hung SC (2011) Enhancement of wound healing by human multipotent stromal cell conditioned medium: the paracrine factors and p38 MAPK activation. Cell Transplant 20(5):693–706

Zech NH (2005) Plasticity of stem cells: cell-fusion versus transdifferentiation. J Reproduktionsmed Endokrinol 2(4):239–245

Zhang QZ, Su WR, Shi SH, Wilder-Smith P, Xiang AP, Wong A, Nguyen AL, Kwon CW, Le AD (2010) Human gingiva-derived mesenchymal stem cells elicit polarization of m2 macrophages and enhance cutaneous wound healing. Stem Cells 28(10):1856–1868

Zhang J, Lian Q, Zhu G, Zhou F, Sui L, Tan C, Mutalif RA, Navasankari R, Zhang Y, Tse HF, Stewart CL, Colman A (2011) A human iPSC model of Hutchinson Gilford Progeria reveals vascular smooth muscle and mesenchymal stem cell defects. Cell Stem Cell 8(1):31–45

Zhang DY, Pan Y, Zhang C, Yan BX, Yu SS, Wu DL, Shi MM, Shi K, Cai XX, Zhou SS, Wang JB, Pan JP, Zhang LH (2013) Wnt/beta-catenin signaling induces the aging of mesenchymal stem cells through promoting the ROS production. Mol Cell Biochem 374(1–2):13–20

Zou Z, Zhang Y, Hao L, Wang F, Liu D, Su Y, Sun H (2010) More insight into mesenchymal stem cells and their effects inside the body. Expert Opin Biol Ther 10(2):215–230

Zouboulis CC, Adjaye J, Akamatsu H, Moe-Behrens G, Niemann C (2008) Human skin stem cells and the ageing process. Exp Gerontol 43(11):986–997

Retinal Repair Using Stem Cells 15

Mark Gutierrez and Deepak A. Lamba

Abstract

The retina is the main light-sensing layer of the eye. Degenerations of the retinal cells are leading causes of blindness worldwide. This cellular loss in the mammalian retina is not followed by any appreciable cellular regeneration. In this chapter, we will cover various age-associated retinal degenerations. We will provide evidence of retinal regeneration in nonmammalian systems as well as attempts to rejuvenate stem cell phenotype in mammalian retinas. We will finish with recent work using pluripotent stem cells, both embryonic stem cells as well as induced pluripotent stem cells, to generate new retinal cells for replacement purposes.

15.1 Introduction: The Retina

The retina is the innermost layer of the eye. It arises from the ventral diencephalon extending out as an optic placode, which subsequently indents to form a two-layered optic cup. The outer layer differentiates to form the retinal pigment epithelium (RPE) while the inner layer forms the neural retina. The neural retina has three main cellular layers: outer nuclear, inner nuclear, and the ganglion cell layer with two plexiform layers in between where the cells synapse. The outer nuclear layer contains the rod and cone photoreceptors of the eye that mediate our vision (Fig. 15.1). The photoreceptors respond to photons of light by changes in their membrane potential resulting in hyperpolarization. The rod photoreceptors are capable of responding to low levels of light, and hence responsible for our night vision, while cone photoreceptors mediate high-acuity vision in the daylight. They also provide color vision due to the presence of three types based on peak sensitivity to red, green, and blue spectrum.

M. Gutierrez • D.A. Lamba (✉)
Department of Biology, Buck Institute for Research on Aging,
8001 Redwood Blvd Rm: 3-234, Novato, CA 94945, USA
e-mail: dlamba@buckinstitute.org

H. Geiger et al. (eds.), *Stem Cell Aging: Mechanisms, Consequences, Rejuvenation*,
DOI 10.1007/978-3-7091-1232-8_15

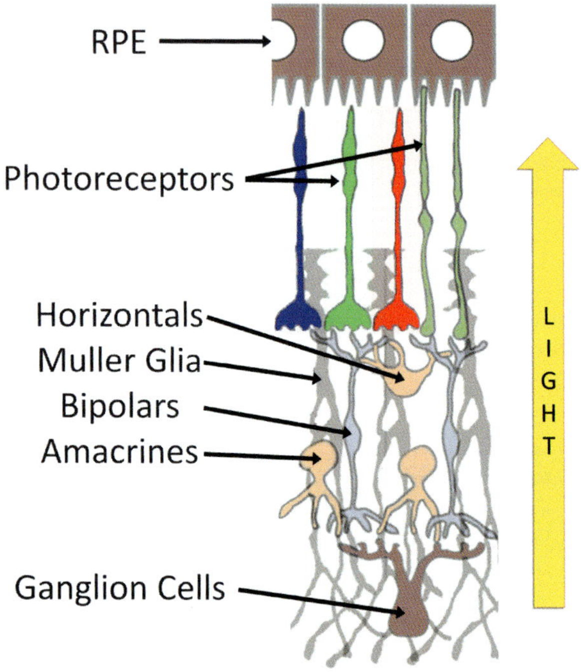

Fig. 15.1 Schematic of the human retina. The figure shows the laminated retinal structure with the various cell types arranged in distinct layers

In humans and nonhuman primates, the central region of the retina is enriched with cones and is called the macula. The central part of the macula called *fovea centralis* has a pit due to the presence of only cone photoreceptors. This is the region of highest density of cone photoreceptors in the eye. The photoreceptors transmit the light information to a group of retinal interneurons called bipolar cells. These cells, in turn, transmit the signals to ganglion cells, either directly or through another class of inhibitory retinal interneurons, called amacrine cells. The ganglion cells are the main output neurons of the retina, whose axons combine to form the optic nerve to send the visual information to central visual centers. Following retinal differentiation, the final cell that comes out of the retinal stem cells is the Müller glial cells.

Outer to the photoreceptor layer is the layer of pigmented cells, called RPE. RPE cells are vital for the maintenance of the rods and cones. The apical end of RPE cells faces the photoreceptors and has microvilli, which completely engulf the outer segments of the photoreceptors. RPE cells perform multiple supportive functions. The RPE layer transports ions, water, and metabolic by-products from the photoreceptor side to the choriocapillaris on the other side. The rod and cone photoreceptors shed approximately 10 % of their outer segments a day. This generates enormous amount of cellular debris into the subretinal space and the RPE cells are responsible for getting rid of photoreceptor disc debris. Additionally, RPE cells are critical for the normal visual cycle. The cells recycle visual cycle component, all-trans retinol, back to 11-cis form. The RPE layer is also part of the blood–retinal barrier, which may play a critical role in immune protection and immune privilege. Finally, the pigmented cells absorb and prevent light scattering thereby aiding in visual acuity.

15.2 Retinal Degenerations

The retina is subject to a number of inherited and acquired forms of retinal degenerative conditions. Age-related macular degeneration (AMD) is a disease of the retina characterized by the degeneration of cells in the RPE and the photoreceptors in the central region of the retina. AMD is one of the leading causes of blindness in the USA. It affects approximately 10 % of the population over 65 and up to 1 in 3 over the age of 80 years. Glaucoma and diabetic retinopathy also lead to retinal neuronal loss. Glaucoma primarily leads to apoptosis of ganglion cells, while diabetic retinopathy leads to loss of most retinal neurons secondary to a vasculopathy. The incidence of both of these diseases increases with age. There are also a number of inherited retinal disorders including Usher's syndrome and retinitis pigmentosa (RP), which do not manifest until much later in life.

Age-related macular degeneration, the leading cause of worldwide blindness in the elderly, affects approximately 15 million people in the USA alone. About 1.75 million people in USA currently have advanced age-related macular degeneration with associated vision loss, and the number is expected to grow to almost three million by 2020. The disease poses a significant financial burden on the society accounting for $68 billion in direct costs in the USA. The pathogenesis of AMD is unclear and multiple factors may be involved in the progression of the disorder including chronic oxidative stress, lifelong light exposure, small RNA processing defects in the RPE, choriocapillaris atrophy, accumulation of lipofuscin, and inflammation and complement activation (Fig. 15.2) (Jarrett and Boulton 2012; Kaneko et al. 2011; Khan et al. 2006). The disease starts in the macula, the region of central retina with highest visual acuity. There are two main forms of the AMD disorder: dry and wet form. The disease is characterized by the development of drusen which are pathological extracellular deposits containing glycolipids, proteins, and cellular

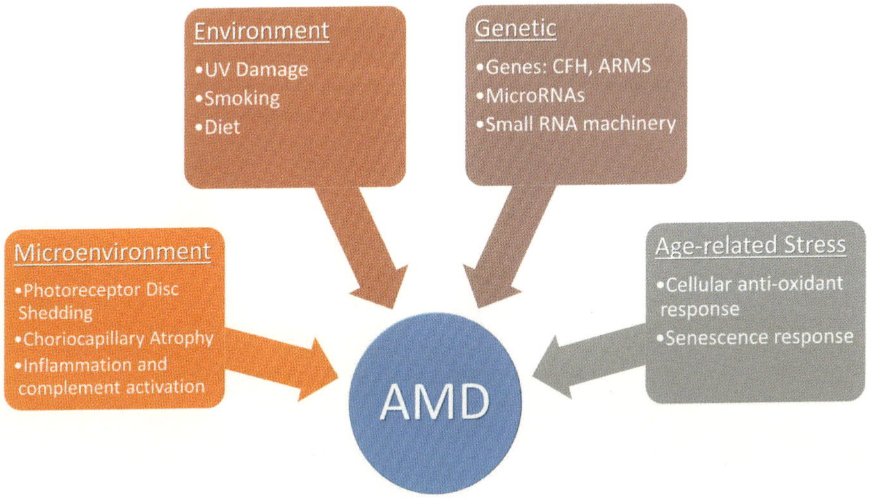

Fig. 15.2 Multifactorial etiology of age-related macular degeneration. A number of genetic, environmental, local microenvironment, and age-associated factors have been linked to the development of the disorder

debris, between the RPE and the overlying Bruch's membrane. Early AMD is associated with more numerous soft drusen in the macula, pigmentary changes in the RPE, and thickening of Bruch's membrane. Advanced forms of AMD can manifest as either geographic atrophy, which is the dry form, or neovascular/wet form of AMD. Geographic atrophy results in dysfunction and loss of the RPE and degeneration of the underlying photoreceptors occurs. Wet AMD is characterized by the development of choroidal neovascularization following disruption of the Bruch's membrane, leading to exudation, bleeding, and scar formation.

A number of studies have associated oxidative stress as the key driver to AMD (Jarrett and Boulton 2012). Evidences pointing to this include (a) decrease in microsomal glutathione S-transferase-1, an enzyme that reduces peroxides, oxidized RPE lipids, and oxidized retinoids, (b) increased lipid peroxidation, and (c) age-associated decrease in catalase activity in the human RPE. The resultant oxidant damage manifests in age-related increases in autofluorescent lipofuscin granules especially those containing A2E, a pyridinium bisretinoid by-product of visual pathway component *all-trans* retinol. A2E is a potent photo-inducible generator of ROS in the RPE (Sparrow and Boulton 2005). Additional cellular changes include 8-oxodG accumulation, lipid peroxidation and glycation of end products, and mitochondrial DNA damage. Linkage and genome-wide association studies have implicated genetic modulators of AMD risk related to many mechanistic pathways, including oxidative stress, complement system dysregulation, DNA repair, mitochondrial dysfunction, neovascularization, and microglial recruitment (Gorin 2012). The Y402H polymorphism in complement factor CFH has been consistently demonstrated as a significant risk factor for AMD (Boon et al. 2008). Single nucleotide polymorphisms (SNPs) in ARMS2 and HTRA1, two genes with strong linkage related to extracellular matrix function, are also associated with AMD susceptibility (Andreoli et al. 2009). Polymorphisms in the apolipoprotein E (apoE) gene modulate AMD risk as well (Thakkinstian et al. 2006). The E4 allelic variant seems to have a protective effect while the E2 allele is associated with increased AMD risk. Some recent SNPs in genes have suggested a role of cholesterol and lipid dysfunction in the development of the pathology. These include hepatic lipase gene (LIPC), cholesterylester transfer protein (CETP), lipoprotein lipase (LPL), and ATP-binding cassette, subfamily A member 1 (ABCA1).

Small RNAs have come into focus for their involvement in a number of developmental and disease states. Recent papers have suggested a role of microRNAs, small RNA processing enzyme, Dicer1, and Alu RNAs in the pathogenesis of AMD (Kaneko et al. 2011; Lukiw et al. 2012). NF-κB-regulated microRNAs are upregulated in both AMD and Alzheimer's disease while other miRNAs including miRNA-9, miRNA-125b, miRNA-146a, and miRNA-155 are known to target 3′-UTR of CFH, thereby downregulating CFH. It has also been found that mir23a decreases in RPE cells from AMD donor eyes and is associated with cellular resistance to oxidative stress (Lin et al. 2011). Recent studies looking at small RNA processing have identified a reduction in the enzyme Dicer1, the small RNA processing protein, in RPE of patients with AMD (Kaneko et al. 2011). They however did not see a huge difference in microRNAs. Instead, they found a concomitant rise in Alu RNA accumulation in the cells. The group found that knocking-down *DICER1* or overexpressing Alu RNA activates the NLRP3 inflammasome and triggers TLR-independent MyD88 signaling via IL18 in RPE cells (Tarallo et al. 2012). These could provide us with novel targets to treat the disease.

Glaucoma is one of the leading causes of vision loss and blindness, affecting as many as 60 million people around the world. It is estimated that nearly 80 million people will have glaucoma by the year 2020 (Quigley and Broman 2006). Like AMD, the pathogenesis of glaucoma is extremely complex and poorly understood. Glaucoma is characterized by progressing cupping of the optic nerve head, thinning of the nerve fiber layer, and retinal gliosis, all of which are associated with concomitant retinal ganglion cell death. The disease is classically associated with rise in the intraocular pressure. Dysfunction of the trabecular meshwork in the angle of the anterior chamber of the eye is linked with the disease development. There are two broad types: angle-closure glaucoma, where the angle of the eye is narrow preventing the outflow of the aqueous fluid, and open-angle glaucoma, in which case the meshwork itself is dysfunctional. In either case, there is rise in the intraocular pressure, which due to some yet unknown mechanism, leads to selective ganglion cell stress and loss. Additionally, there is a subset of patients with, what is referred to as, normal tension glaucoma. These patients have all the changes classically described for glaucoma in the absence of intraocular pressure changes.

A number of mechanisms have been postulated to explain the pathogenesis of glaucoma, including chronic intermittent ischemia due to rise in pressure, accumulation of reactive oxygen species, defective axon transport, loss of retrograde trophic factor support, and excitotoxicity (Calkins 2012). Defective axon transport is one of the earliest changes demonstrated in animal models of glaucoma. This could be due to mechanic pressure on the optic nerve head at the *lamina cribrosa* or be secondary to ischemia. This transport blockage also prevents retrograde transport of neurotrophic factors such as brain-derived neurotrophic factor (BDNF) and likely of other neurotrophic factors from central targets back to ganglion cells, which are critical for survival and dendritic maintenance. All together, these result in programmed cell death of the ganglion cells by upregulation of various proapoptotic markers. Several approaches have been utilized to identify genetic factors that are associated with glaucoma and multiple loci have been linked to open-angle glaucoma. Four genes with strong linkage include myocilin (*MYOC*), optineurin (*OPTN*), WD repeat domain 36 (*WDR36*), and neurotrophin 4 (*NTF4*) (Allingham et al. 2009). Genome-wide association studies have linked additional genes including caveolin 1 and 2 (*CAV1/CAV2*), transmembrane and coiled-coil domains 1 (*TMCO1*), and cyclin-dependent kinase inhibitor 2B (*CDKN2B*).

Diabetic retinopathy is one of the most common complications of diabetes and remains the leading cause of blindness among individuals in developed countries. Population-based studies suggest that approximately one-third of the population with diabetes have early signs of retinopathy and approximately one in ten has vision-threatening stages of retinopathy, such as macular edema and proliferative retinopathy (Chistiakov 2011). The major risk factors for diabetic retinopathy are hyperglycemia, high blood pressure, and duration of diabetes. Patients with diabetic retinopathy include the formation of microaneurysms, retinal hemorrhages, hard exudates, cotton wool spots, venous abnormalities, and neovascularization. The disease is associated with the loss of pericytes and endothelial cells, formation of acellular capillaries, thickening of the basement membrane, formation of microaneurysms, and neovascularization. The underlying mechanism on how hyperglycemia induces these changes is unclear but may involve changes in growth factor expression including VEGF and IGF1, activation of diacylglycerol–protein kinase C pathway, accumulation of advanced glycation end products, RAS activation, and misregulation of polyol pathway (Simo 2011).

Following degeneration in any of the aforementioned disorders, the retina needs to be rebuilt as innate repair mechanisms fall short of any visual improvement. Current avenues of research focus on two broad categories: (1) restoration of stem cell characteristics in host support cells to initiate some host repair and (2) replacement or generation of new cells in vitro which can then be transplanted to take over the function of lost cells.

15.3 Retinal Regeneration: Endogenous Mechanisms

Amphibians, fish, birds, and mammals have all been shown to have varying degrees of retinal regenerative potential. This has been most prominent at the earliest stages of development and diminishes over time as the retina attains maturity (Del Rio-Tsonis and Tsonis 2003). Most retinal regeneration studies employ a process described as transdifferentiation, i.e., the ability of one cell type to take over a different fate. Transdifferentiation of neural retina involves three types of cells, support cells in the retina (called the Müller glial cells), cells from the RPE, and cells located in the ciliary epithelium/ciliary marginal zone at the periphery of the retina (Fig. 15.3). Regeneration via transdifferentiation involves three steps. First is the process of dedifferentiation where the RPE cells/Müller glial cells lose their native characteristics; second, they proliferate; and third and last, some of the cells differentiate into retinal cells (Del Rio-Tsonis and Tsonis 2003; Haynes and Del Rio-Tsonis 2004).

15.3.1 Transdifferentiation from RPE Cells

Amphibians have a remarkable capacity to regenerate various organs and tissues including brain, limbs, and liver. Newt retina has been a great model to study neural retinal regeneration. Following removal of the entire neural retina, the eye can regenerate an entire complete and laminated new retina using the surrounding RPE cells (Stone 1950a, b; Stone and Steinitz 1957). The process involves proliferation and dedifferentiation of the RPE cells into an optic placode-like stem cell. These cells then regenerate a new bilayered structure, which has both a new neural retina and a new RPE cell layer. The newly formed retina has all the different cells including photoreceptors, bipolar cells, and ganglion cells as well as the inhibitory interneurons and glial cells. The only difference lies from the fact that the newly

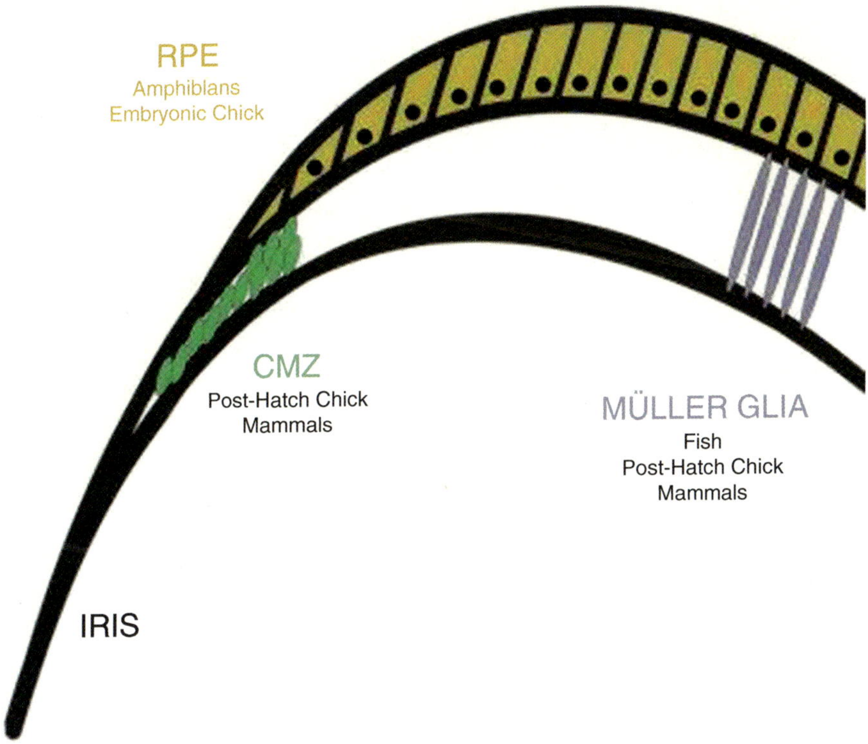

Fig. 15.3 Sources of endogenous retinal regeneration and repair. The figure highlights the three main sources of repair in the various species, i.e., RPE, CMZ, and Müller glia

developed retina is inverted in orientation (Okada 1980). Recent studies have shown that the RPE transdifferentiation into neural cells occurs only in the presence of underlying connective tissue, i.e., choroid (Mitsuda et al. 2005). The results suggest the possibility of choroid playing an essential role in newt retinal regeneration potentially by means of diffusible substances. Two secretory factors that are upregulated in the choroid following newt retina removal include bFGF and IGF1. There is downregulation of Otx2, a key RPE gene, and upregulation of Notch pathway activity (Nakamura and Chiba 2007; Sakami et al. 2005). Some key regulators of this regeneration process include extracellular matrix components, such as laminin and heparin sulfate proteoglycans (Reh et al. 1987), as well as growth factors FGF-2 and IGF1 (Mitsuda et al. 2005; Sakaguchi et al. 1997), among others.

This phenomenon of RPE transdifferentiation also occurs in the embryonic chicken eyes (Coulombre and Coulombre 1965; Park and Hollenberg 1989; Pittack et al. 1991). When the neural retina is dissected out from the chicken embryos within the first 3–4 days of incubation (E3–E4), the RPE dedifferentiates and proliferates to form retinal progenitors, which then differentiate to form a new retina as in the newt. In the chick embryo, this process is stimulated by bFGFs with contribution from BMPs (Haynes et al. 2007) while activin signaling inhibits the regeneration

(Sakami et al. 2008). Both FGFs and activin signaling are known to drive Pax6, a key transcription factor in retinal development. Recent experiments have shown that overexpression of Pax6 is sufficient to stimulate the production of neurons from the embryonic chicken RPE (Azuma et al. 2005). As opposed to the newt, in chickens, this RPE regenerative capacity is restricted to very early embryonic development. The mature chicken RPE does not contribute to retinal repair. So far, the data from mammalian RPE suggests very little contribution, if any, to retinal repair.

15.3.2 Transdifferentiation from Ciliary Margin

The peripheral margin of the retina where it meets the ciliary body has been thought of as another source of retinal stem cells and is called the ciliary marginal zone (CMZ). Retina of fish grows throughout life roughly matching the overall growth of the animal by the addition of new neurons from stem cells in the CMZ (Hollyfield 1968; Straznicky and Gaze 1971). In fact, most of the fish retina is generated from this CMZ as they grow to adult size. The CMZ cells resemble the embryonic retinal progenitor cells. The cells can generate clones of over 100 cells containing all the different retinal cell types (Wetts and Fraser 1988). Teleost fish also retain the ability to regenerate new retinal neurons following neurotoxin damage or surgical lesions throughout life (Hitchcock et al. 2004; Raymond et al. 2006). The chicken retina was largely believed to be completely developed at hatching. However, a study by Fischer and Reh showed that the CMZ is active in posthatch chickens (Fischer and Reh 2000, 2002). These cells at the retinal margin express genes common to embryonic retinal progenitors including Pax6. Injection of a combination of insulin IGF and EGF induces the cells in the CMZ to express progenitor markers and undergo proliferation. Similarly, sonic hedgehog induces proliferation in the adult chicken CMZ (Moshiri et al. 2005). A similar active CMZ has also been demonstrated in adult quails (Kubota et al. 2002).

Several reports have shown that cells isolated from the ciliary epithelium of several mammals, including mice, pig, and humans, can give rise to proliferative clones (neurospheres) when cultured in vitro (Ahmad et al. 2000; Asami et al. 2007; Coles et al. 2004; Kohno et al. 2006; MacNeil et al. 2007; Sun et al. 2006; Tropepe et al. 2000). The cells express a variety of neural progenitor markers and have been shown to undergo extensive proliferation. Pax6, neural progenitor gene, is required for the generation of these neurospheres (Xu et al. 2007). Upon exposure to differentiation conditions, these spheres were shown to express rod photoreceptors, bipolar and Müller glial genes. The sphere proliferation can be enhanced by exposure to Wnts, SCF, and PEDF (Inoue et al. 2006; Das et al. 2004, 2006a; De Marzo et al. 2010). Despite this early hope, some recent reports have suggested that these spheres may not completely resemble retinal stem cells. A detailed analysis by Mike Dyer's group showed that all cells in the CMZ-derived spheres, including those expressing neural progenitor markers, were still pigmented and exhibited other characteristics of normal pigmented cells (Cicero et al. 2009). Thus, the jury is still out on the existence of a CMZ in humans, which can be exploited for native retinal repair.

15.3.3 Transdifferentiation from the Müller Glia

As discussed above, the fish retina contains a CMZ at the retinal margin, and it contributes to retinal growth as the fish increase in size. However, this zone is not thought to contribute to regeneration in central retina. Instead, fish contains two other quiescent progenitor populations, the rod progenitor cells and the Müller glia (Johns 1982). Upon injury, fish retinal Müller glia slowly start dividing to generate the rod progenitor cells (Bernardos et al. 2007). Using transgenic fish with GFP expressed in Müller glia, several labs have demonstrated that these cells are the primary source of regenerated neurons in fish central retina. Following damage to retinal neurons, the retina Müller glia reenter the cell cycle and dedifferentiate toward a progenitor-like phenotype (Raymond et al. 2006; Bernardos et al. 2007; Faillace et al. 2002; Fausett and Goldman 2006; Fimbel et al. 2007; Kassen et al. 2007; Thummel et al. 2008; Vihtelic et al. 2006; Wu et al. 2001; Yurco and Cameron 2005). A number of groups are now attempting to better understand the processes involved in reentry of the glial cells into cell cycle with the hopes of applying it to mammalian regeneration (Raymond et al. 2006; Kassen et al. 2007; Cameron et al. 2005; Yurco and Cameron 2007). It has been demonstrated that not all glial cells have the capacity to regenerate the retina but that these cells are asymmetrically distributed (Julian et al. 1998). Dan Goldman's group has identified *Ascl1*, a basic helix-loop-helix gene expressed highly in retinal progenitors, as one that gets upregulated in Müller glia upon retinal damage (Fausett et al. 2008). They showed using morpholinos that inhibition of *Ascl1* blocks Müller glial dedifferentiation and their subsequent reentry into the cell cycle. This is also interesting as a recent study in mammalian retina demonstrated the Ascl1 expression defines a subpopulation of lineage-restricted progenitors in the mammalian retina (Brzezinski et al. 2011). The process is mediated by EGF/MAPK/Notch cascade (Wan et al. 2012) and Wnt/β--catenin signaling (Ramachandran et al. 2011). A recent study looked for the inducer of Müller glial proliferation following damage. They found that TNF-α was produced by dying retinal neurons and was necessary for expression of ASCL1 in Müller glia (Nelson et al. 2013). Another secreted factor, which has been shown to induce glial proliferation in the absence of injury, is CNTF. It does this through a STAT3-dependent mechanism (Kassen et al. 2009). Injury-induced Ascl1 directly regulates lin-28, a microRNA-binding protein. Lin-28 then represses let-7 miRNA, which is normally involved in repression of regeneration-associated genes including *ASCL1*, *PAX6*, *OCT4*, *KLF4*, and *c-MYC* (Ramachandran et al. 2010). A detailed characterization study by Pam Raymond's group into the genes involved in the zebrafish regenerative process showed that cells upregulated the Notch–Delta pathway as well as key retinal progenitor genes, *Pax6a*, *Rx*, and *Chx10*, upon reentry into the cell cycle (Raymond et al. 2006). These data suggest that regeneration and development use the similar molecular mechanisms and revoking a similar pathway in mammalian glial cells may restore regenerative capacity.

Like the fish retina, chicken retina displays some regenerative capacity (Fischer and Reh 2001). The earliest demonstration of this was using neurotoxic damage of inner retinal neurons with *N*-methyl-D-aspartate (NMDA) in posthatch chicks

(Fischer and Reh 2001). Müller glia reenter the cell cycle and dedifferentiate into retinal progenitors (Fischer and Reh 2001, 2003; Fischer et al. 2002; Hayes et al. 2007). These Müller glial-derived retinal progenitors differentiate into amacrine cells, bipolar cells, and ganglion cells. There was no evidence of photoreceptor cell differentiation. A close examination of the Notch–Delta pathway revealed a dual role in the regenerative response (Hayes et al. 2007). Notch upregulation was indeed necessary for the glia to dedifferentiate and reenter cell cycle. However, after this initial need, Notch inhibition is necessary for differentiation of the Müller glial-derived retinal progenitors to form new neurons.

As opposed to fish and avian retinas, the mammalian Müller glia do not spontaneously reenter the cell cycle following retina damage (Chang et al. 2007; Close et al. 2006; Dayer et al. 2005; Sahel et al. 1990; Zhao et al. 2005). In order to get them to proliferate following damage, they require additional stimulation including growth factors. Ooto et al. used NMDA to induce damage in the rat inner retina (Ooto et al. 2004). They found that Müller glia underwent proliferation, though it was limited. They detected some of the Müller glial-derived progeny, which expressed markers of bipolar cells and photoreceptors. This number could be increased by co-injecting retinoic acid. Alternatively, overexpression of certain transcription factors including Math3, NeuroD1, Pax6, and Crx could generate bipolar and amacrine cells and photoreceptors. A number of additional publications have also reported similar BrdU-expressing photoreceptor cell generation following either retinal damage or growth factor treatment in adult rodents (Close et al. 2006; Das et al. 2006b; Osakada et al. 2007; Wan et al. 2007; Wan et al. 2008; Del Debbio et al. 2010). To gain a better handle on understanding the regenerative process in mouse retinas, Karl et al. conducted a detailed examination of the progeny derived from glia following toxic damage and EGF stimulation in various transgenic strains of mice to track various lineages (Karl et al. 2008). Using this protocol, up to 10 % of Müller glia entered the cell cycle within 3 days. They found that almost all of the Müller glial-derived progenitors differentiated to amacrine cells. Consistent with this observation, they found that FoxN4, a transcription factor necessary for the amacrine cell differentiation, was one of the most highly upregulated genes in both the mouse and chick retina after NMDA-induced damage. Other studies using Wnt3a or Shh following damage have reported photoreceptor cell differentiation (Osakada et al. 2007; Wan et al. 2007, 2008). There have been no studies of functional regeneration in either the chicken or the mammalian retina. Thus, it is still not known whether Müller glial-derived cells functionally integrate into the retinal circuitry.

Recently, the field of direct reprogramming has come into a lot of focus. Direct reprogramming or induced transdifferentiation refers to the process that allows cells to directly progress from one cell fate to another without an intermediate stem cell fate. The earliest landmark paper describing this was the direct conversion of fibroblast cells to myoblasts using a single transcription factor MyoD1 (Davis et al. 1987). More recently, a similar transdifferentiation strategy was used to generate neurons from fibroblast using ASCL1, BRN2, and MYT1L (Vierbuchen et al. 2010). A recent publication from Tom Reh's lab attempted to employ the strategy to restore retinal stem cell characteristics to mammalian Müller glia (Pollak et al. 2013). As described above, Ascl1a is required for retinal regeneration in the fish and is rapidly upregulated in proliferating Müller glia. However, Ascl1 is not upregulated in the mouse retina after *N*-methyl-D-aspartate (NMDA)-induced damage (Karl et al. 2008). Therefore, the authors induced expression of ASCL1 in Müller glia in either dissociated cultures or explant retinas (Pollak et al. 2013). They found that ASCL1 was sufficient to induce neural differentiation of the glial cells. ASCL1 remodeled the chromatin at retinal progenitor gene sites activating their expression while at the same time downregulating glial genes. The reprogrammed cells expressed early postmitotic retinal markers including Otx2, Islet1, Neurod1, Prox1, and Crx. Thus, direct reprogramming could serve as an alternate strategy to induce retinal regeneration in mammalian retinas.

15.4 Retinal Repair: Cell Replacement

An alternative to finding ways to restore innate repair is to transplant fresh cells into the degenerated retina in order to restore visual function. Several potential sources for replacement neurons and photoreceptor cells have been hypothesized and tested. These include (1) dissociated differentiated retinal cells, (2) intact retinal sheets, (3) retinal progenitor cells, (4) neural progenitors derived from the CNS, (5) pluripotent stem cells, and (6) cells derived from CMZ and ciliary epithelium. The earliest published work on rodent retinal transplantation involved transplanting embryonic rat eyes into the brains and skin of newborn rats (Tansley 1946). Upon transplantation in skin pockets, only mesenchyme survived while the retina matured well in case of brain transplantations. Subsequently, a number of other intracranial transplantation studies were carried out by Ray Lund's lab, to look at axon pathfinding. They showed that the axons of the transplanted cells can make contacts along the visual pathway and develop light responses (Lund and Hankin 1995; McLoon et al. 1981; McLoon and Lund 1980). The rats could also use functional visual cue from the graft site in various behavioral assays (Coffey et al. 1989; Klassen and Lund 1990). This early work showed that the mature nervous system is still plastic enough to accept new connectivity and information for functional tasks.

15.5 Transplantation of Retinal Cells

More recent work has concentrated on replacement of cells within the retina. There are two routes that could potentially be used to replace cells: (a) injection into the vitreal cavity close to the ganglion cell side and (b) injection into the subretinal space, potential space between the photoreceptors and overlying RPE cells. Integration following intravitreal transplantation has been limited due to inadequate migration of the transplanted cells into the retina probably due to tight junctions between astrocytes or end feet of Müller glia (Takahashi et al. 1998). Cell migration into the photoreceptor layer is more effective in normal adult retina if a subretinal approach is used (Lu et al. 2002; MacLaren et al. 2006). Developing retinal cells have been transplanted as intact retinal sheets (Seiler and Aramant 1998), retinal aggregates, or retinal spheres (del Cerro et al. 1991; Kwan et al. 1999) or as dissociated cell suspensions (MacLaren et al. 2006; Chacko et al. 2000; Qiu et al. 2005; Barber et al. 2013; Lakowski et al. 2011). Although retinal sheets and spheres survive and differentiate well in the subretinal space, the cells seem to self-aggregate into rosette and do not migrate much into the host retina from the graft site. However, enzymatically dissociating the developing retinas and transplanting cell suspensions results in better integration of the transplanted cells (Qiu et al. 2005).

The best demonstration of photoreceptor integration in a rodent retina was by MacLaren et al. (MacLaren et al. 2006); the authors dissociated retinal cells from a GFP-expressing transgenic mice wherein all the rod photoreceptors are green into the subretinal space of wild-type mice of different ages. They found that GFP-expressing cells migrating into the photoreceptor layer developed the morphology of mature photoreceptors. The cells upon integration elaborated outer segments and made synaptic connections with the host bipolar cells. The transplanted cells expressed markers of mature rod photoreceptors including rhodopsin. In the study, they compared integration from different donor ages varying from embryonic mice to adult mice. Interestingly, they found that the most successful integration was from newborn rod photoreceptors obtained from mice in the first postnatal week. Surprisingly, the most immature donor cells, from embryonic stages, were not effective for transplantation. These cells survived and matured but formed rosettes instead of integrating. A recent study from Tom Reh's lab showed that adult mice can integrate just as well if their survival can be improved (Gust and Reh 2011). They found that the reason or poor integration was cell death due to the stress of dissociation and transplantation. Robin Ali's group recently published a report comparing integration in various models of retinal degeneration ranging from slow to rapid degeneration (Barber et al. 2013). They found that a number of factors can influence integration efficiency including integrative of the outer limiting membrane of the retina formed by tight junction between glia, the gliotic status of the host retina, as well as speed of degeneration. Using approaches to alter the outer limiting membrane integrity just as ZO-1 siRNA or chondroitinase ABC can help enhance integration.

15.6 Transplantation of Non-retinal Cells

As discussed in the previous section, neurospheres can be generated from adult CMZ and ciliary epithelium of mammals (Ahmad et al. 2000; Asami et al. 2007; Coles et al. 2004; Kohno et al. 2006; MacNeil et al. 2007; Sun et al. 2006; Tropepe et al. 2000). In order to assess if these cells can be coaxed to differentiate into mature photoreceptor cells, Coles et al. transplanted ciliary-derived spheres into newborn mice by injecting them into the vitreal cavity of a newborn mouse and analyzing 4 weeks later (Coles et al. 2004). The authors saw migration of the transplanted cells to the retina and some integration into the photoreceptor layer as well as RPE. On account of the limited integration, the group recently tried to enhance integration and maturation by overexpressing two critical photoreceptor and RPE genes, *OTX2* and *CRX*, while at the same time inhibiting *CHX10*, a progenitor gene (Inoue et al. 2010). They reported improved morphological integration with some functional improvement on behavioral and electrophysiological assays. A few groups have looked at neural progenitors as a source of replacement neurons (Takahashi et al. 1998; Young et al. 2000). These progenitors are obtained from the dentate gyrus of the hippocampus, a site of active neurogenesis even in adult mice. Hippocampal-derived neural progenitors exhibit a high degree of migration into all layers of the retina, following intravitreal injection in either newborn mice or adult mice with retinal degeneration or ischemia (Grozdanic et al. 2006; Guo et al. 2003; Akita et al. 2002; Kurimoto et al. 2001). Despite their hippocampal origin, they develop morphologies highly reminiscent of several types of retinal neurons; however, they did not show an ability to differentiate into mature photoreceptors.

15.7 Transplantation of Pluripotent Stem Cell-Derived Retinal Cells

Pluripotent stem cells are most commonly known as cells with "blank slates," having the potential to differentiate into any cell in the developed body. Currently under scientific investigation are two main groups of pluripotent cells: embryonic stem cells (ESCs) and induced pluripotent stem cells (iPSCs). ESCs are derived from the inner cell mass of a developing embryo at the blastocyst stage of development (Thomson et al. 1998), while iPSCs are pluripotent cells generated in vitro by the insertion and subsequent expression of a group of transcription factors: OCT4, KLF4, SOX2, c-MYC, LIN28, and NANOG (Fig. 15.4) (Takahashi et al. 2007; Takahashi and Yamanaka 2006; Yu et al. 2007). By virtue of pluripotency, these cells have the capacity to form the endoderm, mesoderm, and ectoderm germ layers seen in early development. Additionally, when grown under specific differentiation factors that help mimic the cellular environment of specific parts of the embryo, pluripotent cells can be urged to develop into specific cell fates that are seen in the developed body. Such a process that can be studied in vitro presents immense potential of investigating developmental biology, drug/toxicology studies, disease modeling, and aging, among other fields of study. Building on the knowledge of the network of eye-field transcription factors in the development of the vertebrate eye (Bharti et al. 2012; Zuber et al. 2003), scientists have been able to develop protocols with which retinal cells can be derived from pluripotent cells in vitro (Lamba et al. 2006, 2009a; Meyer et al. 2009; Osakada et al. 2008, 2009; Phillips et al. 2012; Hambright et al. 2012). This knowledge may play a further role in the development of practical and safe cell replacement treatments for retinal degenerative disorders.

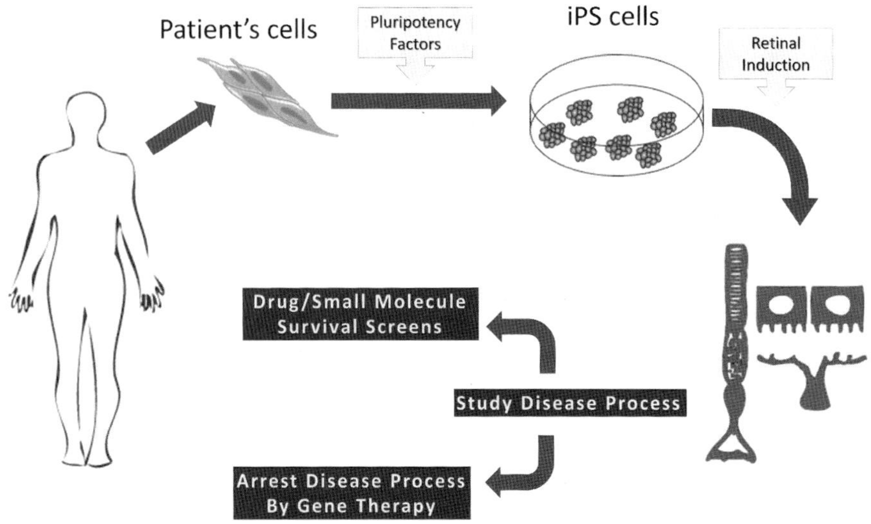

Fig. 15.4 Induced pluripotent stem cell uses for retinal degenerations. The schematic shows the process involved in the use of patient-derived pluripotent stem cells and their potential for either replacement therapy or drug screening

To date, multiple groups have developed protocols with which retinal cell types may be obtained in vitro. The process of the in vitro differentiation that is generally under investigation by many groups is that which mimics the developmental biology of the developing optic vesicle in vivo. This process involves the generation of neural rosette structures in the presence of a retinal induction medium. Pax6 and Rx co-expressing cells indicate the presence of neural retina and RPE progenitor cells (Lamba et al. 2006; Ikeda et al. 2005; Tucker et al. 2013). The derivation of these cells from pluripotent stem cells relies on the inhibition of Wnt/B-catenin signaling and TGFβ/BMP signaling, which is accomplished through the treatment of pluripotent cells with Dickkopf-1 (DKK-1) and Noggin (Lamba et al. 2006) or Lefty-A (Ikeda et al. 2005). Richard-Parpaillon and colleagues have also found in 2002 that the overexpression of insulin-like growth factor 1 in *Xenopus* embryos induces the ectopic formation of eyes, indicating that the IGF pathway is also required for eye formation (Richard-Parpaillon et al. 2002). Currently, the use of small molecules is under investigation in order to potentially replace the use of recombinant protein factors. Such molecules would promise to be more financially favorable to synthesize than recombinant protein factors and easier to handle since small molecules do not easily degrade as proteins do. Others use a nonspecific neural induction media followed by sorting a subpopulation of cells with optic characteristics (Meyer et al. 2009; Sridhar et al. 2013; Zhu et al. 2013). Alternatively, small molecules affecting the pathways are also being used to influence retinal differentiation of pluripotent stem cells (Osakada et al. 2009). Molecules such as LDN-193189 and K02288 have been identified as functional analogues of Dorsomorphin, specific inhibitors of

BMP (Sanvitale et al. 2013). Additionally, IWR1 molecules have been shown to effectively suppress Wnt signaling (Chen et al. 2009). Whether or not these small molecules would effectively lead to the efficient generation of retinal progenitors from pluripotent stem cells has yet to be published. Whether or not there are specific functions that are done by protein factors that are not done by their protein analogues would also be interesting to investigate.

The transplantation of pluripotent stem cell-derived retinal cells has also seen significant advances through continuous investigation, reinforcing the utility of the biotechnology. In 2009, Lamba and colleagues showed that the transplantation of human embryonic stem cell-derived retinal progenitors into *Crx−/−* mice, an animal model for Leber's congenital amaurosis, led to the restoration of light sensitivity in these mice. Upon functional analysis 3–4 weeks posttransplantation in adult mice using an electroretinogram, the transplanted eyes responded to light stimulation. Furthermore, upon histological analysis, the group observed successful integration and expression of differentiated photoreceptor markers, such as rhodopsin, and synaptic markers like synaptophysin (Lamba et al. 2009b). Another group, Tucker and colleagues, found similar results in 2011, using a mouse iPS cell line. In this investigation, the group obtained mouse dermal fibroblasts in which the Yamanaka factors were overexpressed using retroviral vectors (Tucker et al. 2011). Then, they induced the differentiation of these iPSCs with DKK1, noggin, fibroblast growth factor (FGF), and IGF-1 until Pax6 was successfully expressed. These cells were then transplanted into severe combined immunodeficient (SCID) mice by subretinal injection. Posttransplantation analyses included electroretinographical assays and functional anatomical analysis by immunohistochemistry, which showed successful integration of Pax6 cells in the outer nuclear layer, increased retinal function, and functional anatomy. A recent study looked at the need for immunosuppression for long-term integration of human embryonic stem cell-derived retinal cells in mouse retinas (Hambright et al. 2012). They observed that immunosuppression may not be required in mice for long-term survival in the subretinal space as long as the tissue is healthy and intact. They however did not observe much photoreceptor integration. In mice where the blood–brain barrier was damaged during the surgery, the transplanted cells were lost to rejection. An in vivo disease environment, which may not have a well-preserved blood–brain barrier, will likely leak following transplantation surgery and in such cases may need immunosuppression.

In order to influence RPE differentiation of pluripotent stem cells, further knowledge of the developmental biology of the eye has come in handy. Once the expression of Pax6 has been achieved through the antagonism of BMP and Wnt signaling, such progenitors may proceed to develop into cells found in the neural retina, consisting of rod and cone photoreceptors, or the RPE layer. In order to obtain RPE cells from this point, transcription factors, particularly MITF, are required to further direct differentiation (Bharti et al. 2012). Multiple groups have found that inducing the expression of MITF can be achieved by the induction of Wnt signaling after the induction of Pax6 expression through the use of the small molecule CHIR99021 (Bharti et al. 2012; Nakano et al. 2012). Others such as Buchholz and colleagues propose that activin signaling coupled with FGF inhibition is another avenue for consideration in the way of directed RPE fate induction (Buchholz et al. 2013). The same group has used nicotinamide, which acts as a poly(ADP-ribose) polymerase (PARP) inhibitor, enhancing the generation of RPE when used with activin A and FGF inhibitor SU5402. With their methods, the derivation of RPEs from human pluripotent stem cells can be accomplished with 80 % efficiency, with the observation of the cobblestone-like morphology being seen within two weeks. Idelson and colleagues, on the other hand, have proposed that the derivation of RPEs from human pluripotent stem cells can be accomplished through treatment solely with activin A and nicotinamide after 4 weeks, at which point pigmented clusters can be seen in culture (Idelson et al. 2009). These pigmented clusters may then be manually extracted and transferred to a separate culture vessel for expansion and analysis. However, in terms of its practical use for patient transplantation, such a process may be relatively more inefficient, more time-consuming, and more labor-intensive than directed derivation of a specified population.

In cases of retinal degenerative disorders where the primary defect lies in the RPE layer including macular degeneration and Best's disease, replacement of the RPE cells prior to photoreceptor could obviate the need for direct neural retinal repair. Toward this end, a number of groups have developed methods to derive RPE cells from human embryonic stem cells (Idelson et al. 2009; Lund et al. 2006; Vugler et al. 2008; Haruta et al. 2004). Haruta et al. (2004) found that RPE cells, easily identified by their pigmentation, spontaneously developed in monkey embryonic stem cell cultures when exposed to the PA6 stromal cell line. The cells survived transplantation to the subretinal space of a rat model of retinal degeneration and even improved visually mediated behaviors. Similar results were obtained by the groups of Lund and Lanza (Lund et al. 2006) with human embryonic stem cell-derived RPE cells; the RPE cells derived from human embryonic stem cells were from as many as 18 different lines. A recent study also looked at long-term survival of the human ES cell-derived RPE cells in RCS rats (Lu et al. 2009) and reported survival in the subretinal space for over 200 days. These promising animal results have led to one of the first clinical trials of human embryonic stem cell products by Advanced Cell Technology for age-related macular degeneration and Stargardt's disease. The company reported the initial findings in these patients 4 months after the subretinal injection. There was evidence of some survival of the transplanted human embryonic stem cell-derived RPE cells and absence of any untoward effects or immunorejection 4 months after transplantation in some patients (Schwartz et al. 2012). The trial is currently approved in the USA as well as Europe. The use of biodegradable and biocompatible nanofibrous scaffolds is also being considered by researchers. The idea here is to generate polarized RPE cells on substrates, which can then be transplanted as a sheet instead of using a cell suspension. This will allow circumventing issues such as integration and cell attachment on an old and degenerating Bruch's membrane. A recent report looking at attachment of human embryonic stem cell-derived RPE on aged basement membrane in vitro from autopsy samples showed poor resurfacing and function (Sugino et al. 2011) upon coculture on Bruch's membrane. Semipermeable parylene membranes have been tested to be biocompatible with RPE cells (Lu et al. 2012). This group then went on to show that a parylene–RPE coculture sheet can be injected into the subretinal space of rats with survival ranging from a week to a year posttransplantation (Hu et al. 2012; Diniz et al. 2013). The group is currently seeking approval for a phase 1 clinical trial.

15.8 Conclusion and Future Directions

Age-related degenerations in the retina are an increasing health-care problem in the USA and worldwide. There are no effective forms of treatment for most of these degenerations. The field of regenerative medicine has a lot of hope and at the same time a number of challenges to overcome before it is widely applicable. Regeneration of the vertebrate retina is being increasingly studied as a potential mechanism of repair esp. since lower animals are some adept at it. Potential of Müller glia to reenter the stem cell state and regenerate lost cells will likely be the solution, and recent work on transdifferentiation in vitro using a single transcription factor, ASCL1, is very encouraging. Additional genes may, however, be needed to regenerate all the cells in vitro. If successful, the challenge will be on delivery of the genes in vivo. Viral vectors including AAVs and replication-incompetent lentiviruses may be used. A nonviral gene delivery system will be ideal to reduce risks and potential spill out. Reprogramming requires a number of epigenetic changes and stabilizing these will also be critical to maintenance of the newly regenerated cells and phenotype. Finally, small molecules, which have similar effects to the reprogramming genes, as well as small molecule epigenetic modifiers might be alternate tools to induce reprogramming to regenerate retinal neurons.

Cell replacement therapy as an option to restore vision has been the holy grail. A number of early clinical trials have tested the concept using fetal retinal cells in either small aggregates or sheets in patients with macular degeneration or retinitis pigmentosa. The potential of pluripotent stem cells to be able to generate millions or billions of these cells in vitro is tremendous. However, there are still a number of issues to be worked out. Immunogenicity and rejection still is a big possibility especially when transplanting in a disease environment. A number of groups are looking into the potential of generating HLA-matched banks of stem cells. Another alternative is the induced pluripotent stem cells from either patient itself or a matched donor. Induced pluripotent stem cells also have their own drawbacks including epigenetic memory and need to use oncogenes for their generation as well as the regulatory issues that may arise in the entire process not to mention the exorbitant costs that may be involved in generation and characterization of these cells prior to any retinal induction. Finally, diseases where both RPE and photoreceptors are affected require a combined cell therapy approach, and research into the best approach in terms of replacing these cells is very limited. However, in spite of the challenges that lay on the road to clinical trials, studies involving the use of stem cells in the treatment of retinal degenerative disorders continue to grow. Increases in the understanding of underlying pathological mechanisms will allow researchers avenues in which the optimization of retinal cell derivation in clinically compatible methods, the creation of more comprehensive model systems, and the development of new platforms for cell replacement therapy can be achieved.

References

Ahmad I, Tang L, Pham H (2000) Identification of neural progenitors in the adult mammalian eye. Biochem Biophys Res Commun 270(2):517–521
Akita J et al (2002) Neuronal differentiation of adult rat hippocampus-derived neural stem cells transplanted into embryonic rat explanted retinas with retinoic acid pretreatment. Brain Res 954(2):286–293
Allingham RR, Liu Y, Rhee DJ (2009) The genetics of primary open-angle glaucoma: a review. Exp Eye Res 88(4):837–844
Andreoli MT et al (2009) Comprehensive analysis of complement factor H and LOC387715/ARMS2/HTRA1 variants with respect to phenotype in advanced age-related macular degeneration. Am J Ophthalmol 148(6):869–874
Asami M et al (2007) Multipotent cells from mammalian iris pigment epithelium. Dev Biol 304(1):433–446
Azuma N et al (2005) Transdifferentiation of the retinal pigment epithelia to the neural retina by transfer of the Pax6 transcriptional factor. Hum Mol Genet 14(8):1059–1068
Barber AC et al (2013) Repair of the degenerate retina by photoreceptor transplantation. Proc Natl Acad Sci U S A 110(1):354–359
Bernardos RL et al (2007) Late-stage neuronal progenitors in the retina are radial Müller glia that function as retinal stem cells. J Neurosci 27(26):7028–7040
Bharti K et al (2012) A regulatory loop involving PAX6, MITF, and WNT signaling controls retinal pigment epithelium development. PLoS Genet 8(7), e1002757
Boon CJ et al (2008) Basal laminar drusen caused by compound heterozygous variants in the CFH gene. Am J Hum Genet 82(2):516–523
Brzezinski JA 4th et al (2011) Ascl1 expression defines a subpopulation of lineage-restricted progenitors in the mammalian retina. Development 138(16):3519–3531
Buchholz DE et al (2013) Rapid and efficient directed differentiation of human pluripotent stem cells into retinal pigmented epithelium. Stem Cells Transl Med 2(5):384–393
Calkins DJ (2012) Critical pathogenic events underlying progression of neurodegeneration in glaucoma. Prog Retin Eye Res 31(6):702–719
Cameron DA et al (2005) Gene expression profiles of intact and regenerating zebrafish retina. Mol Vis 11:775–791
Chacko DM et al (2000) Survival and differentiation of cultured retinal progenitors transplanted in the subretinal space of the rat. Biochem Biophys Res Commun 268(3):842–846
Chang ML et al (2007) Reactive changes of retinal astrocytes and Müller glial cells in kainate-induced neuroexcitotoxicity. J Anat 210(1):54–65
Chen B et al (2009) Small molecule-mediated disruption of Wnt-dependent signaling in tissue regeneration and cancer. Nat Chem Biol 5(2):100–107
Chistiakov DA (2011) Diabetic retinopathy: pathogenic mechanisms and current treatments. Diabetes Metab Syndr 5(3):165–172
Cicero SA et al (2009) Cells previously identified as retinal stem cells are pigmented ciliary epithelial cells. Proc Natl Acad Sci U S A 106(16):6685–6690
Close JL et al (2006) Epidermal growth factor receptor expression regulates proliferation in the postnatal rat retina. Glia 54(2):94–104
Coffey PJ, Lund RD, Rawlins JN (1989) Retinal transplant-mediated learning in a conditioned suppression task in rats. Proc Natl Acad Sci U S A 86(18):7248–7249
Coles BL et al (2004) Facile isolation and the characterization of human retinal stem cells. Proc Natl Acad Sci U S A 101(44):15772–15777
Coulombre JL, Coulombre AJ (1965) Regeneration of neural retina from the pigmented epithelium in the chick embryo. Dev Biol 12(1):79–92
Das AV et al (2004) Identification of c-Kit receptor as a regulator of adult neural stem cells in the mammalian eye: interactions with Notch signaling. Dev Biol 273(1):87–105

Das AV et al (2006a) Neural stem cells in the adult ciliary epithelium express GFAP and are regulated by Wnt signaling. Biochem Biophys Res Commun 339(2):708–716

Das AV et al (2006b) Neural stem cell properties of Müller glia in the mammalian retina: regulation by Notch and Wnt signaling. Dev Biol 299(1):283–302

Davis RL, Weintraub H, Lassar AB (1987) Expression of a single transfected cDNA converts fibroblasts to myoblasts. Cell 51(6):987–1000

Dayer AG et al (2005) New GABAergic interneurons in the adult neocortex and striatum are generated from different precursors. J Cell Biol 168(3):415–427

De Marzo A, Aruta C, Marigo V (2010) PEDF promotes retinal neurosphere formation and expansion in vitro. Adv Exp Med Biol 664:621–630

del Cerro M et al (1991) Intraretinal grafting restores visual function in light-blinded rats. Neuroreport 2(9):529–532

Del Debbio CB et al (2010) Notch and Wnt signaling mediated rod photoreceptor regeneration by Müller cells in adult mammalian retina. PLoS One 5(8), e12425

Del Rio-Tsonis K, Tsonis PA (2003) Eye regeneration at the molecular age. Dev Dyn 226(2):211–224

Diniz B et al (2013) Subretinal implantation of retinal pigment epithelial cells derived from human embryonic stem cells – improved survival when implanted as a monolayer. Invest Ophthalmol Vis Sci 54:5087–5096

Faillace MP, Julian D, Korenbrot JI (2002) Mitotic activation of proliferative cells in the inner nuclear layer of the mature fish retina: regulatory signals and molecular markers. J Comp Neurol 451(2):127–141

Fausett BV, Goldman D (2006) A role for alpha1 tubulin-expressing Müller glia in regeneration of the injured zebrafish retina. J Neurosci 26(23):6303–6313

Fausett BV, Gumerson JD, Goldman D (2008) The proneural basic helix-loop-helix gene ascl1a is required for retina regeneration. J Neurosci 28(5):1109–1117

Fimbel SM et al (2007) Regeneration of inner retinal neurons after intravitreal injection of ouabain in zebrafish. J Neurosci 27(7):1712–1724

Fischer AJ, Reh TA (2000) Identification of a proliferating marginal zone of retinal progenitors in postnatal chickens. Dev Biol 220(2):197–210

Fischer AJ, Reh TA (2001) Müller glia are a potential source of neural regeneration in the postnatal chicken retina. Nat Neurosci 4(3):247–252

Fischer AJ, Reh TA (2002) Exogenous growth factors stimulate the regeneration of ganglion cells in the chicken retina. Dev Biol 251(2):367–379

Fischer AJ, Reh TA (2003) Potential of Müller glia to become neurogenic retinal progenitor cells. Glia 43(1):70–76

Fischer AJ et al (2002) Insulin and fibroblast growth factor 2 activate a neurogenic program in Müller glia of the chicken retina. J Neurosci 22(21):9387–9398

Gorin MB (2012) Genetic insights into age-related macular degeneration: controversies addressing risk, causality, and therapeutics. Mol Aspects Med 33(4):467–486

Grozdanic SD et al (2006) Morphological integration and functional assessment of transplanted neural progenitor cells in healthy and acute ischemic rat eyes. Exp Eye Res 82(4):597–607

Guo Y et al (2003) Engraftment of adult neural progenitor cells transplanted to rat retina injured by transient ischemia. Invest Ophthalmol Vis Sci 44(7):3194–3201

Gust J, Reh TA (2011) Adult donor rod photoreceptors integrate into the mature mouse retina. Invest Ophthalmol Vis Sci 52:5266–5272

Hambright D et al (2012) Long-term survival and differentiation of retinal neurons derived from human embryonic stem cell lines in un-immunosuppressed mouse retina. Mol Vis 18:920–936

Haruta M et al (2004) In vitro and in vivo characterization of pigment epithelial cells differentiated from primate embryonic stem cells. Invest Ophthalmol Vis Sci 45(3):1020–1025

Hayes S et al (2007) Notch signaling regulates regeneration in the avian retina. Dev Biol 312(1):300–311

Haynes T, Del Rio-Tsonis K (2004) Retina repair, stem cells and beyond. Curr Neurovasc Res 1(3):231–239
Haynes T et al (2007) BMP signaling mediates stem/progenitor cell-induced retina regeneration. Proc Natl Acad Sci U S A 104(51):20380–20385
Hitchcock P et al (2004) Persistent and injury-induced neurogenesis in the vertebrate retina. Prog Retin Eye Res 23(2):183–194
Hollyfield JG (1968) Differential addition of cells to the retina in Rana pipiens tadpoles. Dev Biol 18(2):163–179
Hu Y et al (2012) A novel approach for subretinal implantation of ultrathin substrates containing stem cell-derived retinal pigment epithelium monolayer. Ophthalmic Res 48(4):186–191
Idelson M et al (2009) Directed differentiation of human embryonic stem cells into functional retinal pigment epithelium cells. Cell Stem Cell 5(4):396–408
Ikeda H et al (2005) Generation of Rx+/Pax6+ neural retinal precursors from embryonic stem cells. Proc Natl Acad Sci U S A 102(32):11331–11336
Inoue T et al (2006) Activation of canonical Wnt pathway promotes proliferation of retinal stem cells derived from adult mouse ciliary margin. Stem Cells 24(1):95–104
Inoue T et al (2010) Maximizing functional photoreceptor differentiation from adult human retinal stem cells. Stem Cells 28(3):489–500
Jarrett SG, Boulton ME (2012) Consequences of oxidative stress in age-related macular degeneration. Mol Aspects Med 33(4):399–417
Johns PR (1982) Formation of photoreceptors in larval and adult goldfish. J Neurosci 2(2):178–198
Julian D, Ennis K, Korenbrot JI (1998) Birth and fate of proliferative cells in the inner nuclear layer of the mature fish retina. J Comp Neurol 394(3):271–282
Kaneko H et al (2011) DICER1 deficit induces Alu RNA toxicity in age-related macular degeneration. Nature 471(7338):325–330
Karl MO et al (2008) Stimulation of neural regeneration in the mouse retina. Proc Natl Acad Sci U S A 105(49):19508–19513
Kassen SC et al (2007) Time course analysis of gene expression during light-induced photoreceptor cell death and regeneration in albino zebrafish. Dev Neurobiol 67(8):1009–1031
Kassen SC et al (2009) CNTF induces photoreceptor neuroprotection and Müller glial cell proliferation through two different signaling pathways in the adult zebrafish retina. Exp Eye Res 88(6):1051–1064
Khan JC et al (2006) Age related macular degeneration and sun exposure, iris colour, and skin sensitivity to sunlight. Br J Ophthalmol 90(1):29–32
Klassen H, Lund RD (1990) Retinal graft-mediated pupillary responses in rats: restoration of a reflex function in the mature mammalian brain. J Neurosci 10(2):578–587
Kohno R et al (2006) Sphere formation of ocular epithelial cells in the ciliary body is a reprogramming system for neural differentiation. Brain Res 1093(1):54–70
Kubota R et al (2002) A comparative study of neurogenesis in the retinal ciliary marginal zone of homeothermic vertebrates. Brain Res Dev Brain Res 134(1–2):31–41
Kurimoto Y et al (2001) Transplantation of adult rat hippocampus-derived neural stem cells into retina injured by transient ischemia. Neurosci Lett 306(1–2):57–60
Kwan AS, Wang S, Lund RD (1999) Photoreceptor layer reconstruction in a rodent model of retinal degeneration. Exp Neurol 159(1):21–33
Lakowski J et al (2011) Effective transplantation of photoreceptor precursor cells selected via cell surface antigen expression. Stem Cells 29(9):1391–1404
Lamba DA et al (2006) Efficient generation of retinal progenitor cells from human embryonic stem cells. Proc Natl Acad Sci U S A 103(34):12769–12774
Lamba DA, Gust J, Reh TA (2009a) Transplantation of human embryonic stem cell-derived photoreceptors restores some visual function in Crx-deficient mice. Cell Stem Cell 4:1–7
Lamba DA, Gust J, Reh TA (2009b) Transplantation of human embryonic stem cell-derived photoreceptors restores some visual function in Crx-deficient mice. Cell Stem Cell 4(1):73–79

Lin H et al (2011) Effect of miR-23 on oxidant-induced injury in human retinal pigment epithelial cells. Invest Ophthalmol Vis Sci 52(9):6308–6314

Lu B et al (2002) Transplantation of EGF-responsive neurospheres from GFP transgenic mice into the eyes of rd mice. Brain Res 943(2):292–300

Lu B et al (2009) Long-term safety and function of RPE from human embryonic stem cells in preclinical models of macular degeneration. Stem Cells 27(9):2126–2135

Lu B et al (2012) Mesh-supported submicron parylene-C membranes for culturing retinal pigment epithelial cells. Biomed Microdevices 14(4):659–667

Lukiw WJ et al (2012) Common micro RNAs (miRNAs) target complement factor H (CFH) regulation in Alzheimer's disease (AD) and in age-related macular degeneration (AMD). Int J Biochem Mol Biol 3(1):105–116

Lund RD, Hankin MH (1995) Pathfinding by retinal ganglion cell axons: transplantation studies in genetically and surgically blind mice. J Comp Neurol 356(3):481–489

Lund RD et al (2006) Human embryonic stem cell-derived cells rescue visual function in dystrophic RCS rats. Cloning Stem Cells 8(3):189–199

MacLaren RE et al (2006) Retinal repair by transplantation of photoreceptor precursors. Nature 444(7116):203–207

MacNeil A et al (2007) Comparative analysis of progenitor cells isolated from the iris, pars plana, and ciliary body of the adult porcine eye. Stem Cells 25(10):2430–2438

McLoon SC, Lund RD (1980) Specific projections of retina transplanted to rat brain. Exp Brain Res 40(3):273–282

McLoon LK, McLoon SC, Lund RD (1981) Cultured embryonic retinae transplanted to rat brain: differentiation and formation of projections to host superior colliculus. Brain Res 226(1–2):15–31

Meyer JS et al (2009) Modeling early retinal development with human embryonic and induced pluripotent stem cells. Proc Natl Acad Sci U S A 106(39):16698–16703

Mitsuda S et al (2005) Tissue interaction between the retinal pigment epithelium and the choroid triggers retinal regeneration of the newt Cynops pyrrhogaster. Dev Biol 280(1):122–132

Moshiri A, McGuire CR, Reh TA (2005) Sonic hedgehog regulates proliferation of the retinal ciliary marginal zone in posthatch chicks. Dev Dyn 233(1):66–75

Nakamura K, Chiba C (2007) Evidence for Notch signaling involvement in retinal regeneration of adult newt. Brain Res 1136(1):28–42

Nakano T et al (2012) Self-formation of optic cups and storable stratified neural retina from human ESCs. Cell Stem Cell 10(6):771–785

Nelson CM et al (2013) Tumor necrosis factor-alpha is produced by dying retinal neurons and is required for Müller glia proliferation during zebrafish retinal regeneration. J Neurosci 33(15):6524–6539

Okada TS (1980) Cellular metaplasia or transdifferentiation as a model for retinal cell differentiation. Curr Top Dev Biol 16:349–380

Ooto S et al (2004) Potential for neural regeneration after neurotoxic injury in the adult mammalian retina. Proc Natl Acad Sci U S A 101(37):13654–13659

Osakada F et al (2007) Wnt signaling promotes regeneration in the retina of adult mammals. J Neurosci 27(15):4210–4219

Osakada F et al (2008) Toward the generation of rod and cone photoreceptors from mouse, monkey and human embryonic stem cells. Nat Biotechnol 26(2):215–224

Osakada F et al (2009) In vitro differentiation of retinal cells from human pluripotent stem cells by small-molecule induction. J Cell Sci 122(Pt 17):3169–3179

Park CM, Hollenberg MJ (1989) Basic fibroblast growth factor induces retinal regeneration in vivo. Dev Biol 134(1):201–205

Phillips MJ et al (2012) Blood-derived human iPS cells generate optic vesicle-like structures with the capacity to form retinal laminae and develop synapses. Invest Ophthalmol Vis Sci 53(4):2007–2019

Pittack C, Jones M, Reh TA (1991) Basic fibroblast growth factor induces retinal pigment epithelium to generate neural retina in vitro. Development 113(2):577–588

Pollak J et al (2013) ASCL1 reprograms mouse Müller glia into neurogenic retinal progenitors. Development 140(12):2619–2631

Qiu G et al (2005) Photoreceptor differentiation and integration of retinal progenitor cells transplanted into transgenic rats. Exp Eye Res 80(4):515–525

Quigley HA, Broman AT (2006) The number of people with glaucoma worldwide in 2010 and 2020. Br J Ophthalmol 90(3):262–267

Ramachandran R, Fausett BV, Goldman D (2010) Ascl1a regulates Müller glia dedifferentiation and retinal regeneration through a Lin-28-dependent, let-7 microRNA signalling pathway. Nat Cell Biol 12(11):1101–1107

Ramachandran R, Zhao XF, Goldman D (2011) Ascl1a/Dkk/beta-catenin signaling pathway is necessary and glycogen synthase kinase-3beta inhibition is sufficient for zebrafish retina regeneration. Proc Natl Acad Sci U S A 108(38):15858–15863

Raymond PA et al (2006) Molecular characterization of retinal stem cells and their niches in adult zebrafish. BMC Dev Biol 6:36

Reh TA, Nagy T, Gretton H (1987) Retinal pigmented epithelial cells induced to transdifferentiate to neurons by laminin. Nature 330(6143):68–71

Richard-Parpaillon L et al (2002) The IGF pathway regulates head formation by inhibiting Wnt signaling in Xenopus. Dev Biol 244(2):407–417

Sahel JA, Albert DM, Lessell S (1990) Proliferation of retinal glia and excitatory amino acids. Ophtalmologie 4(1):13–16

Sakaguchi DS, Janick LM, Reh TA (1997) Basic fibroblast growth factor (FGF-2) induced transdifferentiation of retinal pigment epithelium: generation of retinal neurons and glia. Dev Dyn 209(4):387–398

Sakami S et al (2005) Downregulation of Otx2 in the dedifferentiated RPE cells of regenerating newt retina. Brain Res Dev Brain Res 155(1):49–59

Sakami S, Etter P, Reh TA (2008) Activin signaling limits the competence for retinal regeneration from the pigmented epithelium. Mech Dev 125(1–2):106–116

Sanvitale CE et al (2013) A new class of small molecule inhibitor of BMP signaling. PLoS One 8(4), e62721

Schwartz SD et al (2012) Embryonic stem cell trials for macular degeneration: a preliminary report. Lancet 379(9817):713–720

Seiler MJ, Aramant RB (1998) Intact sheets of fetal retina transplanted to restore damaged rat retinas. Invest Ophthalmol Vis Sci 39(11):2121–2131

Simo R (2011) Neurodegeneration as an early event in diabetic retinopathy. Endocrinol Nutr 58(5):211–213

Sparrow JR, Boulton M (2005) RPE lipofuscin and its role in retinal pathobiology. Exp Eye Res 80(5):595–606

Sridhar A, Steward MM, Meyer JS (2013) Nonxenogeneic growth and retinal differentiation of human induced pluripotent stem cells. Stem Cells Transl Med 2:255–264

Stone L (1950a) The role of retinal pigment cells in regenerating neural retinae of adult salamander eyes. J Exp Zool 113:9–31

Stone LS (1950b) Neural retina degeneration followed by regeneration from surviving retinal pigment cells in grafted adult salamander eyes. Anat Rec 106(1):89–109

Stone LS, Steinitz H (1957) Regeneration of neural retina and lens from retina pigment cell grafts in adult newts. J Exp Zool 135(2):301–317

Straznicky K, Gaze RM (1971) The growth of the retina in Xenopus laevis: an autoradiographic study. J Embryol Exp Morphol 26(1):67–79

Sugino IK et al (2011) Comparison of FRPE and human embryonic stem cell-derived RPE behavior on aged human Bruch's membrane. Invest Ophthalmol Vis Sci 52(8):4979–4997

Sun G et al (2006) Retinal stem/progenitor properties of iris pigment epithelial cells. Dev Biol 289(1):243–252

Takahashi K, Yamanaka S (2006) Induction of pluripotent stem cells from mouse embryonic and adult fibroblast cultures by defined factors. Cell 126(4):663–676

Takahashi M et al (1998) Widespread integration and survival of adult-derived neural progenitor cells in the developing optic retina. Mol Cell Neurosci 12(6):340–348

Takahashi K et al (2007) Induction of pluripotent stem cells from fibroblast cultures. Nat Protoc 2(12):3081–3089

Tansley K (1946) The development of the rat eye in graft. J Exp Biol 22:221–224

Tarallo V et al (2012) DICER1 loss and Alu RNA induce age-related macular degeneration via the NLRP3 inflammasome and MyD88. Cell 149(4):847–859

Thakkinstian A et al (2006) Association between apolipoprotein E polymorphisms and age-related macular degeneration: a HuGE review and meta-analysis. Am J Epidemiol 164(9):813–822

Thomson JA et al (1998) Embryonic stem cell lines derived from human blastocysts. Science 282(5391):1145–1147

Thummel R et al (2008) Inhibition of Müller glial cell division blocks regeneration of the light-damaged zebrafish retina. Dev Neurobiol 68(3):392–408

Tropepe V et al (2000) Retinal stem cells in the adult mammalian eye. Science 287(5460):2032–2036

Tucker BA et al (2011) Transplantation of adult mouse iPS cell-derived photoreceptor precursors restores retinal structure and function in degenerative mice. PLoS One 6(4), e18992

Tucker BA et al (2013) Use of a synthetic xeno-free culture substrate for induced pluripotent stem cell induction and retinal differentiation. Stem Cells Transl Med 2(1):16–24

Vierbuchen T et al (2010) Direct conversion of fibroblasts to functional neurons by defined factors. Nature 463(7284):1035–1041

Vihtelic TS et al (2006) Retinal regional differences in photoreceptor cell death and regeneration in light-lesioned albino zebrafish. Exp Eye Res 82(4):558–575

Vugler A et al (2008) Elucidating the phenomenon of HESC-derived RPE: anatomy of cell genesis, expansion and retinal transplantation. Exp Neurol 214(2):347–361

Wan J et al (2007) Sonic hedgehog promotes stem-cell potential of Müller glia in the mammalian retina. Biochem Biophys Res Commun 363(2):347–354

Wan J et al (2008) Preferential regeneration of photoreceptor from Müller glia after retinal degeneration in adult rat. Vision Res 48(2):223–234

Wan J, Ramachandran R, Goldman D (2012) HB-EGF is necessary and sufficient for Müller glia dedifferentiation and retina regeneration. Dev Cell 22(2):334–347

Wetts R, Fraser SE (1988) Multipotent precursors can give rise to all major cell types of the frog retina. Science 239(4844):1142–1145

Wu DM et al (2001) Cones regenerate from retinal stem cells sequestered in the inner nuclear layer of adult goldfish retina. Invest Ophthalmol Vis Sci 42(9):2115–2124

Xu S et al (2007) The proliferation and expansion of retinal stem cells require functional Pax6. Dev Biol 304(2):713–721

Young MJ et al (2000) Neuronal differentiation and morphological integration of hippocampal progenitor cells transplanted to the retina of immature and mature dystrophic rats. Mol Cell Neurosci 16(3):197–205

Yu J et al (2007) Induced pluripotent stem cell lines derived from human somatic cells. Science 318(5858):1917–1920

Yurco P, Cameron DA (2005) Responses of Müller glia to retinal injury in adult zebrafish. Vision Res 45(8):991–1002

Yurco P, Cameron DA (2007) Cellular correlates of proneural and Notch-delta gene expression in the regenerating zebrafish retina. Vis Neurosci 24(3):437–443

Zhao X et al (2005) Growth factor-responsive progenitors in the postnatal mammalian retina. Dev Dyn 232(2):349–358

Zhu Y et al (2013) Three-dimensional neuroepithelial culture from human embryonic stem cells and its use for quantitative conversion to retinal pigment epithelium. PLoS One 8(1), e54552

Zuber ME et al (2003) Specification of the vertebrate eye by a network of eye field transcription factors. Development 130(21):5155–5167

Closing the Circle: Stem Cell Rejuvenation and Longevity

16

Maria Carolina Florian and Hartmut Geiger

Abstract

Many organs with high cell turnover (e.g., intestine and blood as well as the germ line) are composed of short-lived cells that require continuous replenishment by stem cells (Potten and Morris J Cell Sci Suppl 10:45–62, 1988; Morrison et al. Annu Rev Cell Dev Biol 11: 35–71, 1995; Fuchs et al. Cell 100(1):143–155, 2000; Tani et al. Proc Natl Acad Sci U S A 97(20):10960–10965, 2000; Stappenbeck et al. Proc Natl Acad Sci U S A 100(3):1004–1009, 2003). Aging results in the inability of these tissues to maintain homeostasis. A number of theories have been proposed regarding the cellular and molecular mechanisms regulating aging, and genetic, behavioral, and environmental factors may all be involved. Declines in the functional capacity of both somatic organ cells but also adult stem cells impair tissue maintenance and regeneration during aging, which may limit lifespan and thus longevity. Tissues that depend on stem cell activity for long-term tissues maintenance of course will be most vulnerable to stem cell aging. High incidences of anemia, impaired wound healing, and intestinal dysfunction in geriatric patients indicate that such alteration can affect the health status of aging humans (de Craen et al. BMJ 327(7407):131–132, 2003). In addition, an age-dependent impairment of stem cell function and a reduction in regenerative capacity can limit stress reactions in response to diseases. Along

M.C. Florian, PhD (✉)
Institute for Molecular Medicine, University of Ulm,
James Franck-Ring 11c, Ulm 89081, Germany
e-mail: carolina.florian@uni-ulm.de

H. Geiger, PhD
Institute for Molecular Medicine, University of Ulm,
James Franck-Ring 11c, Ulm 89081, Germany

Department of Experimental Hematology and Cancer Biology,
Cincinnati Children Hospital Medical Centre (CCHMC),
3333 Burnet Ave, Cincinnati, OH 45229, USA
e-mail: hartmut.geiger@uni-ulm.de, hartmut.geiger@cchmc.org

H. Geiger et al. (eds.), *Stem Cell Aging: Mechanisms, Consequences, Rejuvenation*,
DOI 10.1007/978-3-7091-1232-8_16

these lines, "old age" represents a major risk factor for the evolution of, for example, liver cirrhosis as a consequence of chronic hepatitis (Poynard et al. J Hepatol 38(3):257–265, 2003) or various forms of leukemia. Vas et al. PLoS One 7(8):e42080, 2012).

Stem cells are defined by their long-term self-renewal as well as multi-lineage differentiation ability, which a priori should lead to "immortal stem cells." Consequently, stem cells were initially thought to be endowed with unlimited self-renewal capacity and thus exempt from aging. However, as it has also been reported in this book, a measurable and successive age-dependent decline in stem cell activity from adulthood to old age exists, ranging from yeast to worms to flies to mice and human beings and include prominent stem cell populations like germ line stem cells, hematopoietic, intestinal, skin, retinoid, and muscle stem cells. Aging of stem cell thus (1) exists and (2) clearly contributes to tissue aging and (3) limits by this means healthy aging.

16.1 Stem Cell Rejuvenation: Mechanistic Prerequisites

The identification of mechanisms under which aged stem cells become functionally similar to young stem cells is an important step toward devising clinical treatments of aging-associated imbalance in tissue homeostasis. Attenuating or even reverting stem cell aging might improve tissue regeneration, lead to healthy aging, and might even contribute to longevity. One central prerequisite to achieve that aims though is that the unwanted or causal changes and also the damage that is associated with aging of stem cells are actually reversible or upon their halt or reversion do not further contribute to stem cell aging. While this is obviously a very basic and logical requirement, if translated into biology and mechanisms, things are getting a little more complicated. In general tough, the observation that cellular differentiation is reversible, which was awarded the Nobel Prize in medicine in 2012, proved that stem cell specification and changes in cells over time are in general not associated with irreversible modifications. So, it will be worthwhile to further explore whether, in contrast to what is frequently suggested, aging of stem cells can also be attenuated or reversed.

Mechanisms of aging have been presented in this book so far on a large variety of organisms and types of stem cells. Such mechanisms of stem cell aging include, but are not restricted to, altered cellular communication; genome and telomere instability; epigenetic alterations; changes in proteostasis; altered nutrient sensing; reactive oxygen species as well as mitochondrial dysfunction, primarily due to mutations in the mitochondrial genome; and, especially important to stem cells, aging of the stem cell niche (Rando 2006). They have been referred to as hallmarks of aging (Lopez-Otin et al. 2013). In general, mechanisms that are irreversible like DNA damage, genomic instability, as well as mitochondrial mutations, which might be able to drive premature aging of stem cells (Rossi et al. 2007; Norddahl et al. 2011; Beerman et al. 2014), will not fulfill this critical requirement of reversibility. Reactive oxygen species might fall here somewhat into the middle, as they list as aging inducers via conferring DNA damage, for example, in skin and for mesenchymal stem cells (Andrade et al. 2012; Borodkina et al. 2014), while at the same time they might also play a more general role in signaling as well as for protein damage (Ray et al. 2012).

Genetically, data derived from studies ranging from yeast, *C. elegans*, *D. melanogaster*, and the mouse have shown that alterations in specific signaling pathways can limit stem cell function and induce aging. This in itself is a remarkable finding, as it implies that there are evolutionary shared mechanisms of stem cell aging, so that not each individual types of stem cell in a distinct organism ages at its own pace and mechanism. Specifically, alterations in developmental pathways – e. g., Wnt signaling (Brack et al. 2007; Liu et al. 2007; Takashima et al. 2008; Florian et al. 2013) or Notch signaling (Ohlstein and Spradling 2006; Biteau et al. 2008; Florian et al. 2013) – have been associated with decreasing stem cell function during aging. Furthermore, Wnt signaling has been linked to Klotho, a factor that is involved in the suppression of various aging phenotypes (Liu et al. 2007). Activation of Notch signaling initiates specific changes in the chromatin-modifying gene regulatory complexes, stem cell maintenance, and cell cycle progression (Campbell et al. 2008; Salat et al. 2008). In muscle precursor cells, activity of the Notch signaling pathway seems to be critical for regeneration (Conboy et al. 2003). In addition, a functional role of p21 has been established in activating a forward loop stabilizing senescence cell cycle arrest (Passos et al. 2010). An upregulation of the $p16^{Ink4a}$ cell cycle inhibitor has also been reported in aging organs, including human skin, and has been associated with reduced self-renewal of organ stem cells (Janzen et al. 2006; Ressler et al. 2006), while the NF-kB pathway has been shown to be closely related to inflammation and aging (Donato et al. 2008).

Another signaling cascade that has a major role in control of longevity is the insulin signaling pathway (Dillin et al. 2002; Bartke 2008), together with its sibling and hallmark of aging, metabolic regulation of (stem) cell function, as has been demonstrated, for example, in the *D. melanogaster* intestinal stem cell model system (Biteau et al. 2010; Guo et al. 2014). There is emerging evidence that these pathways are also affected by DNA damage (Niedernhofer et al. 2006; Mueller et al. 2014) and might be connected to the immune system (Guo et al. 2014) and that they control maintenance of pluripotency of stem cells (Bendall et al. 2007).

Changes in proteostasis (which refers to the control of the biogenesis, folding, trafficking, and degradation of proteins) and autophagy (elimination of proteins from a cell) have been also associated with aging, with the underlying paradigm that an accumulation of nonfunctional proteins over time in a cell will result in senescence or aging. Defects in proteostasis have been linked to diseases that are usually age associated like cardiovascular and neurodegenerative disorders (Warr et al. 2013; Brehme et al. 2014; Zaglia et al. 2014). Changes in proteostasis and autophagy with aging in stem cells though have been not investigated in great detail, with human ES cells, intestinal as well as hematopoietic stem cells being the current focus of investigation (Vilchez et al. 2014).

To further link all these cellular and genetic pathways to age-associated clinical phenotypes, it will be crucial to clearly assess elderly patients with up-to-date geriatric assessment technology to avoid mixing different variables, such as disability, frailty, and comorbidity (Fontana et al. 2014).

These data in aggregation indicate that in general cell extrinsic as well as stem cell-intrinsic mechanisms can contribute to age-related decline in stem cell function and that most listed hallmarks and their underlying molecular pathways might contribute in one or the other way to stem cell aging. The causes leading to molecular alterations in aging stem cells and in the stem cell environment are not completely understood. Yet the questions of which hallmarks of stem cell aging and of which cellular phenotypes are causally affected by which of these primarily genetically tested molecular pathways is still not know in detail, ultimately rendering the search for rational and targeted interventions to achieve attenuation of stem cell aging or even better rejuvenation very difficult. Again, as mentioned before, one central prerequisite to achieve stem cell rejuvenation is that the unwanted changes associated with aging of stem cells are actually reversible or not causally involved in maintaining the aged status.

16.2 Rejuvenation or Attenuation of Stem Cell Aging

The underlying causes leading to molecular alterations in aging stem cells and in the stem cell environment are not completely understood and most likely a combination of factors contribute to stem cell aging, as discussed so far in the different chapters of this book. Nevertheless, as already stated, the number of these distinct factors or pathways seems to be not open-ended but rather limited in number. Amelioration and rejuvenation of stem cells are usually addressed in mammalian systems, due to the translational aspects of this kind of question and the somewhat delayed attention that aging of somatic stem cells attracted in other model organisms. It has been reported that aged muscle stem ("satellite") cells, in which the aged phenotype is a response to Wnt signaling, can be activated to differentiate and regenerate muscle in aged animals as efficiently as young muscle stem cells – either by forced activation of Notch or by factors in serum from young animals supplied by parabiosis (Conboy et al. 2003, 2005; Brack et al. 2007; Liu et al. 2007). In old mice, exposure to a young systemic environment increases activation of the Notch pathway improving satellite cell function and muscle regeneration (Conboy et al. 2003, 2005). Interestingly, parabiosis also increased the dendritic spine density of mature neurons and improved synaptic plasticity in the hippocampus of aged heterochronic parabionts. At the cognitive level, systemic administration of young blood plasma into aged mice improved age-related cognitive impairments in both contextual fear conditioning and spatial learning and memory (Villeda et al. 2014). Especially growth and differentiation factor 11 (GDF11), a member of the TGF-beta superfamily of cytokines, recently draw a lot of attention with respect to parabiosis as one of the factors found in the blood from young mice and reduced upon aging that when given to aged animals reverted both age-related cardiac hypertrophy as well as age-related dysfunction of mouse skeletal muscle. At least in the case of reverting aging-associated function of skeletal muscle, it was shown that GDF11 results, among others, in functional improvement of satellite cells and reduced markers associated with replication fork stalling (Flach et al. 2014; Sinha et al. 2014). In addition in cardiac and skeletal muscle, GDF-11 was also found to enhance vascular remodeling in the brain and by this means enhance neurogenesis in aged mice (Katsimpardi et al. 2014). Parabiosis and GDF-11 though seem to act to a great extent on more differentiated cells, especially in the case of the neuronal and cardiac system, as for these tissues very active contribution of stem cells to tissue homeostasis has not been reported, and as a consequence, rejuvenation should be driven by differentiated cells. The molecular mechanisms though of the action of GDF-11 are not known yet in detail.

It was further reported that deletion of components of the DNA damage checkpoint (Exo1, p21) attenuated loss of stem cell function, organ maintenance, and lifespan of telomere dysfunctional mice without increasing the cancer risk in vivo (Choudhury et al. 2007; Schaetzlein et al. 2007), but pharmacological targeting of these checkpoints is currently difficult and would need to be further developed.

Caloric restriction (CR) has been shown to extend lifespan under certain circumstances in most model organisms, although recently the effect of CR in nonhuman primates has been more critically discussed. Effects of CR can be seen in multiple, also stem cell-based tissues; thus, it most likely also affects aging of somatic stem cells, especially given that the molecular pathways that are linked to CR, like IGF-1, insulin, mTOR, and leptin signaling, have been all shown to affect stem cell behavior (Mazzoccoli et al. 2014). Surprisingly though, analyses on whether CR actually attenuates or rejuvenates stem cell aging are still rare, and only recently it was reported to have rejuvenating effects on murine hematopoietic stem cells (Cheng et al. 2014). In *D. melanogaster*, CR was found to attenuate the age-related decline in the function of the male germ line stem cell, which contributed to an extension of the reproductive period of flies when subjected to CR (Mair et al. 2010), while short-term CR enhanced skeletal muscle stem cell function in mice (Cerletti et al. 2012).

In this context rapamycin (which targets the mTOR pathway) has also been listed as a drug to halt or even revert stem cell aging. In the light of the finding that rapamycin has been one of the few drugs that resulted so far in lifespan extension when given to "normal" mice (Harrison et al. 2009; Neff et al. 2013) and which effects are also already measured in yeast (Powers et al. 2006), rapamycin has been implied in attenuation or rejuvenation of aging of in two distinct types of stem cells (Blagosklonny 2010), while the function of rapamycin on stem cell in vivo in longevity studies has not been performed yet. To note, it is likely that side effects associated with rapamycin treatment might preclude its translation into the clinic (Lamming et al. 2013).

In summary, these studies provide a "proof of principle" that reversion of molecular alterations in aging stem cells represents a promising therapeutic approach to improve organ maintenance and function during aging and that interventions targeting both the molecular and the cellular level might be successful and clinically relevant.

16.3 The Paradigm: Rejuvenation of Hematopoietic Stem Cells (HSCs)

HSCs from young and aged mice differ in their function. Aging exerts a deleterious effect on HSCs self-renewal and differentiation ability, and HSC aging is driven by both intrinsic and extrinsic factors (Geiger and Van Zant 2002; Geiger et al. 2005; Rossi et al. 2005; Rando 2006; Ju et al. 2007; Mayack et al. 2010). Aged HSCs show reduced self-renewal activity determined in serial transplant assays (Janzen et al. 2006). When aged HSCs are transplanted together with young HSCs into lethally irradiated young recipients, aged HSCs are on average twofold less efficient in contributing to hematopoiesis compared to young HSCs (Morrison et al. 1996; Chen et al. 2000) and exhibit a twofold reduced ability to home to the bone marrow (BM) (Liang et al. 2005). Aging also affects the differentiation potential of HSCs. Aged HSCs are deficient in their ability to support erythropoiesis, and aged HSCs do not efficiently generate B-lymphoid progeny at the expense of increased myeloid cell lineage output. Aged HSCs exhibit distinct whole genome expression signatures (Rossi et al. 2005; Chambers et al. 2007) and increased double-strand breaks as detected by increased levels of gammaH2AX staining, a surrogate marker for DNA double-strand breaks (Rossi et al. 2007). Our data show that aged HSCs are less efficient in their ability to adhere to stroma cells and exhibit significantly elevated cell protrusion activity in vivo, reducing the time for effective interaction with the microenvironment (Xing et al. 2006; Geiger et al. 2007; Kohler et al. 2009). Thus, a defined canonical set of features phenotypically separates young from aged HSCs. It might be for this reason that research to improve aging of hematopoietic stem cells has gained momentum, as based on these phenotypes the level of attenuation of stem cell aging can be quantified in more detail and determined in more standard experimental settings compared to other stem cell systems.

While multiple cell-intrinsic mechanisms for aging of HSCs have been discussed and summarized extensively elsewhere (refer to reviews by several authors (Geiger et al. 2013) (Snoeck 2005; Kamminga and de Haan 2006; Geiger et al. 2007; Rossi et al. 2008; Waterstrat et al. 2008; Beerman et al. 2010; Wang et al. 2011), the causative molecular mechanisms of HSC aging remain still largely unclear. We have recently identified a novel candidate mechanism of HSC aging the loss of stem cell polarity. This loss of polarity is induced by elevated Cdc42 activity driven by stem cell-intrinsic elevated expression of Wnt5a upon aging (Florian et al. 2012, 2013) and affects both proteins within the cytoplasm like Cdc42 itself and tubulin but also epigenetic markers in the nucleus, for example, the polar distribution of the acetylated form of histone 4 on lysine 16 (H4K14Ac). Cell polarity has been well characterized in epithelial cells (planar cell polarity) and neuronal stem cells, but not so far in HSCs. One intensely debated paradigm holds that asymmetric distribution of cellular components at HSC division and, subsequently, in the daughter cells determines their fate. Such an asymmetric distribution of cellular components has been shown by single-cell immunostaining (Beckmann et al. 2007; Rajendran et al. 2009) and in a fluorescent Notch-activity indicator system (Wu et al. 2007). It is not clear though whether polarity directly determines the mode of division and/or cell fate. Supporting a determining role for polarity in stem cell aging is a recent study showing that a loss of proper polarity in aged *Drosophila* germ line stem cells correlates with their reduced functions (Cheng et al. 2008). Given the established role of the small Rho GTPase Cdc42 in HSC polarity and differentiation, we postulate that Cdc42 may coordinate cell polarity and division symmetry in HSCs and that an altered mode of division contributes to stem cell aging. Most interestingly, pharmacological inhibition of the aging-associated increase in Cdc42 activity ex vivo resulted in rejuvenation of aged HSCs, as indicated by the reversal of almost all the phenotypic parameters associated with aged HSCs in transplantation experiments, including the level and position of H4K16 acetylation and cell polarity (Florian et al. 2012). The data thus support an important role for a correct cytosol and epigenetic polarity for stem cell function. Along the same line, genetic inhibition of the elevated levels of Wnt5a, a member of the non-canonical Wnt signaling network, in aged HSCs resulted, via reduced activity of Cdc42, in rejuvenation of aged HSCs and restored the youthful dependence on canonical Wnt signaling of HSCs (Florian et al. 2013).

Besides these approaches, attenuation of HSC aging has been achieved by lifelong caloric restriction in Balb/C inbred mice (Chen et al. 2003) or antioxidative therapy with NAC (Chen et al. 2009), which improved the self-renewal capacity of aged HSCs as tested in serial transplantation assays, thereby leading to partial or segmental HSC rejuvenation. Inhibition of mTOR activity, which has been reported to be elevated in HSCs from aged mice, both in vivo and ex vivo by rapamycin restored the self-renewal and hematopoiesis of HSCs, most likely via suppression of stem cell senescence, and enabled effective vaccination against a lethal challenge with influenza virus. These results imply that rapamycin, via inhibition of mTOR, has the ability to rejuvenate aged HSCs (Chen et al. 2009; Luo et al. 2014). In this context an interesting finding is also that iPS cells generated from aged HSCs, when reintroduced into blastocyst, give rise to young HSCs, establishing that aging of HSCs might be fully reversible (Wahlestedt et al. 2013). Therefore, aged HSCs can be fully rejuvenated, and the causal mechanisms of HSC aging do not irreversibly alter HSCs. Very recently also proteostasis was identified as a valid target to genetically rejuvenate aged HSCs. A regulatory branch of the mitochondrial unfolded protein response (UPR(mt)) is mediated by the interplay of SIRT7 and NRF1 and is coupled to cellular energy metabolism and proliferation. While SIRT7 expression was reduced in aged HSCs, SIRT7 upregulation improved the regenerative capacity of aged HSCs defining the deregulation of a UPR(mt)-mediated metabolic checkpoint as a reversible contributing factor for HSC aging (Mohrin et al. 2015). Whether these pharmacological and genetic pathways of rejuvenation are interdependent, as one would assume, has to be further investigated.

In aggregation, these observations support the notion that attenuation or even rejuvenation of aging of HSCs can be achieved by genetic as well as pharmacological targeting of cell-intrinsic mechanisms. Although such observations imply that reversal of the cell-intrinsic parameters of aging may be sufficient to achieve rejuvenation, HSC extrinsic factors, such as circulating cytokine levels and niche-specific factors, should perhaps also be targeted, especially in the context of rejuvenation in vivo. Mechanisms of HSC rejuvenation might also serve as a blueprint or template for testing similar rejuvenation interventions in other stem cell systems.

16.4 Rejuvenation of Stem Cells and Longevity

A key challenge in aging research is to confirm that molecular mechanisms of aging in model organisms are relevant for human aging. Another obvious question stemming from stem cell aging research is whether rejuvenation of stem cells will influence longevity. So far, there are no reports published that unequivocally determine that aging of a distinct type of stem cells limits lifespan, with the exception of some work on intestinal stem cells in *D. melanogaster* (Wang et al. 2014). Lifespan is most likely not limited by aging of just one single type of stem cell, and multiple parameters could be involved in determining longevity of the whole organism. On the other hand, drugs like rapamycin, which have been able to extent lifespan in mammals under certain circumstances, seem to affect multiple stem cell systems at once. Thus, targeting a critical and common pathway of stem cell aging, if that exists, might indeed serve as our fountain of youth.

Acknowledgements The authors are supported by grants from the Deutsche Forschungsgemeinschaft (SFB 1074 and 1149) and the GRK 1079 CEMMA (Deutsche Forschungsgemeinschaft), the BMBF-funded Program SyStaR, and the Exzellenz-Programm of the Baden-Württemberg-Stiftung and from the National Institute of Health, HL076604, DK077762, and AG040118.

References

Andrade LN, Nathanson JL, Yeo GW, Menck CF, Muotri AR (2012) Evidence for premature aging due to oxidative stress in iPSCs from Cockayne syndrome. Hum Mol Genet 21(17): 3825–3834

Bartke A (2008) Impact of reduced insulin-like growth factor-1/insulin signaling on aging in mammals: novel findings. Aging Cell 7(3):285–290

Beckmann J, Scheitza S, Wernet P, Fischer JC, Giebel B (2007) Asymmetric cell division within the human hematopoietic stem and progenitor cell compartment: identification of asymmetrically segregating proteins. Blood 109(12):5494–5501

Beerman I, Maloney WJ, Weissmann IL, Rossi DJ (2010) Stem cells and the aging hematopoietic system. Curr Opin Immunol 22(4):500–506

Beerman I, Seita J, Inlay MA, Weissman IL, Rossi DJ (2014) Quiescent hematopoietic stem cells accumulate DNA damage during aging that is repaired upon entry into cell cycle. Cell Stem Cell 15(1):37–50

Bell DR, Van Zant G (2004) Stem cells, aging, and cancer: inevitabilities and outcomes. Oncogene 23(43):7290–7296

Bendall SC, Stewart MH, Menendez P, George D, Vijayaragavan K, Werbowetski-Ogilvie T, Ramos-Mejia V, Rouleau A, Yang J, Bosse M, Lajoie G, Bhatia M (2007) IGF and FGF cooperatively establish the regulatory stem cell niche of pluripotent human cells in vitro. Nature 448(7157):1015–1021

Biteau B, Hochmuth CE, Jasper H (2008) JNK activity in somatic stem cells causes loss of tissue homeostasis in the aging Drosophila gut. Cell Stem Cell 3(4):442–455

Biteau B, Karpac J, Supoyo S, Degennaro M, Lehmann R, Jasper H (2010) Lifespan extension by preserving proliferative homeostasis in Drosophila. PLoS Genet 6(10), e1001159

Blagosklonny MV (2010) Rapamycin and quasi-programmed aging: four years later. Cell Cycle 9(10):1859–1862

Borodkina A, Shatrova A, Abushik P, Nikolsky N, Burova E (2014) Interaction between ROS dependent DNA damage, mitochondria and p38 MAPK underlies senescence of human adult stem cells. Aging (Albany NY) 6(6):481–495

Brack AS, Conboy MJ, Roy S, Lee M, Kuo CJ, Keller C, Rando TA (2007) Increased Wnt signaling during aging alters muscle stem cell fate and increases fibrosis. Science 317(5839):807–810

Brehme M, Voisine C, Rolland T, Wachi S, Soper JH, Zhu Y, Orton K, Villella A, Garza D, Vidal M, Ge H, Morimoto RI (2014) A chaperome subnetwork safeguards proteostasis in aging and neurodegenerative disease. Cell Rep 9(3):1135–1150

Campbell C, Risueno RM, Salati S, Guezguez B, Bhatia M (2008) Signal control of hematopoietic stem cell fate: Wnt, Notch, and Hedgehog as the usual suspects. Curr Opin Hematol 15(4):319–325

Cerletti M, Jang YC, Finley LW, Haigis MC, Wagers AJ (2012) Short-term calorie restriction enhances skeletal muscle stem cell function. Cell Stem Cell 10(5):515–519

Chambers SM, Shaw CA, Gatza C, Fisk CJ, Donehower LA, Goodell MA (2007) Aging hematopoietic stem cells decline in function and exhibit epigenetic dysregulation. PLoS Biol 5(8), e201

Chen J, Astle CM, Harrison DE (2000) Genetic regulation of primitive hematopoietic stem cell senescence. Exp Hematol 28(4):442–450

Chen J, Astle CM, Harrison DE (2003) Hematopoietic senescence is postponed and hematopoietic stem cell function is enhanced by dietary restriction. Exp Hematol 31(11):1097–1103

Chen C, Liu Y, Zheng P (2009) mTOR regulation and therapeutic rejuvenation of aging hematopoietic stem cells. Sci Signal 2(98):ra75

Cheng J, Turkel N, Hemati N, Fuller MT, Hunt AJ, Yamashita YM (2008) Centrosome misorientation reduces stem cell division during ageing. Nature 456(7222):599–604

Cheng CW, Adams GB, Perin L, Wei M, Zhou X, Lam BS, Da Sacco S, Mirisola M, Quinn DI, Dorff TB, Kopchick JJ, Longo VD (2014) Prolonged fasting reduces IGF-1/PKA to promote hematopoietic-stem-cell-based regeneration and reverse immunosuppression. Cell Stem Cell 14(6):810–823

Choudhury AR, Ju Z, Djojosubroto MW, Schienke A, Lechel A, Schaetzlein S, Jiang H, Stepczynska A, Wang C, Buer J, Lee HW, von Zglinicki T, Ganser A, Schirmacher P, Nakauchi H, Rudolph KL (2007) Cdkn1a deletion improves stem cell function and lifespan of mice with dysfunctional telomeres without accelerating cancer formation. Nat Genet 39(1):99–105

Conboy IM, Conboy MJ, Smythe GM, Rando TA (2003) Notch-mediated restoration of regenerative potential to aged muscle. Science 302(5650):1575–1577

Conboy IM, Conboy MJ, Wagers AJ, Girma ER, Weissman IL, Rando TA (2005) Rejuvenation of aged progenitor cells by exposure to a young systemic environment. Nature 433(7027):760–764

de Craen AJ, Gussekloo J, Teng YK, Macfarlane PW, Westendorp RG (2003) Prevalence of five common clinical abnormalities in very elderly people: population based cross sectional study. BMJ 327(7407):131–132

Dillin A, Crawford DK, Kenyon C (2002) Timing requirements for insulin/IGF-1 signaling in C. elegans. Science 298(5594):830–834

Donato AJ, Black AD, Jablonski KL, Gano LB, Seals DR (2008) Aging is associated with greater nuclear NF kappa B, reduced I kappa B alpha, and increased expression of proinflammatory cytokines in vascular endothelial cells of healthy humans. Aging Cell 7(6):805–812

Flach J, Bakker ST, Mohrin M, Conroy PC, Pietras EM, Reynaud D, Alvarez S, Diolaiti ME, Ugarte F, Forsberg EC, Le Beau MM, Stohr BA, Mendez J, Morrison CG, Passegue E (2014) Replication stress is a potent driver of functional decline in ageing haematopoietic stem cells. Nature 512(7513):198–202

Florian MC, Dorr K, Niebel A, Daria D, Schrezenmeier H, Rojewski M, Filippi MD, Hasenberg A, Gunzer M, Scharffetter-Kochanek K, Zheng Y, Geiger H (2012) Cdc42 activity regulates hematopoietic stem cell aging and rejuvenation. Cell Stem Cell 10(5):520–530

Florian MC, Nattamai KJ, Dorr K, Marka G, Uberle B, Vas V, Eckl C, Andra I, Schiemann M, Oostendorp RA, Scharffetter-Kochanek K, Kestler HA, Zheng Y, Geiger H (2013) A canonical

to non-canonical Wnt signalling switch in haematopoietic stem-cell ageing. Nature 503(7476): 392–396
Fontana L, Kennedy BK, Longo VD, Seals D, Melov S (2014) Medical research: treat ageing. Nature 511(7510):405–407
Fuchs E, Segre JA (2000) Stem cells: a new lease on life. Cell 100(1):143–155
Geiger H, Van Zant G (2002) The aging of lympho-hematopoietic stem cells. Nat Immunol 3(4):329–333
Geiger H, Rennebeck G, Van Zant G (2005) Regulation of hematopoietic stem cell aging in vivo by a distinct genetic element. Proc Natl Acad Sci U S A 102(14):5102–5107
Geiger H, Koehler A, Gunzer M (2007) Stem cells, aging, niche, adhesion and Cdc42: a model for changes in cell-cell interactions and hematopoietic stem cell aging. Cell Cycle 6(8):884–887
Geiger H, de Haan G, Florian MC (2013) The ageing haematopoietic stem cell compartment. Nat Rev Immunol 13(5):376–389
Guo L, Karpac J, Tran SL, Jasper H (2014) PGRP-SC2 promotes gut immune homeostasis to limit commensal dysbiosis and extend lifespan. Cell 156(1–2):109–122
Harrison DE, Strong R, Sharp ZD, Nelson JF, Astle CM, Flurkey K, Nadon NL, Wilkinson JE, Frenkel K, Carter CS, Pahor M, Javors MA, Fernandez E, Miller RA (2009) Rapamycin fed late in life extends lifespan in genetically heterogeneous mice. Nature 460(7253):392–395
Janzen V, Forkert R, Fleming HE, Saito Y, Waring MT, Dombkowski DM, Cheng T, Depinho RA, Sharpless NE, Scadden DT (2006) Stem-cell ageing modified by the cyclin-dependent kinase inhibitor p16(INK4a). Nature 443(7110):421–426
Ju Z, Jiang H, Jaworski M, Rathinam C, Gompf A, Klein C, Trumpp A, Rudolph KL (2007) Telomere dysfunction induces environmental alterations limiting hematopoietic stem cell function and engraftment. Nat Med 13(6):742–747
Kamminga LM, de Haan G (2006) Cellular memory and hematopoietic stem cell aging. Stem Cells 24:1143–1149
Katsimpardi L, Litterman NK, Schein PA, Miller CM, Loffredo FS, Wojtkiewicz GR, Chen JW, Lee RT, Wagers AJ, Rubin LL (2014) Vascular and neurogenic rejuvenation of the aging mouse brain by young systemic factors. Science 344(6184):630–634
Kohler A, Schmithorst V, Filippi MD, Ryan MA, Daria D, Gunzer M, Geiger H (2009) Altered cellular dynamics and endosteal location of aged early hematopoietic progenitor cells revealed by time-lapse intravital imaging in long bones. Blood 114(2):290–298
Lamming DW, Ye L, Sabatini DM, Baur JA (2013) Rapalogs and mTOR inhibitors as anti-aging therapeutics. J Clin Invest 123(3):980–989
Liang Y, Van Zant G, Szilvassy SJ (2005) Effects of aging on the homing and engraftment of murine hematopoietic stem and progenitor cells. Blood 106(4):1479–1487
Liu H, Fergusson MM, Castilho RM, Liu J, Cao L, Chen J, Malide D, Rovira II, Schimel D, Kuo CJ, Gutkind JS, Hwang PM, Finkel T (2007) Augmented Wnt signaling in a mammalian model of accelerated aging. Science 317(5839):803–806
Lopez-Otin C, Blasco MA, Partridge L, Serrano M, Kroemer G (2013) The hallmarks of aging. Cell 153(6):1194–1217
Luo Y, Li L, Zou P, Wang J, Shao L, Zhou D, Liu L (2014) Rapamycin enhances long-term hematopoietic reconstitution of ex vivo expanded mouse hematopoietic stem cells by inhibiting senescence. Transplantation 97(1):20–29
Mair W, McLeod CJ, Wang L, Jones DL (2010) Dietary restriction enhances germline stem cell maintenance. Aging Cell 9(5):916–918
Mayack SR, Shadrach JL, Kim FS, Wagers AJ (2010) Systemic signals regulate ageing and rejuvenation of blood stem cell niches. Nature 463(7280):495–500
Mazzoccoli G, Tevy MF, Borghesan M, Delle Vergini MR, Vinciguerra M (2014) Caloric restriction and aging stem cells: the stick and the carrot? Exp Gerontol 50:137–148
Mohrin M, Shin J, Liu Y, Brown K, Luo H, Xi Y, Haynes CM, Chen D (2015) Stem cell aging. A mitochondrial UPR-mediated metabolic checkpoint regulates hematopoietic stem cell aging. Science 347(6228):1374–1377

Morrison SJ, Uchida N, Weissman IL (1995) The biology of hematopoietic stem cells. Annu Rev Cell Dev Biol 11:35–71
Morrison SJ, Wandycz AM, Akashi K, Globerson A, Weissman IL (1996) The aging of hematopoietic stem cells [see comments]. Nat Med 2(9):1011–1016
Mueller MM, Castells-Roca L, Babu V, Ermolaeva MA, Muller RU, Frommolt P, Williams AB, Greiss S, Schneider JI, Benzing T, Schermer B, Schumacher B (2014) DAF-16/FOXO and EGL-27/GATA promote developmental growth in response to persistent somatic DNA damage. Nat Cell Biol 16(12):1168–1179
Neff F, Flores-Dominguez D, Ryan DP, Horsch M, Schroder S, Adler T, Afonso LC, Aguilar-Pimentel JA, Becker L, Garrett L, Hans W, Hettich MM, Holtmeier R, Holter SM, Moreth K, Prehn C, Puk O, Racz I, Rathkolb B, Rozman J, Naton B, Ordemann R, Adamski J, Beckers J, Bekeredjian R, Busch DH, Ehninger G, Graw J, Hofler H, Klingenspor M, Klopstock T, Ollert M, Stypmann J, Wolf E, Wurst W, Zimmer A, Fuchs H, Gailus-Durner V, Hrabe de Angelis M, Ehninger D (2013) Rapamycin extends murine lifespan but has limited effects on aging. J Clin Invest 123(8):3272–3291
Niedernhofer LJ, Garinis GA, Raams A, Lalai AS, Robinson AR, Appeldoorn E, Odijk H, Oostendorp R, Ahmad A, van Leeuwen W, Theil AF, Vermeulen W, van der Horst GT, Meinecke P, Kleijer WJ, Vijg J, Jaspers NG, Hoeijmakers JH (2006) A new progeroid syndrome reveals that genotoxic stress suppresses the somatotroph axis. Nature 444(7122):1038–1043
Norddahl GL, Pronk CJ, Wahlestedt M, Sten G, Nygren JM, Ugale A, Sigvardsson M, Bryder D (2011) Accumulating mitochondrial DNA mutations drive premature hematopoietic aging phenotypes distinct from physiological stem cell aging. Cell Stem Cell 8(5):499–510
Ohlstein B, Spradling A (2006) The adult Drosophila posterior midgut is maintained by pluripotent stem cells. Nature 439(7075):470–474
Passos JF, Nelson G, Wang C, Richter T, Simillion C, Proctor CJ, Miwa S, Olijslagers S, Hallinan J, Wipat A, Saretzki G, Rudolph KL, Kirkwood TB, von Zglinicki T (2010) Feedback between p21 and reactive oxygen production is necessary for cell senescence. Mol Syst Biol 6:347
Potten CS, Morris RJ (1988) Epithelial stem cells in vivo. J Cell Sci Suppl 10:45–62
Powers RW 3rd, Kaeberlein M, Caldwell SD, Kennedy BK, Fields S (2006) Extension of chronological life span in yeast by decreased TOR pathway signaling. Genes Dev 20(2):174–184
Poynard T, Mathurin P, Lai CL, Guyader D, Poupon R, Tainturier MH, Myers RP, Muntenau M, Ratziu V, Manns M, Vogel A, Capron F, Chedid A, Bedossa P, PANFIBROSIS Group (2003) A comparison of fibrosis progression in chronic liver diseases. J Hepatol 38(3):257–265
Rajendran L, Beckmann J, Magenau A, Boneberg EM, Gaus K, Viola A, Giebel B, Illges H (2009) Flotillins are involved in the polarization of primitive and mature hematopoietic cells. PLoS One 4(12), e8290
Rando TA (2006) Stem cells, ageing and the quest for immortality. Nature 441(7097):1080–1086
Ray PD, Huang BW, Tsuji Y (2012) Reactive oxygen species (ROS) homeostasis and redox regulation in cellular signaling. Cell Signal 24(5):981–990
Ressler S, Bartkova J, Niederegger H, Bartek J, Scharffetter-Kochanek K, Jansen-Durr P, Wlaschek M (2006) p16INK4A is a robust in vivo biomarker of cellular aging in human skin. Aging Cell 5(5):379–389
Rossi DJ, Bryder D, Zahn JM, Ahlenius H, Sonu R, Wagers AJ, Weissman IL (2005) Cell intrinsic alterations underlie hematopoietic stem cell aging. Proc Natl Acad Sci U S A 102(26):9194–9199
Rossi DJ, Bryder D, Seita J, Nussenzweig A, Hoeijmakers J, Weissman IL (2007) Deficiencies in DNA damage repair limit the function of haematopoietic stem cells with age. Nature 447(7145):725–729
Rossi DJ, Jamieson CH, Weissman IL (2008) Stems cells and the pathways to aging and cancer. Cell 132(4):681–696
Salat D, Liefke R, Wiedenmann J, Borggrefe T, Oswald F (2008) ETO, but not leukemogenic fusion protein AML1/ETO, augments RBP-Jkappa/SHARP-mediated repression of notch target genes. Mol Cell Biol 28(10):3502–3512

Schaetzlein S, Kodandaramireddy NR, Ju Z, Lechel A, Stepczynska A, Lilli DR, Clark AB, Rudolph C, Kuhnel F, Wei K, Schlegelberger B, Schirmacher P, Kunkel TA, Greenberg RA, Edelmann W, Rudolph KL (2007) Exonuclease-1 deletion impairs DNA damage signaling and prolongs lifespan of telomere-dysfunctional mice. Cell 130(5):863–877

Sinha M, Jang YC, Oh J, Khong D, Wu EY, Manohar R, Miller C, Regalado SG, Loffredo FS, Pancoast JR, Hirshman MF, Lebowitz J, Shadrach JL, Cerletti M, Kim MJ, Serwold T, Goodyear LJ, Rosner B, Lee RT, Wagers AJ (2014) Restoring systemic GDF11 levels reverses age-related dysfunction in mouse skeletal muscle. Science 344(6184):649–652

Snoeck HW (2005) Quantitative trait analysis in the investigation of function and aging of hematopoietic stem cells. Methods Mol Med 105:47–62

Stappenbeck TS, Mills JC, Gordon JI (2003) Molecular features of adult mouse small intestinal epithelial progenitors. Proc Natl Acad Sci U S A 100(3):1004–1009

Takashima S, Mkrtchyan M, Younossi-Hartenstein A, Merriam JR, Hartenstein V (2008) The behaviour of Drosophila adult hindgut stem cells is controlled by Wnt and Hh signalling. Nature 454(7204):651–655

Tani H, Morris RJ, Kaur P (2000) Enrichment for murine keratinocyte stem cells based on cell surface phenotype. Proc Natl Acad Sci U S A 97(20):10960–10965

Vas V, Senger K, Dorr K, Niebel A, Geiger H (2012) Aging of the microenvironment influences clonality in hematopoiesis. PLoS One 7(8), e42080

Vilchez D, Simic MS, Dillin A (2014) Proteostasis and aging of stem cells. Trends Cell Biol 24(3):161–170

Villeda SA, Plambeck KE, Middeldorp J, Castellano JM, Mosher KI, Luo J, Smith LK, Bieri G, Lin K, Berdnik D, Wabl R, Udeochu J, Wheatley EG, Zou B, Simmons DA, Xie XS, Longo FM, Wyss-Coray T (2014) Young blood reverses age-related impairments in cognitive function and synaptic plasticity in mice. Nat Med 20(6):659–663

Wahlestedt M, Norddahl GL, Sten G, Ugale A, Frisk MA, Mattsson R, Deierborg T, Sigvardsson M, Bryder D (2013) An epigenetic component of hematopoietic stem cell aging amenable to reprogramming into a young state. Blood 121(21):4257–4264

Wang J, Geiger H, Rudolph KL (2011) Immunoaging induced by hematopoietic stem cell aging. Curr Opin Immunol 23(4):532–536

Wang L, Karpac J, Jasper H (2014) Promoting longevity by maintaining metabolic and proliferative homeostasis. J Exp Biol 217(Pt 1):109–118

Warr MR, Binnewies M, Flach J, Reynaud D, Garg T, Malhotra R, Debnath J, Passegue E (2013) FOXO3A directs a protective autophagy program in haematopoietic stem cells. Nature 494:323–327

Waterstrat A, Oakley E, Miller A, Swiderski C, Liang Y, Van Zant G (2008) Mechanisms of stem cell aging. In: Rudolph KL (ed) Telomeres and telomerase in ageing, disease and cancer. Springer Publishing, Berlin, Germany, pp 111–140

Wu M, Kwon HY, Rattis F, Blum J, Zhao C, Ashkenazi R, Jackson TL, Gaiano N, Oliver T, Reya T (2007) Imaging hematopoietic precursor division in real time. Cell Stem Cell 1(5):541–554

Xing Z, Ryan MA, Daria D, Nattamai KJ, Van Zant G, Wang L, Zheng Y, Geiger H (2006) Increased hematopoietic stem cell mobilization in aged mice. Blood 108(7):2190–2197

Zaglia T, Milan G, Ruhs A, Franzoso M, Bertaggia E, Pianca N, Carpi A, Carullo P, Pesce P, Sacerdoti D, Sarais C, Catalucci D, Kruger M, Mongillo M, Sandri M (2014) Atrogin-1 deficiency promotes cardiomyopathy and premature death via impaired autophagy. J Clin Invest 124(6):2410–2424

Printed by Printforce, the Netherlands